Sustainable Green Nanotechnology

The book places a strong emphasis on the significance of synthesizing nanomaterials and explores various methods of achieving this, with particular attention given to the production of environmentally friendly or "green" nanomaterials. This book illustrates the current gaps in the agriculture and food industry, and explores the potential applications of nanomaterials within this field. The integration of nanotechnology into the fields of biology and biochemistry holds the potential to exert a significant impact on the food industry in the future. This book comprehensively explores the various facets of design, utilization, and implementation of green nanomaterials within the agriculture and food industry. It delves into their applications in areas such as fertilizers, sensors, food processing, food packaging, and sustainable agricultural practices. Additionally, this book explores the current limitations and potential solutions in the field of green nanomaterials, highlighting their role in fostering a sustainable economy.

Features:

- Addressing the global food crisis and presenting a picture of depleting food reserves
- Importance of nanomaterials and their efficient utilization
- Green synthesis of nanomaterials
- Utilization of green nanomaterials for development of smart and sustainable agriculture
- Emergence and growing application of green nanomaterials in the food industry
- Green nanomaterials towards a sustainable economy
- Existing challenges and limitations, with future prospects of using green nanomaterials in food and agricultural conversion devices

Yugal Kishore Mohanta is now working as Assistant Professor and Group Leader of Nanobiotechnology and Translational Knowledge Laboratory (NanoBioTraK_Lab) in the Department of Applied Biology, School of Biological Sciences, University of Science and Technology, Meghalaya (USTM), Meghalaya, India.

Ishani Chakrabartty is working as a medical writer at Indegene Pvt. Ltd. Bengaluru, India. She has served as an Assistant Professor in the Department of Applied Biology, School of Biological Sciences, University of Science and Technology, Meghalaya (USTM), Meghalaya, India. She earned her PhD from the Department of Biosciences and Bioengineering, IIT Guwahati, India. She is a regular columnist in *The Assam Tribune*.

Saravanan Muthupandian is currently working as a Professor and Group Leader of the AMR and Nanomedicine Laboratory, Department of Pharmacology, SIMATS, Saveetha University, Chennai, Tamil Nadu, India and Prince Fahad bin Sultan Chair for Biomedical Research, University of Tabuk, Tabuk, Saudi Arabia.

Sustainable Green Nanotechnology

Food and Agriculture Paradigm

Edited by
Yugal Kishore Mohanta, Ishani Chakrabartty, and
Saravanan Muthupandian

CRC Press
Taylor & Francis Group
Boca Raton London New York

CRC Press is an imprint of the
Taylor & Francis Group, an **informa** business

Designed cover image: Shutterstock

First edition published 2025
by CRC Press
2385 NW Executive Center Drive, Suite 320, Boca Raton FL 33431

and by CRC Press
4 Park Square, Milton Park, Abingdon, Oxon, OX14 4RN

CRC Press is an imprint of Taylor & Francis Group, LLC

ISBN: 9781032485072 (hbk)
ISBN: 9781032485089 (pbk)
ISBN: 9781003389408 (ebk)

DOI: 10.1201/9781003389408

Typeset in Times
by Newgen Publishing UK

Contents

Preface

With the existing advances and technology, improvement in crop production and agricultural practices has reached its threshold; that is, a stage has been achieved where it would be nearly impossible to feed the projected population of the near future. There is a huge demand for food worldwide, particularly in the African nations, and every country is looking for suitable options to solve this food crisis. There is an urgent need for smart agricultural practices that use advanced and cutting-edge technologies for sustainable agriculture. This includes the need for new and innovative fertilizers that have a very high fertilizer-use efficiency and minimal loss. As such, designing nanomaterials, particularly biosynthesized nanomaterials, can serve as a possible solution to this ongoing search.

This book will emphasize the importance and different modes of synthesis of nanomaterials, with detailed emphasis on green nanomaterials. It will identify the existing lacunas in the agriculture and food industry and how nanomaterials can be utilized in this area. Nanotechnological advances in biology and biochemistry can eventually influence the food industry, and this book will address the different aspects of design, exploitations, and application of green nanomaterials in the agriculture and food industry—for fertilizers, sensors, food processing, food packaging, and sustainable agricultural practices. It will also try to address current limitations and the question of how green nanomaterials can be utilized as future prospects towards a sustainable economy.

A lot of literature is available on the applications of green nanomaterials, but there are limited studies on this specific subject. This book in general will be viewed as a boon for engineers (mechanical, agricultural, food, etc.), nanotechnologists, biologists, economists, researchers, agriculture scientists, as well as business strategists and management students, and it will be an asset for survival strategies. It will address the global food demand and how nanotechnology plays a vital role in providing a possible solution by involving both the food industry and the agriculture sector. It will highlight the role of nanotechnology in a sustainable economy in the future. Furthermore, this book will recognize the innovative measures taken by different countries to overcome the existing limitations in solving the problem of food demand and the application of green nanomaterials for sustainable growth, for the environment, and for a future biobased economy.

Acknowledgments

We begin by thanking our Creator, the Almighty God, the most benevolent, the most merciful, because from Him we begin, and to Him is our end. Next, we are thankful to our families—parents, spouses, relatives—whose support has been very important in the completion of this book.

This book would be nothing without the contributing authors—we thank each and every author who has shown interest and made their valuable contribution to this book, which has enabled its completion on time. At the same time, we are immensely grateful to the publisher CRC Press for choosing to publish our book. We express our gratitude to Ms. Renu Upadhyay, Commissioning Editor at CRC Press, and Ms. Jyotsna Jangra, Editorial Assistant at CRC Press, for their unwavering editorial support and valuable suggestions in refining the book to its current form. The editors also heartily acknowledge Mr. Jibanjyoti Panda (Project Associate, SERB-DST, Govt. of India), Nano-biotechnology and Translational Knowledge Laboratory, Department of Applied Biology, University of Science and Technology Meghalaya, for his assistance in editing and formatting the chapters.

Lastly, the three editors thank each other for the coordination, communication, and cooperation that has been instrumental in realizing the idea of this book and making it a success.

Contributors

Sumel Ashique
Department of Pharmaceutical Sciences, Bengal College of Pharmaceutical Sciences & Research, Durgapur 713212, West Bengal, India

Devanshi Baghla
UIAS Chandigarh University

Meghali Barua
Assam Agricultural University, Jorhat, Assam, India

Vishwa Jyoti Baruah
Dibrugarh University, Dibrugarh, Assam, India

Indrani Bhattacharyya
Department of Pharmaceutical Sciences, Pandaveswar School of Pharmacy, Pandaveswar, West Bengal, India

Subhra Prakash Bhattacharyya
Department of Pharmaceutical Sciences, Pandaveswar School of Pharmacy, Pandaveswar, West Bengal, India

Tanmayi Boreda
Chaitanya Bharathi Institute of Technology, Gandipet, Hyderabad, India

Garima Chauhan
Chaudhary Sarwan Kumar Himachal Pradesh Krishi Vishvavidyalaya, Palampur, India

Arif Hassan Dar
Institute of NanoScience and Technology (INST), Mohali, India

Ajaz Ahmad Dar
University of Cincinnati, Ohio, USA

Subhankar Das
Mangalore University, Mangalagangotri, Mangalore Karnataka, India

Sweety Das
University of Science and Technology Meghalaya (USTM), Ri-Bhoi, Meghalaya, India

Archana Deka
ICMR-Regional Medical Research Centre, N.E. Region, Dibrugarh, Assam, India

Vinicius Marx Silva Delgado
Federal University of São João del Rei, Divinópolis, Brazil

Juliana Espindola
KROM-Center of Research and Industrial revolution, Campinas, São Paulo, Brazil

Nikhat Farhana
Bharat Institute of Technology, Mangalpally (V), Ibrahimpatnam (M) Hyderabad, Telangana, India; Department of Pharmaceutical Chemistry, Vidy, Siri College Of Pharmacy, Off Sarjapur Road, affiliated to RGUHS, Bengaluru, Karnataka-560035

Arshad Farid
Gomal Center of Biochemistry and Biotechnology, Gomal University, D.I. Khan, KPK, Pakistan

Mohd Azhardin Ganayee
Indian Institute of Technology Madras, Chennai, India

Ashish Garg
Department of Pharmaceutics, Guru Ramdas Khalsa Institute of Science and Technology (Pharmacy), Jabalpur, Madhya Pradesh

Chandan Kumar Gautam
University of Miami, Coral Gables, FL, USA

Amol D. Gholap
St. John Institute of Pharmacy and Research, Palghar, Maharashtra, India

T Gourav
Chaitanya Bharathi Institute of Technology, Gandipet, Hyderabad, India

Khalid Rehman Hakeem
Department of Biological Science, Faculty of Science, King Abdulaziz University, Jeddah, Saudi Arabia

Navnath T. Hatvate
Institute of Chemical Technology, Mumbai, Maharashtra, India

Aabid Hussain
Sher-e-Kasmir University of Agricultural Sciences and Technology of Kashmir (SKUAST-K), Jammu and Kashmir, India

Afzal Hussain
Department of Pharmaceutics, College of Pharmacy, King Saud University, Riyadh, Saudi Arabia

Manjula Ishwara Kalyani
Mangalore University, Chikka Aluvara Kodagu, Mangalore, Karnataka India

Pooja Kashid
Navsahyadri Institute of Pharmacy, Pune, Maharashtra, India

Yashasvi Kambhampati
Chaitanya Bharathi Institute of Technology, Gandipet, Hyderabad, India

Sehijpreet Kaur
University of Florida, Gainesville, FL, USA

Pankaj R. Khuspe
College of Pharmacy, Paniv, Solapur, Maharashtra, India

Neeraj Kumar
MARES, Salooni, Chamba, Chaudhary Sarwan Kumar Himachal Pradesh Krishi Vishvavidyalaya, Palampur, India

Nitesh Kumar
Chaudhary Charan Singh Haryana Agriculture University, Hisar

Sachin Kumar
Centre for Geo-Informatics Research and Training CSKHPKV, Palampur

Shubneesh Kumar
Department of Pharmacy, Bharat Institute of Technology (BIT), School of Pharmacy, Meerut, Uttar Pradesh, India

Barasa Malakar
University of Science and Technology Meghalaya (USTM), Ri-Bhoi, Meghalaya, India

Akhtar Hussain Malik
Govt Degree College Sopore, Jammu & Kashmir, India

Moon Mandal
Pandit Deendayal Upadhyaya Adarsha Mahavidyalaya, Dalgaon, Darrang, Assam, India

Sanjeeb Kumar Mandal
Chaitanya Bharathi Institute of Technology, Gandipet, Hyderabad, India

Trushali A. Mandhare
Navsahyadri Institute of Pharmacy, Pune, Maharashtra, India

Luciana Gama de Mendonça
Federal University of Ceará, Fortaleza, Brazil

Bishwambhar Mishra
Chaitanya Bharathi Institute of Technology, Gandipet, Hyderabad, India

Neeraj Mishra
Department of Pharmaceutics, Amity Institute of Pharmacy, Amity University Madhya Pradesh (AUMP), Gwalior, India

Yugal Kishore Mohanta
University of Science and Technology Meghalaya (USTM), Ri-Bhoi, Meghalaya, India and Chettinad Academy of Research and Education, Kelambakkam, Tamil Nadu, India

Rachid Mrabet
National Institute of Agricultural Research, INRA, Rabat, Morocco

Amilia Nongbet
University of Science and Technology Meghalaya (USTM), Ri-Bhoi, Meghalaya, India

Lakshmayya NSV
Chaitanya Bharathi Institute of Technology, Gandipet, Hyderabad, India

Sanjana Reddy Pailla
Chaitanya Bharathi Institute of Technology, Gandipet, Hyderabad, India

Jibanjyoti Panda
University of Science and Technology Meghalaya (USTM), Ri-Bhoi, Meghalaya, India

Geicielly da Costa Pinto
Federal University of São João del Rei, Divinópolis, Brazil

Majid Sharifi-Rad
University of Zabol, Iran

Adil Majeed Rather
North Carolina State University, USA

Bharat Bhushan Rana
Chaudhary Sarwan Kumar Himachal Pradesh Krishi Vishvavidyalaya, Palampur, India

Manny Saluja
KWS Gateway Research Center, St. Louis, MO, USA

Bhaswati Sarmah
Assam Agricultural University, Jorhat, Assam, India

Tarun Sharma
Chaudhary Sarwan Kumar Himachal Pradesh Krishi Vishvavidyalaya, Palampur, India

Christopher Santos Silva
Federal University of São João del Rei, Divinópolis, Brazil

Akashdeep Singh
Chaudhary Sarwan Kumar Himachal Pradesh Krishi Vishvavidyalaya, Palampur, India

Ekta Sirohi
Department of Pharmacy, Bharat Institute of Technology (BIT), School of Pharmacy, Meerut, Uttar Pradesh, India

B Sumithra
Chaitanya Bharathi Institute of Technology, Gandipet, Hyderabad, India

Farzad Taghizadeh-Hesary
ENT and Head and Neck Research Center and Department, The Five Senses Health Institute, School of Medicine, Iran University of Medical Sciences, Tehran, Iran; Clinical Oncology Department, Iran University of Medical Sciences, Tehran, Iran

Mousham Talukdar
University of Science and Technology Meghalaya (USTM), Ri-Bhoi, Meghalaya, India

Fernanda Maria Policarpo Tonelli
Federal University of São João del Rei, Divinópolis, Brazil

Flávia Cristina Policarpo Tonelli
Federal University of São João del Rei, Divinópolis, Brazil

Aakash Upadhyay
Department of Pharmacy, Bharat Institute of Technology (BIT), School of Pharmacy, Meerut, Uttar Pradesh, India

Kiran Yellappa Vajanthri
Chaitanya Bharathi Institute of Technology, Gandipet, Hyderabad, India

Vanga Sri Varsha
Chaitanya Bharathi Institute of Technology, Gandipet, Hyderabad, India

Bhushan Vishal
Nanyang Technology University, Singapore, Singapore

Lalitkumar K Vora
School of Pharmacy, Queen's University Belfast, UK

Saima Masood Wani
Sher-e-Kasmir University of Agricultural Sciences and Technology of Kashmir (SKUAST-K), Jammu and Kashmir, India

Mira Landep Widiastuti
National Research and Innovation Agency, Republic of Indonesia

Rafael Audino Zambelli
Federal University of Ceará, Fortaleza, Brazil

About the Editors

Yugal Kishore Mohanta has more than 12 years of post-masters teaching and research experience and currently holds the position of Assistant Professor at the Department of Applied Biology within the School of Biological Sciences at the University of Science and Technology, Meghalaya (USTM), India. Dr. Mohanta's current area of research pertains to the study of metal and composite nanoparticles from bioresources and their potential applications in the fields of food, agriculture, and biomedicine. Dr. Mohanta has international research experience both as a PhD visiting scholar and as a postdoctoral fellow at KU Leuven, Belgium and University of Nizwa, Sultanate of Oman, respectively. Dr. Mohanta has authored 90 research publications in peer-reviewed journals which are indexed in the SCIE. These papers have high impact factors and have garnered a total of 2438 citations, with an *h*-index of 24 and an i-10 index of 52. He has served as a reviewer for over 30 international peer-reviewed journals from Springer, Elsevier, Wiley, Frontiers, and MDPI and holds the position of associate editor for four SCIE-indexed journals. Dr. Mohanta has also made contributions to 16 book chapters, edited books for CRC Press and Elsevier-Academic Press, and authored numerous articles on scientific topics for national science publications. Dr. Mohanta currently holds the position of group leader at the Nano-biotechnology and Translational Knowledge Laboratory and serves as the principal investigator for the SERB-DST project, a start-up research grant funded by the Government of India in New Delhi. Dr. Mohanta is also working as State President of the Microbiologist Society India for Assam and Meghalaya, India.

Ishani Chakrabartty recently made a shift from academia to industry, and for the past year has been working as a medical writer in Indegene Pvt. Ltd. She has served as an Assistant Professor at the Department of Applied Biology, School of Biological Sciences, University of Science and Technology, Meghalaya (USTM), India, accredited "A" Grade by the NAAC for two years, and Assistant Professor and Head of the Department of Sciences, P.A. First Grade College, Mangalore. She completed her PhD at the Department of Biosciences and Bioengineering, IIT Guwahati in 2019. She specializes in biotechnology, particularly in Natural products, Microbiology, and Nanotechnology. She has authored around 16 papers in international journals of repute (all Scopus- and WoS-indexed) and several book chapters with national and international publishers, and has an edited volume in press (CRC Press). She has a number of national awards to her credit and holds life memberships of some national organizations. She was awarded "Best Science Story" in 2018 by AWSAR-DST, Government of India. She has around 200 citations and an *h*-index of 7. Teaching is her passion, but writing is her first love — she is a regular columnist in *The Assam Tribune*, and has authored a book, *Symphony of Catharsis*.

Saravanan Muthupandian has over 21 years of teaching and research experience, and he is ranked in the top 2% of scientists worldwide by Stanford University in 2021 and 2022. He has worked as a professor in the AMR and Nanotherapeutics Lab, Department of Pharmacology, Saveetha University, SIMATS, Chennai, India since January 2021. He has published more than 200 research papers, including in high-impact journals *The Lancet*, *Nature* and *Nature Medicine*, with more than 27,000 citations, an *h*-index of 60, and an i-10 index of 135. He has published seven books and 35 book chapters. He has participated in more than 100 national and international conferences, reviewed in more than 100 international peer-reviewed journals, and been guest editor for various reputed PubMed- and Scopus-indexed journals. In particular, he has been an associate editor for *Frontiers in Pharmacology*, *Frontiers in Oncology*, and MDPI's *Functional Biomaterials* and *Medicina*. He has received many fellowships and awards, notably two consecutive IET Nanobiotechnology Premium Awards (in 2019 and 2020), an international fellowship "Advanced Course on Diagnostics" sponsored by LSH&TM and Fondation Mérieux, in France 2013, an international conference "Pertussis: Biology, Epidemiology and Prevention" sponsored by Foundation Mérieux and the World Health Organization in France 2014, an International Union of Microbiological Societies travel grant in 2015 to Canada, and an international fellowship "Advanced Course on Antibiotics" (AdCAb) sponsored by the Institute of Pasteur and Foundation Mérieux, 2016.

1 Green Nanomaterials
Why Are They Important?

Nikhat Farhana and Khalid Rehman Hakeem
Corresponding author: Nikhat Farhana

1.1 INTRODUCTION

Green nanotechnology is the branch of green technology that utilizes the concepts of green chemistry and green engineering [1], where the word *green* refers to plant products (Figure 1.1). Nano refers to metric measures of one billionth of a meter and covering a width of 10 atoms; in terms of comparison with real objects, a useful example is a hair, which is approximately 150,000 nanometers wide. Technology is the application of knowledge to achieve practical goals in a reproducible way; as such, technology plays a critical role in science and technology. Green nanotechnology is a way to synthesize nanoparticles from plant metabolites with inorganic metals in translation research [2].

The evolution of metallic nanoparticles on an ecofriendly approach is one of the remarkable milestones in research and development [3]. Nanotechnology requires basic knowledge of many domains, including chemistry, physics, biological science, engineering, materials science, and computational science, for the formulation of nanostructure [4].

Nanostructures have different applications attributable to their new or increased properties such as size, distribution, and morphology [5]. They have a wide range of applications in various fields including biomedical, catalysis, chemical industries, cosmetics drug delivery, electronic environment energy science, food and feed, health care, mechanics, optics, the space industry, nonlinear optical devices, single electron transistors, and photoelectrochemicals [6]. Metallic nanoparticles are considered one of the most promising systems for all the aforementioned applications [7].

Global warming is due to rapid industrialization and urbanization. Toxic gases are released into the environment, resulting in increased temperatures over the past 50 years. Global warming occurs when CO_2, and other pollutants and greenhouse gases, collect in the atmosphere and absorb sunlight and solar radiation which get bounced from the Earth's surface into the atmosphere. The pollutants are then present in the atmosphere for centuries [8], trapping heat and causing the planet to get hotter [9], a process known as the greenhouse effect. Combustion of fossil fuels generates H_2SO_4, CO_2, and HNO_3, which falls as acid rain, impacting both natural areas and the built environment. Sculptures and monuments made from marble and limestone are vulnerable, as acid dissolves $CaCO_3$. Fossil fuels contain radioactive materials like uranium and thorium which, when released into the atmosphere, have negative health impacts on human beings including premature death, acute respiratory illness, asthma, and bronchitis due to lung infections [10].

1.2 GREEN ENGINEERING: SUSTAINABLE CHEMISTRY PRINCIPLES

An example of green technology is a solar cell that converts energy to electrical energy by photovoltaics. The generation of electricity from solar energy means less consumption of fossil fuels, which results in scarcity of many natural resources, thereby reducing pollution and greenhouse

DOI: 10.1201/9781003389408-1

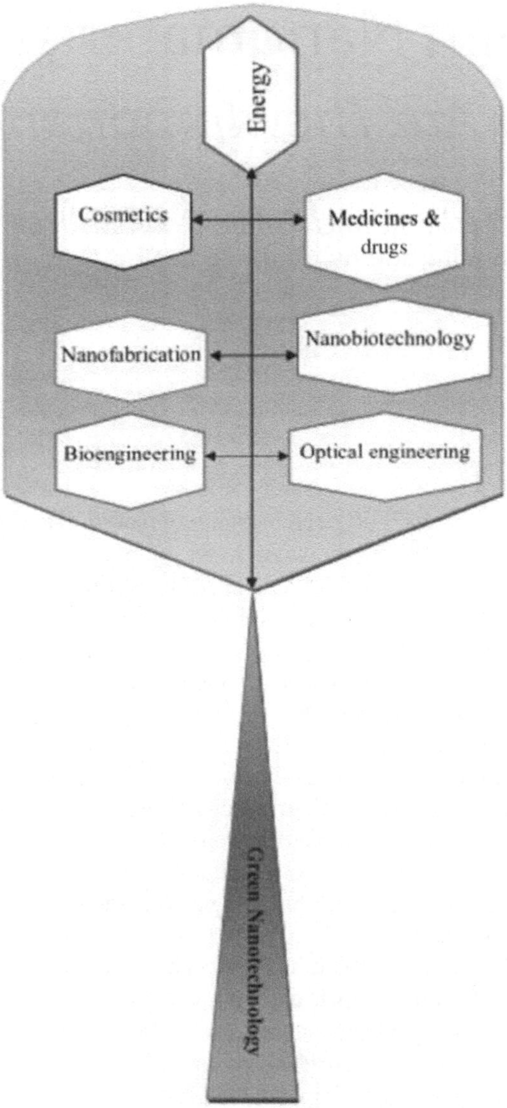

FIGURE 1.1 Branches of green nanotechnology.

gas emissions. Hence, utilization of renewable natural resources will not harm the planet, and by applying the green chemistry principle (productivity) and the green engineering principle (improvement), we can construct a healthier and better environment [11].

1.2.1 Branches of Green Technology

Some popular green technology and practices are as follows:

- Green chemistry: based on the 12 principles of green chemistry minimizing hazardous substances
- Green nanotechnology: production of nanomaterials without harming the environment or human health

- Green building: ecofriendly architectural life cycle in terms of design, construction and maintenance, renovation, and demolition
- Green information and technology: reduction of carbon footprint to reduce negative effects on human activity
- Green energy: produced from natural resources (sunlight, wind, rain, tides, plants, algae, and geothermal heat) which have a much smaller impact on the environment than fossil fuels (i.e., oil, gas, coal, etc.)

This chapter considers green nanotechnology, which is the emerging field of science currently focusing on commercial applications like scratch-free paints, surface coatings, electronics, cosmetics [12], environmental remediation, sports equipment, sensors, and energy storage devices. In this chapter, we focus on the importance of green nanotechnology on a single platform; however, it is challenging to elaborate in detail the background to and current trends in this technology [13]. Hence, we take snapshots of the available literature to discuss in one place some different types, classifications, properties, methods of synthesis, possible opportunities, and applications related to green nanotechnology [14].

1.2.2 Types of Nanoparticles

Background analytical survey captures some distinct forms of different types of inorganic metallic nanoparticles: core–shell, photochromic polymer NPs, polymer-coated magnetite NPs, AgNPs, CuNPs, AuNPs, PtNPs, PdNPs, SiNPs, and NiNPs [15], while others include metallic oxide and metallic dioxide NPs like ZnONPs, CuONPs, FeONPs [16], MgONPs, TiO_2NPs, CeO_2NPs, and ZrO_2NPs [17]. As well as therapeutic importance, these have applications in the fields of science and technological research and development, and the distinct sets can be synthesized according to conventional and unconventional protocols.

1.2.3 Classification of Nanoparticles

Nanomaterials are classified into two types, based on their dimensions or on their synthesis.

1.2.3.1 Classification Based on Dimensions

Nanomaterials are further subdivided into three types based on their dimensions and size: one-dimensional (1D), two-dimensional (2D), and three-dimensional (3D) nanomaterials.

1.2.3.1.1 One-dimensional (1D) Nanomaterials

The number 109, which represents one billionth of any unit that leads to the production of one-dimensional nanomaterials such as thin films, is denoted by the word nano. Such nanoparticles are used in a wide range of scientific disciplines, including chemistry, pharmaceuticals, electronics, and engineering. The thickness of the thin films or monolayers ranges from 1 to 100 nm. The fabrication of electronics, storage systems, nanoscale LEDs, optoelectronics, chemicals, biosensors, magneto-optics, fiberoptic systems, and optical devices all make use of these nanomaterials, which are highly significant in research. One-dimensional nanomaterials are used to create significant materials at nanoscales.

1.2.3.1.2 Two-dimensional (2D) Nanomaterials

Key components of nanodevices are constructed from 2D nanomaterials, which have two dimensions and a distinctive shape outside the range of the nanometric scale [18]. Applications for two-dimensional nanomaterials include nanocontainers, sensor photocatalysts, nanoreactors, and templates for two-dimensional structures. Two-dimensional nanoparticles include carbon nanotubes.

1.2.3.1.3 Three-dimensional (3D) Nanomaterials

Shape, size, and morphology, which are the fundamental factors governing the use and effectiveness of nanostructures, determine how nanomaterials behave [19–22]. Three-dimensional nanomaterials have attracted attention in medical science and research over the past ten years. These nanoparticles have numerous uses in the fields of catalysis, batteries, and the transport of reactants and products through magnetic materials. Fullerenes, dendrimers, and quantum dots are examples of three-dimensional nanoparticles.

1.2.3.2 Classification Based on Synthesis

Due to nanomaterials' high surface-to-volume ratio and extremely small size, they differ significantly from bulk materials in terms of their physicochemical characteristics. Additionally, the size, shape, and crystal structure of nanoparticles have a significant impact on their optical, electrical, and catalytic characteristics. Due to overwhelming demand, the creation of nanomaterials has been accelerated through a variety of techniques. Since the field's inception, two methods have been used to synthesize nanoparticles (see Figure 1.2).

The two primary categories of the bottom-up approach are the liquid phase method and the gaseous phase approach. The wet and dry (grinding) processes are further separated in the top-down approach. Although the dry approach is inexpensive, monodispersity and surface chemistry control are still issues. By employing a wet method, it is possible to avoid nanoparticle condensation and produce highly dispersed nanoparticles. However, neither the wet nor the dry processes for making nanoparticles is ecofriendly. Their downsides include the use of hazardous solvents, contamination from precursor chemicals, and the production of hazardous byproducts. These production techniques are therefore costly, labor-intensive, and potentially harmful to the environment and living things. The development of ecologically friendly nanoparticle production processes through approaches that are also cost-effective is therefore desirable.

1.3 TRENDS IN GREEN TECHNOLOGY: UPTAKE AND ACCUMULATION WITH INORGANIC METALLOIDS

The species of plant, as well as the size, type, chemical composition, and stability of the nanoparticles, determines uptake, translocation, and accumulation. A model plant can absorb, biotransform, and translocate a variety of nanoparticles [23–25]. Studies of the interaction of ryegrass with nanoparticles confirmed the nanoparticles' adsorption and aggregation on the root surface. TEM images of ryegrass root cross-sections also demonstrated the presence of particles in the apoplast, cytoplasm, and nuclei of the endodermal cells [26–28]. In a hydroponic culture, Fe_3O_4 nanoparticles were taken up by pumpkin seedlings. The plants' roots, stalks, and leaves all had nanoparticles. It was proposed that the growing media influence the uptake of the nanoparticles, which may be related to the Fe_3O_4 nanoparticles' affinity for soil and sand particles. Cucurbita pepo plants treated with AgNPs had shoots with an Ag content that was 4.7 times greater than plants treated with bulk Ag powder exposed to 10–1000 mg L1 AgNPs [29–32].

1.4 GREEN SYNTHETIC APPROACH TO DEVELOPING NANOPARTICLES

Many people consider nanoparticles to be the essential building components of nanotechnology. Numerous approaches, including physical, chemical, biological, and hybrid approaches, can be used to create them. They serve as the foundation for numerous bottom-up methods used to create nanostructured materials and electronics. As a result, their synthesis plays a crucial role in the rapidly expanding fields of nanoscale science and engineering [33–38]. Numerous techniques can be used to create nanoparticles made of a wide variety of materials. Instead of going into detail about

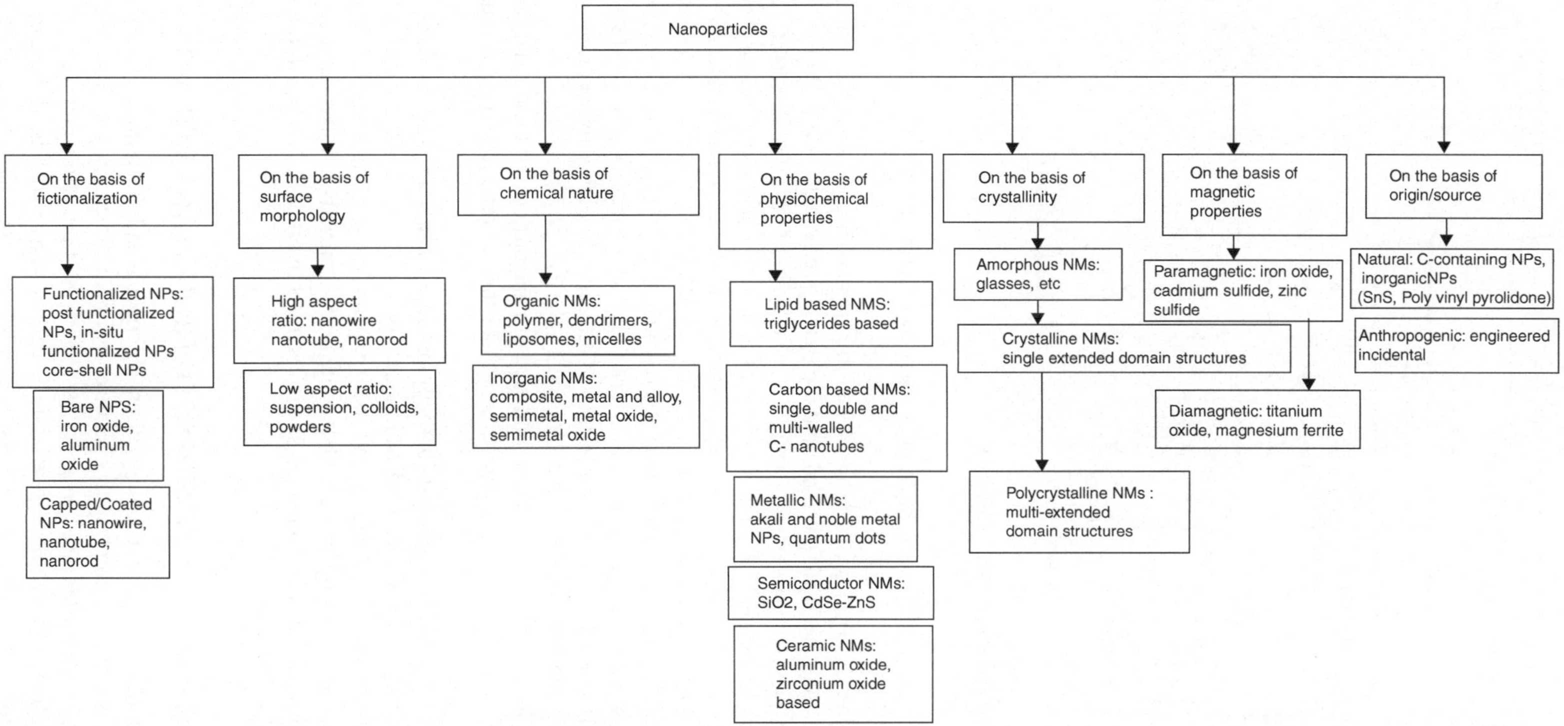

FIGURE 1.2 Bottom-up and top-down approaches for the production of nanoparticles.

(A) top-down approach: a decomposition method in which a large component is decomposed into smaller desired sizes; (B) bottom-up approach: an atom-based folding method based on atomic transformations and molecular condensation.

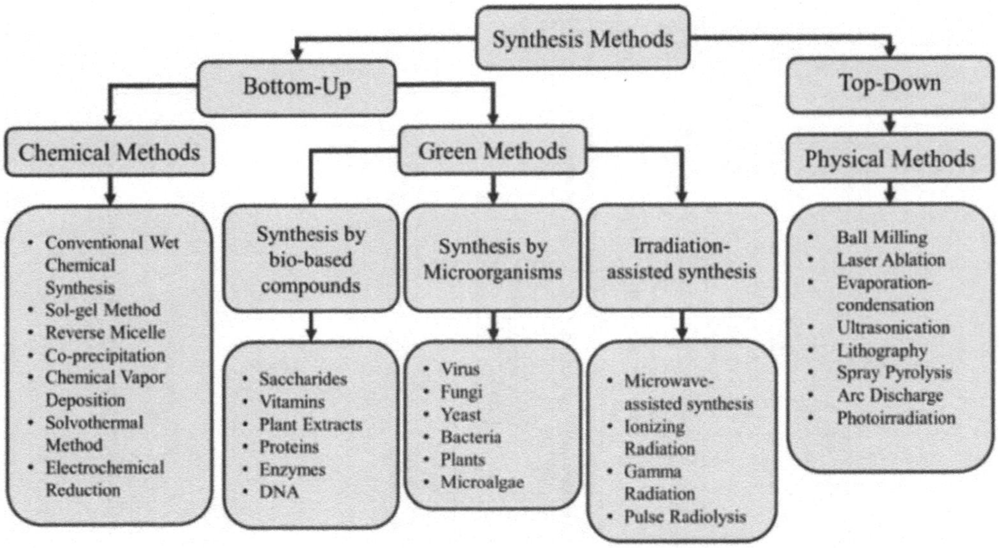

FIGURE 1.3 Classification based on synthetic sources.

any specific methodology, an effort has been made here to describe advances in a variety of synthesis methods [39–42].

According to Figure 1.3, the production of metallic nanoparticles involves top-down and bottom-up strategies using chemical, physical, and biological methods [43–48]. On the bottom-up approach, biologically produced silver nanoparticles are categorized [49–50]. The creation of silver nanoparticles has been accomplished using a variety of techniques, including chemical reduction, microwave-assisted synthesis, ultrasonic-assisted synthesis, electrochemical reduction, the template method, photo-induced or photocatalytic reduction, irradiation reduction, the microemulsion method, and biochemical reduction. The creation of silver nanoparticles uses a variety of techniques [51–53]. After a thorough search of the literature, diverse processes for producing silver nanoparticles were discovered: (1) room temperature synthesis, (2) higher temperature synthesis, (3) microwave-assisted synthesis, and (4) light-induced synthesis.

Green technology, in general, refers to the production of nanoparticles or nanomaterials through biological processes that involve bacteria, plants, viruses, or their byproducts, such as those shown in Figure 1.4.

Based on a variety of variables, green technology-produced nanoparticles are vastly superior to those produced via physical and chemical techniques for the study of proteins and lipids [54]. For instance, green practices use less energy and more expensive chemicals to generate goods and byproducts that are environmentally benign. Green chemistry concepts are used as a guide today by researchers, scientists, chemical technologists, and chemists to produce less hazardous chemical products and byproducts [54–55].

1.5 GREEN NANOPARTICLES SYNTHESIZED FROM PLANT EXTRACTS

The rising incidence of diseases caused by virus pathogens and bacteria, in developed and developing countries, is a challenging task that requires the medical fraternity to search for alternative therapeutics to combat resistant microorganisms. Belief and faith in ancient medicine are still directly or indirectly utilized by people to overcome diseases and ailments, in all countries regardless of their level of modernization. Researchers from several nations have reported synthesizing

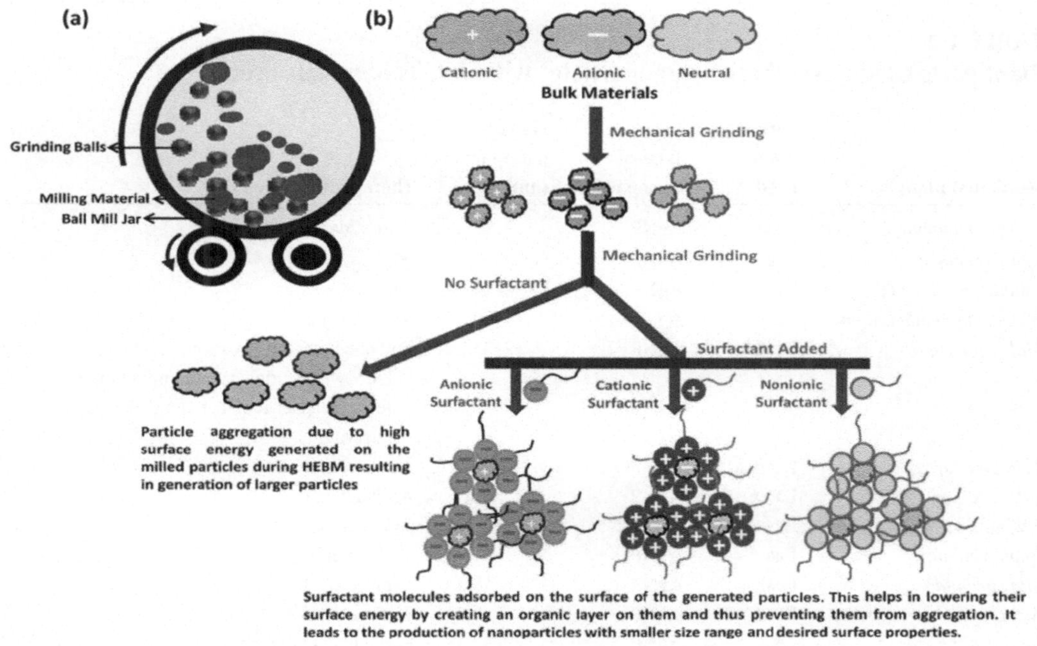

FIGURE 1.4 Mechanisms of nanoparticle formation based on dimensions and ionization.

distinct metal nanoparticles from medicinal plants that exhibit anticancer, antioxidant, antibacterial, wound-healing, insecticidal, hepatoprotective, and antidiabetic properties. As the therapeutic actions and characteristics summarized in Table 1.1 indicate, the increase in so-called nanomedicine can be considered epoch-making. Research, application, and use of metal nanoparticles will benefit sustainable development [56].

To decrease and stabilize nanoparticles, a variety of plants can be employed. Numerous researchers have used biological techniques to create metal or metal oxide nanoparticles utilizing a variety of plant parts, including leaves, stems, roots, and fruits, as shown in Figure 1.5. Plants include a variety of biomolecules that turn metal ions into nanoparticles, including proteins, coenzymes, and carbohydrates. Different nanoparticles are created using various methods [57].

1.6 IMPORTANCE AND AWARENESS OF GREEN NANOTECHNOLOGY

A large number of theoretical and experimental literature reviews on nanotechnology and nanomaterials have recently been published. The efficiency with which nanoscale materials can be modified for different applications will define future technologies. However, the production and efficient use of nanomaterials pose many challenges. Some of the major challenges are included in the list below:

- The fundamental properties of nanomaterials can be compromised and their performance can be hampered by defects. For example, one of the strongest known materials is carbon nanotubes. However, the tensile strength of carbon nanotubes can be significantly degraded by impurities, discontinuous tube lengths, defects, and random orientation [58].
- Another critical problem is finding an affordable way to synthesize nanomaterials. The production of high-quality nanomaterials usually requires special equipment and difficult environmental conditions, limiting their use in mass production. This issue becomes even more

TABLE 1.1
Plant parts used to synthesize nanoparticles with details of metals used

Medicinal plant	Part of plant used	Type of nanoparticle	Size of nanoparticle (nm)	Therapeutic activity
Syzygium cumini	Seed	AgNP	93	Antioxidant
Ixora coccinea	Leaf	AgNP	13–57	ND
Ananas comosus L.	Fruit	AgNP	12	Antioxidant
Memecylon umbellatum	Leaf	AgNP	15–20	Antimicrobial
Malva parviflora	Leaf	AgNP	19–25	Treatment of inflammation, abscesses, pimples, kidney infections, vaginal inflammation, fever, headache, spider stings, bronchitis, pharyngitis, tuberculosis
Jatropha curcas	Latex	AgNP	73	Antibacterial
Jatropha gossypiifolia	Latex	AgNP	62	Antibacterial
Pedilanthus tithymaloides	Latex	AgNP	123	Antibacterial
Euphorbia milii	Latex	AgNP	105	Antibacterial
Vitex negundo	Leaf	AgNP	18.2	Antibacterial
Artocarpus heterophyllus	Seed	AgNP	10.78	Antibacterial
Euphorbia hirta	Leaf	AgNP	20–30	Larvicidal and pupicidal
Coleus aromaticus	Leaf	AgNP	40–50	Bactericidal
Cycas	Leaf	AgNP	2–6	Antioxidant
Iresine herbstii	Leaf	AgNP	44–64	Antioxidant, antimicrobial, and antitumor
Melia azedarach	Leaf	AgNP	78	Antitumor
Eclipta prostrata	Leaf	AgNP	35–65	Larvicidal
Tinospora cordifolia Miers	Leaf	AgNP	55–80	Pediculocidal and larvicidal
Boswellia ovalifoliolata	Stem	AgNP	30–40	Rheumatic pains
Andrographis paniculata	Leaf	AgNP	35–55	Antiplasmodial
Nelumbo nucifera	Leaf	AgNP	45	Larvicidal against malaria and filariasis vectors
Ocimum	Leaf, stem, root	AgNP	3–20	Treatment of headaches, coughs, diarrhea, constipation, worms, kidney malfunctions
Catharanthus roseus	Leaf	AgNP	35–55	Antiplasmodial
Azadirachta indica	Leaf	AgNP	43	Antimicrobial
Olive	Leaf	AgNP	20–25	Antibacterial
Hibiscus rosa sinensis	Leaf	AgNP	5000	Antibacterial against fish pathogen *A. hydrophila*
Aloe vera	Leaf	AgNP	70–192	Antibacterial
Acalypha indica	Leaf	AgNP	20–30	Antibacterial against fish pathogens
Garcinia mangostana	Leaf	AgNP	35	MDR human pathogens
Phyllanthus niruri	Leaf	AgNP	32–53	Antimicrobial and cytotoxic
Argemone mexicana	Leaf	AgNP	20	Antimicrobial
Hibiscus sabdariffa	Leaf	ZnONP	12–46	Antibacterial and antidiabetic
Aloe barbadensis Miller	Leaf	ZnONP	25–40	Cosmetics
Trifolium pratense	Flower	ZnONP	100–190	Antibacterial
Nyctanthes arbor-tristis	Flower	ZnONP	12–32	Antifungal
Solanum nigrum	Leaf	ZnONP	29.79	Antibacterial
Azadirachta indica	Leaf	ZnONP	18	Antimicrobial
Passiflora caerulea	Leaf	ZnONP	37.67	Antibacterial
Murraya koenigii	Leaf	ZnONP	30–35	Antimicrobial
Hibiscus subdariffa	Leaf	ZnONP	16–60	Antibacterial and antidiabetic
Berberis aristata	Root	ZnONP	20–40	Urinary tract pathogens

Source: Singh et al. [31]

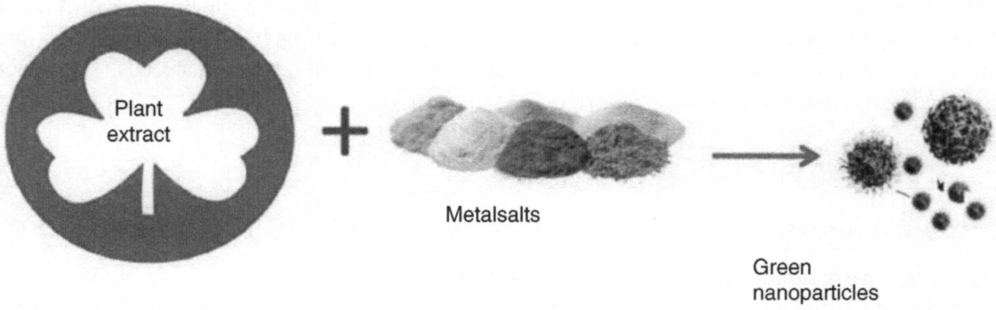

FIGURE 1.5 Green methods of synthesis of nanoparticles.

important in the fabrication of 2D nanomaterials. Most of the low-cost production techniques are widely used, but they tend to produce defective products. Controlled synthesis of nanomaterials remains a challenging task. For example, achieving chiral selectivity, electrical conductivity, and precisely controlled diameters are major challenges in carbon nanotube fabrication. The theoretically predicted properties outlined in the literature can only be achieved by using structurally pure nanomaterials [59].

- Synthesis of nanomaterials in an inexpensive manner is another major challenge. High-quality nanomaterials are generally manufactured using advanced equipment and harsh conditions, limiting large-scale production. This issue is more important for the synthesis of 2D nanomaterials. Most of the processes that have been used for large-scale production are inexpensive, and these processes generally produce poor-quality defective material. Controlled synthesis of nanomaterials remains a challenge. For example, key challenges associated with the synthesis of carbon nanotubes are achieving chiral selectivity, conductivity, and precisely controlled diameters. More focused efforts are needed to develop new synthetic techniques that overcome the shortcomings of existing techniques [60].
- Performance in related fields is severely impacted by the inherent problem of nanoscale particle agglomeration. When nanomaterials interact with each other, most start to aggregate. Physical entanglement, electrostatic interactions, or high surface energies can all be contributing factors to the aggregation process. Van der Waals interactions cause He CNTs to cluster together, making it difficult for He to align or evenly disperse them within the polymer matrix. Similarly, p–p interactions and van der Waals forces initiate graphene aggregation from the basal plane of the graphene layer. Graphene's enormous surface area and other properties are compromised by strong aggregation. These issues make it difficult to use high-throughput electrode materials or composites in practical applications.
- Both the use of nanomaterials in industry and the production rate of nanoscale materials are increasing.

The field of nanotechnology is already very wide, and will become wider in future as new nanomaterials with attractive properties are explored. One of the major concerns associated with nanomaterials that cannot be ignored is toxicity, which is still poorly understood and a significant concern given their use in the home, workplace, and environment. It is unknown to what extent compounds from nanoparticles cause cytotoxicity [61]. Science must work to bridge the information gap between the rapid development of nanomaterials and their potential for harm to living organisms. A thorough understanding of how they interact with organisms, tissues, and proteins is essential [62].

Nanotechnology breakthroughs are related to the future of modern technology. The development of technologies based on nanomaterials will realize the goal of generating clean energy. As a result of the advances made possible by these materials, new types of solar cell and hydrogen fuel cell technologies, effective water-splitting catalysts, and superior hydrogen storage capabilities have been developed. The topic of nanomedicine indicates a bright future for nanomaterials. Therapeutic compounds can be delivered via nanocarriers [63].

1.7 APPLICATIONS AND LIMITATIONS

1.7.1 Applications

Because they are naturally superparamagnetic, iron and iron oxide nanoparticles are frequently used in biomedical applications. AuNPs have shown to be useful tools in a variety of possible medicinal applications, including development of cancer nanotechnology alternatives, DNA modeling, treatment of life-threatening disorders, and biosensing applications. Nanoparticles are now being considered potential antibacterial agents. AgNPs are potential cancer cell growth inhibitors that can be used in a variety of molecular ways, including virtual devices and chemical substances derived from plants, like quinine [64]. *Plasmodium falciparum* resistant strains have been effectively treated with artemisinin and aromatic compounds. Immunoreactive substances are produced via the anti-inflammatory cascade mechanism. The therapy of malignant deposits has been transformed by newly developed biobased nanoparticles that do not harm healthy cells. The HeLa cell was significantly cytotoxic when silver nanoparticles made by green synthesis were used compared to other chemical-based drug synthesis [65].

1.7.2 Limitations

Greater attempts are being made as the technology is developed to establish ways to evaluate or measure the impact of nanotechnology on particular policy goals like green growth. This is a really difficult task. In comparison to the hazards associated with using existing technologies, such as nanotechnology for green cars, the risks associated with employing future green nanotechnologies must be taken into account when compared to the human and environmental costs of not resolving important global issues [66]:

- high cost of implementation
- lack of information (uncertainty on the extent to which research groups, universities, and businesses are pursuing this)
- no known substitute chemicals or raw materials
- no alternative process technology available
- uncertainty regarding effects on performance
- shortage of talent and resources.

1.8 APPLICATIONS IN HERBAL MEDICINE

The therapeutic value of herbal medicine depends on the overall function of bioactive constituents present in specific herbs for particular diseases, with each constituent released from the herbal medicine playing a significant role (see Table 1.1). The insoluble bioactive molecules of some herbal drugs decreases the bioavailability, which affects the ADME properties, reducing the therapeutic efficacy and causing the drug to be classified as less effective for therapeutic use [67].

In phytoformulation research, designing drug forms based on nanotechnology, such as proliposomes, liposomes, nanoemulsions, solid lipid nanoparticles (SLNs), and polymeric nanoparticles (nanospheres and nanocapsules) [68], has several advantages for herbal medicines. These include increasing solubility and bioavailability, increasing stability, reducing toxicity, increasing pharmacological effect, delivering the substance for a longer period of time, optimizing tissue macrophage circulation, and protecting substances from physical and chemical damage [69]. As a result, herbal medicines with nanoscale drug delivery systems (NDDS) have the potential to improve their activity while solving problems with herbal medicine. Consequently, it would be important to incorporate nanocarriers such as NDDS into conventional health systems for the use of herbal medicines to treat chronic diseases such as diabetes, cancer, and asthma [70].

1.8.1 Herbal Nanoparticles

1.8.1.1 Gold and Silver Nanoparticles

Due to their customizable surface attributes and adaptable size, shape, and composition, nanoparticles have drawn much attention, and gold nanoparticles have been examined in particular. For prospective applications in fields including biosensing, hyperthermia therapy [71], and antibiotics, a platform for the delivery of pharmaceuticals and genetic engineering is required, as well as ecologically sound sources. Plants are used to produce AuNPs, since they are biological factories that use sustainable chemistry methods. Studies on nanoparticle synthesis also show that, depending on the pH of the reaction media, different shapes, such as rod-shaped, irregular, decahedral, icosahedral, and hexagonal, can be created. Additionally, gold nanoparticles can be created using leaf extract of *Eucalyptus macrocarpa* (Table 1.2). As the primary result, spherical particles with diameters ranging from 20 to 80 nm were produced. Several studies have described alloying Au and Ag to examine the characteristics of the resulting bimetallic nanoparticles and the synthesis of pure metal nanoparticles by plants in this way. Two aq solutions compete with one another in a reduction process to produce bimetallic nanoparticles, each of which has a unique metal ion precursor that reacts with the plant extract [72]. Bimetallic nanoparticles occur, with gold forming first and becoming the center of the resulting core–shell structure because gold has a greater reduction potential than silver. Similar to this, the reduction of Ag ions causes the Ag to condense on the core to form the shell. Some plants, including cashew nuts, neem, and West Indian mahogany, yield bimetallic (Au–Ag) nanoparticles [72].

1.8.1.2 Copper and Copper Oxide Nanoparticles

Plant extracts have been used to create a range of copper (Cu) and copper oxide (CuO) nanoparticles. Cu nanoparticles were biologically synthesized from magnolia leaf extract to produce stable

TABLE 1.2
Herbal drug-loaded nanoparticles

Formulation	Method of preparation	Active ingredient	Biological activity
Curcumin-loaded solid lipid nanoparticles	Micro-emulation	Curcuminoids	Anticancer and antioxidant
Glycyrrhizic acid-loaded nanoparticles	Rotary-evaporated film ultrasonication	Glycyrrhizic acid	Anti-hypertensive and anti-inflammatory
Nanoparticles of *Cuscuta chinenisis*	Nanosuspension	Flavonoids and lignans	Heptoprotective and antioxidant
Artemisinin nanocapsules	Self-assembled	Artemisinin	Anticancer
Berberine-loaded nanoparticles	Ionic gelation	Berberine	Anticancer
CPT-encapsulated nanocapsules	Dialysis	Camptothecin	Anticancer
Taxol-loaded nanoparticles	Emulsion solvent evaporation	Taxol	Anticancer

nanoparticles of 40 to 100 nm. Furthermore, Cu nanoparticles demonstrated potential antibacterial activity against *Escherichia coli* cells. Cu nanoparticles with a spherical to granular shape and an average particle size of 40 nm have been synthesized using *Syzygium aromaticum* (clove) extract. Cu nanoparticles were created utilizing the stem latex of *Euphorbia nivulia*, often known as common milk hedge. These nanoparticles are stabilized by peptides and terpenoids found in latex. Furthermore, human adenocarcinomas of alveolar-origin epithelial cells have been shown to be hazardous to these nanoparticles [73].

1.8.1.3 Palladium and Platinum Nanoparticles

Cinnamon bark extract produces palladium nanoparticles. The temperature, concentration, and reaction pH of the bark extract changed during synthesis, but morphology and particle size (15 to 20 nm) were unaffected. Palladium nanoparticles with sizes of 75 to 85 nm were also synthesized using custard apple peel extract. Tea and coffee extracts were used to produce palladium nanoparticles with sizes ranging from 20 to 60 nm and cubic crystal symmetry at the center, as well as platinum nanoparticles with sizes ranging from 2 to 12 nm from *Diospyros kaki* leaf extract. Recently, biological production of platinum nanoparticles with particle size and shape control has been reported. Plant wood has been used to extract lignin from red pine (*Pinus resinosa*) to produce palladium and platinum nanoparticles at the nanoscale.

1.8.1.4 Titanium Dioxide and Zinc Oxide Nanoparticles

Oxide nanomaterials of this important metal have been synthesized from several plant extracts, establishing that TiO_2 nanoparticles can be efficiently synthesized from *Annona squamosa* bark. Meanwhile, spherical particles ranging from 100 to 150 nm were detected in *Nyctanthes arbor-tristis* leaf extract. *Eclipta prostrata* leaf extract can also yield particles with sizes ranging from 36 to 68 nm. Velutham et al. fabricated TiO_2 nanoparticles using *Catharanthus roseus* leaf extract with irregular shape and size ranging from 25 to 110 nm. TiO_2 suspension was larvicidal and adulticidal against *Bovicola ovis* (sheep louse) and *Hippobosca maculata* (hematophagous fly). TiO_2 nanoparticles were tested for antioxidant and antibacterial effects against *Pseudomonas aeruginosa*, *Staphylococcus aureus*, *Proteus mirabilis*, *Aeromonas hydrophila*, *E. coli*, and other pathogens. Furthermore, the antioxidant and antibacterial activities of TiO_2 nanoparticles were investigated at nanometer and bulk scales.

1.8.1.5 Nanoparticles of Indium Oxide (In_2O_3), Iron Oxide, Lead, and Selenium

Several new metal oxides and metal nanoparticles have been organically synthesized utilizing different plants. Aloe vera (*Aloe barbadensis Miller*) leaf extract was used to create indium oxide nanoparticles. The precipitates were heat-treated at 400–600 °C after the initial synthesis for the creation of nanoparticles. The size of the synthesized spherical nanoparticles ranged from 5 to 50 nm and was temperature dependent. Several environmental cleanup technologies rely on iron nanoparticles. As a result, several studies have concentrated on green chemistry to synthesize these iron (Fe) nanoparticles. Aqueous extracts of sorghum bran, for example, synthesized Fe nanoparticles. Recently, spherical iron nanoparticles with a particle size of 100 nm were synthesized from *Azadirachta indica* (neem) leaf extract. Iron nanoparticles were synthesized from plant extracts including *Cymbopogon citratus* (lemongrass tea), *Datura innoxia*, *Tridax procumbens*, *Calotropis procera*, *Tinospora cordifolia*, and *Euphorbia milii*. The smaller ones were created using the stem extract. Spherical nanoparticles with a size range of 43 to 342 nm were synthesized from *Cymbopogon citratus* leaf extract. Lead (Pb) and selenium (Se) are two other important biologically synthesized nanoparticles. Joglekar et al. [73] were able to synthesize spherical Pb nanoparticles with diameters ranging from 10 to 12.5 nm from the latex of *Jatropha curcas*. Sasidharan et al. [74] recently synthesized spherical selenium (Se) nanoparticles with a size of 70 nm from citrus reticulata bark extract.

1.9 CONCLUSIONS AND FUTURE PROSPECTS

Green nanotechnology contributes quickly to the release of contaminants from wastewater by making use of the principles of green chemistry. This chapter looked at a variety of green nanomaterial categories. The inability of conventional technologies to entirely remove organic pollutants from wastewater makes them uneconomical and energy intensive. However, using green technology is a very effective strategy, since it produces nanoparticles that are simple to separate without the need for harmful chemicals. Photocatalysts with various nanostructures include CuO, NiO, ZnO, TiO_2, Fe_3O_4, ZnO-Ag, ZnS, and ZnS-Gd. This chapter discussed environmentally friendly synthetic methods that make use of plant extracts and microorganisms.

Plant extract use is regarded as a successful strategy since plants are easier to utilize and less expensive than microorganisms. There has been discussion about photocatalyst selectivity. The decomposition of hazardous substances generated by various dyeing companies is accomplished using photocatalysis. Solar energy is being harnessed to ease the world's energy issues through a variety of energy storage and conversion technologies.

Nanotechnology has thereby expanded its potential applications to include environmental preservation and restoration. Every day, demand across numerous industries is increasing dramatically for nanoscale materials, sensors, and equipment. However, there are several issues that need to be addressed, including bioavailability, bioaccumulation, transport, toxicity, and the effects of nanoparticle manufacturing on the environment and human health. By using environmentally friendly nanotechnology manufacturing techniques, tainted waste can be eliminated at source.

To use nanoscale particles for purification or in catalysis for environmental remediation, more research must be done on how they are produced. Nevertheless, the future for green nanotechnology appears promising and offers a lot of potential to achieve a balance between harmful materials and sustainable practices. Green nanotechnology will be crucial in this area going forward.

ACKNOWLEDGMENTS

We would like to thank all the authors who helped to build the repository based on background analysis of nanomaterials.

CONFLICT OF INTEREST

The authors declare no conflict of interest.

REFERENCES

1. Gadekar JM, Kadam KD. Green innovation. *Nanotechnol.* 2014;3(1):1–4.
2. Nakum J, Bhattacharya D. Various green nanomaterials used for wastewater and soil treatment: A mini-review. *Front Environ Sci.* 2022;9(January):1–9.
3. Patra JK, Baek K. Factors affecting synthesis and characterization techniques. *Green Nanobiotechnol.* 2014; 20(4):101–118.
4. Bharti B, Kumar R, Kumar H, Li H. Advanced applications and current status of green nanotechnology in the environmental industry. *J Pharmacol.* 2021:303–340.
5. Adhikary K, Mandal T, Majumder J, Chandra B, Vishwavidyalaya K, Chandra B, et al. Applications of green nano materials and its importance: A review. *Int J Creat Res Thoughts* 2020;8(6):2136–2143.
6. Kanwar R, Rathee J, Salunke DB, Mehta SK. Green nanotechnology-driven drug delivery assemblies. *ACS Omega* 2019;2(1):111–118.
7. Ahmad S, Munir S, Zeb N, Ullah A, Ali J, Bilal M, et al. A review on green synthesis of silver nanoparticles: An ecofriendly approach. *Green Nanobiotechnol.* 2019;5(3):5087–6107.
8. Mahajan G, Kaur M, Gupta R. Green functionalized nanomaterials: Fundamentals and future opportunities. *Nano and Microtechnologies* 2022;4(1):18–21.

9. Gurtu MG, Sharma M. Major applications of green nanotechnology in sustainable development and its impact on global environment: A review. *Nat Vol Essent Oil.* 2021;8(4):12823–12851.
10. Kammakakam I, Bara J. As featured in: Materials advances. J. Nanosci. 2020;2(1):11–15.
11. Okafor F, Janen A, Kukhtareva T, Edwards V. Green synthesis of silver nanoparticles, their characterization, application and antibacterial activity. *J Pharmacol.* 2013:5221–5238.
12. Nikhat F, Satyanarayana D, Joshia AB. Phytological and pharmacological investigation of roots of *Syzygium cuminii (L) Skeel. Asian J Res Chem.* 2008:22–25.
13. Nikhat F, Satyanarayana D, Subhramanyam EVS. Isolation, characterization and screening of antioxidant activity of the roots of *Syzygium cuminii (L) Skeel. Asian J Res Chem.* 2009:218–221.
14. Farhana N et al. Fundamental principles to address green chemistry and green engineering for sustainable future. InTechOpen. 2022;456–465.
15. Muhammad N, Khan AU, Tahir K. Biomedical applications of green-synthesized Nobel metal nanoparticles. *J Photochem Photobiol B Biol.* 2017;10(4):10–15.
16. Rani K, Sridevi V. An overview on role of nanotechnology in green and clean technology. *Pharm. Nanosci.* 2017;2(3):245–250.
17. Kumari MM, Jacob J, Philip D. Green synthesis and applications of Au–Ag bimetallic nanoparticles. *Spectrochim ACTA PART A Mol Biomol Spectrosc.* 2015;137:185–192.
18. Manuel C, Toni U, Allan P, Katrin L. Challenges in determining the size distribution of nanoparticles in consumer products by asymmetric flow field-flow fractionation coupled to inductively coupled plasma-mass spectrometry: The example of Al_2O_3, TiO_2, and SiO_2 nanoparticles in toothpaste. *Separations* 2018;5:56.
19. Hua M, Zhang S, Pan B, Zhang W. Heavy metal removal from water/wastewater by nano-sized metal oxides: A review. *J Hazard Mater.* 2012;211:317–331.
20. Lu Y, Ozcan S. Green nanomaterials: On track for a sustainable future. *Future J Pharm Sci.* 2015;2(1):41–51.
21. Datta JK. Green synthesis of silver nanoparticles and its application for mosquito control. *Green Nanotechnol.* 2014;4(3):10–18.
22. Yousaf H, Mehmood A, Shafique K, Raffi M. Green synthesis of silver nanoparticles and their applications as an alternative antibacterial and antioxidant agents. *Mater Sci Eng C [Internet].* 2020;112(October 2019):110901.
23. Lu Y, Ozcan S. Green nanomaterials: On track for a sustainable future. *Nano Today [Internet]* 2015;10:1–4.
24. Karthik KV, Raghu AV, Raghava K, Ravishankar R, Sangeeta M, Shetti NP, et al. Green synthesis of Cu-doped ZnO nanoparticles and its application for the photocatalytic degradation of hazardous organic pollutants. *Chemosphere [Internet].* 2022;287(P2):132081.
25. Verma A, Gautam SP, Bansal KK, Prabhakar N, Rosenholm JM. Advancement in green nanotechnology. *J Ethnopharmacol.* 2019;4(1):101–118.
26. Devatha CP, Thalla AK. Chapter 7—Green synthesis of nanomaterials. In Mohan Bhagyaraj, S, Oluwafemi, OS, Kalarikkal, N, Thomas, S (Eds.), Synthesis of Inorganic Nanomaterials, pp. 169–184, Woodhead Publishing, Sawston, UK, 2018.
27. Chandra H, Kumari P, Bontempi E, Yadav S. Medicinal plants: Treasure trove for green synthesis of metallic nanoparticles and their biomedical applications. *Biocatal Agric Biotechnol [Internet].* 2020;24(January):101518. Available from: https://doi.org/10.1016/j.bcab.2020.101518
28. Hoseinpour V, Ghaemi N. Green synthesis of manganese nanoparticles: Applications and future perspective: A review. *J Photochem Photobiol B Biol [Internet].* 2018;189(August):234–243.
29. Shankar S, Rakhi D. Saraca indica bark extract mediated green synthesis of polyshaped gold nanoparticles and its application in catalytic reduction. *Green Nanoparticles* 2014;485–40.
30. Ijaz I, Gilani E, Nazir A, Bukhari A. Detail review on chemical, physical and green synthesis, classification, characterizations and applications of nanoparticles. *Green Chem Lett Rev.* 2020; 185–190.
31. Singh J, Dutta T, Kim KH, Rawat M, Samddar P, Kumar P. 'Green' synthesis of metals and their oxide nanoparticles: Applications for environmental remediation. *J Nanobiotechnol.* 2018;3(2):1–24.
32. Benelli G, Lukehart CM. Special issue: Applications of green-synthesized and entomology. *J Clust Sci.* 2017;28(1):1–2.

33. Rana A, Yadav K, Jagadevan S. A comprehensive review on green synthesis of nature-inspired metal nanoparticles: Mechanism, application and toxicity. *J Clean Prod [Internet]*. 2020;272:122880. Available from: https://doi.org/10.1016/j.jclepro.2020.122880
34. Afique M, Sadaf I, Rafique MS, Tahir MB. A review on green synthesis of silver nanoparticles and their applications. *Artif Cell Nanomed Biotechnol*. 2017;3(1):127–135.
35. Kamran U, Bhatti HN, Iqbal M, Nazir A. Green synthesis of metal nanoparticles and their applications in different fields. *J Green Technol*. 2019;7(4):123–133.
36. Narayanan M, Divya S, Natarajan D. Green synthesis of silver nanoparticles from aqueous extract of Ctenolepis garcini L. and assess their possible biological applications. *Process Biochem [Internet]*. 2021;107(May):91–9.
37. Mousavi SM, Hashemi SA, Ghasemi Y, Amani AM, Dashtaki AS, et al. Green synthesis of silver nanoparticles toward bio and medical applications: Review study. *Artif Cell Nanomed Biotechnol [Internet]*. 2018;46(S3):S855–872.
38. Tarannum N, Gautam YK. Nanoparticles: A state-of-the-art review. *J Nanotechnol*. 2019;34926–34948.
39. Davar F, Majedi A, Mirzaei A. Green synthesis of ZnO nanoparticles and its application in the degradation of some dyes. *J Am Ceram Soc*. 2015;1746:1739–1746.
40. Princy G. A review on green synthesis of silver nanoparticles. *Asian J Pharmaceut Clin Res*. 2013;(March):10–15.
41. Bin AM. Cerium oxide nanoparticles: Green synthesis and biological applications. *J Nanosci*. 2017:1401–1413.
42. Samuel MS, Ravikumar M, Ashwini J, Selvarajan E, Patel H, Chander PS, et al. A review on green synthesis of nanoparticles and their diverse biomedical and environmental applications. *Ethnopharmacol*. 2022:245–250.
43. Khan AK, Rashid R, Murtaza G, Zahra A. Gold nanoparticles: Synthesis and applications in drug delivery. *Trop J Pharm Res*. 2014; 13:1169–1177.
44. Centi M, Perathoner S. Carbon nanotubes for sustainable energy applications. *ChemSusChem* 2011; 4:913–925.
45. Qiao Y, Li CM. Nanostructured catalysts in fuel cells. *J Mater Chem*. 2011; 21:4027–4036.
46. Cardellina JH. Challenges and opportunities confronting the botanical dietary supplement industry. *J Nat Prod*. 2000; 65:1073–1084.
47. Williamson EM. Synergy and other interactions in phytomedicines. *Phytomedicine* 2001; 8:401–409.
48. Samaligy MS, Afif NN, Mahmoud EA. Increasing bioavailability of silymarin using a buccal liposomal delivery system: Preparation and experimental design investigation. *Int J Pharm*. 2006; 308:140–148.
49. Li Y, Dong L, Jia A, Chang X, Xue H. Preparation and characterization of solid lipid nanoparticles loaded traditional Chinese medicine. *Int J Biol Macromol*. 2006; 38:296–299.
50. Chen H, Chang X, Du D, Liu W, Yang X. Podophyllotoxin-loaded solid lipid nanoparticles for epidermal targeting. *J Control Release*. 2006; 110:296–306.
51. Lawrence MJ, Rees GD. Microemulsion-based media as novel drug delivery systems. *Adv Drug Deliv Rev*. 2000; 45:89–121.
52. Zhang H, Cui Y, Zhu S, Feng F, Zheng X. Characterization and antimicrobial activity of a pharmaceutical microemulsion. *Int J Pharm*. 2010; 395:154–160.
53. Goel A, Kunnumakkara AB, Aggarwal BB. Curcumin as curecumin: From kitchen to clinic. *Biochem Pharmacol*. 2008;75:787–809.
54. Lin YH, Lin JH, Chou SC, Chang S, Chung CC. Berberine-loaded targeted nanoparticles as specific Helicobacter pylori eradication therapy: In vitro and in vivo study. *Nanomedicine* 2015;10:157–171.
55. Bhatt RL, Vries P, Tulinsky J, Bellamy G, Baker B, Singer JW. Synthesis and in vivo antitumor activity of poly(L-glutamic acid) conjugates of 20(S)-camptothecin. *J Med Chem*. 2003;46:190–19.
56. Fonseca C, Simões S, Gaspar R. Paclitaxel-loaded PLGA nanoparticles: Preparation, physicochemical characterization and in vitro anti-tumoral activity. *J Control Release* 2002;4:273–286.
57. Harikumar PS, Aravind A. Antibacterial activity of copper nanoparticles and copper nanocomposites against Escherichia coli bacteria. *Int J Sci*. 2016;2:84–90.
58. Tran QH, Van Quy N, Le AT. Silver nanoparticles: Synthesis, properties, toxicology, applications and perspectives. *Adv Nat Sci*. 2013;4:1–21.

59. Vijay Kumar PPN, Pammi SVN, Kollu P, Satyanarayana KV, Shameem VU. Green synthesis and characterization of silver nanoparticles using Boerhaavia diffusa plant extract and their anti-bacterial activity. *Ind Crops Prod.* 2014;52:562–566.
60. Dhuper S, Panda D, Nayak PL. Green synthesis and characterization of zero valent iron nanoparticles from the leaf extract of Mangifera indica. *Nano Trends J Nanotechnol Appl.* 2012;13:16–22.
61. Kalishwaralal K, Deepak V, Pandian SRK, Kottaisamy M, Barath Mani Kanth S, Kartikeyan B, Gurunathan S. Biosynthesis of silver and gold nanoparticles using Brevibacterium casei. *Colloids Surf B Biointerfaces* 2010;77:257–262.
62. Narayanan KB, Park HH. Antifungal activity of silver nanoparticles synthesized using turnip leaf extract (Brassica rapa L.) against wood rotting pathogens. *Eur J Plant Pathol.* 2014;140:185–192.
63. Mondal SN, Roy RA, Laskar I, Sk S, Basu D. Biogenic synthesis of Ag, Au and bimetallic Au/Ag alloy nanoparticles using aqueous extract of mahogany (Swietenia mahogani JACQ) leaves. *Colloids Surf B Biointerfaces* 2011;82:497–504.
64. Sharma R, Rana V. Effect of carboxymethylation on rheological and drug release characteristics of Terminalia catappa gum. *Carbohydr Polym.* 2017;175:728–738.
65. Chandran SP, Chaudhary M, Pasricha R, Ahmad A, Sastry M. Synthesis of gold nanotriangles and silver nanoparticles using Aloe vera plant extract. *Biotechnol Prog.* 2006;22:577–583.
66. Petros RA, DeSimone JM. Strategies in the design of nanoparticles for therapeutic applications. *Nat Rev Drug Discov.* 2010;9:615–627.
67. Brewer E, Coleman J, Lowman A. Emerging technologies of polymeric nanoparticles in cancer drug delivery. *J Nanomater.* 2011; 2011: 1–10.
68. Lim ZZJ, Li JEJ, Yung LYL, Bay BH. Gold nanoparticles in cancer therapy. *Acta Pharmacol Sin.* 2011;32:983–990.
69. Boisselier E, Astruc D. Gold nanoparticles in nanomedicine: Preparations, imaging, diagnostics, therapies and toxicity. *Chem Soc Rev.* 2009;38:1759–1782.
70. Kundu S, Ghosh SK, Mandal M, Pal T. Silver and gold nanocluster catalyzed reduction of methylene blue by arsine in micellar medium. *Mater Sci.* 2002;25:577–579.
71. Justin T, Seil JT, Webster TJ. Antimicrobial applications of nanotechnology: Methods and literature. *Int J Nanomed.* 2012;7:2767–2781.
72. Cao YW, Jin R, Mirkin CA. DNA-modified core–shell Ag/Au nanoparticles. *J Am Chem Soc.* 2001;123:7.
73. Wegner K, Schimmöller B, Thiebaut B, Fernandez C, Rao TN. Pilot plants for industrial nanoparticle production by flame spray pyrolysis. *KONA Powder Part. J.* 2011;29:251–265. [Google Scholar] [Green Version].
74. Nikhat F. Meticulous endorsement of Cassia fistula Linn. *Asian J. Res. Chem.* 2019;12(3):169–181.

2 Current Trends in Green Nanotechnology

Recent Advances, Sustainable Future Prospects

Shubneesh Kumar, Ekta Sirohi, Aakash Upadhyay, Sumel Ashique, Subhra Prakash Bhattacharyya, Indrani Bhattacharyya, Neeraj Mishra, Arshad Farid, Afzal Hussain, Ashish Garg, and Farzad Taghizadeh-Hesary
Corresponding authors: Sumel Ashique, Neeraj Mishra

2.1 INTRODUCTION

Materials science, physics, chemistry, and biology are just a handful of the disciplines that nanotechnology unites. Green nanotechnology (GN) is a key component of clean technologies intended to clean up the environment and turn extra-bioactive items into green nanomaterials that are more profitable and environmentally benign. Along with population expansion, human biowaste is rising, posing a serious hazard to environmental purity. Accordingly, it is crucial to progress a safe as well as practical solution to this issue, which has a substantial environmental cost (Figure 2.1).

Green nanotechnology has been proposed as a solution to societal issues because of its extensive range of applications, which include nanoelectronics, nanomedicine, food, packaging, energy, composite materials, construction, environmental remediation, coatings, water treatment, and agriculture. It is not surprising that the application of nanostructured supplies has raised concerns regarding environmental and human health wellbeing, also fostering the development of a subdiscipline devoted to safe-by-design and ecofriendly solutions. Innovative strategies could, for instance, replace damaging processes or goods with acceptable ecofriendly ones to lessen the risks that nanomaterials represent to the atmosphere and to human wellbeing in the span of their lifetime. According to several studies [1–6], nanotechnology is a vibrant technology that unites scientists from several fields, including biology, chemistry, materials science, and physics. Owing to its varied range of applications, including nanomedicine, packaging, food, nanoelectronics, composite materials, coatings, energy, construction, environmental remediation, water treatment, and agriculture [7,8], it has been suggested as a solution to societal problems. It should come as no surprise that the usage of nanostructured materials has led to safety issues for human wellbeing and the atmosphere [5,9,10], backing the extension of a subdiscipline devoted to safe-by-design and green solutions [1,2,6]. Innovative approaches could reduce the dangers that nanomaterials pose to the troposphere and human wellbeing throughout the span of their use, for instance by substituting acceptable ecofriendly processes or products for harmful ones [11].

In order to create, produce, utilize, and dispose of nanomaterials in a sustainable manner [12], green nanotechnology relies on the concepts of green synthesis. While environmental studies may

DOI: 10.1201/9781003389408-2

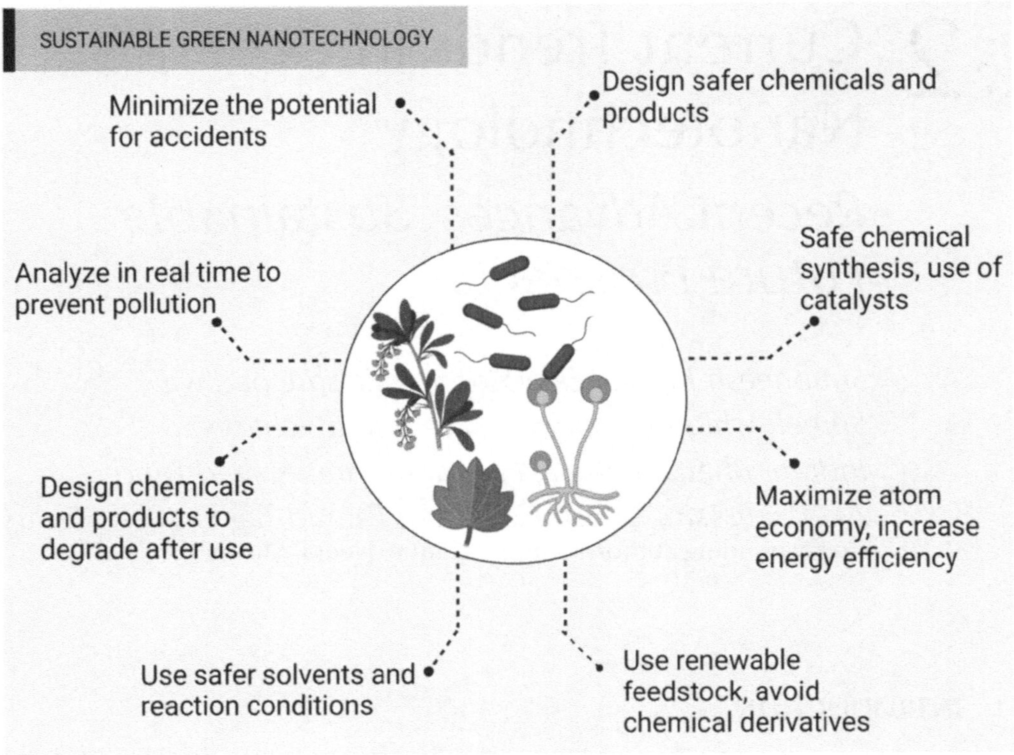

FIGURE 2.1 Various advantages of green sustainable nanomaterials.

only take a few hundred years or so into account, future thinking requires planning on a hundred-year time scale. Whether it is hereditarily modified stem cell research, food, or nanotechnology, the long-term usefulness of new technologies is a major concern nowadays. This presentation also includes future perspectives, inspiration to adopt GN industries globally, and the implementation of advanced technology within the United Nations sustainable development goals (SDGs). According to this chapter, using GC choices, merging life cycles as well as risk evaluation, applying feasible manufacturing techniques, and using life-cycle thinking to inform product design are all viable alternatives for the global ecosystem.

2.2 OBJECTIVES OF THIS CHAPTER

The methodical visions of green nanotechnology, green synthesis, and wide scientific boundaries are now being surpassed. The cornerstones of modern logical investigation in the domain of nanotechnology and of green nanotechnology are technological vision and systematic justification. The activities discussed in this chapter include a great deal of in-depth scientific reflection. The weightiness of green nanotechnology, green synthesis, and green chemistry is an effect of the scientific prospect being confronted by the extensive implementation of nanotechnology besides the health risks they create. Specifically, attention to up-to-the-minute scientific bids in green nanoscience include the hefty field of groundwater remediation as well as its anticipation for the future. Ingenuities in the field of nanotechnology are currently focused on sustainable chemistry, green engineering, and green chemistry. In this review, the immense scientific latent of green nanoadvancement is precisely outlined. The fundamental tenets of this chapter are deep scientific comprehension, broad scientific reflection, and the revelation of green technology and discipline [13,14].

2.3 THE SCIENTIFIC DOCTRINE AND TRUTH BEHIND NANOTECHNOLOGY APPLICATIONS

At the moment, nanodisciplines require vast technical foresight and profound technical patience. The focus of modern scientific study is on green science and green engineering. Future-oriented thinking, scientific philosophy, and scientific truth are precursors to a more complete deliverance of green nanoscience. The contemporary building of science is being substantiated by the water scarcity state of affairs. Groundwater poisoning with heavy metals and arsenic is transforming human culture and science. At every single step of the modern human lifespan, the philosophy of the arts of nanotechnology and green nanotechnology needs to be rethought and re-explained. The pitches of sustainable chemistry and green nanotechnology are being refashioned by hi-tech revelations and scientific corroboration. Environmental engineering science is currently being more fully realized thanks to the concept of conservational and energy sustainability along with its instantaneous necessities. The two sides of the coin are environmental engineering and green nanotechnology.

Movement in the direction of today's innovative scientific rejuvenation and reconstruction is through scientific doctrine as well as cognizance. Ecofriendly engineering problems and comprehensive aquatic constraints are genuinely altering the nature of scientific research. Groundwater contamination from heavy metals and arsenic are today's perplexing problems brought on by scientific empowerment. The definite guidelines for conducting scientific research today are technological validation and scientific revelation. The vast scientific rigor of hominoid evolution, the academic rigor behind current environmental engineering research, and the numerous nanotechnology applications are all milestones along a lengthy and challenging road to the actual deliverance and fulfillment of green chemistry and nanotechnology. The tactic of prodigious reconstruction and temperance is veiled by technical modernization and perfection. This chapter places special emphasis on the effective use of green nanoadvancement for the sole motive of evolving engineering to science [13,14].

2.4 SCIENTIFIC ENDEAVORS IN THE FIELD OF GREEN NANOTECHNOLOGY

The road to weighty scientific revitalization and scientific forbearance is through green technology and sustainable engineering. Today's scientific advances are ushering in a more progressive and imaginative era of green nanotechnology. The gigantic scientific perimeter is being hard pressed by the wide-reaching water problematic, the scarcity of drinking water, and the heavyweight metal toxic condition of pounded water. This chapter deliberately emphasizes the enormous scientific tolerance and vision required in the quest of green expertise. The OECD's Science, Technology and Industry Policy Paper No. 5 (2013) significantly redefined nanotechnology designed for use in green modernization [15]. The OECD Working Party on Nanotechnology (WPN) has engaged in talks and projects that have yielded a wealth of information that is crucial to the advancement and application of nanotechnology intended for mammoth revolution, but also environmentally sustainable growth [15]. The motive of this study is to provide contextual knowledge for impending work on nanotechnology's use in green engineering and innovation by the WPN.

A new epoch in the progression of technology and science is being shepherded in by means of green nanotechnology [15]. A grander manumission of nanoadvancement and nanotechnology is headed by hi-tech visualization, systematic want, and all-embracing academic diligence. Due to enormous difficulties, humanity is currently in a position of great scientific crisis. The authors' research highlights the enormous hurdles, serious scientific crises, and current triumph of nanoadvancement and nanotechnology. The overall argument is fragmented into three elements: a complete instigation in nanotechnology, techniques for promoting green modernization and green engineering by means of nanotechnology, and the effects of green nanotechnology [15]. There has never been a time when the demand for affordable and safe sustainable solutions to the world's problems in the areas of energy, the environment, and health has been more urgent. The foundation and lynchpin of contemporary scientific and engineering pursuits are energy and environmental sustainability [15]. Between 2010 and 2035, the demand for energy is projected to intensify by more than 30% [15]. Currently, more than 8,000 lakhs individuals globally lack adequate ingress to clean drinkable water. The scientific worldview of scientists, engineers, and policymakers, as well as that of politicians in industrialized and emerging economies, has changed as a result of these issues. The foundation of a higher revelation for a superior delivery of green nanoadvancement is scientific restraint and profound insight [15]. By improving energy proficiency, lowering waste or greenhouse gas emissions, and reducing the usage of unrenewable raw inventory, green innovation aims to lessen its negative effects on the environment.

Green nanotechnology and engineering currently represent cutting-edge technology and science. Today's environmental engineering research will go a long way toward environmental sustainability if scientific accomplishment, a visionary approach, and overcoming difficulties are all factors. The OECD WPN has created a figure of workable responsibilities since it started in 2007 that target new and expansive technology, knowledge, and modernization policies relevant to the sensible upgrading of nanotechnology [15]. Green nanotechnology, also known as nanotechnology of green innovation, attempts to create a wide array of scientific progressions that are innocuous, energy-conserving, waste-reducing, and significantly lower in greenhouse gas emissions. Technological validation is crucial to the advance of academic and scientific diligence. In their 2009 book, Karn and Bergeson et al. give clear insights into the enormous opportunities and risks existing in the current, wide field of green nanotechnology. A new technological era has begun thanks to the highly developed engineering skill and expertise of nanotechnology and nanoscience [16]. Today, green nanotechnology is redefining what it means to be successful in the field of applying nanotechnology to goods for use in human civilization. In their work [16], Karn and Bergeson investigate green nanotechnology and the numerous conceivable reasons why old chemical evaluation and managing systems would not be adequate for the advance of logical exploration. Green nanotechnology is prominent in a venture report by the Woodrow Wilson International Center for Scholars (WWIC-S) [17] with boundless aspects. The truthful discernment of nanotechnology for implementation in green technologies will be fueled by scientific goals, technological developments, and a forward-looking vision [17]. Global green nanopolicy, clean and green nanotechnology, and nano-enhanced green nanotechnology are all well observed in this report. The fundamentals of green chemistry, according to this paper, include waste prevention, better chemical design, less hazardous material design, and safer process synthesis. The use of catalytic agents, avoiding organic elements, minimizing costs, and envisioning reuse are also part of this vast scientific frontier.

Green chemistry correspondingly encompasses reducing the likelihood of quirks and successfully implementing chemical process safety. The scientific merit and profound scientific patience of green nanotechnology are pointed out with rigor in this work [17]. The current status of scientific vision is extremely distressing given advances in research and technology. The difficulty of this study project goes beyond what is imaginable for scientists and opens up new horizons for endeavor. This incisive analysis examines the activities of North American academic and research institutes [17].

Nath and Banerjee's [18] detailed analysis of the practice of green nanotechnology and its crossing point with health biology sheds light on the field's triumph. To expand their biological and medicinal uses, it is crucial to develop ecofriendly material synthesis technologies. Due to fresh ideas, technological depth and splendor are currently undergoing a massive regeneration [18]. They emphasize the classification of nanoparticles and the critical role played by living organisms in the creation of metallic nanoparticles in their investigation [18]. Knowledge of nanoparticles is constantly expanding and breaking new ground. Today, biomedical engineering is likewise pushing the boundaries of cutting-edge science. For green nanotechnology leaders in a new stage of development in the fields of nanotechnology and nanoscience, technological validity and depth are of the utmost importance. This learning also concentrates on the prospective benefits of biologically created nanoparticles in health biology fields, such as the management of cancer, targeted drug delivery, catalysis, the development of antibacterial mediators, and the building of biosensors [18].

Nanomaterials are at the forefront of scientific examination in nanotechnology, with distinctive diameters in the limit of 1–100 nm [18]. Numerous areas of applied research, from materials science to biotechnology, have shown tremendous interest in nanotechnology, principally in metallic nanomaterials. The physicochemical characteristics of materials containing nanoparticles differ significantly from those of majority constituents owing to the tremendously insignificant size and the high surface-to-volume ratio related to nanoparticles. The two pillars of recent scientific research are scientific tolerance and scientific revelation. The same authors discussed the sorting of nanoparticles [18], together with liposomes, superparamagnetic (SP) nanoparticles, fullerenes, and buckyballs, as well as carbon nanotubes, quantum dots, dendrimers, and liquid crystals. The blend of metallic nanoparticles using orthodox physical as well as chemical practices was meticulously covered by the authors. Techniques include hydrothermal synthesis, sol-gel synthesis, inert gas condensation, laser ablation, and a variety of colloidal techniques. The authors [18] also discussed the synthesis of green nanomaterials with bio-inspired design.

In their detailed discussion of green metal nanoparticle manufacturing, Virkutyte and Varma [19] undertook supplementary examination of green polymers in addition to enzymes in equilibrium and surface functionalization. A new epoch of scientific revelation and scientific erudition is being ushered in by engineering knowledge and green fusion technology. New advances in green nanotechnology have the potential to alter drastically a variety of currently used methods and goods that progress ecofriendly values, reduce waste, and preserve natural resources. At the moment, learning must navigate challenging terrain. New guidelines in scientific and engineering learning are being opened up by hi-tech improvements, new ways of thinking, and careful academic ethical consideration of green polymers and metallic nanoparticles [19].

With amazing vision and scientific patience, Schwarz [20] gave a clear overview of green nanotechnology. Recently, sustainability and the so-called green agenda have been linked to nanotechnology [20]. In their research, Schwarz and colleagues claim that deeper cognitive and societal systems are at work to identify all the possibilities of green nanotechnologies. At the moment, green nanotechnology is seen as a borderline conception where vastly different discourses are handled with a great amount of scientific insight. In line with its original purpose of evolving engineering and science, this work also gradually stimulates green sustainability and nanotechnology. Schwarz's research provides a broader perspective on the pertinency of green nanotechnology in German-speaking countries; primed green nanotechnology is crammed with precise revelation and embraces both technological legitimacy and scientific firmness.

Yehia et al. [21] debated the magnetic and structural landscapes of nanocrystalline spinel ferrite powders with great clarity. These are created with a cutting-edge, organically approachable sol-gel derivative. The inverse spinel structure is a feature of nickel ferrite ($NiFe_2O_4$). This treatise [21] credentials the discipline of nanotechnology besides its far-reaching scientific perception with persistence as well as a thorough comprehension of science. *Citrullus colocynthis* aqueous extracts were used in Shawkey et al.'s highly interesting examination of the antitumor efficacy of silver (Ag) nanoparticles [22]. There are two sides to today's innovative coin: biotechnology and green nanotechnology. Attributable to their conceivable use in nanomediscience, green production of metallic nanoparticles is a burgeoning area of research [22]. The study of nanomedicines is a complex and interdisciplinary field. In comparison to chemical synthesis, the green fusion of silver (Ag) nanoparticles (SNPs) is a more practical, affordable, and biologically friendly method [22]. Modern technology has made significant scientific advances. Today's engineering discipline for green nanotechnology is overflowing with deep systematic consideration and scientific discovery. Hydrated extracts of the leaves, roots, seeds, and fruits of *C. colocynthis* were used as capping and reducing agents to synthesize SNPs from $AgNO_3$ [22]. Due to their diverse physical characteristics, biological labeling, chemical reactivity, biosensing, antibacterial, drug transport, antiviral activity, exposure of genomic oddities, gene therapy, and DNA sequencing, nanoparticles of free metals have received substantial research [22]. Green nanotechnology is inaugurating innovative spaces of scientific modernization and deep systematic predisposition for the years ahead. Current exploration focuses on the antineoplastic activity of greenly created silver (Ag) NPs on melanoma cells [22]. The lengthy and difficult path to the accurate liberation of nanomediscience is being paved by green nanotechnology, the ultramodern discoveries of malignancy biology, and deep speculative and scientific diligence [22].

Hutchison [23] emphasized greener nanoscience by means of a positive strategy for evolving applications and curtailing the ramifications of nanoscientific skills with incredible vision and clarity. The antecedents of modern research include scientific grasp, scientific erudition, and deep scientific comprehension [23]. To truly realize and use green nanotechnology, one must have a broad view of the future and overcome technical and scientific obstacles. Although there is concern around the undesirable effects on human health and the environment of the manufacture and widespread use of nanoproducts, nanotechnology continues to propose novel constituents and implementations which will provide humanity with serious improvements [23]. Through coordinated study of the uses and effects of novel materials, the complexity of nanomaterials is the main topic in this field. At the cutting edge of nanoscience today is greener nanoscience. The sophistication of technology as well as the technical considerations related to green nanotech science are undeniably altering the nature of methodical investigation. The goal of addressing global nanomaterials and green technology challenges is a clear focus of Hutchison's thesis [23].

Albrecht et al.'s study provides lucid clarification of greener chemistry and the impacts of nanoparticles on social healthiness [24]. Amazing advances in nanotechnology have occurred despite the fact that there are serious health risks associated with it. Except for current legislation covering the same material in bulk form, there are essentially no regulations that specifically apply to nanoparticles [24]. Today's technology and engineering fields must overcome significant scientific challenges. This review focuses on the possible negative consequences of nanoparticles on human health as well as the medicinal uses of nanoparticles, such as imaging, medicine administration, cleaning, and tissue restoration. Science and expertise are progressively moving in the direction of a more futuristic, greener nanotechnology and green chemistry [24].

Greener nanotechnology was defended influentially by Fagan et al. [25], along with the creation of nanoparticles for environmental and energy purposes. Today's green nanotechnology expertise and engineering disciplines go beyond conventional limitations. The comprehensive debate concerning the fusion of some nanomaterials for use in greener nanotechnology is explained by Fagan et al., whose study gives special attention to development in growing fields like the environment and energy [25]. As human society moves into a new, systematic epoch, the discoveries of green nanotechnology are reaching incredible scientific heights. This treatise goes into great detail about various techniques for creating nanostructured photocatalysts, plus sol-gel, microwave-assisted, hydrothermal, and sonochemical procedures [25]. These approaches include methods for producing materials like titanium (Ti) dioxide, iron oxide, zinc oxide, and metallic sulfides. The realization of nanotechnology research today will come at the end of a long and visionary journey, as will the ultramodern interpretation of nanotechnology, the target of systematic diligence and scientific keenness and grasp [25].

2.5 LATEST INNOVATIONS AND INSIGHTS INTO THE DOMAIN OF GREEN NANOTECHNOLOGY

This distinct observation, consisting of five original research articles [2–6], combines cutting-edge implementation, goods, machineries, and high-tech methods in a number of technical domains related to green nanotechnology; research on safe-by-design methods for nanomaterial synthesis besides (eco)toxicologic assessment of pioneering nano substances were also offered (Figure 2.1).

Gemini surface-active agents are suggested in terms of potential recyclable alternatives to cutting-edge surfactants, for example, to synthesize more environmentally friendly nanomaterials. Brycki et al. [3] offered an environmentally friendly synthesis of gemini surfactants manufactured using a solvent-free technique to stabilize AgNPs. The smallest AgNPs were produced with a molar ratio of nAg:nGemini = 5, a surplus of reductant, and the surfactant 16-6-16 as a stabilizing agent [3]. In the same vein, Kaczerewska et al. proposed that greener silica mesoporous nanocapsules (SiNC) might be synthesized using innovative gemini surfactants [2]. SiNC (silica nanocapsule) is frequently utilized in numerous applications owing to its distinctive characteristics. Nevertheless, its association with environmental issues arises from the presence of the cationic surface-active agent N-hexadecyl-N, N,N-trimethylammonium bromide (CTAB), which is encapsulated within the nanostructure. This compound is slowly released when SiNCs are dispersed in saltwater, with potential environmental implications [5,9,10,26–29]. Consequently, 1,4-bis-[N-(1-dodecyl)-N,N-dimethyl ammonium methyl]benzene dibromide (QSB2-12) was used by Kaczerewska et al. [2] as a minimally harmful prototype agent to replace CTAB. Silicate nanocapsules have shown promise in significantly reducing the harmful effects of nanoparticles on aquatic microalgae and micro crawfishes. These capsules act as a protective barrier, similar to older methods, shielding the organisms from direct exposure to the nanoparticles. This advance in nanotechnology holds great potential for safeguarding our aquatic ecosystems [2]. Future lines of research are routinely determined by existing societal problems, such as ecological restrictions, rather than simply by innovative scientific concepts.

The creation of risk-free, multifunctional nanoconstructed phospholipid transporters, which have been proposed for usage in medicinal domains, is another fascinating subject. It is preferable to employ natural lipids in medication delivery systems. In order to create more environmentally friendly nanostructured lipid carriers, Galvão et al. [4] evaluated the upshot of carvacrol on the crystallinity of compacted ordinary phospholipids (beeswax, stearic acid, and carnauba wax). The authors proved the auspicious properties related to this monoterpenoid hydroxybenzene near the progress of green drug transport systems constructed on fatty nanoparticles by indicating that the higher the carvacrol extent, the lower the crystallinity of the solid substances of aimed fats [4]. The coatings industry has also used green nanotechnology to substitute hazardous substances or immobilize them for gradual controlled release [5,6,9,26]. Mennucci et al. [6] illustrated the robust anti-rust properties of nanostructured nickel coatings, suggesting a novel alternative to chromium textures typically utilized in ornamental, optical, and solar applications. On the other hand, new research has shown that antifouling biocides, which are commonly used in maritime coatings and include DCOIT, Zn, and Cu pyrithiones, can be nanoencapsulated to dramatically reduce their toxicity and risk to marine species [9,10,26–29]. Santos et al. [5] showed that SiNC-DCOIT is less hazardous than unrestricted DCOIT throughout the larval expansion period and has a great antifouling efficiency toward aim primary life steps of the mussel *Perna*. This original insight highlights the advantages of encapsulating harmful substances in nanocarriers.

2.5.1 Natural and Renewable Sources of Reducing Agents

Plant extracts, vitamins, biopolymers, proteins, sugars (like glucose and fructose), and peptides (like glutathione) are just some of the common chemical compounds found in nature that can operate as superlative reducing agents for the creation of nanoparticles [30,31]. The most extensively researched category so far is plant extracts. Plant extracts are thought to be highly proficient natural reducing groups because of how common they are [32]. The manufacture of metallic nanoparticles, valuable in computer electronics and medicinal applications, and the use of extracts from plants as reducing agents is one area of remarkable success [33]. The use of gold and silver nanoparticles in biomedical applications, including the delivery of drugs and genes, has recently gained significant research attention. Plants, bacteria, algae, and fungi are employed as relatively harmless greener reduction groups to increase biocompatibility.

Elia et al. used extracts of four distinct plants as stabilizing and reducing agents to create golden nanoparticles [34]. These extracts included *Lippia citriodora*, *Salvia officinalis*, *Punica granatum*, and *Pelargonium graveolens*. A different group of natural deposits employed as stabilizing and reducing agents for the manufacture of metallic nanoparticles are biopolymers. Due to their widespread industrial application, these polymeric carbohydrate molecules are easily accessible for the mass manufacturing of nanoparticles. Chitosan, cellulose, and dextran, which are extracted from plants, crustaceans, sugar cane, and exoskeletons, are examples of biopolymers used for nanoparticle engineering [35–37]. A familiar natural antioxidant or reductant is vitamin C. Vitamin C can reduce metallic ions in a hydric solution to form metallic nanoparticles, much like other naturally occurring reducing agents [38]. It is interesting to note that by decreasing colloidal hydrated oxide of iron under hydrothermal surroundings, Fe_3O_4 nanoparticles may be produced [39]. Vitamins B and E are also employed in the creation of metal nanoparticles [39]. Despite vitamins being widely available on the market, their low cost may be a barrier to their use in commerce [40]. The cost-related difficulties faced by further natural deposits, such as peptides and proteins, make them unsuitable for the extensive manufacturing of nanostuffs [41–43].

2.5.2 The Importance of Sustainability for the Global Community

The impacts of biodiversity forfeiture, desertification, irresponsible dumping of waste, air pollution, toxic loads, water insecurity, plastic pollution, climate warming from fossil fuels, and macroclimate transformation are currently being experienced both separately and communally throughout the world. In many areas, social stressors result in ill health, insufficient access to healthcare, poor education, and inadequate skills development, all of which contribute to growing disparities. The future of civilization is unknown, but it is traveling down a terrible path. The UN SDGs were formulated to address the pressing political, economic, and environmental issues facing society [44]. These 17 goals are interconnected, and achieving one goal can help achieve the others. For example, science and technology are working hard to advance the following goals: good health (goal 3), clean water (goal 6), energy and the environment (goals 7, 13, 14, and 15), industrial modernization and sustainable communities (goals 9 and 11), and sustainable consumption and production (goal 12). Success here will result in less hunger and poverty (goals 1 and 2), increased equality, education, and decent work (goals 4, 5, 8, and 10), which will lead to peace, justice, and strong institutions (goal 16), the ultimate universal aim. Putting together partnerships (goal 17) to address up-to-date demands with incipient and experienced comprehensive capacity is the best method of achieving these goals [45]. Nanotechnology delivers disruptive, paradigm-shifting breakthroughs and inventions that allow quick responses to and explanations of difficulties upsetting our social order, the atmosphere, and the planet in the form of boosts to already-existing technologies. By creating appropriate supplies and allied strategies, breakthroughs in nanotechnology are transforming the fields of energy, environmental protection, supply management, and health care. Furthermore, advances in nanotechnology and nanoscience have enhanced communication capabilities, facilitating the integration of systematic, technical, therapeutic, and other communities, thereby impacting a range of interconnected fields [47].

2.5.3 Funding Allocated for Nanotechnology by Various National Funding Agencies

Many nations have made significant investments in nanotechnology during the past decade, including Canada, Australia, Japan, the Netherlands, the nations of the contributors to this chapter, and several others. For the past two years, federal funding in the USA for nanotechnology has exceeded USD 1800 billion [48]. Between 2010 and 2019 [49], the Natural Sciences and Engineering Research Council (NSERC, the prime national subsidy agency for STEM disciplines) funded CAD

0.432 billion in nanotechnology in Canada. The Netherlands' national nanotechnology program, NanoNextNL, received funding from the Dutch government totaling EUR 0.250 billion between 2011 to 2017 [50]. In 2018, Japanese institutions in Japan allocated USD 79 million, while academic institutions and national research laboratories invested JPY 58 billion yen, representing a total of 164,000 million (approximately USD 1.5 billion) in funding for nanotechnology research [51].

2.5.4 Natural Sources as Precursors of Various Nanomaterials

The status of greener path measures depends on the extent to which they make it possible to produce nanomaterials in precise and hygienic conditions, ensuring their recyclability. Supplementary measure for going green include using plant extracts and making nanoparticles with the help of fungi, bacteria, viruses, and algae (Figure 2.2).

2.5.4.1 Natural and Renewable Sources of Reducing Agents

Nature affords us with plentiful chemical matters that are suitable for use as reducing agents for the amalgamation of nanoparticles, proteins, plant extracts, vitamins, biopolymers, peptides (such as glutathione), and sugars (such as glucose and fructose) [30,33]. Plant extracts have received the most attention so far. Given their copious availability, they are regarded as the natural reducing agents with the greatest potential [31]. One particularly successful area is the amalgamation of metallic nanoparticles, advantageous in integrated circuit expertise and therapeutic claims, via plant extracts as reducing agents [32].

Recent advances in biomedical research have sparked interest in the use of gold and silver nanoparticles for a range of applications. These tiny particles show potential in areas such as drug delivery and facilitation of inheritable modifications. To enhance their biocompatibility, researchers are exploring the use of safe and environmentally friendly reduction agents derived from plants, microbes, algae, and fungi. Elia et al. produced gold nanoparticles using four different types of plant extracts (*Lippia citriodora*, *Salvia officinalis*, *Punica granatum*, *Pelargonium graveolens*) as

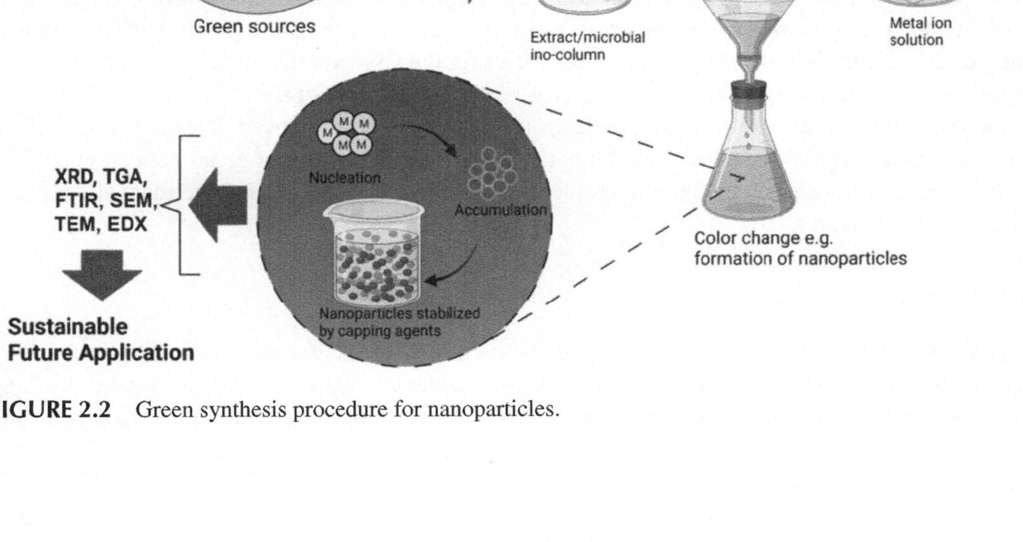

FIGURE 2.2 Green synthesis procedure for nanoparticles.

a reducing and stabilizing agent [34]. Biopolymers are natural sources used as reducing and stabilizing agents for metallic nanoparticle amalgamation. These polymeric simple carbohydrates have already been used in innumerable commercial contexts and consequently are regarded as having great potential for the construction of nanoparticles. Examples of biopolymers for nanoparticle amalgamation are chitosan, cellulose, and dextran from plants, crayfish exoskeleton, and sugar cane [35–37]. Vitamin C is a well-acknowledged natural reducing agent or antioxidant. Like natural reducing agents, vitamin C can reduce metallic ions in an aqueous solution to harvest metallic nanoparticles [38]. Interestingly, it can also be used to fashion Fe_3O_4 nanoparticles by reducing colloidal iron hydrated oxide in hydrothermal settings [39]. Other vitamin uses for metallic nanoparticle amalgamation involve vitamins B and E [40]. Notwithstanding the easy availability of vitamins, their low price may present difficulties for marketable demand. Other natural bases, for instance proteins and peptides, face challenges in relation to price and consequently are inappropriate for the large-scale construction of nanomaterials [41–43].

2.5.4.2 Natural Sources as Precursors for Carbon Nanomaterials

One evolving area of green nanoexpertise research concerns natural sources of carbon nanostuffs [44–46], which are advantageous in a host of contexts due to their unique characteristics. For example, vegetable oil has been shown to be a suitable source of high-superiority carbon nanotubes via a spray pyrolysis slant [47–49]. Consuming diverse catalytic agents, both single-fenced and multi-fenced carbon nanotubes can be fashioned by this method [50–52]. In another example, sugars and biopolymers have been used as sources of carbon quantum dots in microwave-assisted synthesis. Given the abundance of carbon-rich natural bases, it is anticipated that their use as a source of carbon nanostuffs will intensify rapidly in the next few years [52].

2.5.4.3 Nanocellulose: A Green Nanomaterial

As discussed above, a noteworthy example of green nanoexpertise is the extension of green interaction to manufactured nanostuffs. Indeed, there has been a growing interest and emphasis on natural nanomaterials as alternatives to synthetic products. This shift reflects a broader movement towards sustainability and ecofriendliness in various industries. By harnessing the properties of natural nanomaterials, such as nanoparticles derived from plants or other organic sources, researchers and industries aim to create more environmentally friendly products and processes. This trend aligns with the increasing awareness of the potential risks associated with synthetic materials and the desire to mitigate their negative impacts on the environment and human health. One example is nanocellulose, which is nanosized cellulose fibers or crystals fashioned from microbes or resulting from vegetation. Nanocellulose has unique characteristics in terms of being lightweight, transparent, and biocompatible. Like some other nanostuffs, nanocellulose is abundant and relatively easy to harvest. Because of this, nanocellulose is regarded as a highly suitable auxiliary to artificial nanostuffs in fields such as polymer nanocomplexes and transport of medications [53–55]. Nanocellulose aerogels have indeed found applications as templates for the fabrication of metallic oxide nanotubes. In this process, the nanocellulose aerogel acts as a support structure or scaffold onto which metallic oxide precursors are deposited. After the deposition, the aerogel is removed, leaving behind hollow metallic oxide nanotubes with a well-defined structure [56]. Interest in green nanocomplexes is snowballing, as evinced by the mounting quantity of papers in this field and the innovativeness of its commercialization.

2.5.5 Green Processing

Green nanotechnology is not only about utilizing natural materials for nanoparticle synthesis; it is also about developing ecologically friendly and sustainable processes. This encompasses everything from the materials used to the energy consumption and waste produced during the manufacturing process. By focusing on sustainability and environmental impact, green nanotechnology aims to minimize harm to the environment while still harnessing the unique properties of nanoparticles for various applications. As alternatives to organic solvents, aquatic and supercritical (SC) carbon dioxide have been explored in this area [57–58]. Different alternatives to traditional heating are particularly intriguing [59]. The so-called hydrothermal technique, which uses water as the reaction medium, is the most popular green approach. Another interesting development involves the use of the Leidenfrost effect by Abdelaziz et al. to produce nanoparticles and coatings on complex objects by overheating and charging a green chemical reactor [60]. As sources of heat for the fusion of nanoparticles, microwave energy [59] and concentrated sunlight [61] have also been investigated. Although these methods work well in the lab, putting them to use in commercial production comes with a lot of difficulties. For instance, microwave heating's quick reaction times are only applicable to small-scale reactions. The inconsistent intensity of sunshine and the need for a magnifier restrict the use of intensive sunlight. Nevertheless, recent developments in flow chemical processes have improved the financial viability of green chemistry [61].

From harnessing natural resources for initial stages to the synthesis of nanoparticles, one facet of green nanotechnology involves utilizing natural materials and processes. By leveraging these resources, researchers aim to develop ecofriendly methods for nanoparticle production. In this connection, liquid and supercritical CO_2 have been inspected as substitutions to carbon-based diluters [58]. Of specific interest are the many alternatives to orthodox heating [62]. In another exciting development, Abdelaziz et al. recently used the Leidenfrost effect to generate an enflamed and charged green biochemical receptacle to engineer nanoparticles and glazes on composite items [60].

The use of microwave oven energy and engrossed sunbeams [61] as auxiliary warmth sources for the amalgamation of nanoparticles offers considerable potential. These unconventional methods provide alternative ways to control and manipulate the synthesis process of nanoparticles. Microwave irradiation, for instance, can provide rapid and uniform heating, leading to efficient nanoparticle formation [62]. On the other hand, harnessing sunbeams can be environmentally friendly and cost-effective, especially in regions with abundant sunlight. These approaches showcase the versatility and creativity in nanoparticle synthesis, opening doors to innovative applications in various fields such as materials science, medicine, and environmental engineering. The auxiliary warmth sources are circumscribed by their inconsistency and the need for a lens. Recently, however, flow biochemical progressions have been identified that may make green interaction more cost-effective [63].

2.5.6 Nanotechnology and Materials

Materials and nanotechnology are predicted to be crucial to the implementation of Society 5.0. By producing a variety of nanodevices for use in Society 5.0, including IoT sensors, self-directed automobiles, and intelligent automatons, nanotechnology will be the driving force behind the digital transition. In addition, nanotechnology and its constituents are projected to help form a sustainable humanity by safeguarding water purity [59], reducing CO_2 emissions, and encouraging reuse and recycling. Wearable biosensors and biostuffs for renewing medication are two additional ways in which nanotechnology and materials can assist human health and wellbeing, despite the technological difficulties.

2.5.7 Society 5.0 and the SDGs

We acknowledge six technological issues in nanotechnology and resources for Society 5.0. In order to bring about innovations in computing, these encounters include IoT edge and quantum devices and AI chips; high-level security and short impression carriage; nano-biotechnology drug delivery and medical care; service automatons concomitant with and auxiliary individuals; smart resources to empower sustainability of air, water, and supplies; and energy resources and diplomacies to empower renewable energy techniques and energy reserves. Addressing these complications will help Society 5.0 become a reality in terms of technology [51,64]. From the standpoint of human development, Society 5.0 is central not only to digital advances but also to building a new knowledge society that is sustainable and leaves no one behind. Because the SDGs take a broad view of our future society as a human-focused society, there is a lot of overlap and consistency between the idea of Society 5.0 and the SDGs.

2.5.8 Opportunities for Technology and Sustainability

Nanotechnology has indeed opened up a vast array of possibilities, and the advances in this field continue to amaze. From medicine to electronics, energy to materials science, nanotechnology has the potential to revolutionize various industries, and there is much untapped potential waiting to be explored. While many green nanotechnologies show promise at the lab scale, scaling up production for industrial applications can be challenging. The regulatory landscape for nanomaterials is still evolving, with different standards and guidelines across different regions. Clear and consistent regulations are needed to ensure the safe and responsible development and deployment of green nanotechnologies. Green nanotechnology often involves interdisciplinary approaches that require integration with existing technologies and industries. Overcoming compatibility issues and fostering collaboration between different sectors is crucial for successful implementation. However, there may be risks as well as advantages associated with nanotechnology research and development. The understanding of the effects of nanotechnology on human health and the atmosphere still has gaps [65,66]. Ensuring the safe and sustainable use of nanotechnology is crucial for its continued progress and successful application in various fields. Hence, in spite of the potential for nanotechnologies to improve the likelihood of achieving each of the nine ecofriendly planetary boundaries (climate variation, biosphere integrity, land-system transformation, freshwater use, biogeochemical streams, ocean acidification, atmospheric aerosol smog, stratospheric ozone weakening, and release of innovative elements) [67], the ultimate goal for nanotechnology is to overcome conventional hazards. This chapter examines the primary uses of GN in sustainable development (SD) and their effects on the environment at large. Currently, nanotechnology (NT) has a lot of potential to help solve sustainability problems, but it is important to remember that nanomaterials can have damaging impacts on the atmosphere and human wellbeing. Even though nanoremediation technology has increased effectiveness and decreased costs, more research is still required to understand and prevent potential adverse environmental consequences (Figure 2.3).

2.5.9 Ongoing Challenges

Green nanotechnology, the application of nanotechnology to enhance environmental sustainability, is an evolving field facing several challenges even as it progresses [68]. Green nanotechnology holds significant promise for creating a more sustainable future by integrating environmentally friendly principles into the design, production, and application of nanomaterials and nanotechnology-enabled products. Green nanotechnology aims to minimize resource consumption and waste generation. [68]. In recent years, legally binding conservation protocols have had a significant impact on chemical innovativeness. Environmental protocols focus on reducing health risks and environmental hazards by minimizing exposure. However, many multinational companies are investing in delegated arrangements and end-of-pipe solutions rather than engaging in greener practices that would inherently make their products and processes safer. This is evidenced by the fact that biochemical commerce has spent more on cutting pollution than on new engineering projects [68]. Undeniably, the law has endorsed such alliances by providing financial backing for the commercialization of green nanoskills. Relevant administration supervisory codes and procedures for the

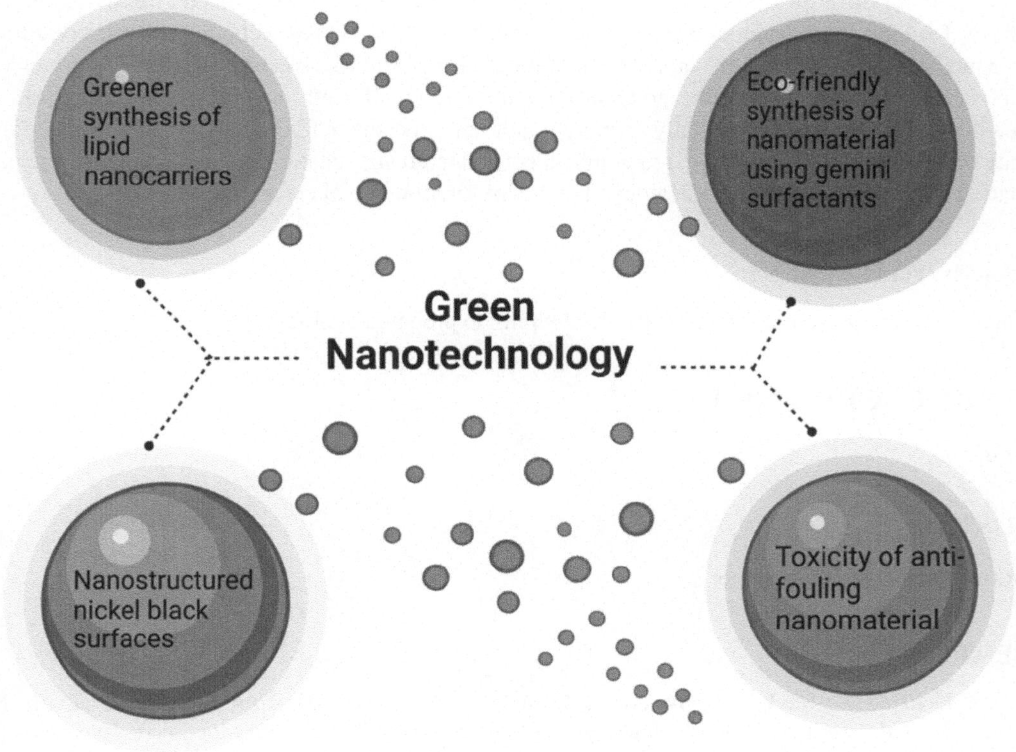

FIGURE 2.3 Main contributions to the field of green nanotechnology.

green construction of nanostuffs could boost the progress and enactment of green nanoskills on a commercial scale.

2.6 CONCLUSIONS AND FUTURE PERSPECTIVES

The engineering and science of green nanotechnology are escalating the frontiers of vision and advancing towards newer knowledge dimensions. According to Palit et al. [69–71], ecological sustainability and ecological engineering science are genuinely shifting the scientific field. Science and academic rigor will advance in the future thanks to nanotechnology, particularly green nanotechnology. Sustainable advance in green nanotechnology has its roots in hi-tech methods, scientific excellence, and ultramodern discoveries. Up-to-date skills are being applied to greener nanotechnology, sustainable chemistry, and greener chemistry. By emphasizing green engineering and workable development, this chapter offers a broader view of the science behind green nanotechnology. Today's pursuit of scientific inquiry is evolving due to scientific accomplishment, discovery, and moderation. As science progresses and time passes, biochemical process engineering and ecological engineering science must be continually re-envisioned. As we proceed towards a new era of planetary technology and nuclear-powered research, scientific resilience and achievements are on the way to rejuvenation. Today's knowledge is a goliath with its own size, vision, and scope. Given the indistinct visualization of the nanotechnology, ecological engineering, and green engineering disciplines today, technological validation and scientific perspicacity are prerequisites. With the progression of scientific and academic rigor, nanotechnology's scientific achievements will undoubtedly bear new and distinct fruits. The difficulties associated with green nanotechnology are equally significant and extensive. The enormous triumph of implementations of nanotechnology will undeniably stimulate innovation for the improvement of the discipline in the years to come. This chapter discusses the use of GN for long-term development. The development of clean-energy technologies and pollution-free methods of pollution removal is essential for the long-term survival of humanity. NT-grounded innovative technologies have the potential to contribute significantly towards the development of healthier and more environmentally friendly goods and procedures [67]. By 2030, GN applications may offer solutions for resolving, reducing, and cleaning up air, aquatic, and land-dwelling contamination and for increasing the efficiency of orthodox environmental cleanup machineries.

ACKNOWLEDGMENTS

The authors are grateful to BIT, Meerut for helping to prepare the draft.

CONFLICT OF INTEREST

None.

FUNDING

Not applicable for this chapter.

REFERENCES

1. Gottardo, S.; Mech, A.; Drbohlavová, J.; Małyska, A.; Bøwadt, S.; Riego Sintes, J.; Rauscher, H. Towards Safe and Sustainable Innovation in Nanotechnology: State-of-Play for Smart Nanomaterials. *NanoImpact* 2021, *21*, 100297. https://doi.org/10.1016/j.impact.2021.100297
2. Kaczerewska, O.; Sousa, I.; Martins, R.; Figueiredo, J.; Loureiro, S.; Tedim, J. Gemini Surfactant as a Template Agent for the Synthesis of More Eco-Friendly Silica Nanocapsules. *Appl. Sci.* 2020, *10* (22), 8085. https://doi.org/10.3390/app10228085
3. Brycki, B.; Szulc, A.; Babkova, M. Synthesis of Silver Nanoparticles with Gemini Surfactants as Efficient Capping and Stabilizing Agents. *Appl. Sci.* 2021, *11* (1), 154. https://doi.org/10.3390/app11010154
4. Galvão, J. G.; Santos, R. L.; Lira, A. A. M.; Kaminski, R.; Sarmento, V. H.; Severino, P.; Dolabella, S. S.; Scher, R.; Souto, E. B.; Nunes, R. S. Stearic Acid, Beeswax and Carnauba Wax as Green Raw Materials for the Loading of Carvacrol into Nanostructured Lipid Carriers. *Appl. Sci.* 2020, *10* (18), 6267. https://doi.org/10.3390/app10186267
5. Santos, J. V. N. dos; Martins, R.; Fontes, M. K.; Campos, B. G. de; Silva, M. B. M. do P. e; Maia, F.; Abessa, D. M. de S.; Perina, F. C. Can Encapsulation of the Biocide DCOIT Affect the Anti-Fouling Efficacy and Toxicity on Tropical Bivalves? *Appl. Sci.* 2020, *10* (23), 8579. https://doi.org/10.3390/app10238579
6. Mennucci, M. M.; Montes, R.; Bastos, A. C.; Monteiro, A.; Oliveira, P.; Tedim, J.; Ferreira, M. G. S. Nanostructured Black Nickel Coating as Replacement for Black Cr(VI) *Finish. Appl. Sci.* 2021, *11* (9), 3924. https://doi.org/10.3390/app11093924
7. Janczarek, M.; Endo-Kimura, M.; Wei, Z.; Bielan, Z.; Mogan, T. R.; Khedr, T. M.; Wang, K.; Markowska-Szczupak, A.; Kowalska, E. Novel Structures and Applications of Graphene-Based Semiconductor Photocatalysts: Faceted Particles, Photonic Crystals, Antimicrobial and Magnetic Properties. *Appl. Sci.* 2021, *11* (5), 1982. https://doi.org/10.3390/app11051982
8. Tiwari, E.; Singh, N.; Khandelwal, N.; Monikh, F. A.; Darbha, G. K. Application of Zn/Al Layered Double Hydroxides for the Removal of Nano-Scale Plastic Debris from Aqueous Systems. *J. Hazard. Mater.* 2020, *397*, 122769. https://doi.org/10.1016/j.jhazmat.2020.122769
9. Figueiredo, J.; Oliveira, T.; Ferreira, V.; Sushkova, A.; Silva, S.; Carneiro, D.; Cardoso, D. N.; Gonçalves, S. F.; Maia, F.; Rocha, C.; Tedim, J.; Loureiro, S.; Martins, R. Toxicity of Innovative Anti-Fouling Nano-Based Solutions to Marine Species. *Environ. Sci. Nano.* 2019, *6* (5), 1418–1429. https://doi.org/10.1039/C9EN00011A

10. Figueiredo, J.; Loureiro, S.; Martins, R. Hazard of Novel Anti-Fouling Nanomaterials and Biocides DCOIT and Silver to Marine Organisms. *Environ. Sci. Nano.* 2020, *7* (6), 1670–1680. https://doi.org/10.1039/D0EN00023J
11. Kaczerewska, O.; Martins, R.; Figueiredo, J.; Loureiro, S.; Tedim, J. Environmental Behaviour and Ecotoxicity of Cationic Surfactants towards Marine Organisms. *J. Hazard. Mater.* 2020, *392*, 122299. https://doi.org/10.1016/j.jhazmat.2020.122299
12. Gałuszka, A.; Migaszewski, Z.; Namieśnik, J. The 12 Principles of Green Analytical Chemistry and the SIGNIFICANCE Mnemonic of Green Analytical Practices. *TrAC Trends Anal. Chem.* 2013, *50*, 78–84. https://doi.org/10.1016/j.trac.2013.04.010
13. Wikipedia, Green Nanotechnology, Wikipedia, the free encyclopedia, 2017.
14. Wikipedia, Green Chemistry, Wikipedia, the free encyclopedia, 2017.
15. OECD, Nanotechnology for Green Innovation, OECD Science, Technology and Industry Policy Papers, No.5, OECD Publishing, 2013.
16. Karn, B. P.; Bergeson, L. L. Green Nanotechnology: Straddling Promise and Uncertainty. *Nat. Resour. Env.* 2009, *24*, 9.
17. Woodrow Wilson International Center for Scholars, *Project for Emerging Technologies, Green Nanotechnology: Its Easier Than You Think*, April 2007.
18. Nath, D.; Banerjee, P. Green Nanotechnology – A New Hope for Medical Biology. *Environ. Toxicol. Pharmacol.* 2013, *36* (3), 997–1014. https://doi.org/10.1016/j.etap.2013.09.002
19. Vaseghi, Z.; Nematollahzadeh, A.; Tavakoli, O. Green Methods for the Synthesis of Metal Nanoparticles Using Biogenic Reducing Agents: A Review. *Rev. Chem. Eng.* 2018, *34* (4), 529–559. https://doi.org/10.1515/revce-2017-0005
20. Schwarz, A. E. Green Dreams of Reason. Green Nanotechnology between Visions of Excess and Control. *Nanoethics* 2009, *3* (2), 109–118. https://doi.org/10.1007/s11569-009-0061-3
21. Yehia, M.; Labib, Sh.; Ismail, S. M. Structural and Magnetic Properties of Nano-NiFe2O4 Prepared Using Green Nanotechnology. *Phys. B: Conden. Matter.* 2014, *446*, 49–54. https://doi.org/10.1016/j.physb.2014.04.032
22. Shawkey, A. M.; Rabeh, M. A.; Abdulall, A. K.; Abdellatif, A. O. Green nanotechnology: anticancer activity of silver nanoparticles using Citrullus colocynthis aqueous extracts. *Adv. Life Sci. Technol.* 2013, *13*, 60–70.
23. Hutchison, J. E. Greener Nanoscience: A Proactive Approach to Advancing Applications and Reducing Implications of Nanotechnology. *ACS Nano.* 2008, *2* (3), 395–402. https://doi.org/10.1021/nn800131j
24. Albrecht, M. A.; Evans, C. W.; Raston, C. L. Green Chemistry and the Health Implications of Nanoparticles. *Green Chem.* 2006, *8* (5), 417–432. https://doi.org/10.1039/B517131H
25. Han, C.; Andersen, J.; Pillai, S. C.; Fagan, R.; Falaras, P.; Byrne, J. A.; Dunlop, P. S. M.; Choi, H.; Jiang, W.; O'Shea, K.; Dionysiou, D. D. Chapter Green Nanotechnology: Development of Nanomaterials for Environmental and Energy Applications. In *Sustainable Nanotechnology and the Environment: Advances and Achievements*; ACS Symposium Series; American Chemical Society, 2013; Vol. 1124, pp. 201–229. https://doi.org/10.1021/bk-2013-1124.ch012
26. Avelelas, F.; Martins, R.; Oliveira, T.; Maia, F.; Malheiro, E.; Soares, A. M. V. M.; Loureiro, S.; Tedim, J. Efficacy and Ecotoxicity of Novel Anti-Fouling Nanomaterials in Target and Non-Target Marine Species. *Mar. Biotechnol.* 2017, *19* (2), 164–174. https://doi.org/10.1007/s10126-017-9740-1
27. Gutner-Hoch, E.; Martins, R.; Oliveira, T.; Maia, F.; Soares, A. M. V. M.; Loureiro, S.; Piller, C.; Preiss, I.; Weis, M.; Larroze, S. B.; Teixeira, T.; Tedim, J.; Benayahu, Y. Antimacrofouling Efficacy of Innovative Inorganic Nanomaterials Loaded with Booster Biocides. *J. Mar. Sci. Eng.* 2018, *6* (1), 6. https://doi.org/10.3390/jmse6010006
28. Gutner-Hoch, E.; Martins, R.; Maia, F.; Oliveira, T.; Shpigel, M.; Weis, M.; Tedim, J.; Benayahu, Y. Toxicity of Engineered Micro- and Nanomaterials with Antifouling Properties to the Brine Shrimp Artemia Salina and Embryonic Stages of the Sea Urchin Paracentrotus Lividus. *Environ. Pollut.* 2019, *251*, 530–537. https://doi.org/10.1016/j.envpol.2019.05.031
29. Jesus, É. P. S. de; Figueirêdo, L. P. de; Maia, F.; Martins, R.; Nilin, J. Acute and Chronic Effects of Innovative Antifouling Nanostructured Biocides on a Tropical Marine Microcrustacean. *Mar. Pollut. Bull.* 2021, *164*, 111970. https://doi.org/10.1016/j.marpolbul.2021.111970

30. Virkutyte, J.; Varma, R. S. Green Synthesis of Metal Nanoparticles: Biodegradable Polymers and Enzymes in Stabilization and Surface Functionalization. *Chem. Sci.* 2011, *2* (5), 837–846. https://doi.org/10.1039/C0SC00338G
31. Iravani, S. Green Synthesis of Metal Nanoparticles Using Plants. *Green Chem.* 2011, *13* (10), 2638–2650. https://doi.org/10.1039/C1GC15386B
32. Kumar, V.; Yadav, S. K. Plant-Mediated Synthesis of Silver and Gold Nanoparticles and Their Applications. *J. Chem. Technol. Biotechnol.* 2009, *84* (2), 151–157. https://doi.org/10.1002/jctb.2023
33. *Sustainable Nanotechnology and the Environment: Advances and Achievements*; Shamim, N., Sharma, V. K., Eds.; ACS symposium series; American Chemical Society, 2013.
34. Elia, P.; Zach, R.; Hazan, S.; Kolusheva, S.; Porat, Z.; Zeiri, Y. Green Synthesis of Gold Nanoparticles Using Plant Extracts as Reducing Agents. *Int. J. Nanomed.* 2014, *9*, 4007–4021. https://doi.org/10.2147/IJN.S57343
35. He, F.; Zhao, D. Manipulating the Size and Dispersibility of Zerovalent Iron Nanoparticles by Use of Carboxymethyl Cellulose Stabilizers. *Environ. Sci. Technol.* 2007, *41* (17), 6216–6221. https://doi.org/10.1021/es0705543
36. Kharlampieva, E.; Slocik, J. M.; Tsukruk, T.; Naik, R. R.; Tsukruk, V. V. Polyaminoacid-Induced Growth of Metal Nanoparticles on Layer-by-Layer Templates. *Chem. Mater.* 2008, *20* (18), 5822–5831. https://doi.org/10.1021/cm801475v
37. Jiang, W.; Yang, H. C.; Yang, S. Y.; Horng, H. E.; Hung, J. C.; Chen, Y. C.; Hong, C.-Y. Preparation and Properties of Superparamagnetic Nanoparticles with Narrow Size Distribution and Biocompatible. *J. Magn. Magn. Mater.* 2004, *283* (2), 210–214. https://doi.org/10.1016/j.jmmm.2004.05.022
38. Nadagouda, M. N.; Varma, R. S. A Greener Synthesis of Core (Fe, Cu)-Shell (Au, Pt, Pd, and Ag) Nanocrystals Using Aqueous Vitamin C. *Cryst. Growth Des.* 2007, *7* (12), 2582–2587. https://doi.org/10.1021/cg070554e
39. Xiao, L.; Li, J.; Brougham, D. F.; Fox, E. K.; Feliu, N.; Bushmelev, A.; Schmidt, A.; Mertens, N.; Kiessling, F.; Valldor, M.; Fadeel, B.; Mathur, S. Water-Soluble Superparamagnetic Magnetite Nanoparticles with Biocompatible Coating for Enhanced Magnetic Resonance Imaging. *ACS Nano.* 2011, *5* (8), 6315–6324. https://doi.org/10.1021/nn201348s
40. Nadagouda, M. N.; Varma, R. S. Green and Controlled Synthesis of Gold and Platinum Nanomaterials Using Vitamin B2: Density-Assisted Self-Assembly of Nanospheres, Wires and Rods. *Green Chem.* 2006, *8* (6), 516–518. https://doi.org/10.1039/B601271J
41. Xie, J.; Zheng, Y.; Ying, J. Y. Protein-Directed Synthesis of Highly Fluorescent Gold Nanoclusters. *J. Am. Chem. Soc.* 2009, *131* (3), 888–889. https://doi.org/10.1021/ja806804u
42. Scott, D.; Toney, M.; Muzikár, M. Harnessing the Mechanism of Glutathione Reductase for Synthesis of Active Site Bound Metallic Nanoparticles and Electrical Connection to Electrodes. *J. Am. Chem. Soc.* 2008, *130* (3), 865–874. https://doi.org/10.1021/ja074660g
43. Baruwati, B.; Polshettiwar, V.; Varma, R. S. Glutathione Promoted Expeditious Green Synthesis of Silver Nanoparticles in Water Using Microwaves. *Green Chem.* 2009, *11* (7), 926–930. https://doi.org/10.1039/B902184A
44. United Nations. Department of Economic and Social Affairs Sustainable Development. The 17 Goals; https://sdgs.un.org/goals (accessed 2021-05-14)
45. International Network 4 Sustainable Nanotechnology. International Network 4 Sustainable Nanotechnology; https://network4sustainablenano.org/ (accessed 2021-05-14)
46. Sustainable Development Solutions Network (SDSN) Canada. Sustainable Development Solutions Network (SDSN) Canada; https://uwaterloo.ca/sustainable-development-solutions-network-canada/ (accessed 2021-05-14)
47. Zhao, Y.; Bai, C.; Brinker, C. J.; Chi, L.; Dawson, K. A.; Gogotsi, Y.; Halas, N. J.; Lee, S.-T.; Lee, T.; Liz-Marzán, L.; Miller, J. F.; Mitra, S.; Nel, A. E.; Nordlander, P.; Parak, W. J.; Rowan, A.; Rogach, A. L.; Rotello, V. M.; Tang, B. Z.; Wee, A. T. S.; Weiss, P. S. Nano as a Rosetta Stone: The Global Roles and Opportunities for Nanoscience and Nanotechnology. *ACS Nano* 2019, *13* (10), 10853–10855. https://doi.org/10.1021/acsnano.9b08042
48. United States National Nanotechnology Initiative. *NNI Budget*; www.nano.gov/about-nni/what/funding (accessed 2021-08-26)

49. Fortin, M. Nanotechnology: Advancing Foundational Technologies through Horizontal Initiatives. International Workshop on Nanotechnology for a Sustainable Future, Waterloo Canada, 2020; www.youtube.com/watch?v=BVJIlegMkxk (accessed 2021-11-03)
50. Rijksdienst voor Ondernemend Nederland. NanoNextNL; www.rvo.nl/subsidies-regelingen/projecten/nanonextnl (accessed 2021-09-20)
51. Pokrajac, L.; Abbas, A.; Chrzanowski, W.; Dias, G. M.; Eggleton, B. J.; Maguire, S.; Maine, E.; Malloy, T.; Nathwani, J.; Nazar, L.; Sips, A.; Sone, J.; van den Berg, A.; Weiss, P. S.; Mitra, S. Nanotechnology for a Sustainable Future: Addressing Global Challenges with the International Network4Sustainable Nanotechnology. *ACS Nano* 2021;*15* (12), 18608–18623. https://doi.org/10.1021/acsnano.1c10919
52. Titirici, M. M.; White, R. J.; Brun, N.; Budarin, V. L.; Su, D. S.; Del Monte, F.; Clark, J. H.; MacLachlan, M. J. Sustainable Carbon Materials. *Chem. Soc. Rev.* 2015;*44* (1):250–290. https://doi.org/10.1039/C4CS00232F
53. Habibi, Y.; Lucia, L. A.; Rojas, O. J. Cellulose Nanocrystals: Chemistry, Self-Assembly, and Applications. *Chem. Rev.* 2010 Jun 9;*110* (6):3479–3500. https://doi.org/10.1021/cr900339w
54. Moon, R. J.; Martini. A.; Nairn, J.; Simonsen, J.; Youngblood, J. Cellulose Nanomaterials Review: Structure, Properties and Nanocomposites. *Chem. Soc. Rev.* 2011;*40* (7):3941–3994. https://doi.org/10.1039/C0CS00108B
55. Nadagouda, M. N.; Varma, R. S. Synthesis of Thermally Stable Carboxymethyl Cellulose/Metal Biodegradable Nanocomposites for Potential Biological Applications. *Biomacromolecules* 2007 Sep 10;*8* (9):2762–2767. https://doi.org/10.1021/bm700446p
56. Korhonen, J. T.; Hiekkataipale, P.; Malm, J.; Karppinen, M.; Ikkala, O.; Ras, R. H. Inorganic Hollow Nanotube Aerogels by Atomic Layer Deposition onto Native Nanocellulose Templates. *ACS Nano* 2011 Mar 22;*5* (3):1967–1974. https://doi.org/10.1021/nn200108s
57. Ji, S.; Shi, T.; Qiu, X.; Zhang, J.; Xu, G.; Chen, C.; Jiang, Z.; Ye, C. A Route to Phase Controllable Cu2ZnSn(S1−xSex)4 Nanocrystals with Tunable Energy Bands. *Sci. Rep.* 2013, *3* (1), 2733. https://doi.org/10.1038/srep02733
58. Sui, Z.; Meng, Q.; Zhang, X.; Ma, R.; Cao, B. Green Synthesis of Carbon Nanotube–Graphene Hybrid Aerogels and Their Use as Versatile Agents for Water Purification. *J. Mater. Chem.* 2012, *22* (18), 8767–8771. https://doi.org/10.1039/C2JM00055E
59. Nadagouda, M. N.; Speth, T. F.; Varma, R. S. Microwave-Assisted Green Synthesis of Silver Nanostructures. *Acc. Chem. Res.* 2011, *44* (7), 469–478. https://doi.org/10.1021/ar1001457
60. Abdelaziz, R.; Disci-Zayed, D.; Hedayati, M. K.; Pöhls, J.-H.; Zillohu, A. U.; Erkartal, B.; Chakravadhanula, V. S. K.; Duppel, V.; Kienle, L.; Elbahri, M. Green Chemistry and Nanofabrication in a Levitated Leidenfrost Drop. *Nat. Commun.* 2013, *4* (1), 2400. https://doi.org/10.1038/ncomms3400
61. Vinayan, B. P.; Nagar, R.; Ramaprabhu, S. Solar Light Assisted Green Synthesis of Palladium Nanoparticle Decorated Nitrogen Doped Graphene for Hydrogen Storage Application. *J. Mater. Chem. A.* 2013, *1* (37), 11192–11199. https://doi.org/10.1039/C3TA12016C
62. Nagar, A.; Pradeep, T. Clean Water through Nanotechnology: Needs, Gaps, and Fulfillment. *ACS Nano* 2020, *14* (6), 6420–6435. https://doi.org/10.1021/acsnano.9b01730
63. Deadman, B. J.; Battilocchio, C.; Sliwinski, E.; Ley, S. V. A Prototype Device for Evaporation in Batch and Flow Chemical Processes. *Green Chem.* 2013, *15* (8), 2050–2055. https://doi.org/10.1039/C3GC40967H
64. Sato, F. *Japan's Policy Package for Society 5.0 and STI for SDGs: Expert Group Meeting on STI for SDGs Roadmaps – Paving the Pathways for Sustainable Recovery and Future Resilience*; UN Inter-Agency Task Team on Science, Technology and Innovation for the SDGs, 2020; https://sustainabledevelopment.un.org/content/documents/ 26879Joint_key_messages_HLPF_STI_for_SDGs_consultations.pdf (accessed 2021-11-03)
65. Meng, H.; Xia, T.; George, S.; Nel, A. E. A Predictive Toxicological Paradigm for the Safety Assessment of Nanomaterials. *ACS Nano* 2009, *3* (7), 1620–1627. https://doi.org/10.1021/nn9005973
66. Nel, A. E.; Parak, W. J.; Chan, W. C. W.; Xia, T.; Hersam, M. C.; Brinker, C. J.; Zink, J. I.; Pinkerton, K. E.; Baer, D. R.; Weiss, P. S. Where Are We Heading in Nanotechnology Environmental Health and Safety and Materials Characterization? *ACS Nano* 2015, *9* (6), 5627–5630. https://doi.org/10.1021/acsnano.5b03496

67. Stockholm Resilience Centre. The Nine Planetary Boundaries, Stockholm University; www.stockholmresilience.org/ research/planetary-boundaries/the-nine-planetary-boundaries.html (accessed 2021-08-25)
68. Matus, K. J.; Clark, W. C.; Anastas, P. T.; Zimmerman, J. B. Barriers to the Implementation of Green Chemistry in the United States. *Environ. Sci. Technol.* 2012 Oct 16;46(20):10892–10899. https://doi.org/10.1021/es3021777
69. Palit, S. Nanofiltration and Ultrafiltration – The Next Generation Environmental Engineering Tool and a Vision for the Future. *Int. J. ChemTech Res.* 2016, 848–856.
70. Palit, S. Filtration: Frontiers of the Engineering and Science of Nanofiltration-a far-Reaching Review. *CRC Concise Encycl. Nanotechnol.* 2016:205–214.
71. Palit, S. Advanced Oxidation Processes, Nanofiltration, and Application of Bubble Column Reactor. In *Nanomaterials for Environmental Protection*; John Wiley & Sons, Ltd, 2014; pp 205–215. https://doi.org/10.1002/9781118845530.ch13

3 Are Biogenic Nanomaterials Biocompatible?

Fernanda Maria Policarpo Tonelli, Vinicius Marx Silva Delgado, Christopher Santos Silva, Geicielly da Costa Pinto, and Flávia Cristina Policarpo Tonelli
Corresponding author: Fernanda Maria Policarpo Tonelli

3.1 INTRODUCTION

Due to the rapid development that the world has been experiencing, relevant problems are being faced by humanity. Environmental pollution (especially as a consequence of industrial activities performed ignoring sustainability principles and discharging toxic residues into air, water, and soil; Figure 3.1) and food scarcity are important examples.

Since the 1950s, different protocols have been proposed to obtain materials in nanoscale presenting a high surface area to volume ratio to perform different tasks efficiently (with or without undergoing a step of chemical surface modification, called functionalization, to optimize the achievement of the target), such as remediating environmental pollution (Figure 3.2) and eliminating food pathogens. However, only in the 1970s did the term nanotechnology became available to refer to the field of expertise dedicated to the study of these materials [1–4].

Nanomaterials, in the food segment, can contribute not only to improving quality control but also to offering safer food to the world's growing population. They can, for example, serve as additives to enhance nutritional aspects or to extend the shelf life of food. During food processing they can improve aspects such as the food's appearance, consistency, taste, and texture, and also act as nanocarriers of food additives. Besides that, materials in nanoscale can act as components in nanosensors that indicate an expired product or can be added to packages to improve thermal resistance, among other functions [5,6].

Regarding food packaging that aims to avoid the growth of food pathogens, due to their antimicrobial characteristics even at low concentrations, nanoparticles (NPs), especially metallic ones, are notable candidates [7,8]. The incorporation of silver nanoparticles (AgNPs) into a film of tragacanth, hydroxypropyl methylcellulose, and beeswax contributed to the inhibition of food pathogens *Bacillus cereus*, *Escherichia coli*, *Klebsiella pneumonia*, *Pseudomonas aeruginosa*, *Staphylococcus aureus*, and *Salmonella typhimurium* [9]. Titanium dioxide nanoparticles (TiO_2NPs) applied to polyethylene-based films promoted the inhibition of the bacteria *E. coli*, *Listeria monocytogenes*, and *S. aureus* when exposed to UVA light [10]. Copper oxide nanoparticles (CuONPs) applied to carbohydrate-based composite films exhibited antibacterial effects against *E. coli* and *L. monocytogenes* besides other important features to the food industry field such as improved thermal stability [11].

With the aim of obtaining an increase in the shelf life of products, several techniques involving nanomaterials are being applied and studied. Through the nanoencapsulation of tarragon essential oil, combined with the use of chitosan–gelatin-based coating, it was possible to obtain a decrease

DOI: 10.1201/9781003389408-3

FIGURE 3.1 Industries are primarily responsible for environmental problems currently faced by humanity.

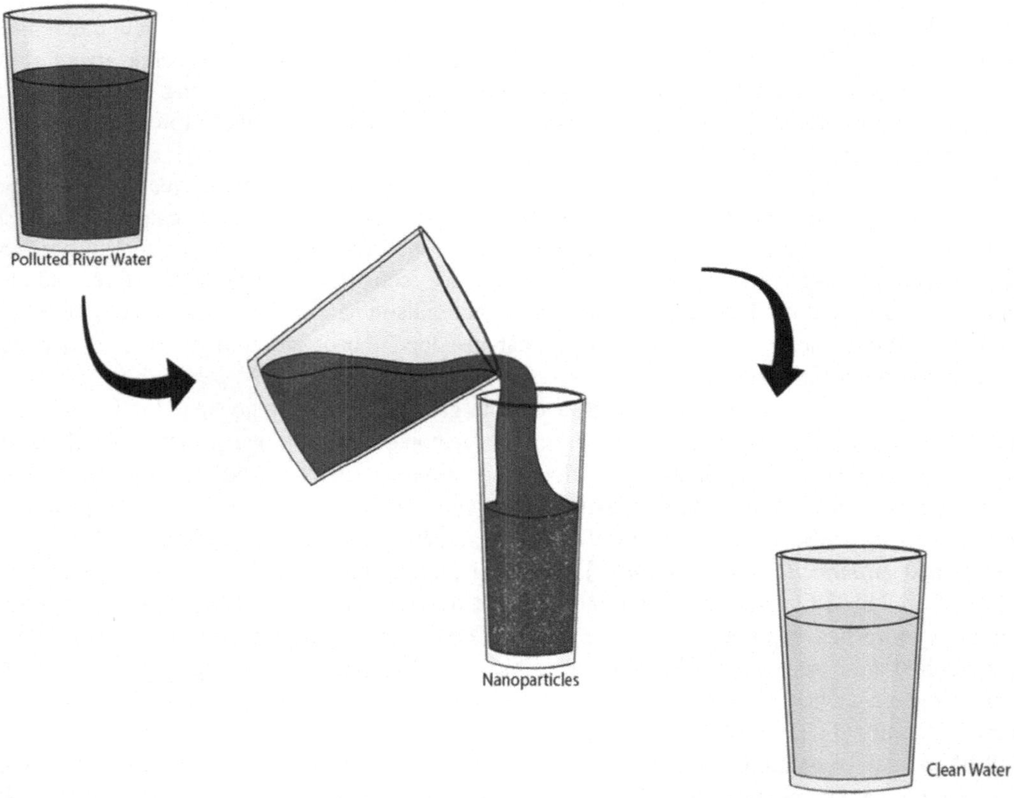

FIGURE 3.2 Nanomaterials can be used as nanoremediators to deal with pollutants present in samples, such as aqueous ones, restoring their quality.

in pH change and oxidation processes in pork slices, delaying the deterioration process [12]. Beer bottles manufactured from nanocomposites/nanoclay promoted the restriction of oxygen entrance and carbon dioxide exit, which increased by approximately 6 months the shelf life of the product; it also offered increased strength in the bottle's structure, reducing the chance of it breaking and making it safer to handle [13]. Sodium caseinate-based edible coating nanoemulsion containing ginger essential oil displayed antioxidant and antibacterial properties that could extend the shelf life of products such as chicken breast fillets [14].

In the field of agriculture, nanomaterials can be applied from the initial stages of planting to the crop's harvesting, aiming not only to guarantee the quality of the grains through adequate control of nutrients but also to offer greater security in the final product through extermination of pests and microorganisms. Iron oxide nanoparticles, for example, used in association with EDTA functionalized iron oxide nanoparticles, demonstrated the potential to improve the nutritional qualities of plants [15]. Unlike conventional herbicides that may eliminate only the superficial part of weeds, nanoherbicides are able to act at the core of the pest, eliminating its seed bank and preventing its proliferation, while exhibiting low risk to the workers dealing with the product [16,17]. Another important aspect is that nanomaterials can offer the possibility of reducing the amounts of conventional substances necessary to promote plant protection and that are harmful to other living beings; by doing that, they favor a reduction in costs and also favor safety in agriculture-related activities [18].

To improve the agronomic quality of the crop, carbon nanotubes can be used to promote a better rate of growth and development of plants, both in the aerial part and in the subsoil portions [19–21]. Extracts obtained from pomegranate peel and coffee grounds were successfully used to produce phosphorous-containing hydroxyapatite nanoparticles that worked as nanofertilizers [22]. It is also possible to use nanomaterials to prevent soil nutrient loss [18].

Remediation of environmental contaminants, biological and nonbiological (organic and inorganic substances), can also be performed by nanomaterials. Metal nanoparticles, such as zinc oxide NPs could be successfully applied to promote the remediation of fungal pathogens in soybeans and coffee [23,24]. Nanosensors are capable of being applied for the purpose of detecting pests and, simultaneously, monitoring the nutritional condition of the soil [25]. The nanoencapsulation of chemical compounds is able to promote the gradual release of compounds according to the plant's needs, regulating its growth and development; nanomaterials can act as nanocarriers of important agrochemicals aimed at improving the final crop yield [26].

Furthermore, the application of nanomaterials, if they are biogenic, being synthesized from sustainable protocols, in the field of agriculture presents a low risk of environmental contamination [27,28]. In fact, they offer an elegant and strategic solution to the problem of dealing with waste products of agriculture. Wheat straw, for example, can serve as raw material to generate nanostructures that can be useful [29].

The synthesis of nanomaterials can be performed through different protocols: chemical, physical, and/or biological. Of these, chemical protocols commonly require the use of toxic substances that can contribute to worsening the problem of environmental pollution and also damage the health of living beings. Physical protocols commonly demand the use of expensive pieces of equipment and may be time-consuming. Therefore, these traditional methods present important limitations that have motivated the search for more sustainable protocols [30–32].

Biological protocols emerged as an interesting alternative offering biogenic ecofriendly nanomaterials produced using as raw materials: algae, bacteria, fungi, and plants as well as their bioderived products. These biobased materials provide metabolites that can participate in the synthesis and also act as stabilizing and/or capping agents for the nanomaterials [33–41].

3.2 BIOGENIC NANOMATERIALS: GREEN SYNTHESIS

This type of strategy to generate ecofriendly nanomaterials applies the principles of green chemistry, aiming to prevent waste production and reduce derivatives by using benign chemicals, preferentially obtained from renewable sources optimizing energy efficiency associated with the process [33,42,43].

Before advances in the green chemistry field, synthesis of nanomaterials was commonly performed by applying physical and chemical protocols, divided in two different approaches: bottom-up and top-down (Figure 3.3).

The bottom-up approach involves complexation of smaller molecules into nanometric materials through many different techniques such as spray pyrolysis, laser pyrolysis, sol-gel processes, and molecular condensation. Top-down techniques involve processing larger structures of the desired material into its nanometric scale using laser ablation, chemical etching, mechanical milling, and electro-explosion, for example. Those techniques might require expensive materials and equipment, high temperature and pressure, and skilled operators [44]. However, those protocols are still widely used nowadays due to their advantages such as scalability, the possibility of exerting control over the particle's morphology (which favors reproducibility), and the development of materials designed for targeted disease therapy. Nevertheless, most traditional synthesis uses toxic agents that worsen environmental pollution and are dangerous to human beings [45]. In that regard, green chemistry presents a convenient alternative when it comes to nanoparticles. Biogenic protocols might use living beings such as bacteria, algae, and fungi, or plant extracts, for example, to reduce and stabilize nanomaterials through a bottom-up approach (Figure 3.4).

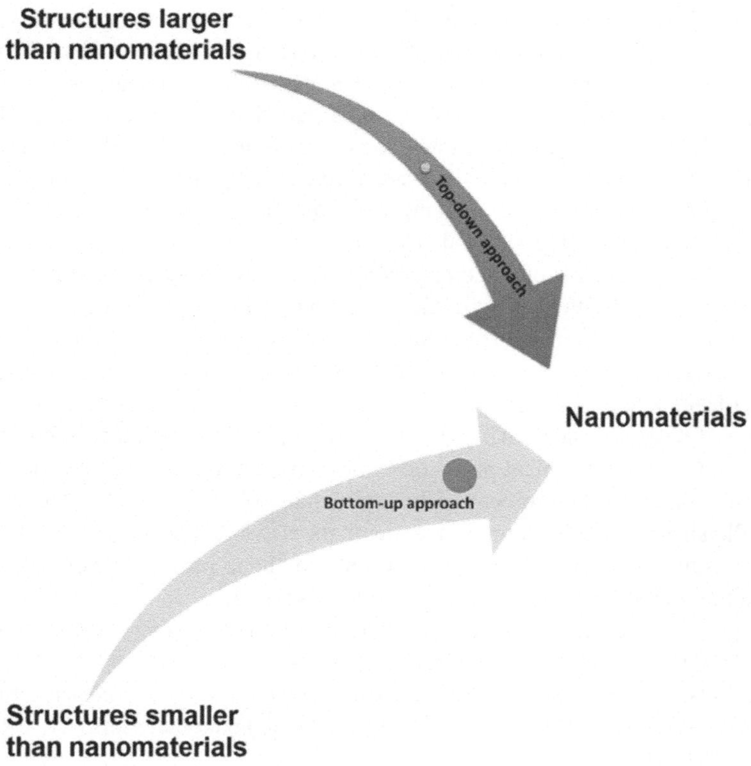

FIGURE 3.3 Nanomaterials can be produced through top-down or bottom-up approaches.

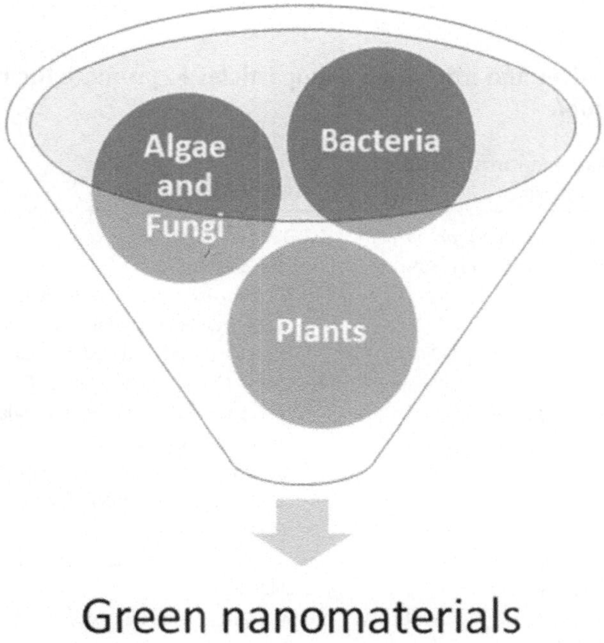

FIGURE 3.4 Important examples of biobased sources of green nanomaterials.

Most procedures demand room temperature and neutral pH and are simpler to perform than physical and chemical conventional protocols of synthesis. The morphology and characteristics of those nanoparticles can be altered by varying the pH, temperature, and other variables during the synthesis [46], and each biogenic nanomaterial has different properties depending on the raw biomaterial used. For example, metallic nanoparticles generated using the same ion might have different properties depending on which organism is the source of the raw material applied in synthesis. Moreover, those particles might be activated or modified with other components to increase the desirable properties [47]. In terms of sustainability, green synthesis offers little or no danger to the environment and to human beings, owing to the lack of toxic components applied during production. Instead, many studies have shown that a large array of metallic nanoparticles present remediation properties, such as silver and iron NPs that can efficiently adsorb heavy metals [48,49] and degrade dyes [50,51] (Table 3.1).

AgNPs are also particularly efficient when applied against bacteria [92], which can be seen as biological environmental contaminants (Table 3.1). These nanomaterials present interesting properties associated with the metal silver that facilitate their use in different applications; among these characteristics it is possible to highlight antimicrobial and catalytic ones, and also good chemical and thermal stability, which broadens the array of applications [93–95]. When it comes to iron oxide, Fe_3O_4NPs present magnetic properties, which facilitate its recovery after use by means of a magnet [96].

Some studies have already reported the effectiveness of green NPs on food and agriculture issues. Many studies, as previously mentioned, have already shown that AgNPs exhibit antibacterial and antifungal activity, which indicates a potential for application in food industries to prevent contamination by food pathogens [97]. TiO_2NPs have presented effectiveness on increasing spinach growth, inducing the formation of chlorophyll, increasing the photosynthetic rate during the growth stage of spinach, and increasing its RuBisCo activity [98]. There are different studies that present green protocols for generating this nanoparticle [99]. For example, by using the plant extract from

TABLE 3.1
Recent examples of silver and iron green nanoparticles to promote the remediation of environmental pollution

Type of metallic green nanoparticle	Biobased raw material used in the synthesis	Pollutant(s) remediated	Reference
AgNPs	*Trigonella foenum-graecum (TF)*	RB19 and RY186 dyes	[52]
	Bacillus licheniformis M09	Methylene blue	[53]
	Syzygium cumini	*Pseudomonas aeruginosa*, *Escherichia coli*, methylene blue, crystal violet, amaranth, indigo carmine, acid red, vat red, malachite green, Congo red, rhodamine B	[54]
	Zingiber officinale	4-Nitrophenol, 2-nitrophenol, Rhodamine B, methyl red, Congo red	[55]
		Methyl orange, *Pseudomonas aeruginosa*, *Methicillin-resistant Staphylococcus aureus*, *Candida albicans*	[56]
	Trichoderma spp.	*Pseudomonas aeruginosa*, *Escherichia coli*	[57]
	Aloe barbendensis	Naphthalene	[58]
	Crataegus pentagyna fruit extract	Three different organic dyes: eosin, methylene blue, rhodmine B	[59]
	Bacillus amyloliquefaciens MSR5	4-nitrophenol	[60]
	Chlorella ellipsoidea	Methyl orange, methylene blue	[61]
	Sargassum myriocystum	*Escherichia coli*, *Staphylococcus aureus*, *S. epidermidis*, *Proteus vulgaris*, *Klebsiella pneumonia*, *Pseudomonas aeruginosa*, methylene blue	[62]
	Species from *Metschnikowia* genus	*Escherichia coli*, *Pseudomonas aeruginosa*, *Staphylococcus aureus*	[63]
	Anabaena variabilis and *Spirulina platensis*	Malachite green	[64]
	Chloroxylon swietenia	Congo red, coomassie blue, crystal violet, *Staphylococcus nepalensis*, *Staphylococcus gallinarum*, *Bacillus subtilis*, *Enterococcus faecalis*, *Pseudomonas stuteria*	[65]
	Origanum majorana	Multidrug-resistant *Klebsiella pneumoniae*	[66]
	Poria cocos	4-Nitrophenol	[67]
	Smilax aspera	*Klebsiella pneumoniae*, *Streptococcus pneumoniae*, *Escherichia coli*, *Pseudomonas aeruginosa*, *Staphylococcus aureus*, methylene blue, rose Bengal, acridine orange	[68]
	Catharanthus roseus	Cr, Cd	[69]
FeNPs	*Colpomenia sinuosa* and *Pterocladia capillacea*	*Bacillus subtilis*, *Escherichia coli*, *Pseudomonas aeruginosa*, *Salmonella typhi*, *Staphylococcus aureus*, *Vibrio cholera*, *Fusarium oxysporum*, *Aspergillus flavus*	[70]
	Sargassum vulgare, Jania rubens, Ulva fasciata	Biofilm-forming bacteria	[71]
	Aspergillus niger BSC-1	Cr^{6+}	[72]
	Penicillium pimiteouiense	Cr^{6+}	[73]
	Prangos ferulacea, Teucrium polium	Arsenic	[74]
	Hybrid of *E. urophylla* and *E. grandi*	Arsenate	[75]

TABLE 3.1 (Continued)
Recent examples of silver and iron green nanoparticles to promote the remediation of environmental pollution

Type of metallic green nanoparticle	Biobased raw material used in the synthesis	Pollutant(s) remediated	Reference
	Jania rubens, *Sargassum vulgare*, *Ulva fasciata*	Marine bacterial biofilm	[71]
	Algae present in the lichen *Ramalina sinensis*	*Pseudomonas aeruginosa*, *Staphylococcus aureus*	[76]
	Apple peels	Malachite green	[77]
	Penicillium oxalicum	Methylene blue	[78]
	Algae present in the lichen *Ramalina sinensis*	Cd, Pb	[79]
	Fusarium oxysporum	*Bacillus cereus*, *Listeria monocytogenes*, *Enterococcus faecalis*, *Pseudomonas aeruginosa*, *Escherichia coli*, *Salmonella typhi*, *Candida albicans*, Cd, Cr, Ni, Pb, Zn	[80]
	Spirulina platensis	Crystal violet, methyl orange	[81]
	Padina pavonica, *Colpomenia sinuosa*, *Petalonia fascia*	Compounds containing nitrogen and phosphorus, harmful algae species	[82]
	Pleurotus florida	*Candida albicans*, *C. glabrata*, *Candida* sp., *Bacillus cereus*, *Staphylococcus aureus*, *Micrococcus mucilaginosus*, *Klebsiella pneumoniae*, *K. terrigena*, *Pseudomonas aeruginosa*, *Escherichia coli*	[83]
	Aspergillus terreus	The dyes Congo red and direct blue-1	[84]
	Moringa oleifera	$Cd^{2+}$and Pb^{2+}	[85]
	Acacia nilotica	Pb ions	[86]
	Catharantus roseus	Methyl orange, *Escherichia coli*	[87]
	Fusarium proliferatum	Malachite green, methyl violet, crystal violet	[88]
	Spirulina spp.	Crystal violet	[89]
	Penicillium spp.	*Staphylococcus aureus*, *Escherichia coli*, *Klebsiella pneumoniae*, *Shigella sonnei*, *Pseudomonas aeruginosa*	[90]
	Saccharomyces cerevisiae	*Bacillus subtilis*, *Escherichia coli*	[91]

Trachyspermum ammi, it was possible to produce TiO_2NPs that increased germination of *Vigna radiata* by 80% [100].

Green nanomaterials may also present ovicidal and or larvicidal properties. An extract obtained using water as solvent and the leaf powder of *Scadoxus multiflorus* allowed the production of zinc oxide nanoparticles that exhibited activity of eliminating eggs and larvae of *Aedes aegypti* [101]. Silver NPs might present similar properties; AgNP produced using *Bacillus marisflav* as raw material and tested against eggs, larvae, and pupae of *Aedes aegypti*, *Culex quinquefasciatus*, and *Anopheles stephensi* exhibited significant activity to fight them [102].

3.3 THE IMPORTANCE OF BIOCOMPATIBILITY OF NANOMATERIALS

As nanomaterials, even the green ones, when applied to exert a desirable function come into contact with living beings, especially humans, it is important to investigate whether they are harmful: that is, whether they exhibit good biocompatibility or stimulate toxicity and immunogenic response [103].

The response that a living organism elaborates when exposed to a nanomaterial, like the type of transportation it will use to be internalized into cells, depends on different features of the material such as size, shape, charge, presence of functional groups on the surface, and hydrophobicity [104–107]. In order to optimize the interaction of nanomaterials with cells from living beings, it is possible to adjust protocols with the aim of obtaining materials in nanoscale with desirable size, for example, or to perform chemical modification on their surface by functionalizing them. This type of approach can offer a reduction in cytotoxicity or an enhancement, depending on the intended use of the nanomaterials. Aiming at a reduction in cytotoxicity, ZnONPs were covered by organic deposits [108]. However, to be used as antibacterial or anticancer substances, platinum NPs could also be functionalized to increase induced cell death rate [109].

Another strategy to favor biocompatibility is the adoption of a green synthesis protocol to generate nanoparticles. By applying a biobased raw material it is possible to generate nanomaterials avoiding the use of toxic chemicals, favoring a reduction in cytotoxicity. Commonly, gold nanoparticles synthesized using plant material exhibit good biocompatibility [110].

3.4 BIOCOMPATIBILITY OF BIOGENIC NANOMATERIALS

As previously mentioned, ecofriendly protocols can be applied when the intention is to generate biocompatible nanomaterials (Table 3.2).

The cytotoxicity of nanomaterials is commonly accessed through assays such as MTT that involve a change in the color of a reagent reflecting mitochondrial function and that can be measured using a spectrophotometer [111]. These assays generally involve human normal cells rather than cancer cell lines [112]. However, there are also studies that evaluate hemolysis of human red blood cells (RBCs) as an indicator of compatibility [113].

To increase the biocompatibility of Cr_2O_3 NPs chemically synthesized, it was possible to adopt a green protocol using the extract obtained from the leaves of *Abutilon indicum* [114]. Bimetallic nanoparticles core–shell (Au–Ag) and hybrid (Au–ZnO and Ag–ZnO) generated using the extract from the leaves of *Manilkara zapota* proved to be biocompatible when assayed using, for example, human RBCs. This feature is important, as the nanomaterials exhibited antidiabetic activity; if they were used to treat disease in human beings, it would not be desirable that they cause the death of human cells [115]. The extract obtained from the leaves of *Ilex paraguariensis* allowed the production of zinc oxide nanoparticles. These particles exhibited no cytotoxicity when small and presenting uniform shape at concentrations up to 10 g/mL [116]. At 5×10^{-1} mg/mL the decrease in J774 cell viability induced by carbon dots (Cdots) prepared using citric acid was significantly higher than that induced by Cdots synthesized using extract of bamboo leaves [117]. Silver nanoparticles synthesized using aqueous extract obtained from the leaves of *Cistus incanus* proved to be immune-compatible and, at a concentration of 16 ppm, non cytotoxic to L929 mouse fibroblasts [118]. The aqueous extract from the fruits of *Myristica fragrans* allowed the production of biogenic zinc oxide nanoparticles with antibacterial properties, notably against *Klebsiella pneumoniae*; among other interesting features, the particles were biocompatible, being considered only slightly hemolytic at a concentration of 400 µg/mL [119]. ZnO nanoparticles synthesized using *Lactobacillus plantarum* and *Lactobacillus acidophilus* were biocompatible when assayed up to 600 µg/ml through MTT using mouse embryonic fibroblast cell line, like the version chemically modified by gum Arabic-polyvinyl alcohol-polycaprolactone nanofiber matrix [120]. *Hibiscus rosa sinensis* leaves allowed the synthesis of biocompatible biogenic Co_3O_4NPs and MgONPs [121]. The leaves of *Olea europaea* could be used to generate silver nanoparticles that proved to be biocompatible when assayed through MTT using normal human cells (mesenchymal stem cells) [122]. Biogenic gold and silver nanoparticles could be generated using the phytocannabinoid cannabidiol to offer a less cytotoxic effect than cannabidiol itself when assayed on HaCat cell lines [123]. An extract obtained from the leaves of *Azadirachta indica* allowed the production of gold nanoparticles

TABLE 3.2
Recent examples of biogenic and biocompatible nanomaterials

Biogenic nanomaterial	Cytotoxicity assay performed	Reference
Cdots synthesized using an extract of bamboo leaves	MTT assay after 24 hours of exposure in J774.1 cells	[117]
Zinc oxide NPs generated using an extract of the leaves of *Ilex paraguariensis*	MTT using mouse fibroblasts (L929)	[116]
AuNPs generated from an extract of the leaves of *Azadirachta indica*	MTT using HuH-7 and CHANG cells	[124]
SeNPs produced using an aqueous extract of the tuber of *Ceropegia bulbosa*	MTT assay using the epithelial human cell line HBL-100	[131]
Zinc oxide NPs produced using an extract of the fruit of *Myristica fragrans*	Hemolysis of RBCs assay	[119]
AgNPs synthesized using an extract of the leaves of *Cistus incanus*	Immunocompatibility assay: using THP1-XBlue™ human monocytes; cytotoxicity test: MTT	[118]
ZnONPs generated using *L. acidophilus* and *L. plantarum*	MTT assay using mouse embryonic fibroblasts cell line	[120]
Ag and Au nanoparticles generated using cannabidiol	Sulforhodamine B staining	[123]
MgONPs and Co_3O_4NPs from *Hibiscus rosa sinensis*	Hemolysis of RBCs assay	[121]
Cr_2O_3 nanoparticles produced using an extract of the leaves of *Abutilon indicum*	MTT assay using Vero cell lines	[114]
AgNPs generated using an extract of the leaves of *Olea europaea*	MTT assay using human mesenchymal stem cells	[122]
Phosphorus-containing hydroxyapatite NPs produced using coffee grounds and the peel of pomegranate fruit	MTT assay using Vero E6 cells	[22]
Zinc oxide NPs produced using coffee grounds and the peel of pomegranate fruit	MTT assay using Vero E6 cells	[125]
Au nanoparticles generated using *Gynostemma pentaphyllum*	MTT assay using 3T3-L1 and RAW 264.7 cells	[126]
Bimetallic nanoparticles core–shell (Au–Ag) and hybrid (Au–ZnO and Ag–ZnO) generated from an extract of the leaves of *Manilkara zapota*	Hemolysis of RBCs assay	[116]
AgNPs synthesized using fruits from *Conocarpus lancifolius*	Hemolysis of RBCs assay	[127]
AgNPs synthesized using an extract of the leaves of *Barleria buxifolia*	MTT assay using human fibroblast L929 cells	[128]
AuNPs synthesized using extracts of *Ephedra alata* and *Pistacia lentiscus*	MTT using noncancerous human fibroblasts	[129]
Iron and SeNPs synthesized using an extract of *Punica granatum*	MTT using noncancerous human fibroblasts	[129]
AuNPs generated using *Pseudobulbus Cremastraeseu Pleiones*	CCK-8 test using L02 cells	[130]

that offered cytotoxicity in liver cancer HuH-7 cells, but human normal CHANG cells were less sensitive to them [124]. Extracts obtained from the peel of *Punica granatum* and coffee grounds were used to produce zinc oxide nanoparticles that proved to be biocompatible when assayed on Vero cells [125]; this biobased raw material was also employed to produce phosphorus-containing hydroxyapatite nanoparticles biocompatible on a Vero cell line when assayed through MTT [22]. Gold nanoparticles were synthesized using an extract of *Gynostemma pentaphyllum* and proved to be nontoxic to 3T3-L1 and RAW 264.7 cells [126]. Silver nanoparticles generated using the extract of fruits obtained from *Conocarpus lancifolius* were biocompatible when assayed against RBCs [127]. Silver nanoparticles synthesized using material from *Barleria buxifolia* also proved to be biocompatible [128]. Gold nanoparticles synthesized using the extracts from *Ephedra alata* and *Pistacia lentiscus* and iron and selenium nanoparticles synthesized using extracts from *Punica granatum* proved to be biocompatible when assayed through MTT using human fibroblasts as target cells [129]. The extract of *Pseudobulbus Cremastrae seu Pleiones* allowed the synthesis of gold nanoparticles that proved to be biocompatible when assayed using normal liver cells (L02) [130]. Selenium nanoparticles produced using the aqueous extract from the tuber of *Ceropegia bulbosa*, through MTT assay using the epithelial human cell line HBL-100, proved to be biocompatible [131].

However, adopting a green protocol to obtain a biogenic nanomaterial is not a guarantee of absent or low cytotoxicity. Other aspects, such as the shape and size of the nanomaterial, as previously mentioned, also influence cytotoxicity. An extract obtained from green tea was used to generate gold nanoparticles capped by chitosan but presenting different shapes. Nanorods proved to be more toxic than nanostars, and nanospheres were the least cytotoxic shape assayed through MTT assay in different cancer cell lines [132]. Gold nanoparticles of different sizes could be produced using the hydroethanolic extract from the bark of *Mimosa tenuiflora*. After 24 h of exposure to NPs using human umbilical vein endothelial cells and performing MTT assay, at a concentration of 50 µg/mL a significant decrease in cell viability was observed for the smallest nanoparticle; however, for the larger one, this decrease was already observed at the lower concentration of 25µg/mL [133].

Sometimes the intention is that the nanomaterial exhibits cytotoxicity, for example to fight a specific cell type (Table 3.3). Human leukemia cell line K-562 had cell viability significantly reduced by silver and gold nanoparticles produced using aqueous extract of *Lavandula dentata*

TABLE 3.3
Recent examples of green nanomaterials designed to present toxicity on target cells or organisms

Nanomaterial	Target cell or organism	Reference
Zinc oxide NPs generated using grounds of coffee and peel of *Punica granatum*	*Enterobacter aerogenes, Klebsiella pneumoniae, Pseudomonas aeruginosa, Staphylococcus aureus*	[125]
Silver nanoparticles produced using an extract of fruits obtained from *Conocarpus lancifolius*	*Staphylococcus aureus, Streptococcus pneumonia, Aspergillus flavus, Rhizopusus stolonifera*	[127]
AgNPs produced using an extract obtained from the leaves of *Barleria buxifolia*	Biofilm-forming bacteria	[128]
Ag and AuNPs produced using an aqueous extract of *Lavandula dentata*	Human leukemia cell line K-562	[134]
SeNPs produced using *Paenibacillus motobuensis* LY5201	HepG2 cells	[135]
AgNPs produced using an extract of *Acanthospermum australe*	Agents capable of causing skin and soft-tissue infections	[136]
AgNPs produced using an extract of walnut green husk	MCF-7 cancerous cells	[137]
AgNPs generated using an extract of *Achillea millefolium*	Lymphoblastic leukemia cell line MOLT-4, *Aspergillus niger, Fusarium sp.,* E.coli	[138]
SeNPs generated using the tuber of *Ceropegia bulbosa*	*Escherichia coli, Bacillus subtilis*	[131]
ZnONPs generated using *Lactobacillus plantarum* and *Lactobacillus acidophilus*	Biofilm-forming bacteria	[139]
Co_3O_4NPs and MgONPs generated using *Hibiscus rosa sinensis*	Bacteria	[121]

[134]. Green selenium nanoparticles produced using *Paenibacillus motobuensis* LY5201 proved to present anticancer properties by being cytotoxic against HepG2 cells [135]. The aqueous extract of the medicinal plant *Acanthospermum australe* allowed the production of silver nanoparticles exhibiting antimicrobial activity against agents capable of causing skin and soft-tissue infections [136]. Commercial silver nanoparticles, for example, could cause the death of more than 55% of MCF-7 cancerous cells, and the walnut green husk extract exhibited a death rate of 40%. The green silver nanoparticles produced using this extract, however, exhibited 70% cytotoxicity due to a synergism [137]. Silver NPs generated using *Achillea millefolium* extract obtained using water as solvent proved to be more active than the anticancer drug Cisplatin in inducing death of lymphoblastic leukemia cell line MOLT-4; this material also proved to be an important tool to be used, for example, in packing, as it exhibited antimicrobial activity against fungi (*Aspergillus niger* and *Fusarium*) and bacteria (*Escherichia coli*) [138]. The aqueous extract from the tuber of *Ceropegia bulbosa* was used to generate selenium nanoparticles capable of inhibiting the growth of *Escherichia coli* and *Bacillus subtilis* (among other important features such as presenting larvicidal, anticancer, and photocatalytic effects) [131]. *Lactobacillus plantarum* and *Lactobacillus acidophilus* were used to generate ZnO nanoparticles presenting antimicrobial and antibiofilm properties [139]. The extract obtained from the leaves of *Hibiscus rosa sinensis* allowed the synthesis of antibacterial green Co_3O_4NPs and MgONPs [140]. Zinc oxide nanoparticles generated using grounds of coffee and peel of *Punica granatum* exhibited antibacterial activity against *Enterobacter aerogenes*, *Klebsiella pneumoniae*, *Pseudomonas aeruginosa*, and *Staphylococcus aureus* [125]. The extract of fruits obtained from *Conocarpus lancifolius* allowed the production of silver nanoparticles that presented antibacterial activity to fight *Staphylococcus aureus* and *Streptococcus pneumonia* and antifungal activity against *Aspergillus flavus* and *Rhizopusus stolonifera* [127]. The extract obtained from the leaves of *Barleria buxifolia* could be used in a biogenic protocol to produce silver nanoparticles exhibiting higher antibiofilm and antibacterial activity when compared to plant extract [128].

3.5 CONCLUSIONS AND FUTURE PERSPECTIVES

Advances in the nanotechnology field have offered several improvements to human quality of life, especially when it comes to food and agriculture issues. They have, for example, allowed improvements related to crops, food quality, and packaging technologies for guaranteeing food security. In relation to these improvements, biogenic nanomaterials deserve to be highlighted, as they avoid the use of toxic chemicals in their synthesis, contributing to a sustainable development. However, before applying the materials in nanoscale to large-scale exposure on human beings, safety-related aspects need to receive attention. The biocompatibility of these structures can be evaluated through methodologies such as MTT and hemolysis of RBCs, and they are commonly nontoxic to human cells. Nevertheless, toxicity may be an interesting feature when the aim is to eliminate a pathogen, for example; accordingly, there are also green protocols that allow the production of biogenic nanomaterials optimized to eliminate undesirable living beings.

The potential of the nanotechnology field to offer safe and sustainable solutions to problems faced by humanity in different areas is far from being completely explored. It is expected that new protocols to generate nanomaterials biocompatible with human cells will continue to be developed and to offer structures capable of solving problems such as food contamination without damaging human health and the environment.

REFERENCES

1. Taniguchi, N.; On the basic concept of nanotechnology. *Proceedings of the International Conference on Production Engineering* 1974; 18–23.
2. Subbiah, R.; Veerapandian, M.; Yun, K.S.; Nanoparticles: Functionalization and multifunctional applications in biomedical sciences. *Curr Med Chem* 2010; 17, 4559–4577.

3. Albanese, A.; Tang, P.S.; Chan, W.C.W.; The effect of nanoparticle size, shape, and surface chemistry on biological systems. *Annu Rev Biomed Eng* 2012; 14, 1–16.
4. Tonelli, F.M.P.; Tonelli, F.C.P.; Ferreira, D.R.C.; da Silva, K.E.; Cordeiro, H.G.; Ouchida, A.T.; Nunes, N.A.M.; Biocompatibility and Functionalization of Nanomaterials. In: Intelligent Nanomaterials for Drug Delivery Applications, Elsevier, New York, EUA, 2020, 204p.
5. Dimitrijevic, M.; Karabasil, N.; Boskovic, M.; Teodorovic, V.; Vasilev, D.; Djordjevic, V.; Kilibarda, N.; Cobanovic, N.; Safety aspects of nanotechnology applications in food packaging. *Procedia Food Sci* 2015; 5, 57–60.
6. Gokularaman, S.; Cruz, S.A.; Pragalyaashree, M.; Nishadh, A.; Nanotechnology approach in food packaging-review. *J Pharm Sci Res* 2017; 9, 1743–1749.
7. Hoseinnejad, M.; Jafari, S.M.; Katouzian, I.; Inorganic and metal nanoparticles and their antimicrobial activity in food packaging applications. *Crit Rev Microbiol* 2018; 44(2), 161–181.
8. Preethi, R.; Maria Leena, M.; Moses, J.A.; Anandharamakrishnan, C.; Biopolymer nanocomposites and its application in food processing. In: Green Nanomaterials: Processing, Properties, and Applications, Springer, Amsterdam, Netherlands, 2020, 346p.
9. Bahrami, A.; Mokarram, R.R.; Khiabani, M.S.; Ghanbarzadeh, B.; Salehi, R.; Physico-mechanical and antimicrobial properties of tragacanth/hydroxypropyl methylcellulose/beeswax edible films reinforced with silver nanoparticles. *Int J Biol Macromol* 2019; 129, 1103–1112.
10. Hashemabad, Z.N.; Shabanpour, B.; Azizi, H.; Ojagh, S.M.; Alishahi, A.; Effect of TiO2 nanoparticles on the antibacterial and physical properties of low-density polyethylene film. *Polym Plast Technol Eng* 2017; 56(14), 1516–1527.
11. Shankar, S.; Wang, L.F.; Rhim, J.W.; Preparation and properties of carbohydrate-based composite films incorporated with CuO nanoparticles. *Carbohydr Polym* 2017; 169, 264–271.
12. Zhang, H.; Liang, Y.; Li, X.; Kang, H.; Effect of chitosan-gelatin coating containing nano-encapsulated tarragon essential oil on the preservation of pork slices. *Meat Sci* 2020; 166, 108137.
13. Lu, J.; Bowles, M.; How will nanotechnology affect agricultural supply chains?. *Int Food Agribus Manag Rev* 2013; 16(1030-2016-82815), 21–42.
14. Noori, S.; Zeynali, F.; Almasi, H.; Antimicrobial and antioxidant efficiency of nanoemulsion-based edible coating containing ginger (*Zingiber officinale*) essential oil and its effect on safety and quality attributes of chicken breast fillets. *Food Control* 2018; 84, 312–320.
15. Haydar, M.S.; Ghosh, S.; Mandal, P.; Application of iron oxide nanoparticles as micronutrient fertilizer in mulberry propagation. *J Plant Growth Regul* 2021; 1–21.
16. Prasad, R.; Kumar, V.; Prasad, K.S.; Nanotechnology in sustainable agriculture: Present concerns and future aspects. *Afr J Biotechnol* 2014; 13(6), 705–713.
17. Dwivedi, S.; Saquib, Q.; Al-Khedhairy, A.A.; Musarrat, J.;Understanding the Role of Nanomaterials in Agriculture. In: Microbial Inoculants in Sustainable Agricultural Productivity: Vol. 2: Functional Applications. Springer, New Delhi, India, 2016; 271–288.
18. Usman, M.; Farooq, M.; Wakeel, A.; Nawaz, A.; Cheema, S.A.; ur Rehman, H.; Ashraf, I.; Sanaullah, M.; Nanotechnology in agriculture: Current status, challenges and future opportunities. *Sci Total Environ* 2020; 721, 137778.
19. Patel, A.; Tiwari, S.; Parihar, P.; Singh, R.; Prasad, S.M.; Carbon Nanotubes as Plant Growth Regulators: Impacts on Growth, Reproductive System, and Soil Microbial Community. In: Nanomaterials in Plants, Algae and Microorganisms Concepts and Controversies, Academic Press, New York, EUA, 2017, 365p.
20. Joshi, A.; Kaur, S.; Dharamvir, K.; Nayyar, H.; Verma, G.;Multi-walled carbon nanotubes applied through seed-priming influence early germination, root hair, growth and yield of bread wheat (*Triticum aestivum* L.). *J Sci Food Agric* 2018; 98(8), 3148–3160.
21. Juárez-Cisneros, G.; Gómez-Romero, M.; de la Cruz, H.R.; Campos-García, J.; Villegas, J.;Multi-walled carbon nanotubes produced after forest fires improve germination and development of Eysenhardtia polystachya. *PeerJ* 2020; 8:e8634.
22. Abdelmigid, H.M.; Morsi, M.M.; Hussien, N.A.; Alyamani, A.A.; Alhuthal, N.A.; Albukhaty, S.;Green synthesis of phosphorous-containing hydroxyapatite nanoparticles (nHAP) as a novel nano-fertilizer: Preliminary assessment on pomegranate (*Punica granatum* L.). *Nanomaterials* 2022; 12, 1527.

23. Lakshmeesha, T.R.; Murali, M.; Ansari, M.A.; Udayashankar, A.C.; Alzohairy, M.A.; Almatroudi, A.; Alomary, M.N.; Asiri, S.M.M.; Ashwini, B.S.; Kalagatur, N.K.; Nayak, C.S.; Niranjana, S.R.; Biofabrication of zinc oxide nanoparticles from Melia azedarach and its potential in controlling soybean seed-borne phytopathogenic fungi. *Saudi J Biol Sci* 2020; 27(8), 1923–1930.
24. Mosquera-Sánchez, L.P.; Arciniegas-Grijalba, P.A.; Patiño-Portela, M.C.; Guerra–Sierra, B.E.; Muñoz-Florez, J.E.; Rodríguez-Páez, J.E.; Antifungal effect of zinc oxide nanoparticles (ZnO-NPs) on Colletotrichum sp., causal agent of anthracnose in coffee crops. *Biocatal Agric Biotechnol* 2020; 25, 101579.
25. Ingale, A.G.; Chaudhari, A.N.; Biogenic synthesis of nanoparticles and potential applications: An eco-friendly approach. *J Nanomed Nanotechol* 2013; 4(165), 1–7.
26. DeRosa, M.C.; Monreal, C.; Schnitzer, M.; Walsh, R.; Sultan, Y.; Nanotechnology in fertilizers. *NatNanotechnol* 2010; 5(2), 91–91.
27. Khot, L.R.; Sankaran, S.; Maja, J.M.; Ehsani, R.; Schuster, E.W.; Applications of nanomaterials in agricultural production and crop protection: A review. *Crop Prot* 2012; 35, 64–70.
28. Malini, S.; Raj, K.; Madhumathy, S.; El-Hady, K.M.; Islam, S.; Dutta, M.; Bioinspired advances in nanomaterials for sustainable agriculture. *J Nanomater* 2022; 892 6133.
29. Mittal, D.; Kaur, G.; Singh, P.; Yadav, K.; Ali, S.A.; Nanoparticle-based sustainable agriculture and food science: Recent advances and future outlook. *Front Nanotechnol* 2020; 2, 579954.
30. Rafique, M.; Sadaf, I.; Rafique, M.S.; Tahir, M.B.; A review on green synthesis of silver nanoparticles and their applications. *Artif Cells Nanomed Biotechnol* 2017; 45, 1272–1291.
31. Singh, J.; Dutta, T.; Kim, K.H.; Rawat, M.; Samddar, P.; Kumar, P.; Green' synthesis of metals and their oxide nanoparticles: Applications for environmental remediation. *J Nanobiotechnol* 2018; 16, 84.
32. Ahmed, S.F.; Mofijur, M.; Rafa, N.; Chowdhury, A.T.; Chowdhury, S.; Nahrin, M.; Islam, A.; Ong, H.C.; Green approaches in synthesising nanomaterials for environmental nanobioremediation: Technological advancements, applications, benefits and challenges. *Environ Res* 2022; 204, 111967.
33. Anastas, P.T.; Warner, J.C.; Green Chemistry: Theory and Practice. Oxford University Press, New York, NY, 2020.
34. Wadhwani, S.A.; Shedbalkar, U.U.; Singh, R.; Karve, M.S.; Chopade, B.A.; Novel polyhedral gold nanoparticles: Green synthesis, optimization and characterization by environmental isolate of Acinetobacter sp. SW30. *World J Microbiol Biotechnol* 2014; 30, 2723–2731.
35. Babu-Maddinedi, S.; Mandal, B.K.; Ranjan, S.; Dasgupta, N.; Diastase assisted green synthesis of size-controllable gold nanoparticles. *RSC Adv* 2015; 5, 26727–26733.
36. Dhanasekar, N.N.; Rahul, G.R.; Narayanan, K.B.; Raman, G.; Sakthivel, N.; Green chemistry approach for the synthesis of gold nanoparticles using the fungus Alternaria sp. *J Microbiol Biotechnol* 2015; 25, 1129–1135.
37. Saratale, R.G.; Karuppusamy, I.; Saratale, G.D., Pugazhendhi, A.; Kumar, G.; Park, Y.; Ghodake, G.S.; Bharagava, R.N.; Banu, J.R.; Shin, H.S.; A comprehensive review on green nanomaterials using biological systems: Recent perception and their future applications. *Colloids Surf B Biointerfaces* 2018; 170, 20–35.
38. Owaid, M.N.; Rabeea, M.A.; Aziz, A.A.; Jameel, M.S.; Dheyab, M.A.; Mushroom-assisted synthesis of triangle gold nanoparticles using the aqueous extract of fresh Lentinula edodes (shiitake), Omphalotaceae. *Env Nanotechnol Monit Manage* 2019; 12, 100270.
39. Bolade, O.P.; Williams, A.B.; Benson, N.U.; Green synthesis of iron-based nanomaterials for environmental remediation: A review. *Environ Nanotechnol Monit Manag* 2020; 13, 100279.
40. Jameel, M.S.; Aziz, A.A.; Dheyab, M.A.; Green synthesis: Proposed mechanism and factors influencing the synthesis of platinum nanoparticles. *Green Process Synth* 2020; 9, 41.
41. Nasrollahzadeh, M.; Sajjadi, M.; Iravani, S.; Varma, R.S.; Green-synthesized nanocatalysts and nanomaterials for water treatment: Current challenges and future perspectives. *J Hazard Mater* 2021; 401, 123401.
42. Anastas, P.; Eghbali, N.; Green chemistry: Principles and practice. *Chem Soc Rev* 2010; 39, 301–312.
43. Jahangirian, H.; Lemraski, E.G.; Webster, T.J.; Rafiee-Moghaddam, R.; Abdollahi, Y.; A review of drug delivery systems based on nanotechnology and green chemistry: Green nanomedicine. *Int J Nanomed* 2017; 12, 2957–2978.

44. Singh, J.; Dutta, T.; Kim, K.H.; Rawat, M.; Samddar, P.; Kumar, P.; 'Green' synthesis of metals and their oxide nanoparticles: Applications for environmental remediation. *J Nanobiotechnol* 2018; 16, 84.
45. Huston, M.; DeBella, M.; DiBella, M.; Gupta, A.; Green synthesis of nanomaterials. *Nanomaterials* 2021; 11, 2130.
46. Patra, J.K.; Baek, K.H.; Green nanobiotechnology: Factors affecting synthesis and characterization techniques. *J Nanomaterials* 2014; 2014(2015), 219–234.
47. Gomaa, E.Z.; Iron nanoparticles α-chitin nanocomposite for enhanced antimicrobial, dyes degradation and heavy metals removal activities. *J Polym Environ* 2018; 26, 3638–3654.
48. Samrot, A.V., Angalene, J.L.A., Roshini, S.M.; Raji, P.; Stefi, S.M.; Preethi, R.; Selvarani, A.J.; Madankumar, A.; Bioactivity and heavy metal removal using plant gum mediated green synthesized silver nanoparticles. *J Clust Sci* 2019; 30, 1599–1610.
49. Ehrampoush, M.H., Miria, M., Salmani, M.H.; Mahvi, A.H.; Cadmium removal from aqueous solution by green synthesis iron oxide nanoparticles with tangerine peel extract.*J Environ Health Sci Engineer* 2015; 13, 84.
50. Beheshtkhoo, N.; Kouhbanani, M.A.J.; Savardashtaki, A.; Amani, A.M.; Taghizadeh, S.; Green synthesis of iron oxide nanoparticles by aqueous leaf extract of Daphne mezereum as a novel dye removing material. *Appl Phys A* 2018; 124 (2018), 1–7.
51. Nakkala, J.R.; Mata, R.; Raja, K.; Chandra, V.K.; Sadras, S.R.; Green synthesized silver nanoparticles: Catalytic dye degradation, in vitro anticancer activity and in vivo toxicity in rats. *Mater Sci Eng C* 2018; 91, 372–381.
52. Singh, J.; Kumar, V.; Jolly, S.S.; Kim, K-H.; Rawat, M.; Kukkar, D.; Tsang, Y.F.; Biogenic synthesis of silver nanoparticles and its photocatalytic applications for removal of organic pollutants in water. *J Ind Eng Chem* 2019; 247–257.
53. Momin, B.; Rahman, S.; Jha, N.; Annapure, U.S.; Valorization of mutant Bacillus licheniformis M09 supernatant for green synthesis of silver nanoparticles: Photocatalytic dye degradation, antibacterial activity, and cytotoxicity. *Bioprocess Biosyst Eng* 2019, 42, 541–553.
54. Parmar, A.; Kaur, G.; Kapil, S.; Sharma, V.; Choudhary, M.K.; Sharma, S.; Novel biogenic silver nanoparticles as invigorated catalytic and antibacterial tool: A cleaner approach towards environmental remediation and combating bacterial invasion. *Mater Chem Phys* 2019, 238, 121861.
55. Ismail, M.; Khan, M.I.; Khan, M.A.; Akhtar, K.; Asiri, A.M.; Khan, S.B.; Plant-supported silver nanoparticles: Efficient, economically viable and easily recoverable catalyst for the reduction of organic pollutants. *Appl Org Chem* 2019, 33(8), e4971.
56. Riaz, M.; Ismail, M.; Ahmad, B.; Zahid, N.; Jabbour, G.; Khan, M.S.; Mutreja, V.; Sareen, S.; Rafiq, A.; Faheem, M.; Shah, M.M.; Khan, M.I.; Bukhari, S.A.I.; Park, J.; Characterizations and analysis of the antioxidant, antimicrobial, and dye reduction ability of green synthesized silver nanoparticles. *Green Process Synth* 2020, 9, 64.
57. Ramos, M.M.; Morais, E.S.; Sena, I.S.; Lima, A.L.; Oliveira, F.R.; Freitas, C.M.; Fernandes, C.P.; Carvalho, J.C.T.; Ferreira, I.M.; Silver nanoparticle from whole cells of the fungi Trichoderma spp. isolated from Brazilian Amazon. *Biotechnol Lett* 2020, 42(5), 833–843.
58. Abbas, S.; Nasreen, S.; Haroon, A.; Asharf, M.A.; Synthesis of silver and copper nanoparticles from plants and application as adsorbents for naphthalene decontamination. *Saudi J Biol Sci* 2020, 27(4), 1016–1023.
59. Ebrahimzadeh, M.A.; Naghizadeh, A.; Amiri, O.; Shirzadi-Ahodashti, M.; Mortazavi-Derazkola, S.; Green and facile synthesis of Ag nanoparticles using Crataegus pentagyna fruit extract (CP-AgNPs) for organic pollution dyes degradation and antibacterial application. *Bioorg Chem* 2020;94, 103425.
60. Samuel, M.S.; Jose, S.; Selvarajan, E.; Mathimani, T.; Pugazhendhi, A.; Biosynthesized silver nanoparticles using Bacillus amyloliquefaciens; Application for cytotoxicity effect on A549 cell line and photocatalytic degradation of p-nitrophenol. *J Photochem Photobiol B: Biol* 2020, 202, 111642.
61. Borah, D.; Das, N.; Das, N.; Bhattacharjee, A.; Sarmah, P.; Ghosh, K.; Chandel, M.; Rout, J.; Pandey, P.; Ghosh, N.N.; Bhattacharjee, C.R.; Alga-mediated facile green synthesis of silver nanoparticles: Photophysical, catalytic and antibacterial activity. *Appl Organomet Chem* 2020, 34(5): e5597.
62. Balaraman, P.; Balasubramanian, B.; Kaliannan, D.; Durai, M.; Kamyab, H.; Park, S.; Chelliapan, S.; Lee, C.T.; Maluventhen, V.; Maruthupandian, A.; Phyco-synthesis of silver nanoparticles mediated

from marine algae Sargassum myriocystum and its potential biological and environmental applications. *Waste Biomass Valorization* 2020, 11(10), 5255–5271.
63. Liu, X.; Chen, J.L.; Yang, W.Y.; Qian, Y.C.; Pan, J.Y.; Zhu, C.N.; Liu, L.; Ou, W.B.; Zhao, H.X.; Zhang, D.P.; Biosynthesis of silver nanoparticles with antimicrobial and anticancer properties using two novel yeasts. *Sci Rep* 2021, 11, 15795.
64. Ismail, G.A.; Allam, N.G.; El-Gemizy, W.M.; Salem, M.A.; The role of silver nanoparticles biosynthesized by Anabaena variabilis and Spirulina platensis cyanobacteria for malachite green removal from wastewater. *Environ Technol* 2021, 42(2), 4475–4489.
65. Nga, N.T.A.; Raghavendra, V.B.; Sindhu, R.; Alshiekheid, M.; Sabour, A.; Krishnan, R.; Chi, N.T.L.; Pugazhendhi, A.; Green fabrication of silver nanoparticles using Chloroxylon swietenia leaves and their application towards dye degradation and food borne pathogens. *Food Chem Toxicol* 2022, 165, 113192.
66. Yassin, M.T.; Mostafa, A.A.-F.; Al-Askar, A.A.; Al-Otibi, F.O.; Facile green synthesis of silver nanoparticles using aqueous leaf extract of Origanum majorana with potential bioactivity against multidrug resistant bacterial strains. *Crystals* 2022, 12, 603.
67. Doan, VD.; Phan, T.L.; Le, V.T.; Vasseghian, Y.; Evgenievna, L.O.; Tran, D.L.; Le, V.T.; Efficient and fast degradation of 4-nitrophenol and detection of Fe(III) ions by Poria cocos extract stabilized silver nanoparticles. *Chemosphere* 2022, 286, 131894.
68. Negi, A.; Rana, P.; Vishwakarma, R.K.; Negi, D.S.; Multi dye degradation, antibacterial, antidiabetic and antioxidant assessment of silver nanoparticles (Ag-NPs) derived via leaves of *Smilax aspera. Inorg Chem Commun* 2022, 143, 109703.
69. Verma, A.; Bharadvaja, N.; Plant-mediated synthesis and characterization of silver and copper oxide nanoparticles: Antibacterial and heavy metal removal activity. *J Clust Sci* 2022, 33, 1697–1712.
70. Salem, D.M.S.A.; Ismail, M.M.; Aly-Eldeen, M.A.; Biogenic synthesis and antimicrobial potency of iron oxide (Fe3O4) nanoparticles using algae harvested from the Mediterranean Sea, Egypt. *Egypt J Aquat Res* 2019, 45(3), 197–204.
71. Salem, D.M.S.A.; Ismail, M.M.; Tadros, H.R.Z.; Evaluation of the antibiofilm activity of three seaweed species and their biosynthesized iron oxide nanoparticles (Fe3O4-NPs). *Egypt J Aquat Res* 2020, 46(4), 333–339.
72. Chatterjee, S.; Mahanty, S.; Das, P.; Chaudhuri, P.; Das, S.; Biofabrication of iron oxide nanoparticles using manglicolous fungus Aspergillus niger BSC-1 and removal of Cr(VI) from aqueous solution. *Chem Eng J* 2020, 385, 123790.
73. Mahanty, S.; Bakshi, M.; Ghosh, S.; Gaine, T.; Chatterjee, S.; Bhattacharyya, S.; Das, S.; Das, P.; Chaudhuri, P.; Mycosynthesis of iron oxide nanoparticles using manglicolous fungi isolated from Indian sundarbans and its application for the treatment of chromium containing solution: Synthesis, adsorption isotherm, kinetics and thermodynamics study. *Environ Nanotechnol Monit Manag* 2019, 12, 100276.
74. Karimi, P.; Javanshir, S.; Sayadi, M.H.; Arabyarmohammadi, H.; Arsenic removal from mining effluents using plant-mediated, green-synthesized iron nanoparticles. *Processes* 2019, 7(10), 759.
75. Wu, Z.; Su, X.; Lin, Z.; Owens, G.; Chen, Z.; Mechanism of As (V) removal by green synthesized iron nanoparticles. *J Hazard Mater* 2019, 379, 120811.
76. Safarkar, R.; Rajaei, G.E.; Khalili-Arjagi, S.; The study of antibacterial properties of iron oxide nanoparticles synthesized using the extract of lichen Ramalina sinensis. *Asian J Nanosci Mater* 2020, 3, 157–166.
77. Ting, A.S.Y.; Chin, J.E.; Biogenic synthesis of iron nanoparticles from apple peel extracts for decolorization of malachite green dye. *Wat Air Soil Poll* 2020, 231(6), 1–10.
78. Mathur, P.; Saini, S.; Paul, E.; Sharma, C.; Mehtani, P.; Endophytic fungi mediated synthesis of iron nanoparticles: Characterization and application in methylene blue decolorization. *Curr Res Green Sustain Chem* 2021, 4, 100053.
79. Arjaghi, S.K.; Alasl, M.K.; Sajjadi, N.; Fataei, E.; Rajaei, G.E.; Green synthesis of iron oxide nanoparticles by RS lichen extract and its application in removing heavy metals of lead and cadmium. *Biol Trace Elem Res* 2020, 199, 1–6.
80. Darwesh, O.; Shalapy, M.; Abo-Zeid, A.; Mahmoud, Y.; Nano-Bioremediation of municipal wastewater using myco-synthesized iron nanoparticles. *Egypt J Chem* 2021, 64(5), 2499–2507.

81. Shalaby, S.M.; Madkour, F.F.; El-Kassas, H.Y.; Mohamed, A.A.; Elgarahy, A.M.; Green synthesis of recyclable iron oxide nanoparticles using Spirulina platensis microalgae for adsorptive removal of cationic and anionic dyes. *Environ Sci Pollut Res Int* 2021, 28(46), 65549–65572.
82. El-Sheekh, M.M.; Hassan, L.H.S.; Morsi, H.H.; Evaluation of antimicrobial activities of blue-green algae-mediated silver and gold nanoparticles. *Rend Lincei Sci Fis Nat* 2021, 32, 747–759.
83. Manikandan, G.; Raju, R.; Biosynthesis of iron nanoparticles from Pleurotus florida and its antimicrobial activity against selected human pathogens. *Ind J Pharmaceut Sci* 2021, 83(1), 45–51.
84. Singh, G.; Kumar, V.; Dwivedi, S.K.; Comparative investigation of Congo Red and direct blue-1 adsorption on mycosynthesized iron nanoparticle. *J Clust Sci* 2022, 33(5), 1889–1905.
85. Vázquez-Guerrero, A.; Cortés-Martínez, R.; Alfaro-Cuevas-Villanueva, R.; Rivera-Muñoz, E.M.; Huirache-Acuña, R.; Cd (II) and Pb (II) adsorption using a composite obtained from moringa oleifera lam. Cellulose nanofibrils impregnated with iron nanoparticles. *Water* 2021, 13(1), 89.
86. Da'na, E.; Taha, A.; Hassanin, H.A.; Green fabrication of iron nanoparticles decorated with amine functionality for the remediation of lead ions from aqueous solutions. *Surf Interfaces* 2022, 30, 101909.
87. Roy, A.; Singh, V.; Sharma, S.; Ali, D.; Azad, A.K.; Kumar, G.; Emran, T.B.; Antibacterial and dye degradation activity of green synthesized iron nanoparticles. *J Nanomater* 2022, 2022, 1–6.
88. Schuster, S.; Su, Y.; Ting, A.; Decolourisation of triphenylmethane dyes by biogenically synthesised iron nanoparticles from fungal extract. *Mycology* 2021, 13, 56–67.
89. Bhukal, S.; Sharma, A.; Divya, R.; Kumar, S.; Deepak, B.; Pal, K.; Mona, S.; Spirulina based iron oxide nanoparticles for adsorptive removal of crystal violet dye. *Top Catal* 2022, 65(19-20), 1675–1685.
90. Zakariya, N.A.; Majeed, S.; Jusof, W.H.W.; Investigation of antioxidant and antibacterial activity of iron oxide nanoparticles (IONPS) synthesized from the aqueous extract of Penicillium spp. *Sens Int* 2022, 3, 100164.
91. Ranjani, V.A.; Rani, G.T.; Sowjanya, M.; Preethi, M.; Srinivas, M.; Nikhil, M.; Yeast mediated synthesis of iron oxide nano particles: Its characterization and evaluation of antibacterial activity. *Sciences (IRJPMS)* 2022, 5(5), 12–16.
92. Yin, I.X.; Zhang, J.; Zhao, I.S.; Mei, M.L.; Li, Q.; Chu, C.H.; The antibacterial mechanism of silver nanoparticles and its application in dentistry. *Int J Nanomedicine* 2020; 15, 2555–2562.
93. Calderón-Jiménez, B.; Johnson, M.E.; Bustos, A.R.M.; Murphy, K.E.; Winchester, M.R.; Baudrit, J.R.V.; Silver nanoparticles: Technological advances, societal impacts, and metrological challenges. *FrontChem* 2017; 5, 00006.
94. Coetzee, D.; Venkataraman, M.; Militky, J.; Petru, M.; Influence of nanoparticles on thermal and electrical conductivity of composites. *Polymers* 2020, 12(4), 742–750.
95. Monti, G.A.; Correa, N.M.; Falcone, R.D.; Silbestri, G.F.; Moyano, F.; New insights into the catalytic activity and reusability of water-soluble silver nanoparticles. *Chem Select* 2021, 6(29), 7436–7442.
96. Moslehi, A.; Zarei, M.; Application of magnetic Fe3O4 nanoparticles as a reusable heterogeneous catalyst in the synthesis of β-lactams containing amino groups. *New J Chem* 2019; 43, 12690–12697.
97. Gibała, A.; Żeliszewska, P.; Gosiewski, T.; Krawczyk, A.; Duraczyńska, D.; Szaleniec, J.; Szaleniec, M.; Oćwieja, M.; Antibacterial and antifungal properties of silver nanoparticles—Effect of a surface-stabilizing agent. *Biomolecules* 2021; 11(10), 1481.
98. Zheng, L.; Hong, F.; Lu, S.; Liu, C.; Effect of nano-TiO2 on strength of naturally aged seeds and growth of spinach. *Biol Trace Elem Res* 2005; 104, 83–91.
99. Agarwal, H.S.; Kumar, V.; Rajeshkumar, S.; A review on green synthesis of zinc oxide nanoparticles–An eco-friendly approach. *ResourEffic Technol* 2017; 34, 406–413.
100. Sunny, N.E.; Mathew, S.S.; Chandel, N.; Saravanan, P.; Rajeshkannan, R.; Rajasimman, M.; Vasseghian, Y.; Rajamohan, N.; Kumar, S.V.; Green synthesis of titanium dioxide nanoparticles using plant biomass and their applications-A review. *Chemosphere* 2022; 2022, 134612.
101. Al-Dhabi, N.A.; Arasu, M.V.; Environmentally-friendly green approach for the production of zinc oxide nanoparticles and their anti-fungal, ovicidal, and larvicidal properties. *Nanomaterials* 2018; 87, 500.
102. Thelma, J.; Balasubramanian, C.; Ovicidal, larvicidal and pupicidal efficacy of silver nanoparticles synthesized by Bacillus marisflavi against the chosen mosquito species. *Plos One* 2021; 16, e0260253.
103. Soufi, G.J.; Iravani, S.; Eco-friendly and sustainable synthesis of biocompatible nanomaterials for diagnostic imaging: Current challenges and future perspectives. *Green Chem* 2020; 22, 2662–2687.

104. Shang, L.; Nienhaus, K.; Nienhaus, G.U.; Engineered nanoparticles interacting with cells: Size matters. *J Nanobiotechnol* 2014; 12, 5–9.
105. Shi, Y.; Miller, M.L.; DiPasqua, A.J.; Biocompatibility of mesoporous silica nanoparticles? *Comment Inorg Chem* 2015; 36, 61–80.
106. Adabi, M.; Naghibzadeh, M.; Adabi, M.; Zarrinfard, M.A.; Esnaashari, S.S.; Seifalian, A.M.; Faridi-Majidi, R.; Aiyelabegan, H.T.; Ghanbari, H.; Biocompatibility and nanostructured materials: Applications in nanomedicine. *Artif Cells Nanomed Biotechnol* 2016; 45, 833–842.
107. Hu, C.C.; Wu, G.H.; Hua, T.E.; Wagner, O.I.; Yen, T.J.; Uptake of TiO2 nanoparticles into C. elegans neurons negatively affects axonal growth and worm locomotion behavior. *ACS Appl Mater Interfaces* 2018; 10, 8485–8495.
108. Król-Górniak, A.; Rafińska, K.; Monedeiro, F.; Pomastowski, P.; Buszewski, B.; Comparison study of cytotoxicity of bare and functionalized zinc oxide nanoparticles. *Int J Mol Sci* 2021; 22, 9529.
109. Gutiérrez de la Rosa, S.Y.;Muñiz Diaz, R.; Villalobos Gutiérrez, P.T.; Patakfalvi, R.; Gutiérrez Coronado, Ó.; Functionalized platinum nanoparticles with biomedical applications. *Int J Mol Sci* 2022; 23, 9404.
110. Rónavári, A.; Igaz, N.; Adamecz, D.I.; Szerencsés, B.; Molnar, C.; Kónya, Z.; Pfeiffer, I.; Kiricsi, M.; Green silver and gold nanoparticles: Biological synthesis approaches and potentials for biomedical applications. *Molecules* 2021; 26(4), 844.
111. Jeyarania, S.; Vinita, N.M.; Puja, P.; Senthamilselvi, S.; Devan, U.; Velangani, A.J.; Biruntha, M.; Pugazhendhi, A.; Kumar, P.; Biomimetic gold nanoparticles for its cytotoxicity and biocompatibility evidenced by fluorescence-based assays in cancer (MDA-MB-231) and non-cancerous (HEK-293) cells. *J Photochem Photobiol B Biol* 2020; 202, 111715.
112. Präbst, K.; Engelhardt, H.; Ringgeler, S.; Hübner, H.; Basic colorimetric proliferation assays: MTT, WST, and Resazurin. *Methods Mol Biol* 2017; 1601, 1–17.
113. Yedgar, S.; Barshtein, G.; Gural, A.; Hemolytic activity of nanoparticles as a marker of their hemocompatibility. *Micromachines* 2022; 13(12), 2091.
114. Khan, S.A.; Shahid, S.; Hanif, S.; Almoallim, H.S.; Alharbi, S.A.; Sellami, H.; Green synthesis of chromium oxide nanoparticles for antibacterial, antioxidant anticancer, and biocompatibility activities. *Int J Mol Sci* 2021; 22, 502.
115. Anjum, S.; Nawaz, K.; Ahmad, B.; Hano, C.; Abbasi, B.H.; Green synthesis of biocompatible core–shell (Au–Ag) and hybrid (Au–ZnO and Ag–ZnO) bimetallic nanoparticles and evaluation of their potential antibacterial, antidiabetic, antiglycation and anticancer activities. *RSC Adv.* 2022; 12, 23845–23859.
116. Bandeira, M.; Possan, A.L.; Pavin, S.S.; Raota, C.S.; Vebber, M.C.; Giovanela, M.; Roesch-Ely, M.; Devine, D.M.; Crespo, J.S.; Mechanism of formation, characterization and cytotoxicity of green synthesized zinc oxide nanoparticles obtained from Ilex paraguariensis leaves extract. *Nano-Struct. Nano-Objects* 2020; 24, 100532.
117. Thoo, L.; Fahmi, M.Z.; Zulkipli, I.N.; Keasberry, N.; Idris, A.; Interaction and cellular uptake of surface-modified carbon dot nanoparticles by J774.1 macrophages. *Cent Eur J Biol* 2017; 42, 324–330.
118. Florkiewicz, W.; Pluta, K.; Malina, D.; Rudnicka, K.; Żywicka, A.; Guigou, M.D.; Tyliszczak, B.; Sobczak-Kupiec, A.; Investigation on green synthesis, biocompatibility, and antibacterial activity of silver nanoparticles prepared using *Cistus incanus*. *Materials* 2021; 14, 5028.
119. Faisal, S.; Jan, H.; Shah, S.A.; Shah, S.; Khan, A.; Akbar, M.T.; Rizwan, M.; Jan, F.; Wajidullah, Akhtar, N.; Khattak, A.; Syed, S.; Green synthesis of zinc oxide (ZnO) nanoparticles using aqueous fruit extracts of *Myristica fragrans*: Their characterizations and biological and environmental applications. *ACS Omega* 2021, 6, 709–9722.
120. Harandi, F.N.; Khorasani, A.C.; Shojaosadati, S.A.; Hashemi-Najafabadi, S.; Living lactobacillus-ZnO nanoparticles hybrids as antimicrobial and antibiofilm coatings for wound dressing application. *Mater Sci Eng C Mater Biol Appl.*2021, 130, 112457.
121. Kainat, Khan, M.A.; Ali, F.; Faisal, S.; Rizwan, M.; Hussain, Z.; Zaman, N.; Afsheen, Z.; Uddin, M.N.; Bibi, N.; Exploring the therapeutic potential of *Hibiscus rosa sinensis* synthesized cobalt oxide (Co_3O_4-NPs) and magnesium oxide nanoparticles (MgO-NPs). *Saudi J Biol Sci* 2021; 28, 5157–5167.
122. Sellami, H.; Khan, S.A.; Ahmad, I.; Alarfaj, A.A.; Hirad, A.H.; Al-Sabri, A.E.; Green synthesis of silver nanoparticles using *Olea europaealeaf* extract for their enhanced antibacterial, antioxidant, cytotoxic and biocompatibility applications. *Int J Mol Sci* 2021; 22, 12562.

123. Josiah, A.J; Pillai, S.K.; Cordier, W.; Nell, M.; Twilley, D.; Lall, N.; Ray, S.S.; Cannabidiol-mediated green synthesis, characterization, and cytotoxicity of metal nanoparticles in human keratinocyte cells. *ACS Omega* 2021; 6(43), 29078–29090.
124. Bin-Jumah, M.N.; Al-Abdan, M.; Al-Basher, G.; Alarifi, S.; Molecular mechanism of cytotoxicity, genotoxicity, and anticancer potential of green gold nanoparticles on human liver normal and cancerous cells. *Dose-Response* 2020; 18(2), 1–10.
125. Abdelmigid, H.M.; Hussien, N.A.; Alyamani, A.A.; Morsi, M.M.; AlSufyani, N.M.; Kadi, H.A.; Green synthesis of zinc oxide nanoparticles using pomegranate fruit peel and solid coffee grounds vs. chemical method of synthesis, with their biocompatibility and antibacterial properties investigation. *Molecules* 2022; 27, 1236.
126. Akter, R.; Ling, L.; Rupa, E.J.; KyuPark, J.; Mathiyalagan, R.; Nahar, J.; Won, L.J.; Hyun, K.D.; Murugesan, M.; Yang, D.C.; Kang, S.C.; Kwak, G.Y.; Binary effects of gynostemma gold nanoparticles on obesity and inflammation via downregulation of PPARγ/CEPBα and TNF-αgene expression. *Molecules* 2022; 27, 2795.
127. Oves, M.; Ahmar Rauf, M.; Aslam, M.; Qari, H.A.; Sonbol, H.; Ahmad, I.; Sarwar Zaman, G.; Saeed, M.; Green synthesis of silver nanoparticles by *Conocarpus Lancifolius* plant extract and their antimicrobial and anticancer activities. *Saudi J Biol Sci* 2022, 29, 460–471.
128. Sekar, V.; Balakrishnan, C.; Kathirvel, P.; Swamiappan, S.; Alshehri, M.A.; Sayed, S.; Panneerselvam, C.; Ultra-sonication-enhanced green synthesis of silver nanoparticles using Barleria buxifolia leaf extract and their possible application. *Artif Cells Nanomed Biotechnol* 2022, 50, 177–187.
129. Shnoudeh, A.J.; Qadumii, L.; Zihlif, M.; Al-Ameer, H.J.; Salou, R.A.; Jaber, A.Y.; Hamad, I.; Green synthesis of gold, iron and selenium nanoparticles using phytoconstituents: Preliminary evaluation of antioxidant and biocompatibility potential. *Molecules* 2022; 27, 1334.
130. Zhu, J.; Liu, Z.; Pu, Y.; Xu, J.; Zhang, S.; Bao, Y.; Green synthesized gold nanoparticles from *Pseudobulbus Cremastrae seu Pleiones* show efficacy against hepatic carcinoma potentially through immunoregulation. *Drug Delivery* 2022, 29, 1983–1993.
131. Cittrarasu, V.; Kaliannan, D.; Dharman, K.; Maluventhen, V.; Easwaran, M.; Liu, W.C.; Balasubramanian, B.; Arumugam, M.; Green synthesis of selenium nanoparticles mediated from Ceropegia bulbosa Roxb extract and its cytotoxicity, antimicrobial, mosquitocidal and photocatalytic activities. *Sci Rep* 2021; 11, 1032.
132. Lee, Y.J.; Ahn, E.Y.; Park, Y.; Shape-dependent cytotoxicity and cellular uptake of gold nanoparticles synthesized using green tea extract. *Nanoscale Res Lett* 2019; 14, 129.
133. Rodríguez-León, E.;RodrRodríguez-Leíguez-Vázquez, B.E.; Martínez-Higuera, A.; Rodríguez-Beas, C.; Larios-Rodríguez, E.; Navarro, R.E.; López-Esparza, R.; Iñiguez-Palomares, R.A.; Synthesis of gold nanoparticles using Mimosa tenuiflora extract, assessments of cytotoxicity, cellular uptake, and catalysis. *Nanoscale Res Lett* 2019; 14, 334.
134. Justus, B.; Arana, A.F.M.; Gonçalves, M.M.; Wohnrath, K.; Boscardin, P.M.D.; Kanunfre, C.C.; Budel, J.M.; Farago, P.V.; dePaula, J.F.P.; Characterization and cytotoxic evaluation of silver and gold nanoparticles produced with aqueous extract of Lavandula dentata L. in relation to K-562 cell line. *Braz Arch Biol Technol* 2019; 62, 125.
135. Long, Q.; Cui, L.; He, S.; Sun, J.; Chen, Q.; Bao, H.; Liang, T.; Liang, B.; Cui, L.; Preparation, characteristics and cytotoxicity of green synthesized selenium nanoparticles using Paenibacillus motobuensis LY5201 isolated from the local specialty food of longevity area. *Sci Rep* 2023; 13, 53.
136. Mussin, J.; Robles-Botero, V.; Casañas-Pimentel, R.; Rojas, F.; Angiolella, L.; San Martín-Martínez, E.; Giusiano, G.; Antimicrobial and cytotoxic activity of green synthesis silver nanoparticles targeting skin and soft tissue infectious agents. *Sci Rep* 2021; 11, 14566.
137. Khorrami, S.; Zarrabi, A.; Khaleghi, M.; Danaei, M.; Mozafari, M.R.; Selective cytotoxicity of green synthesized silver nanoparticles against the MCF-7 tumor cell line and their enhanced antioxidant and antimicrobial properties. *Int J Nanomed* 2018; 13, 8013–8024.
138. Karimi, S.; Shahri, M.M.; Medical and cytotoxicity effects of green synthesized silver nanoparticles using Achillea millefolium extract on MOLT-4 lymphoblastic leukemia cell line. *J Med Virol* 2021; 93(6), 3899–3906.

139. Harandi, F.N.; Khorasani, A.C.; Shojaosadati, S.A.; Hashemi-Najafabadi, S.; Living lactobacillus–ZnO nanoparticles hybrids as antimicrobial and antibiofilm coatings for wound dressing application. *Mater Sci Eng C* 2021; 130, 112457.
140. Kainat; Khan, M.A.; Ali, F.; Faisal, S.; Rizwan, M.; Hussain, Z.; Zaman, N.; Afsheen, Z.; Uddin, M.N.; Bibi, N.; Exploring the therapeutic potential of Hibiscus rosa sinensis synthesized cobalt oxide (Co3O4-NPs) and magnesium oxide nanoparticles (MgO-NPs). *Saudi J. Biol. Sci.* 2021; 28(9), 5157–5167.

4 Nanostructure in Foods

Rafael Audino Zambelli and Luciana Gama de Mendonça
Corresponding authors: Luciana Gama de Mendonça, Rafael Audino Zambelli

4.1 INTRODUCTION

Nanostructures have been explored in the food area for the development of color additives, auxiliaries in the production and manufacture of polymers, development of preservatives, flavor carriers, quality markers in fruits and vegetables, production of anti-caking agents, and nutritional dietary supplements (He et al., 2019).

The two main routes for the development of nanostructures are top-down and bottom-up, based on the properties of materials and their response to chemical, thermal, and mechanical factors (Niinivaara and Cranston, 2020). According to Gonçalves et al. (2022), the main applications are to improve the nutritional and functional value of foods, as well as to extend their shelf life. Nanostructures make it possible to control the interaction of bioactive components with food, regulate their release, and ensure availability in food formulations during their shelf life (Sahani and Sharma, 2021).

Tang (2021) points out that nanostructures can be produced from food-quality molecules such as polypeptides, polysaccharides, minerals, and lipids, as they have characteristics of being available, renewable, biocompatible, biodegradable, and accessible, which are interesting and desirable in food processes.

Nanostructures can be produced by several methods, such as high shear and homogenization (Yi et al., 2014), spraying and atomization (Moghaddam et al., 2015), milling (Burmeister and Kwade, 2013), extrusion and microfluidics (McClements, 2015), coating (Oxley, 2012), polymeric formation (Shpigelman et al., 2014), and supercritical fluid (McClements, 2015).

Despite the versatility and applications that nanotechnology can provide through nanostructures, only a small part of the food industry has introduced this technology into their processing lines (Magro et al., 2016).

Within this context, the chapter aims to present the main developed nanostructures, their applications, and potential health risks.

4.2 LIPID NANOSTRUCTURE IN FOOD APPLICATIONS

Lipid-based nanostructures can improve the solubility, bioavailability, and stability of foods, preventing unwanted interactions with other food components (Mozafari et al., 2008), and have advantages such as high surface-to-volume ratio, load capacity, and high stability in the environment (Rafiee et al., 2018). Mohammadi et al. (2016) emphasized that lipid nanostructures have differential advantages over other molecules, such as encapsulation efficiency, low cost, absence of organic solvent in the formulation, targeted targeting, biocompatibility, and continuous production.

 DOI: 10.1201/9781003389408-4

Nanoliposomes are usually formed by phospholipids that are organized in one or more layers. However, this class of nanostructure has disadvantages, such as lack of chemical and physical stability, which limit its application (García-Pinel et al., 2019).

According to Müller et al. (2016), lipids of different structures can be used for the formation of SLNs, such as chain triglycerides, fatty acids, and waxes. The most traditional methods of producing SLNs are high pressure, hot or cold homogenization, ultrasonication, emulsification, and microemulsion (Ghasemiyeh and Mohammadi-Samani, 2018).

Weber et al. (2014) pointed out that SLNs reorganize their crystalline structure to achieve stabilization and form a gel in the dispersed phase. SLNs have some disadvantages, such as a tendency to gelation, polymorphic transition, and low encapsulation efficiency (Müller et al., 2002). They also have a low diffusion rate and require a long release time (Tapeinos et al., 2017). These limitations of SNLs allowed new studies to be carried out that ended in the development of the second generation of lipid transporters, the NLCs, where solid and liquid lipids are used together to maximize their properties. In this way, NLCs were developed to overcome the limitations associated with SLNs (Patil et al., 2019).

The aqueous dispersion of NLCs has the presence of other colloidal structures, such as mixed micelles, micelles, liposomes, and nanoemulsions. These structures can affect system stability. Thus, during storage, changes may occur to obtain greater thermodynamic stability and imply important changes in the matrix, such as an increase in the size of the particles caused by the phenomenon of coalescence or by the gelation of the dispersion (Nguyen et al., 2022).

A study developed by Rohmah et al. (2020) shows that the use of palm stearin and palm olein together with a surfactant agent can improve the delivery of nutrients through the NLC system.

4.3 NANOLIPOSOMES

Phospholipids are the major components of nanoliposomes, which are amphiphilic substances with hydrophilic head section (water-soluble) and hydrophobic tail section (lipid-soluble) (Aguilar-Pérez et al., 2020). Nanoliposomes are nanoparticles whose structures are spherical or oval vesicles composed of membranes containing two or more lipid bilayers (phospholipids and/or sterols) and have a hydrophobic lipid membrane and hydrophilic core structure (Zarrabi et al., 2020).

Silva et al. (2023) defines nanoliposomes as vesicular transporters formed by phospholipids through self-assembly involving nuclei that have the presence of water. They are composed of water molecules that are hydrophilic in nature, while the membrane is amphiphilic because of the long chain of fatty acids (Large et al., 2021). Nanoliposome encapsulation reduces the oxidative decomposition of loads caused by photothermal conditions during food storage and transport, which can extend shelf life (Li et al., 2022).

4.4 NANOEMULSIONS

Nanoemulsions are colloidal dispersions with sizes between 50 and 1000 nm. They can be applied to produce foods such as salad dressings, custom drinks, sweeteners, and flavored oils (Garti, 2008), and can be of the oil-in-water, water-in-oil, or bicontinuous type.

The two phases are often combined by a homogenization process, in which the use of energy is necessary to increase the surface area of the dispersed phase and provide for the creation of droplets. Because the free energy of formation is positive and the surface tension between the phases is relatively large, nanoemulsions are thermodynamically unstable (Komaiko and McClements, 2015).

According to Ziani et al. (2011) nanoemulsions are formed using methods involving high or low amounts of energy. High-energy methods are more effective. In this way, the sonication method is the most used for productions. Abbas et al. (2013) stated that the main factors that can affect the performance of the formation of nanoemulsions via sonication are the frequency, amplitude, and treatment time. Methods involving high pressure or ultrasonics are applied to form intensive disruptive forces such as collision, compression, and cavitation, so that better energy transfer occurs (Mehrnia et al., 2017).

The phase inversion temperature method was used to produce cinnamon oil nanoemulsions by Chuesiang et al. (2020) with the aim of studying their antimicrobial and antioxidant properties to increase the shelf life of seabass fillet during refrigerated storage. This method was used to produce essential oil nanoemulsions of lemon, key lime, kaffir lime, and calamansi lime by Liew et al. (2020). The phase was composed of lemon essential oil and corn oil, in the proportion of 8:2 and mixed with the aid of a stirrer at room temperature.

Nanoemulsions containing the essential oils carvacrol, cinnamaldehyde, thymol, and a mixture of them were developed and studied by Sanei-Dehkordi et al. (2022). They were used to load thyme essential oil to promote the preservation of ultrafiltered cheese by Bedoya-Serna et al. (2018). The nanoemulsions carrying thyme essential oil managed to eliminate microbial contamination and maintained the product's characteristics for four weeks.

4.5 INORGANIC-BASED NANOSTRUCTURES IN FOOD APPLICATIONS

Inorganic nanostructures, especially metallic ones, have advantages when compared to organic nanostructures. These particles are more effective and resistant to the severe conditions of food thermal processing (Hoseinnejad et al., 2018).

Metallic oxides have detection properties through the mechanism of adsorption and desorption with different gaseous compounds on the surface of the material, causing a change in electrical conductivity, making it interesting in the detection of some gases of interest in the food industry, such as ethylene (Galstyan et al., 2018).

Several nanoparticles are produced from metals or metal oxides of inorganic nanomaterials such as copper, zinc oxide, and titanium dioxide. The main application is in the formation of nanocomposites to compose packaging material (Shankar and Rhim, 2019).

4.6 ZINC OXIDE (ZNO) NANOPARTICLES

Zinc oxide consists of a material that has distinct properties, such as a large specific area, high sensitivity, non-toxicity, and a high isoelectric point, and is considered the most desirable group of nanostructures in terms of morphology, structure, and properties (Theerthagiri et al., 2019).

ZnO nanostructures are produced to be used in electrochemical sensors and biosensors for the detection of analytes, food safety, and monitoring of environmental pollutants, especially in wastewater (Beitollahi et al., 2020). Currently, the main methods of producing zinc nanoparticles are as follows:

(1) Processing by physical vapor deposition: this process involves the use of high temperatures to generate steam from a solid precursor that is subjected to plasma arc energy (Hessien et al., 2019).
(2) Mechanochemical processing involves crushing and mixing starting materials such as zinc chloride and sodium carbonate to produce zinc carbonate and sodium chloride.
(3) Sputtering: this method allows flexibility in growing the size of nanostructures on substrates that have a larger area. The target material is fed energetic ions to perform collisions. Plasma is generated inside a sputtering chamber (Krishna et al., 2023).
(4) Sol-gel: This consists of a low-cost deposition technique with a large available area. The method involves three steps: solution preparation from starting materials; deposition of the solution on a substrate; and thermal treatment of the formed gel. During the sol-gel synthesis

technique, parameters can affect the properties of nanostructures (Milea et al., 2011; Arya et al., 2021).

(5) Hydrothermal: this process generally involves temperatures between 90 °C and 100 °C of one-dimensional zinc oxide nanomaterials and their reaction with hexamethylenetetramine (HMT) (Ashfold et al., 2007).

Nanocomposites composed of aluminum oxide, bentonin, and activated carbon were developed by González-Jartín et al. (2019) and were able to eliminate 87% of mycotoxins. The nanostructures were used in the detoxification of beer.

Copper oxide nanostructures have been used in fungal disinfection processes in food and plants. The particles promote the release of Cu^{2+} ions that damage the cell membrane and lead to the death of fungi (Liu et al., 2020). Zhu et al. (2022) demonstrated that alginate nanogel containing copper nanoparticles more significantly destroyed fungal cell membranes than commercial fungicides. Magnetic iron oxide (Fe_3O_4) was introduced into chitosan and pectin films to produce magnetically active materials based on biopolymers by Zarandona et al. (2022).

One of the advantages of metallic nanoparticles is the modification of their surface, which allows several applications. Limthin et al. (2019) modified the surface of iron oxide nanoparticles with molecularly imprinted polymer and obtained an electrode for the detection of gluten.

4.7 SILVER NANOPARTICLES (AGNPS)

Silver nanoparticles are used in diverse areas such as food, health, industrial purposes, and medicine because of their specific physical and chemical properties, morphology, distribution, shape, size, and high surface area. They can be applied in optical, electrical, and thermal devices, packaging, sensors, cosmetics, and pharmaceuticals (Ghojavand et al., 2020).

The production methods of silver nanoparticles can be classified into chemical, biological and physical methods.

Chemical synthesis is the most widely used method for manufacturing silver nanoparticles. The reagents added in this process have the function of reducing silver ions and stabilizing the nanoparticles. However, due to problems related to toxicity, the biogenic synthesis method has been an alternative (Zhang et al., 2016).

Biogenic synthesis presupposes the use of organisms such as plants, fungi, bacteria, and/or by-products of their metabolism that can act as stabilizing and reducing agents. Nanoparticles are coated with biomolecules derived from organisms to improve stability and may exhibit biological activity (Ballotin et al., 2016). Guilger-Casagrande and de Lima (2019) point out that, compared to chemical synthesis, biogenic synthesis produces a greater number of metabolites.

Physical methods include milling, pyrolysis, and spark discharge processes. These methods produce silver nanoparticles with uniform size and high purity (Iravani et al., 2014). The synthesis of silver nanoparticles by chemical or physical means is generally more laborious and involves greater risks compared to biological methods, which generate nanoparticles with greater yield, solubility, and stability (Zhang et al., 2016). Figure 4.1 shows some applications of silver nanoparticles in food.

Raveendran and Docoslis (2021) used silver nanostructures as scaffolds for bimetallic structures (Ag–Au) for the identification of toxicants in water and food. The toxic components thiabendazole, malachite green, and biphenyl-4-thiol were identified and detected. The nanostructure was validated in apple peel.

Silver nanoparticles were inserted into a biodegradable matrix and incorporated into wheat gluten hydrolysates to be used in thin films using polymers as crosslinking agents. The objective was to confer antibacterial and antioxidant potential to the films from the controlled release of silver nanoparticles (Sajjad et al., 2023).

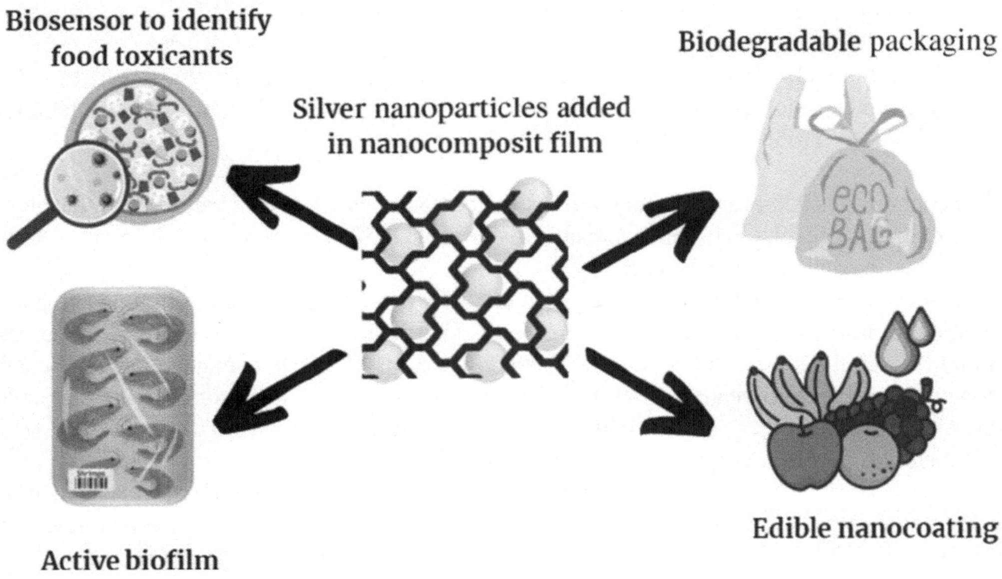

FIGURE 4.1 Applications of silver nanoparticles in food.

4.8 CARBON NANOSTRUCTURES

Carbon nanostructures are generally used with biosensors due to the number of nano allotropes with different structures, hybridization states, and dimensions of each carbon (Dridi et al., 2017). Carbon nanotubes to produce electrochemical sensors have attracted strong attention from researchers in several areas due to their mechanical and electronic properties (Pete et al., 2021). These nanostructures were used by Qian et al. (2020) for the development of an ultrasensitive electrode for the detection of bisphenol A in real food packaging.

Graphene oxide was modified with UiO-66 and Fe_3O_4 nanoparticles for the extraction of food dyes prior to high-performance liquid chromatography analysis by Arabkhani et al. (2023). The modification promotes the hybridization process that produces functional oxygen groups, reduces aggregation, confers magnetic properties, and increases porosity.

4.9 PROTEIN-BASED NANOSTRUCTURE IN FOOD APPLICATIONS

Proteins are biopolymers formed by different amino acids connected through peptide bonds and can have multiple functions in food such as providing unique sensory aspects, foaming, and other functional properties (Zheng et al., 2021). These sets of peptides obtained from different protein sources can present electrochemical properties, environmental response, thermal stability, optical, and electrical properties such as fluorescence, used, in many cases, as a biosensor for the identification of components of interest to the industry (Xu al., 2011; Carneiro et al., 2021). Protein-based nanostructures can be obtained from different sources, as shown in Table 4.1.

Nanostructures manufactured from animal or vegetable proteins are biocompatible with food formulations, which favors the guarantee of providing a safe food for the consumer. Other important characteristics for the production process of protein-based nanostructures are considered important: solubility, stability, and bioaccessibility (Chen et al., 2016).

TABLE 4.1
Protein sources used for the construction of nanostructures and nanoparticles

Protein-based nanostructures source	
Plant protein	**Animal protein**
Gliadin	Casein
Glutenin	Albumin
Albumin	Collagen
Globulin	Gelatin
Zein	Silk protein
Soy protein	

4.10 SOY PROTEIN

According to Utsumi et al. (1997), the structure of soy protein can vary significantly according to its composition, processing, and storage conditions. Its use is associated with characteristics of biodegradability, metabolization, easy handling, possible surface changes, and/or modification.

Nanoparticles structured from soy protein are extracted from soy protein isolate and require the application of some pre-treatments such as shear homogenization, ultrasonication, or microfluidization. The production of nanostructured particles can be obtained through top-down or bottom-up approaches (Tang, 2019), and other ways of producing nanoparticles from proteins and soy are used, such as heat-induced aggregation, emulsification-evaporation, crosslinking, self-assembly, electrospraying, nanoemulsions, and interfacially nanostructured emulsions (Tang, 2019). The author points out that soy-based nanoparticles have considerable advantages for acting as nanocarriers of bioactive components compared to other sources and types of proteins:

(1) Soy proteins have relevant nutritional and functional values, with beneficial effects associated with health.
(2) Proteins from soy products, such as protein isolates, are present in an aggregated state and can interact directly with bioactive components to form nanostructures.
(3) Soy proteins have a balance between hydrophobicity and hydrophilicity properties.
(4) Soy proteins are available in large quantities due to their global production; they are cheaper and more renewable than proteins of animal origin.

In this context, a study was conducted by Li et al. (2022) which focused on obtaining nanoliposomes from soy lecithin due to its phospholipid content. Medium purity soy lecithin (70%) was more stable and had a smaller droplet size.

4.11 GLUTEN

Gluten, as one kind of hydrophobic plant protein, has the capability of yielding sustained bioactive release (Kong et al., 2019). It is one of the most promising materials of natural origin due to its abundant availability, low cost, and hydrophilic characteristics (Ullsten et al., 2006). Gliadins and glutenins make up approximately 30 and 50%, respectively, of the total protein in wheat grain. It is believed that gliadin contributes to the flow properties of dough rheology, while glutenin contributes to its elasticity and strength (Zambelli et al., 2017).

Wheat gluten presents some technological difficulties, mainly due to its low solubility in water. In this way, several modifications can be made to gluten to improve its properties. One of these modifications is the manufacture of nanoparticles, which can promote improvements in gluten functionality and increase commercial application, for example, to promote the encapsulation of bioactive components (Yang et al., 2014).

The existence of a set of hydrated gliadin nanoparticles in high concentrations can be obtained from the disintegration of gluten, and these are directly linked to the resistance of bread doughs (Urade et al., 2018).

Wheat gliadin nanoparticles have been used as carriers of bioactive components and drugs for several years. Ezpeleta et al. (1996) obtained gliadin nanoparticles through the desolvation of wheat macromolecules which allowed nanoparticles with a yield of 90% of the total initial protein to be obtained. It was found that these nanoparticles have low water-solubility and high hydrophobicity.

Türe et al. (2013) evaluated the physicochemical properties of nanostructured hybrid composites of wheat gluten and silica. It was found that the protein phase is involved in the silica matrix due to hydrogen bonds, and silica induces a greater degree of restriction in relation to the wheat gluten matrix, despite less aggregation.

Wheat gluten nanoparticles modified with rice flour were used as stabilizers in soybean oil Pickering emulsion by Wang et al. (2022). The nanoparticles exhibited spherical morphology with an average particle size of 20 nm and satisfactory wettability, in addition to presenting thermal stability.

4.12 CARBOHYDRATE-BASED NANOSTRUCTURE IN FOOD APPLICATIONS

Different types of carbohydrates can be used to produce nanostructures: glycogen, starch, cellulose, hemicelluloses, lignin, chitosan, and chitin. One of the differentials of polysaccharides compared to other biomolecules is their diversity of functional groups, allowing their properties to be adjustable (Luo, 2020).

Nanogels can be produced from polysaccharides because they are biocompatible, biodegradable, nontoxic, and have a lower immunogenic probability (Zhang et al., 2021a). Some adjustments may be necessary to produce nanostructures, such as the hydrophilic/hydrophobic balance of the carbohydrate molecule (Piorkowski and McClements, 2014P). The different sources for obtaining nanostructures from carbohydrates are presented below.

4.12.1 Starch

Starch has been extensively studied as a promising biomaterial for application in the food industry. Several physical, chemical, and enzymatic treatments have been applied to different starch sources to modify their properties and increase the range of applications (Zambelli et al., 2018; Silva et al., 2021). The crystalline portions of starch molecules can be converted into nanostructures from physical disintegration or hydrolysis, and are commonly called nanocrystalline starch or starch nanocrystals (Rajisha et al., 2014).

Methods for producing starch-based nanostructures can be produced by different routes, the main ones being the formation of complexes, emulsion crosslinking, hydrolysis coupled with ultrasonication, and acid or enzymatic hydrolysis (Mukurumbira et al., 2017).

Ganje et al. (2019) proposed the encapsulation of orange peel oil containing 92% amylose nanostructures through a combination of sonication and thermal processes. Amylose nanocarriers with limonene showed stability in acidic conditions and controlled release.

Nano-sized starches with fantastic properties can be readily prepared from starch granules (Rostamabadi et al., 2019). Within this context, starch nanogels have been developed with a size range of 10 to 1000 nm that have been shown to be efficient in the encapsulation and controlled release of bioactive compounds (Ahmad et al., 2020).

4.12.2 Nanogels

Carbohydrate-based nanogels can be applied directly in the development of functional foods or through fortification (Silva and Fabi, 2022), and they have excellent capacity for expansion and water retention without compromising their three-dimensional structure (Bourbon et al., 2016).

Zhu et al. (2013) studied the development of sodium carboxymethylcellulose and lysozyme nanogels through the self-assembly method for drug loading. The nanogel ensured slower release and more efficient delivery to the target organ.

Carbohydrate-based nanogel was developed to encapsulate folic acid, an essential compound in important biochemical processes in the human body. The encapsulation of folic acid is necessary because of its easy degradation due to light, temperature, pH, and oxygen (Gazzali et al., 2016).

4.12.3 Cellulose and Derivates

Cellulose is the most abundant polysaccharide in nature and can be obtained from plants, animals, or bacteria (Trache et al., 2020). It can be used to produce cellulose nanocrystals, which have attractive characteristics such as high abundance and availability, high surface aspect ratio, mechanical properties, renewability, and biocompatibility (González and Gastelú, 2019; Wu et al., 2021).

According to Xu et al. (2013), nanocelluloses are obtained through chemical or mechanical disintegration of cellulosic raw materials and can be classified into two categories: cellulose nanocrystals and cellulose nanofibrils.

Nanofibrillated cellulose consists of long and flexible nanoloads composed of amorphous and crystalline regions alternately (Zhong et al., 2020), with soft and long chain.

Nanostructures have been produced from lignocellulosic biomass of rice straw. This raw material has a variety of aromatic hydrocarbons and phenolic compounds from the carbohydrate and can be used as reducing agents for the reduction of any metallic ion and to aid in the formation of nanomaterials (Zoghlami and Paës, 2019).

Cellulose nanocrystals have been used as a safe and renewable rheological modifier capable of allowing 3D printing of foods such as tomato, spinach, and apple purees. A study conducted by Armstrong et al. (2022) demonstrated that inks loaded with cellulose nanocrystals can be useful in printing multilateral structures and that they are a great rheological additive to promote viscous behavior.

Fotie et al. (2023) evaluated the gas barrier performance of plastic films coated with cellulose nanocrystals obtained from cotton. The results showed that the coatings added from the nanocrystals were more effective in blocking oxygen molecules.

Cellulose nanoparticles have been combined with gold nanoparticles encapsulated on the surface of citric polyacid for the display of fluorescence properties. The biosensor was developed to detect gluten proteins (Mousavizadeh and Sarlak, 2023).

4.12.4 Chitosan

Chitosan is another important carbohydrate; its structure is derived from deacetylated chitin with the presence of an amino group and two hydroxyl groups as a repetitive subunit. Its composition allows its properties to be adjusted more easily than chitin (Qu and Luo, 2020). Chitosan can be coupled to various nanostructures by inclusion in the material or deposited on the surface. The presence of surface amine and hydroxyl groups can promote the formation of inter and intramolecular hydrogen bonds (Silva et al., 2021).

Das et al. (2021) studied the encapsulation of *Pimpinella anisum* essential oil in a chitosan nanobiopolymer to improve the effectiveness and sensory parameters of the essential oil for application in food matrix. The nanobiopolymer showed antifungal and antiaflatoxigenic activity.

Hadidi et al. (2020) encapsulated clove essential oil in chitosan nanoparticles using the emulsion and ionic gelling technique. The oil retention rate was considered good, and the nanoparticles enhanced the antioxidant capacity of the essential oil, in addition to showing bactericidal capacity against *Listeria monocytogenes*.

Nisin-loaded carboxymethyl chitosan nanogels were prepared by the combined method of electrostatic self-assembly and chemical crosslinking by Wu et al. (2023) to enhance the antibacterial activity of nisin.

Chitosan and carboxymethyl konjac glucomannan nanogels were developed by Wu et al. (2021) to be a curcumin delivery vehicle. Nanogels showed stability and efficiency in curcumin encapsulation, showing promise in nutrient delivery.

4.13 PROTEIN- AND CARBOHYDRATE-BASED NANOSTRUCTURES

Nanostructures developed from conjugated proteins and carbohydrates are usually spheres or hollow aggregated particles that have a more effective potential about functional properties (Zhang et al., 2021b).

According to Yan et al. (2017), pectin nanostructures for the food industry can be obtained through nanoemulsions in the presence of whey protein. The main application is the encapsulation of essential oils and lipid-based aromas. Pectin was also studied in the development of nanogel in the presence of bovine serum albumin to perform vitamin C encapsulation by Peng et al. (2016). The conjugated nanostructure was prepared by self-assembly and showed efficiency in the release of vitamin C.

Whey protein nanofibrils and gum arabic were used to create an encapsulation system for curcumin by Mohammadian et al. (2019). The system showed encapsulation efficiency greater than 98% having high affinity with the compound.

4.14 NANOSTRUCTURE FOR FOOD PACKAGING APPLICATIONS

Food packaging is an important structure for maintaining the quality and extending the life of food products. However, just storing the food and protecting it from the external environment filled with contaminants and pathogens is no longer enough. Packaging needs to interact with food.

Active packaging involves the use of nanostructures from the most different sources in packaging materials to extend shelf life, monitor quality, and ensure food safety. The inclusion of nanostructures in packages proposes the creation of an intelligent system with multifunctional possibilities such as sensing, recording, and communication of eventual problems with the health of the food.

Nanostructured materials have been shown to be effective in improving water and water vapor barriers, mechanical, thermal, and functional properties for food packaging, especially bionanocomposite materials such as polysaccharides and proteins (Sothornvit, 2019).

UV light protection films were produced from modified lignin and cellulose nanocrystals extracted from Argania nutshell by Halloub et al. (2022).

Cellulose nanocrystals were extracted from *Enteromorpha prolifera* for the development of films reinforced with nanocrystals based on alginate, carrageenan, and shellac for application in food packaging by Zhang et al. (2022).

Cellulose nanocrystals were extracted from agricultural waste to be used as a reinforcing agent of material for sustainable food packaging by Andrade et al. (2022). The films showed hydrophobic character and thermal stability from the inclusion of cellulose nanocrystals.

Lignin nanoparticles were prepared from an ethylene glycol solution to be incorporated into a biocomposite with wheat gluten to be used in food packaging by Yang et al. (2015).

Films based on wheat gluten and zinc oxide nanocomposites were developed for use in food products by Rezaei et al. (2019). The results showed that the pure gluten film did not promote antibacterial and antifungal effects; however, when zinc nanoparticles were inserted, there was an inhibitory effect and there was antioxidant activity.

4.15 PROPERTIES OF NANOSTRUCTURES IN FOOD APPLICATIONS

The use of nanostructures has been an alternative for the protection and controlled release of antimicrobial agents and natural components such as bacteriocins and antimicrobial proteins, in addition to plant-derived natural antimicrobials. Studies suggest that nanostructures, when used alone, are not effective in eliminating bacteria, requiring the application of photocatalytic methods (Truong et al., 2022).

The use of nanoparticles with antimicrobial potential has several technical advantages, as they provide a greater surface area for interaction with biological substrates (Bahrami et al., 2020).

The adhesion of silver ions from the nanoparticles to the cytoplasmic membrane increases the permeability of the membrane, which causes the rupture of the bacterial envelope (Durán et al., 2016). From the moment the nanoparticle enters the cell, interaction occurs with proteins that contain sulfur and/or phosphorus in their composition, such as DNA (Sondi and Salopek-Sondi, 2007).

The efficiency of silver nanoparticles against cultures of *Escherichia coli*, *Staphylococcus aureus*, *Aspergillus niger*, and *Penicillium phoeniceum* were studied by Khaydarov et al. (2009).

Zinc oxide nanostructures have been increasingly incorporated into food formulations with the aim of protection. The efficiency occurs in gram-positive and gram-negative bacteria (Espitia et al., 2012). Zinc oxide improved the antimicrobial activity of pectin nanofilm against *Staphylococcus aureus* (Dong et al., 2022).

Rai et al. (2016) reported that metallic nanoparticles may have antiviral properties due to a competitive binding with the cell receptor, which promotes the rupture of the viral envelope.

4.16 POSSIBLE HEALTH RISKS ASSOCIATED WITH NANOSTRUCTURES IN FOOD

Although nanostructures have potential benefits, there is still a lack of information about their behavior and interactions with complex food matrices (Rahman et al., 2020). At a global level, there are no specific laws regarding the regulation of nanomaterials in food. Many countries still do not even have regulations for risk assessment. Figure 4.2 presents the main risks that nanostructures can cause in humans.

The European Food Safety Authority (EFSA, 2009) warned that, despite the small number of studies involving nanostructures and their relationship with health, some in vitro studies have shown that nanostructures can induce oxidative stress at high concentrations and that some data suggest the possibility of particles showing genotoxicity and inflammatory response in vitro.

Nanostructures fall into different categories according to the EFSA and the Food and Drug Administration (FDA). They can be framed in the categories of exempt from certification, authorized and recognized as safe (GRAS), or in research and development (He et al., 2019). According to Kim et al. (2022), zinc oxide nanostructures are considered GRAS (generally recognized as safe) by the FDA. Raynes et al. (2014) state that there is interest from food regulatory bodies, mainly in the United States, Europe, Canada, Australia, and New Zealand, which promote multilateral discussions and review of legislation as new studies prove possible health risks.

Silver nanoparticles can contaminate the environment from their production process in an industrial environment, mainly in the steps of drying the solution, mechanical grinding, mixing, and packaging (Park et al., 2009).

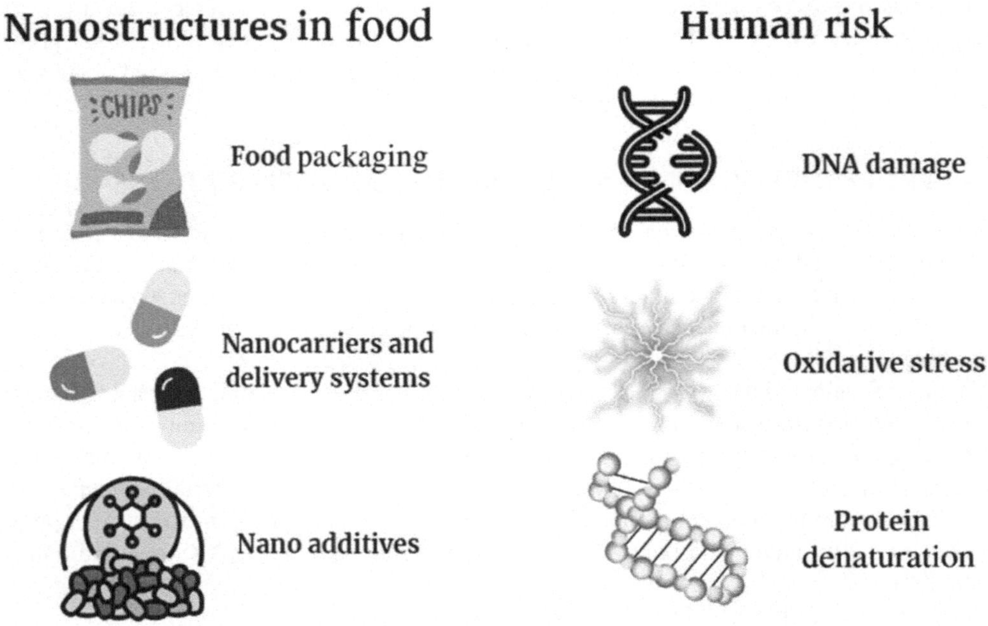

FIGURE 4.2 The main risks that nanostructures can cause in humans.

The presence of metallic nanoparticles in the diet has been a cause for concern regarding the health of consumers. Studies show that the ingestion of these nanoparticles can increase the potential of gluten peptides to activate an immune response and increase the likelihood of developing celiac disease (Mancuso et al., 2021). The risk lies in the toxicity mechanisms induced by the overproduction of reactive oxygen species, which causes oxidative stress, and therefore can cause cell damage that can affect DNA, cell motility, cytotoxicity, and apoptosis, in addition to cancer initiation (Fu et al., 2014; He et al., 2014).

In relation to protein-based amyloid fibrils nanostructures, evidence suggests that the use of these structures in food requires caution. In a study developed by Bucciantini et al. (2002) and Xue et al. (2009), the fibrils demonstrated toxicity to cells, especially the shorter and more fragmented fibrils.

Therefore, there is a need for further studies to be produced to assess the potential toxicity of nanostructures or nanomaterials with a focus on their use in the food industry.

4.17 CONCLUSIONS

Nanostructures are increasingly present in people's daily lives. They can be produced from different raw materials (carbohydrates, proteins, lipids, and minerals). Within this context, nanoemulsions, nanoliposomes, nanostructures based on metallic oxides, and nanogels have been developed and studied by different authors.

Depending on the raw material used and the production methods, nanostructures can have several applications, the main ones being in food packaging, sensors for detecting analytes of interest to the industry, and carriers of bioactive components. In addition, nanostructures can promote the preservation of thermolabile compounds and present antioxidant and antibacterial properties. However, it is necessary to consider the potential environmental and health risks for consumers, especially about the toxicological potential of nanostructures.

REFERENCES

Abbas, S.K., Karangwa, E., Bashari, M., Zhang, X. An overview of ultrasound-assisted food-grade nanoemulsions. *Food Engineering Review*, 2013, 5, 139–157.

Aguilar-Pérez, K.M., Avilés-Castrillo, J.I., Medina, D.I., Parra-Saldivar, R., Iqbal, H.M.N. Insight into nanoliposomes as smart nanocarriers for greening the twenty-first century biomedical settings. *Frontiers in Bioengineering and Biotechnology*, 2020, 8, 536–542.

Ahmad, M., Gani, I., Hassan, I., Huang, Q., Shabbir, H. Production and characterization of starch nanoparticles by mild alkali hydrolysis and ultra-sonication process. *Scientific Reports*, 2020, 10(1), 3533–3544.

Andrade, M.S., Ishikawa, O.H., Costa, R.S., Seixas, M.V.S., Rodrigues, R.C.L.B., Moura, E.A.B. Development of sustainable food packaging material based on biodegradable polymer reinforced with cellulose nanocrystals. *Food Packaging and Shelf Life*, 2022, 31, 807–815.

Arabkhani, P., Sadegh, N., Asfaram, A. Nanostructured magnetic graphene oxide/UIO-66 sorbent for ultrasound-assisted dispersive solid-phase microextraction of food colorants in soft drinks, candies, and pastilles prior to HPLC analysis. *Microchemical Journal*, 2023, 184, 149–162.

Armstrong, C.D., Yue, L., Deng Y., Qi, H.J. Enabling direct link to write edible 3D printing of food purees with cellulose nanocrystals. *Journal of Food Engineering*, 2022, 330, 111–119.

Arya, S., Mahajan, P., Mahajan, S., Khosla, A., Datt, R., Gupta, V., Young, S.J., Oruganti, S.K. Review—Influence of processing parameters to control morphology and optical properties of sol-gel synthesized ZnO nanoparticles. *ECS Journal of Solid State and Technology*, 2021, 10, 300–310.

Ashfold, M.N.R., Doherty, R.P., Ndifor-Angwafor, G., Riley, D.J., Sun, Y. The kinetics of the hydrothermal growth of ZnO nanostructures. *Thin Solid Films*, 2007, 515(24), 8679–8683.

Bahrami, A., Delshadi, R., Assadpour, E., Jafari, S.M., Williams, L. Antimicrobial-loaded nanocarriers for food packaging applications. *Advances in Colloid and Interface Science*, 2020, 278, 140–147.

Ballotin, D., Fulaz, S., Souza, M.L., Corio, P., Rodrigues, A.G., Souza, A.O. Elucidating protein involvement in the stabilization of the biogenic silver nanoparticles. *Nanoescale Research Letters*, 2016, 11, 313–319.

Bedoya-Serna, C.M., Dacanal, G.C., Fernandes, A.M., Pinho, S.C. Antifungal activity of nanoemulsions encapsulating oregano (origanum vulgare) essential oil: In vitro study and application in Minas Padrão cheese. *Brazilian Journal of Microbiology*, 2018, 49(4), 929–935.

Beitollahi, H., Tajik, S., Nejad, F.G., Safaei, M. Recent advances in ZnO nanostructure-based electrochemical sensors and biosensors. *Journal of Materials Chemistry B*, 2020, 8, 5826–5844.

Bourbon, A.I., Cerqueira, M.A., Vicente, A.A. Encapsulation, and controlled release of bioactive compounds in lactoferrin- glycomacropeptide nanohydrogels: Curcumin and caffeine as model compounds. *Journal of Food Engineering*, 2016, 180, 110–119.

Bucciantini, M., Giannoni, E., Chiti, F., Taddei, N., Ramponi, G., Dobson, C.M. Inherent toxicity of aggregates implies a common mechanism for protein misfolding diseases. *Nature*, 2002, 416(6880), 507–511.

Burmeister, C.F., Kwade, A. Process engineering with planetary ball mills. *Chemical Society Reviews*, 2013, 42, 7660–7667.

Carneiro, S.V., Holanda, M.H.B., Cunha, H.O., Oliveira, J.J.P., Pontes, S.M.A., Cruz, A.A.C., Fechine, L.M.U.D., Moura, T.A., Paschoal, A.R., Zambelli, R.A., Freire, R.M., Fechine, P.B.A. Highly sensitive sensing of food additives based on fluorescent carbon quantum dots. *Journal of Photochemistry and Photobiology A: Chemistry*, 2021, 411, 113–119.

Chen, F.P., Ou, S.Y., Tang, C.H. Core-shell soy protein-soy polysaccharide (nano)particles as carriers for improved stability and sustained-release of curcumin. *Journal of Agricultural and Food Chemistry*, 2016, 64, 5053–5059.

Chuesiang P., Sanguadeekul, R., Siripatrawan, U. Phase inversion temperature-fabricated cinnamon oil nanoemulsion as a natural preservative for prolonging shelf-life of chilled Asian seabass (*Lates calcarifer*) fillets. *LWT — Food Science and Technology*, 2020, 12, 109–122.

Das, S., Singh, V.P., Dwivedy, A.K., Chaudhari, A.K., Deepika, A., Dubey, N.W. Nanostructured *Pimpinella anisum* essential oil as novel green food preservative against fungal infestation, aflatoxin B1 contamination and deterioration of nutritional qualities. *Food Chemistry*, 2021, 344, 574–582.

Dong, Z., Du, Z., Wu, X., Zhai, K., Wei, Z., Rashed, M.M.A. Fabrication and characterization of ZnO nanofilms using extracted pectin of Premna microphylla Turcz leaves and carboxymethyl cellulose. *International Journal of Biological Macromolecules*, 2022, 209, 525–532.

Dridi, F., Marrakchi, M., Gargouri, M., Saulnier, J., Jaffrezic-Renault, N., Lagarde, F. Nanomaterial-based electrochemical biosensors for food safety and quality assessment. *Nanobiosensor*, 2017, 1, 167–204.

Durán, N., Nakazato, G., Seabra, A. Antimicrobial activity of biogenic silver nanoparticles, and silver chloride nanoparticles: an overview and comments. *Applied Microbiology Biotechnology*, 2016, 100(15), 6555–6570.

Espitia, P.J.P., Soares, N.F.F., Coimbra, J.S.R., Andrade, N.J., Cruz, R.S., Medeiros, E.A.A. Zinc oxide nanoparticles: synthesis, antimicrobial activity, and food packaging applications. *Food and Bioprocessing Technology*, 2012, 5, 1447–1464.

European Food Safety Authority (EFSA). The potential risks arising from nanoscience and nanotechnologies on food and feed safety. *EFSA Journal*, 2009, 958, 1–39.

Ezpeleta, I., Irache, J.M., Stainmesse, S., Chabenat, C., Gueguen, J., Popineau, Y., Orecchioni, A.M. Gliadin nanoparticles for the controlled release of all-trans-retionic acid. *International Journal of Pharmaceutics*, 1996, 131(2), 191–200.

Fotie, G., Gazzotti, S., Ortenzi, M.A., Limbo, S., Piergiovanni, L. Performance comparison of coatings based on cellulose nanocrystals and microfibrillated cellulose for food packaging. *Carbohydrate Polymer Technologies and Applications*, 2023, 5, 264–269.

Fu, P.P., Xia, Q., Hwang, H.M., Ray, P.C., Yu, H. Mechanisms of nanotoxicity: generation of reactive oxygen species. *Journal of Food and Drug Analysis*, 2014, 22, 64–75.

Galstyan, V., Bhandari, M.P., Sberveglieri, V., Sberveglieri G., Comini, E. Metal oxide nanostructures in food applications: quality control and packaging. *Chemosensors*, 2018, 6(2), 16–31.

Ganje, M., Jafari, S.M., Tamadon, A.M., Niakosari, M., Maghsoudlou, Y. Mathematical and fuzzy modeling of limonene release from amylose nanostructures and evaluation of its release kinetics. *Food Hydrocolloids*, 2019, 95, 186–194.

García-Pinel, B., Porras-Alcalá, C., Ortega-Rodriguez, A., Saraiba, F., Prados, J., Melguizo, C., López-Romero, J.M. Lipid-based nanoparticles: application and recent advances in cancer treatment. *Nanomaterials*, 2019, 9(4), 638–649.

Garti, N. Delivery and controlled release of bioactives in foods and nutraceuticals. Elsevier, Woodland Publishing Co, Cambridge, England, 2008.

Gazzali, A.M., Lobry, M., Colombeau, L., Acherar, S., Azais, H., Mordon, S. Stability of folic acid under several parameters. *European Journal of Pharmaceutical Science*, 2016, 93, 419–430.

Ghasemiyeh, P., Mohammadi-Samani, S. Solid lipid nanoparticles and nanostructured lipid carriers as novel drug delivery systems: applications, advantages, and disadvantages. *Research Pharmaceutical Science*, 2018, 13, 288–303.

Ghojavand, S., Madani, M., Karimi, J. Green synthesis, characterization and antifungal activity of silver nanoparticles using stems and flowers of felty germander. *Journal of Inorganic and Organometallic Polymers and Material*, 2020, 30, 2987–2997.

Gonçalves, R.F.S., Madalena, D.A., Fernandes, J.M., Marques, M., Vicente, A.A., Pinheiro, A.C. Application of nanostructured delivery systems in food: from incorporation to detection and characterization. *Trends in Food Science & Technology*, 2022, 129, 111–125.

González, A., Gastelú, G. Preparation and characterization of soy protein films reinforced with cellulose nanofibers obtained from soybean by-products. *Food Hydrocolloids*, 2019, 89, 758–764.

González-Jartín, Alves., L.C., Alfonso, A., Piñeiro, Y., Vilar, S.Y., Gomez, M.G., Osorio, Z.V., Sainz, M.J., Vieytes, M.R., Rivas, J., Botana, L.M. Detoxification agents based on magnetic nanostructured particles as a novel strategy for mycotoxin mitigation in food. *Food Chemistry*, 2019, 294, 60–66.

Guilger-Casagrande, M., de Lima, R. Synthesis of silver nanoparticles mediated by fungi: a review. *Frontiers in Bioengineering and Biotechnology*, 2019, 7, 287–299.

Hadidi, M., Puramin, S., Adinepour, F., Haghani, S., Jafari, S.M. Chitosan nanoparticles loaded with clove essential oil: characterization, antioxidant, and antibacterial activities. *Carbohydrate Polymers*, 2020, 236, 116–124.

Halloub, A., Raji, M., Essabir, H., Chackchak, H. Boussen, R., Bensalah, M.O., Bouhfid, R., Qaiss, A.E.K. Intelligent food packaging film containing lignin and cellulose nanocrystals for shelf-life extension of food. *Carbohydrate Polymers*, 2022, 296, 972–980.

He, W., Liu, Y., Wamer, W.G., Yin, J.J. Electron spin resonance spectroscopy for the study of nanomaterial-mediated generation of reactive oxygen species. *Journal of Food and Drug Analysis*, 2014, 22, 49–63.

He, X., Deng, H., Hwang, H. The current application of nanotechnology in food and agriculture. *Journal of Food and Drug Analysis*, 2019, 27(1), 1–21.

Hessien, M., Da'na, E., Kawther, A.L., Khalaf, M.M. Nano ZnO (Hexagonal Wurtzite) of different shapes under various conditions: fabrication and characterization. *Materials Research Express*, 2019, 6(8), 85–93.

Hoseinnejad, M., Jafari, S.M., Katouzian, I. Inorganic and metal nanoparticles and their antimicrobial activity in food packaging applications. *Critical Reviews in Microbiology*, 2018, 44, 161–181.

Iravani, S., orbeandi, H., Mirmohammadi, S.V., Zolgafari, B. Synthesis of silver nanoparticles: chemical, physical, and biological methods. *Research Pharmaceutical Science*, 2014, 9, 385–406.

Khaydarov, R.R., Khaydarov, R.A., Estrin, I. Evgrafova, S., Scheper, T., Endres, C., Cho, S.Y. Silver nanoparticles. InLinkov, I., Steevens, J. (Eds), Nanomaterials: Risks and Benefits. NATO Science for Peace and Security Series C: Environmental Security, Springer, Dordrecht, 2009.

Kim, I., Viswanathan, K., Kasi, G., Thanakkasaranee, S., Sadeghi, K., Seo, J. ZnO Nanostructures in active antibacterial food packaging: preparation methods, antimicrobial mechanisms, safety issues, future prospects, and challenges. *Food Reviews International*, 2022, 38(4), 537–565.

Kong, X., Wu, W., Hua, Y., Zhang, C., Chen, Y. Nanostructures of gluten for encapsulation of food ingredients. *Nanoencapsulation in the Food Industry*, 2019, 1, 287–303.

Krishna, K.G., Umadevi G., Parne, S., Pothukanuri, N. Zinc oxide based gas sensors and their derivatives: a critical review. *Journal of Materials Chemistry C*, 2023, 11(2), 3906–3925.

Large, D.E., Abdelmessih, R.G., Fink, E.A., Auguste, D.T. Liposome composition in drug delivery design, synthesis, characterization, and clinical application. *Advanced Drug Delivery Reviews*, 2021, 176, 851–859.

Li, J., Nan, J., Wu, H., Park, H.J., Zhao, Q., Yang, L. Middle purity soy lecithin is appropriate for food grade nanoliposome: preparation, characterization, antioxidant, and anti-inflammatory ability. *Food Chemistry*, 2022, 389, 931–940.

Liew, S.N., Utra, U., Alias, A.K., Tan, T.B., Tan, C.P., Yossof, N.S. Physical, morphological, and antibacterial properties of lime essential oil nanoemulsions prepared via spontaneous emulsification method. *LWT—Food Science and Technology*, 2020, 128, 388–396.

Limthin, D., Klamchuen, A., Phromyothin, D. Surface modification of superparamagnetic iron oxide nanoparticles and methyl methacrylate molecularly imprinted polymer for gluten detection. *Ferroelectrics*, 2019, 552(1), 97–107.

Liu, H.F., Xue, X.J., Yu, Y., Xu, M.M., Lu, C.C., Meng, X.L., Zhang, B.G., Ding, X.H., Chu, Z.H. Copper ions suppress abscisic acid biosynthesis to enhance defence against *Phytophthora* infestans in potato. *Molecular Plant Pathology*, 2020, 21(5), 636–651.

Luo, Y. Food colloids binary and ternary nanocomplexes: innovations and discoveries. *Colloids and Surfaces B: Biointerfaces*, 2020, 196, 309–315.

Magro, M., Moritz, D.E., Bonaiuto, E., Baratella, D., Terzo, M., Jakubec, P., Malina, O., Cepe, K., Aragao, G.M.F., Zboril, R., Vianello, F. Citrinin mycotoxin recognition and removal by naked magnetic nanoparticles. *Food Chemistry*, 2016, 203, 505–512.

Mancuso, C., Re, F., Rivolta, I., Elli, L., Gnodi, E., Beaulieu, J., Barisani, D. Dietary nanoparticles interact with gluten peptides and alter the intestinal homeostasis increasing the risk of celiac disease. *International Journal of Molecular Science*, 2021, 22(11), 6102–6110.

McClements, D.J. *Nanoparticle- and Microparticle-based Delivery Systems. Encapsulation, Protection and Release of Active Compounds*. CRC Press, Taylor & Francis Group LLC, 2015.

Mehrnia, M.A., Jafari, S.M., Makhmal-Zadeh, B.S., Maghsoudlou, Y. Rheological and release properties of double nano-emulsions containing crocin prepared with Angum gum, Arabic gum and whey protein. *Food Hydrocolloids*, 2017, 66, 259–267.

Milea, C.A., Bogatu, C., Duta, A. The influence of parameters in silica sol-gel process. *Bulletin of the Transilvania University of Brasov*, 2011, 4, 59–64.

Moghaddam, M.K., Mortazavi, S.M., Khayamian, T. Preparation of calcium alginate microcapsules containing n-nonadecane by a melt coaxial electrospray method. *Journal of Eletrostatics*, 2015, 73, 56–64.

Mohammadi, A., Jafari, S.M., Assadpour, E., Esfanjani, A.F. Nanoencapsulation of olive leaf phenolic compounds through WPC–pectin complexes and evaluating their release rate. *International Journal of Biological Macromolecules,* 2016, 82, 816–822.

Mohammadian, M., Salami, M., Alavi, F. Fabrication and characterization of curcumin-loaded complex coacervates made of gum Arabic and whey protein nanofibrils. *Food Biophysics*, 2019, 14, 425–436.

Mousavizadeh, F.S., Sarlak, N. A sensitive dual mode turn-on fluorescence and colorimetric nanosensor for ultrasensitive detection of trace amount of gluten proteins in bread products based on crystalline nano cellulose and gold nanoparticles. *Spectrochimica Acta Part A: Molecular and Biomolecular Spectroscopy*, 2023, 287, 122–129.

Mozafari, M.R., Kosravi-Darani, K., Gocke Borazan, G., Cui, J., Pardakhty, A., Yurdugul, S. Encapsulation of food ingredients using nanoliposome technology. *International Journal of Food Properties*, 2008, 11(4), 833–844.

Müller, R.H., Radtke, M., Wissing, S.A. Nanostructured lipid matrices for improved microencapsulation of drugs. *International Journal of Pharmaceutics*, 2002, 242, 121–128.

Müller, R.H., Alexiev, U., Sinambela, P., Keck, C.M. Nanostructured lipid carriers (NLC): the second generation of solid lipid nanoparticles. In: Percutaneous Penetration Enhanc. Chem. Methods Penetration Enhanc. Nanocarriers, Springer Berlin Heidelberg, 2016, 161–185.

Mukurumbira, A., Mariano, M., Dufresne, A., Mellem, J.J., Amonsou, E.O. Microstructure, thermal properties, and crystallinity of amadumbe starch nanocrystals. *International Journal of Biological Macromolecules*, 2017, 102, 241–247.

Nguyen, V.H., Thuy, V.N., Dao, A.H., Lee, B.J. Nanostructured lipid carriers and their potential applications for versatile drug delivery via oral administration. *OpenNano*, 2022, 8, 64–78.

Niinivaara, E., Cranston, E.D. Bottom-up assembly of nanocellulose structures. *Carbohydrate Polymers*, 2020, 247, 116–124.

Park, J., Kwak, B.K., Bae, E., Lee, J., Kim, Y., Choi, K., Yi, J. Characterization of exposure to silver nanoparticles in a manufacturing facility. *Journal of Nanoparticles Research*, 2009, 11, 1705–1712.

Patil, D., Pattewar, S., Palival, S., Patil, G., Sharma, S. Nanostructured lipid carriers: a platform to lipophilic drug for oral bioavailability enhancement. *Journal of Drug Delivery and Therapeutics*, 2019, 9, 758–764.

Peng, H., Chen, S., Luo, M., Ning, F., Zhu, X., Xiong, H. Preparation and self-assembly mechanism of bovine serum albumin–citrus peel pectin conjugated hydrogel: a potential delivery system for Vitamin C. *Journal of Agricultural and Food Chemistry*, 2016, 64(39), 7377–7384.

Pete, S., Kattil, R.A., Thomas, L. Polyaniline-multiwalled carbon nanotubes (PANI- MWCNTs) composite revisited: an efficient and reusable material for methyl orange dye removal. *Diamond and Related Materials*, 2021, 117, 108–116.

Piorkowski, D.T., McClements, D.J. Beverage emulsions: recent developments in formulations, production, and applications. *Food Hydrocolloids*, 2014, 42, 5–41.

Qian, H., Wang, J., Yan, L. Synthesis of lignin-poly (N-methylaniline)-reduced graphene oxide hydrogel for organic dye and lead ions removal. *Journal of Bioresources and Bioproducts*, 2020, 5, 204–210.

Qu, B., Luo, Y. Chitosan-based hydrogel beads: preparations, modifications and applications in food and agriculture sectors–a review. *International Journal of Biological Macromolecules*, 2020, 152, 437–448.

Rafiee, Z., Nejatian, M., Daeihamed, Jafari, S.M. Application of different nanocarriers for encapsulation of curcumin. *Critical Reviews in Food Science and Nutrition*, 2018, 21, 1–77.

Rahman, U.U., Sahar, A., Ishaq, A., Khalil, A.A. Design of nanoparticles for future beverage industry. In: Grumezescu, A.M., Holban, A.M., (Eds.), *Nanoengineering in the Beverage Industry*, Elsevier, 2020, 105–136.

Rai, M., Deshmukh, S.D., Ingle, A.P., Gupta, I.R., Galdiero, M., Galdiero, S. Metal nanoparticles: the protective nanoshield against virus infection. *Critical Reviews in Microbiology*, 2016, 42(1), 46–56.

Rajisha, K.R., Maria, H.J., Pothan, L.A., Ahmad, Z., Thomas, S. Preparation, and characterization of potato starch nanocrystal reinforced natural rubber nanocomposites. *International Journal of Biological Macromolecules*, 2014, 67, 147–153.

Raveendran, J., Docoslis A. Detection, and quantification of toxicants in food and water using Ag–Au core-shell fractal SERS nanostructures and multivariate analysis. *Talanta*, 2021, 231, 383–391.

Raynes, J.K., Carver, J.A., Gras, S.L., Gerrard, J.A. Protein nanostructures in food — should we be worried?. *Trends in Food Science & Technology*, 2014, 37(1), 42–50.

Rezaei, M., Pirsa, S., Chavoshizadeh, S. Photocatalytic/antimicrobial active film based on wheat Gluten/ZnO nanoparticles. *Journal of Inorganic and Organometallic Polymers and Materials*, 2020, 30, 2654–2665.

Rohmah, M., Raharjo, S., Hidayat, C., Martien, R. Application of response surface methodology for the optimization of β-carotene-loaded nanostructured lipid carrier from mixtures of palm stearin and palm olein. *Journal of the American Oil Chemists' Society*, 2020, 97(2), 213–223.

Rostamabadi, H., Falsafi, S.R., Jafari, S.M. Nanostructures of starch for encapsulation of food ingredients. *Biopolymer Nanostructures for Food Encapsulation Purposes*, 2019, 1, 419–462.

Sahani, S., Sharma, Y.C. Advancements in applications of nanotechnology in global food industry. *Food Chemistry*, 2021, 342, 128–135.

Sajjad, A., Zia, M., Xiao, X., Olsson, R.T., Capezza, A.J., Rasheed, F. Wheat gluten hydrolysates with embedded Ag-nanoparticles; a structure-function assessment for potential applications as wound sorbents with antimicrobial properties. *Polymer Testing*, 2023, 118, 896–905.

Sanei-Dehkordi, A., Hatami, S., Zarenezhad, E., Montaseri, Z., Osanloo, M. Efficacy of nanogels containing carvacrol, cinnamaldehyde, thymol, and a mix compared to a standard repellent against Anopheles stephensi. *Industrial Crop and Products*, 2022, 189, 833–840.

Shankar, S., Rhim, J.W. Bionanocomposite films for food packaging applications. In: G. Smithers (Ed.), *Reference Module in Food Science*, 2018, 1–10.

Shpigelman, A., Shoham, Y., Israeli-Lev, Livney, Y.D. β-lactoglobulin–naringenin complexes: nano-vehicles for the delivery of a hydrophobic nutraceutical. *Food Hydrocolloids*, 2014, 40, 214–224.

Silva, A.O., Cunha, R.S., Hotza, D., Machado, R.A.F. Chitosan as a matrix of nanocomposites: a review on nanostructures, processes, properties, and applications. *Carbohydrate Polymers*, 2021, 272, 472–489.

Silva, M.P., Fabi, J.P. Food biopolymers-derived nanogels for encapsulation and delivery of biologically active compounds: a perspective review. *Food Hydrocolloids for Health*, 2022, 2, 79–88.

Silva, S., Jin, J.O., Kim, M. Nanoliposome based biosensors for probing mycotoxins and their applications for food: a review. *Biosensors and Bioelectronics*, 2023, 219, 845–859.

Sondi, I., Salopek-Sondi, B. Silver nanoparticles as antimicrobial agent: a case study on *E. coli* as a model for gram-negative bacteria. *Journal of Colloid Interface*, 2007, 275, 177–182.

Sothornvit, R. Nanostructured materials for food packaging systems: new functional properties. *Current Opinion in Food Science*, 2019, 25, 82–87.

Tang, C.H. Nanostructured soy proteins: Fabrication and applications as delivery systems for bioactives (a review). *Food Hydrocolloids*, 2019, 91, 92–116.

Tang, C.H. Assembled milk protein nano-architectures as potential nanovehicles for nutraceuticals. *Advances in Colloid and Interface Science*, 2021, 292, 102–109.

Tapeinos, C., Battaglini, M., Ciofani, G. Advances in the design of solid lipid nanoparticles and nanostructured lipid carriers for targeting brain diseases. *Journal of Controller Release*, 2017, 264, 306–332.

Theerthagiri, J., Salla, S., Senthil, R.A., Nithyadharseni, P., Mandankumar, A., Arunachalam, P., Maiyalagan, T., Kim, H.S. A review on ZnO nanostructured materials: energy, environmental and biological applications. *Nanotechnology*, 2019, 30, 392–410.

Trache, D., Tarchoun, A.F., Derradji, M., Hamidon, T.S., Masruchin, N., Brosse, N., Hussin, M.H. Nanocellulose: from fundamentals to advanced applications. *Frontier Chemistry*, 2020, 8, 392–400.

Truong, V.K., Kobaisi, M.A., Vasilev, K., Cozzolino, D., Chapman, J. Current perspectives for engineering antimicrobial nanostructured materials. *Current Opinion in Biomedical Engineering*, 2022, 23, 399–412.

Türe, H., Blomfeldt, T.O.J., Gallstedt, M., Hedenqvist, M.S., Farris, S. Nanostructured silica/wheat gluten hybrid materials prepared by catalytic sol-gel chemistry. *Macromolecular Chemistry and Physics*, 2013, 214(10), 1131–1139.

Ullsten, N.H., Gallstedt, M., Johansson, E., Graslund, A., Hedenqvist, M.S. Enlarged processing window of plasticized wheat gluten using salicylic acid. *Biomacromolecules*, 2006, 7(3), 771–776.

Urade, R., Sato, N., Sugiyama, M. Gliadins from wheat grain: an overview, from primary structure to nanostructure of aggregates. *Biophysical Reviews*, 2018, 10, 435–443.

Utsumi, S., Matsumura, Y., Mori, T. Structure-function relationships of soy proteins. In: Damodaran, S., Paraf, A. (Eds.), *Food proteins and their applications*, Marcel Dekker, New York, USA, 1997, 257–292.

Wang, Z., Ma, Y., Chen, H. Deng, Y., Wei, Z., Zhang, Y., Tang, X., Zhao, Z., Zhou, P., Liu, G., Zhang, M. Rice bran-modified wheat gluten nanoparticles effectively stabilized Pickering emulsion: an interfacial antioxidant inhibiting lipid oxidation. *Food Chemistry*, 2022, 387, 132–140.

Weber, S., Zimmer, A., Pardeike, J. Solid lipid nanoparticles (SLN) and nanostructured lipid carriers (NLC) for pulmonary application: a review of the state of the art. *European Journal of Pharmaceutics and Biopharmaceutics*, 2014, 86(1), 7–22.

Wu, C., Sun, J., Jiang, H., Li, Y., Pang, J. Construction of carboxymethyl konjac glucomannan/chitosan complex nanogels as potential delivery vehicles for curcumin. *Food Chemistry*, 2021, 362, 242–250.

Xu, L.P., Liu, Y., Zhang, X. Interfacial self-assembly of amino acids and peptides: scanning tunneling microscopy investigation. *Nanoscale*, 2011, 3, p. 4901–4909.

Xu, X., Liu, F., Jiang, L., Zhu, J.Y., Haagenson, D., Wiesenborn, D.P. Cellulose nanocrystals vs. cellulose nanofibrils: a comparative study on their microstructures and effects as polymer reinforcing agents. *ACS Applied Materials & Interfaces*, 2013, 5(8), 2999–3009.

Xue, W.F., Hellewell, A.L., Gosal, W.S., Homans, S.W., Hewitt, E.W., Radford, S.E. Fibril fragmentation enhances amyloid cytotoxicity. *Journal of Biological Chemistry*, 2009, 284(49), 34272–34282.

Yan, J., Qiu, W., Wang, Y., Wu. J. Biocompatible polyelectrolyte complex nanoparticles from lactoferrin and pectin as potential vehicles for antioxidative curcumin. *Journal of Agricultural and Food Chemistry*, 2017, 65, 5720–5730.

Yang, J.Q., Zhou, Y., Chen, L.Y. Elaboration and characterization of barley protein nanoparticles as an oral delivery system for lipophilic bioactive compounds. *Food & Function*, 2014, 5(1), 92–101.

Yang, W., Kenny, J.M., Puglia, D. Structure and properties of biodegradable wheat gluten bionanocomposites containing lignin nanoparticles. *Industrial Crops and Products*, 2015, 74, 348–356.

Yi, J., Lam, T.I., Yokoyama, W., Cheng, L.W., Zhong, F. Cellular uptake of beta-carotene from protein stabilized solid lipid nanoparticles prepared by homogenization–evaporation method. *Journal of Agricultural and Food Chemistry*, 2014, 62, 1096–1104.

Zambelli, R.A., Galvão, A.M.M.T., Mendonça, L.G., Leão, M.V.S., Carneiro, S.V., Lima, A.C.S., Melo, C.A.L. Effect of different levels of acetic, citric and lactic acid in the cassava starch modification on physical, rheological, thermal and microstructural properties. *Food Science and Technology Research*, 2018, 24(4), 747–754.

Zambelli, R.A., Mendonça, L.G., Carneiro, S.V., Rodrigues, B.G. Effect of Açaí powder and chitosan incorporation on bread quality. *International Journal of Food Engineering*, 2017, 3(2), 89–94.

Zarandona, I., Correia, D.M., Moreira, J., Costa, C.M., Lanceros-Mendez, S., Guerrero, P., Caba, K.L. Magnetically responsive chitosan-pectin films incorporating Fe_3O_4 nanoparticles with enhanced antimicrobial activity. *International Journal of Biological Macromolecules*, Available online, In Press, Corrected Proof, 2022.

Zarrabi, A., Abadi, M.A.A., Khorasani, S., Mohammadabadi, M.R., Jasmshidi, A., Torkaman, S., Taghavi, E., Mozafari, M.R., Rasti, B. Nanoliposomes and tocosomes as multifunctional nanocarriers for the encapsulation of nutraceutical and dietary molecules. *Molecules*, 2020, 25(3), 638–645.

Zhang, X., Liu, F., Shen, W., Gurunathan, S. Silver nanoparticles: synthesis, characterization, properties, applications, and therapeutic approaches. *International Journal of Molecular Sciences*, 2016, 17, 153–159.

Zhang, Q., Zhou, Y., Yue, W., Qin, W., Dong, H., Vasanthan, T. Nanostructures of protein-polysaccharide complexes or conjugates for encapsulation of bioactive compounds. *Trends in Food Science & Technology*, 2021a, 109, 169–196.

Zhang, Z., Hao, G., Liu, C., Fu, J., Hu, D., Rong, J., Yang, X. Recent progress in the preparation, chemical interactions, and applications of biocompatible polysaccharide-protein nanogel carriers. *Food Research International*, 2021b, 147, 564–679.

Zhang, Y., Man, J., Li, J., Xing, Z., Zhao, B., Ji, M., Xia, H., Li, J. Preparation of the alginate/carrageenan/shellac films reinforced with cellulose nanocrystals obtained from *enteromorpha* for food packaging. *International Journal of Biological Macromolecules*, 2022, 218, 519–532.

Zheng, B., Qiao, D., Zhao, S., Lin, Q., Wang, J., Xie, F. Starch-based food matrices containing protein: recent understanding of morphology, structure, and properties. *Trends in Food Science & Technology*, 2021, 114, 212–231.

Zhong, Y., Godwin, P., Jin, Y., Xiao, H. Biodegradable polymers and green-based antimicrobial packaging materials: a mini-review. *Advanced Industrial and Engineering Polymer Research*, 2020, 3, 27–35.

Zhu, K., Ye, T., Liu, J., Peng, Z., Xu, S., Lei, J. Nanogels manufactured by lysozyme and sodium carboxymethyl cellulose for 5-fluorouracil controlled release. *International Journal of Pharmaceutics*, 2013, 441(1-2), 721–727.

Zhu, X., Ma, X., Gao, C., Mu, Y., Pei, Y., Liu, C., Zou, A. Sun, X. Fabrication of CuO nanoparticles composite ε-polylysine-alginate nanogel for high-efficiency management of Alternaria alternate. *International Journal of Biological Macromolecules*, 2022, 223, 1208–1222.

Ziani, K., Chang, Y., McLandsborough, L., McClements, D.J. Influence of surfactant charge on antimicrobial efficacy of surfactant-stabilized thyme oil nanoemulsions. *Journal of Agricultural and Food Chemistry*, 2011, 59, 6247–6255.

Zoghlami, A., Paës, G. Lignocellulosic biomass: understanding recalcitrance and predicting hydrolysis. *Food Chemistry*, 2019, 7, 874–888.

5 Conducting Polymer Nanocomposite-Based Sensors and Their Applications in Food Contaminant Detection

A Review

Archana Deka and Moon Mandal
Corresponding author: Archana Deka

5.1 INTRODUCTION

Food-borne diseases pose a global threat, reinforcing the need for governments, the food sector, and individuals to do more to ensure food safety. As per a World Health Organization (WHO) report, approximately 600 million people suffer from food-borne diseases, and there are 420,000 deaths due to food-borne illnesses every year [1,2]. The global estimate of food-borne diseases is comparable to that of the major infectious diseases HIV/AIDS, malaria, and tuberculosis [3]. The Center for Disease Control and Prevention (CDC) revealed that every year unsafe contaminated food causes approximately 48 million cases of food-borne diseases in the United States alone [4]. Though these estimates have promoted awareness among people, the actual incidence of food-borne diseases is far higher than reported [5]. The low-income countries of Africa and Asia bear a considerably greater burden of food-borne diseases, which is closely linked to poor maintenance of hygiene; lack of access to clean water; inadequate conditions for food production and storage; low levels of education; high prevalence of food-borne pathogens; widespread misuse of chemicals; utilization of obsolete, unduly hazardous, and banned chemicals; and inadequate implementation of food safety and food-borne disease control regulations [6,7]. Understanding the nature of contamination, its sources, the risks to consumers, and approaches to eliminate or reduce contamination levels are critical to food safety management.

Due to the complexity of food matrices and trace levels of contaminants in foods, developing reliable and efficient detection techniques is challenging. Various analytical methods and techniques are utilized to monitor the levels of contaminants and for quantitative assessment of food spoilage. In recent years, rapid and efficient detection methods such as sensors have been developed in order to detect the presence of contaminants in food products immediately. Due to the advantages of low cost, speed, portability, and potential for field-testing, sensors are indispensable in their applications for food quality and safety. Sensors can be suitably modified for their varied applications in detection of food contaminants. Conducting polymer nanocomposites have shown promising potential for a range of sensing applications. Conducting polymer nanocomposite-based sensors offer high

DOI: 10.1201/9781003389408-5

flexibility, sensitivity, robustness, and stability for detection of analytes present in trace amounts. The objective of this review chapter is to provide overview on food contamination and sensing technology, and to illustrate details regarding the fabrication of conducting polymer nanocomposite-based sensors for rapid monitoring of food contaminants.

5.2 FOOD CONTAMINATION AND TYPES OF CONTAMINANTS

5.2.1 Food Contamination and Food Safety

Access to safe and nutritious food is crucial for the well-being of people all over the world. Food safety is of the utmost importance, as it helps to protect the health of consumers from food-borne illnesses. The objective of food safety is to ensure food stays safe at every stage, and it involves safeguarding a nation's food supply chain from the introduction, growth, or survival of dangerous microorganisms and chemicals [8,9]. Many biochemical changes occur in foods due to the presence of inherent enzymes and several intrinsic factors such as pH, redox potential, and water content. These factors have the ability to deteriorate the quality, sensory characteristics, and shelf life of food products [10]. The presence of contaminants further leads to food spoilage, making it unacceptable for human consumption. Food contamination is a serious matter of concern, as high concentrations of contaminants in food have serious implications for human health. There are numerous chances of food contamination during food production, processing, packaging, storage, and transportation. Hence, food contamination and food spoilage are inextricably linked and may cause food-borne diseases in consumers.

Food-borne diseases are typically characterized by short-term symptoms including nausea, vomiting, and diarrhea, but can also display long-term health conditions such as cancer, kidney or liver failure, and brain and neural disorders. Children below 5 years are more susceptible to food-borne diseases, and the detrimental effects post-recovery may impact their quality of life permanently. A food contaminant can be a biological agent (bacteria, viruses, parasites, fungi) or chemical agent (pesticides, fertilizers, heavy metals, small molecule toxins, chemical residues). According to a WHO report (2015), food-borne diseases are caused by 31 agents, including bacteria, viruses, parasites, toxins, and chemicals [11]. Table 5.1 lists the most common contaminants

TABLE 5.1
Food-borne contaminants responsible for major food-borne diseases

Food-borne contaminants	Potential food-borne sources	Diseases/health effects
Pathogenic bacteria		
Staphylococcus aureus	Raw animal meat products and poultry, eggs, unpasteurized milk, dairy products, bakery products	Diarrheal disease, abdominal pain
Campylobacter spp.	Poultry, raw meat, seafood	Diarrhea (often bloody), abdominal pain
Escherichia coli	Unpasteurized milk, vegetables, raw meat products and poultry	Diarrheal disease
Brucella spp.	Unpasteurized milk, animal meat, dairy products	Brucellosis
Salmonella spp.	Eggs and egg products	Diarrheal disease, typhoid, abdominal pain
Shigella spp.	Raw vegetables, fruits, oysters, milk, meat	Diarrheal disease, abdominal pain
Vibrio cholerae	Seafood, fruits, vegetables, poultry, meat, dairy products	Diarrheal disease, abdominal cramps
Entamoeba histolytica	Contaminated vegetables, fruit	Diarrheal disease

(*continued*)

TABLE 5.1 (Continued)
Food-borne contaminants responsible for major food-borne diseases

Food-borne contaminants	Potential food-borne sources	Diseases/health effects
Listeria monocytogenes	Raw vegetables and fruit, processed food products, unpasteurized milk, dairy products, seafood	Sepsis, infection in central nervous system
Clostridium botulinum	Raw vegetables, fish, meat products, canned foods	Respiratory paralysis, diarrhea, botulism
Clostridium perfringens	Meat, poultry	Diarrheal disease, abdominal pain
Viruses		
Norovirus	Shellfish, leafy vegetables, raw fruit	Watery nonbloody diarrhea
Hepatitis A virus	Shellfish, raw vegetables and fruit	Jaundice, abdominal cramps, hepatitis
Hepatitis E virus	Pork and other animal meat, milk	Neurological disorders, haemolytic anemia
Rotavirus	Raw vegetables and fruit	Watery dehydrating diarrhea, abdominal pain
Astrovirus	Molluscs, fruit, vegetables	Watery diarrhea, gastroenteritis
Enteric adenovirus	Raw shellfish, contaminated vegetables and fruit	Gastrointestinal disease, diarrhea
Parasites		
Cryptosporidium spp.	Fruit, vegetables, shellfish, beef and pork	Diarrheal disease
Entamoeba histolytica	Fruit, vegetables	Diarrheal disease
Giardia spp.	Shellfish, fruit, vegetables	Diarrheal disease
Toxoplasma gondii	Fruit, vegetables, shellfish, beef and pork	Abnormalities in central nervous system
Echinococcus spp.	Fruit, vegetables	Abdominopelvic problems, cystic echinococcosis, infection in central nervous system
Taenia solium	Fruit, vegetables, pork	Cysticercosis, neurocysticercosis
Ascaris spp.	Fruit, vegetables	Ascariasis, abdominopelvic problems
Trichinella spp.	Pork and other meat	Acute clinical trichinellosis, gastroenteritis
Fasciola spp.	Fruit, vegetables, beef	Fascioliasis, abdominopelvic problems
Paragonimus spp.	Freshwater crab, crayfish	Pulmonary problems, paragonimiasis
Biotoxins		
Mycotoxins	Cereals, fruit, vegetables, oilseeds	Hepatocellular carcinoma, gastrointestinal diseases, nephrotoxicity
Marine biotoxins	Seafood	Diarrhea, abdominal cramps, respiratory distress, paralysis
Plant secondary metabolites	Fruit and vegetables	Intoxication symptoms, liver toxicity
Other contaminants		
Heavy metals	Cereals, Fruits, vegetables, processed food products	Allergy, gastrointestinal diseases, pulmonary diseases, reproductive system defects, central nervous systems disorders, immunological diseases, cancers
Dioxins and other byproducts	Dairy food products, meat, seafood, poultry	Infertility, carcinogenic diseases
Pesticide residues	Fruit and vegetables	Diarrhea, allergy, leukemia, reproductive, immunological, and neurological cancers

found in the different food sources that are responsible for major food-borne diseases. Diarrheal disease agents represent more than half of the global burden of food-borne diseases. Infectious agents such as norovirus, nontyphoidal *Salmonella*, pathogenic *E. coli*, and *Campylobacter spp.* account for the vast majority of diarrheal diseases [12]. Other agents that contribute substantially to the global burden of food-borne diseases include helminth *Taenia solium*, helminth *Paragonimus spp.*,

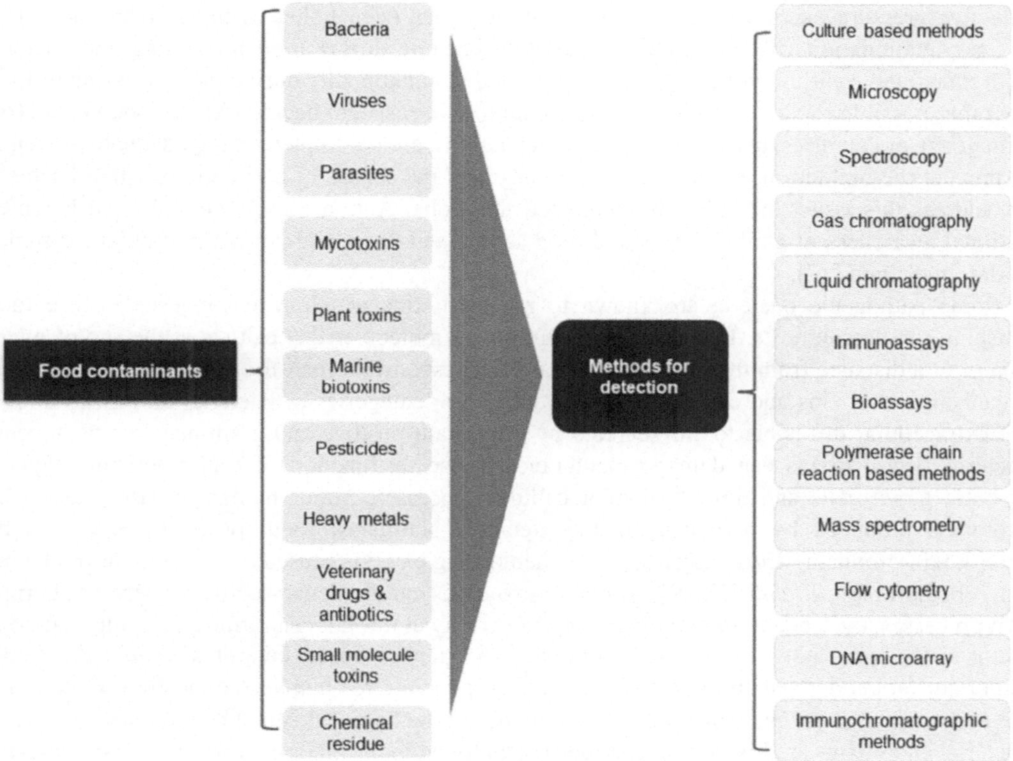

FIGURE 5.1 Types of food contaminants and their detection methods.

hepatitis A virus, and alfatoxin (produced by mold) [13]. Figure 5.1 summarizes different types of food contaminants and the detection methods commonly used. Thus, contaminated foods have the potential to result in serious consequences for human health and can pose economic consequences if certain precautions are not taken.

5.2.2 Food-Borne Pathogenic Bacteria

Foods with high protein and moisture content, especially animal-source food products, are often a breeding ground for the growth of microorganisms and are associated with bacterial food-borne illnesses. *Bacillus cereus*, *Campylobacter jejuni*, enteropathogenic *Escherichia coli*, *Staphylococcus aureus*, *Clostridium perfringens*, *Clostridium botulinum*, *Listeria monocytogenes*, *Salmonella spp.*, *Shigella spp.*, and *Vibrio spp.* are some of the pathogenic bacteria responsible for food-borne illness worldwide. Food-borne diarrheal diseases usually occur when a human host consumes food contaminated with these pathogenic agents. Of approximately 600 million cases of food-borne illnesses, diarrheal diseases particularly caused by *Campylobacter spp.* accounted for the vast majority (96 million cases) [12]. Invasive infectious diseases characterized by nonenteric symptoms occur when pathogenic bacteria penetrate the epithelial cells of the gastrointestinal tract or pass to other internal organs. *Listeria monocytogenes* causing invasive infectious disease is responsible for the majority of deaths due to food-borne diseases. Among the others, the typhoid bacterium *Salmonella typhi* is a frequent cause of food-borne diseases.

The occurrence of antibiotic-resistant bacteria in food products has gained attention as an emerging contaminant and a potential threat to food safety [14]. There is evidence for the emergence of antibiotic resistance in *L. monocytogenes* strains isolated from food products, and the selective pressure exerted by the use of antibiotics has been linked with the resistance [15]. Cross-contamination with antimicrobial-resistant bacteria during food processing and bacteria with transferable antimicrobial resistance genes added intentionally during the processing of food are other possible routes for the emergence of antibiotic-resistant bacteria in the food chain [16]. A majority of countries, particularly developing nations, are not implementing adequate measures to prevent the buildup of antimicrobial resistance and the spread of antibiotic-resistant bacteria. To address this issue, the WHO implemented its Global Action Plan, which aims to heighten national awareness of antibiotic use and food safety, and knowledge of antimicrobial resistance in the food chain [17].

Some pathogenic bacteria are known to release toxins which, when ingested, cause food poisoning in human hosts. Bacterial toxins are potent molecules that can act at the site of infection or at a distance from the infectious site and are responsible for various health effects in host organisms. Exotoxins and endotoxins are two different categories of bacterial toxins. Exotoxins are extracellular diffusible toxins secreted by both gram-positive and gram-negative pathogenic bacteria. These toxins can damage membranes, cause dysfunction of transmembrane signals, or enter target cells and directly disrupt cellular processes. Major human diseases caused by exotoxins produced by pathogens include tetanus, botulisms, food-borne illness caused by enterotoxin-producing *Clostridium spp.*, and hemorrhagic colitis caused by Shiga toxin-producing enterohemorrhagic *E. coli* [18,19]. Although exotoxins can be denatured by modern pasteurization processes, their potent immunogenicity and toxicity at low concentrations is an important risk factor in food poisoning. Botulinum neurotoxins and staphylococcal enterotoxins are some of the important toxins derived from bacteria that are responsible for human intoxication. Endotoxins are cell-associated toxins which are located mainly in the bacterial cell. The most widely known bacterial endotoxins are lipopolysaccharides (LPS) found in the outer membrane of gram-negative bacteria released after death and lysis of the bacterial cell. The endotoxin contamination of food and food products has been associated with a number of diseases, including chronic inflammation, sepsis, liver diseases, and diabetes [20,21]. Both bacterial endotoxins and exotoxins have the potential to damage the health of consumers, and therefore food processing and production departments should take the necessary steps to reduce the level of bacterial toxins and measure the content of bacterial toxins in food samples.

Food-borne pathogenic bacteria exist at very low levels and are difficult to detect among millions of other bacteria. In general, culture-based detection methods are used for detection of viable food-borne pathogens. However, the conventional method is laborious, time-consuming, and slow to provide results. Several methods have been developed in recent years for rapid detection of food-borne pathogens, including simple polymerase chain reaction (PCR), multiplex PCR, real-time PCR, nucleic acid sequence-based amplification (NASBA), loop-mediated isothermal amplification (LAMP) and oligonucleotide DNA microarray, and phage-based methods. Additionally, biosensor-based and immunological-based assays have been developed for rapid detection of food-borne pathogens [22,23]. The standard method usually employed for detection of endotoxins secreted by microbes is the rabbit pyrogen test [24]. Since that method is time-consuming, expensive, and involves animal experimentation, the limulus amoebocyte lysate (LAL) test is the currently preferred method [25]. In this test, an endotoxin-contaminated food sample is mixed with horseshoe-crab blood to derive the amoebocyte extraction, and the reaction between amoebocyte-endotoxin is characterized by one of four techniques: chromogenic, colorimetric, gel-clot, or spectrophotometric. Although the LAL test is highly sensitive, it requires expert personnel and is expensive. Mass spectrometry-based methods have been shown to offer superior performance for detection of microbial toxins; however, the requirement of sophisticated instruments and protocol optimization seems to limit its broader applicability in food analyses [26]. In recent years, several new approaches, such as nanomaterial-based biosensors, electrochemical biosensors, optical biosensors, and aptasensors, have been developed to detect pathogens and microbial toxins [27]. These methods are suitable for fieldwork, easy to use, low cost, generate a quick response, and display greater sensitivity and accuracy.

5.2.3 Food-borne Enteric Viruses

Food-borne enteric viruses are frequently transmitted to human hosts via contaminated food and are known to contribute significantly to the burden of food-borne diseases. Hepatitis A virus, hepatitis E, rotavirus, noroviruses, astroviruses, sapoviruses, adenoviruses, and parvoviruses are known to be transmitted to humans through food [28,29]. Among several types of enteric viruses, noroviruses are responsible for the majority of food-borne illnesses worldwide. Besides norovirus, hepatitis A and rotoviruses are also considered as major causative agents of food-borne diseases. The most commonly reported viral food-borne disease is gastroenteritis, and its symptoms include nausea, vomiting and abdominal pain, fever, and headache. Another frequently reported viral food-borne illness is hepatitis, which is characterized by fever, headache, nausea, vomiting, diarrhea, abdominal pain, and jaundice. Food-borne viruses can often be transmitted to susceptible human consumers through the ingestion of virus-contaminated food [30]. Contamination of foods can occur at different stages of food production and processing. Raw foods, foods that do not require processing, or ready-to-eat foods are at risk of being contaminated with food-borne viruses. Environmental pollution, unhygienic practices, improper handling of food products, and contact with infected persons are associated with the bulk of viral contamination of foods.

Advances in food processing technologies have helped to control viral contamination of raw food products and lessen the risk of cross-contamination. Traditionally, diagnostic methods based on antigen detection and electron microscopy have been used for samples where the viral load is high. These methods cannot be directly applied to routine viral diagnosis in the food processing industry, since food samples contain a low number of infectious viruses. Moreover, enteric viruses are difficult to grow in a laboratory for diagnosis and study. Current methods employed for detection of viruses in food samples involve extraction and concentration of viruses followed by detection using nucleic acid amplification methods and serological assays. Techniques used for detection, quantification, and identification of viruses in food matrices include real-time quantitative reverse transcription PCR (RT-qPCR) of viral RNA, molecular virus detection from intact virus capsids, and nucleic acid aptamers [30]. Serological tests demonstrate high specificity for detection of viruses but often display low sensitivity. Given the ability of the viruses to withstand a wide range of environmental conditions, inadequate systems for routine diagnosis make it difficult to control them. Validated rapid diagnostic methods are essential and remain in high demand for detection of food-borne viruses.

5.2.4 Food-borne Parasites

Parasites are a category of food-borne pathogens that is generally neglected compared to bacteria and viruses [31]. Food-borne products contaminated or infected with a variety of parasites, including nematodes, trematodes, cestodes, and protozoa, can cause zoonotic infections in humans. Some of the common food-borne parasites that are a risk to human populations are *Cryptosporidium parvum*, *Cyclospora cayetanensis*, *Trichinella spp.*, *Taenia spp.*, *Paragonimus spp.*, and *Toxoplasma gondii* [32]. These parasites have complex development cycles, transmission routes, and survival characteristics, and use a wide variety of hosts. Transmission of the parasites to humans occurs when raw meat or meat products infected with parasites are consumed [33]. These parasites often cause a lifelong latent infection in the host's tissues (such as skeletal and heart muscle or the central nervous system) and may go undetected due to a lack of definite pathognomonic symptoms. The outcomes of parasitic infections in humans vary greatly depending on the type of parasite. Protozoa such as *Cryptosporidium spp.* and *Giardia spp.* most commonly cause diarrheal disease [34], while

Toxoplasma gondii causes toxoplasmosis-associated illnesses [35]. A wide variety of insidious and chronic symptoms are displayed in food-borne helminthic infections caused by tapeworms (*Taenia spp.* and *Echinococcus spp.*), roundworms (*Ascaris spp.* and *Trichinella spp.*), and flukes (*Fasciola spp.* and *Paragonimus spp.*) [36]. The major clinical manifestations of helminthic infections in humans include abdominopelvic problems, muscle pain, pulmonary problems, malnutrition, weight loss, and neurological symptoms.

Globalization of the food supply, rise in consumption of animal proteins, increase of highly susceptible individuals, changes in lifestyles, and varied cooking habits are some of the reasons associated with the upsurge in the incidence of food-borne parasitic infections globally. Despite the various health hazards that are caused by food-borne parasites, there are no standardized methods for detection of parasites in the food industry. Food matrices such as undercooked meat, fish, and shellfish, untreated milk and fruit juices, and contaminated fruit or vegetables are usually analyzed for detection of parasites. Unlike pathogenic bacteria, laboratory culture is not a feasible option for detection of food-borne parasites. For several animal-derived foods, visual inspection of carcasses is employed as a validation procedure for detecting food-borne parasites [37]. Monitoring and visual inspection of animal carcasses gives a holistic assessment of meat quality and robust safety information for specific parasites. However, visual inspection lacks both diagnostic sensitivity and specificity. Hence, detection methods rely greatly on morphological identification and microscopy to provide confirmation of clinical infection. Polymerase chain reaction (PCR)-based molecular methods, agglutination, western blot assay, antibody detection by enzyme-linked immunosorbent assay (Ab-ELISA), and antigen detection by enzyme-linked immunosorbent assay (Ag-ELISA) are other widely used methods for rapid and sensitive diagnosis of parasitic infections [38]. Additionally, immuno-magnetic separation (IMS), immuno-fluorescence microscopy, and mass spectrometry have been successfully applied for detection of parasites in different food matrices. Bioassays (animal experimentation) are considered the most sensitive methods for detection of *Toxoplasma gondii* but are not a practical approach for routine screening of food products. Molecular methods, microscopy, serological diagnostic tests, and bioassays are preferred detection methods, but these methods have their own sets of issues, and therefore lack practical applicability in the food industry. Hence, there is an urgent need for use of new technologies for the development of detection methods for monitoring and control of food-borne parasites.

5.2.5 Biotoxins

Biotoxins are toxins derived from natural sources such as plants, fungi, and marine organisms. These usually include peptides, nonproteinaceous small molecules, and proteins. Biotoxins lack the ability to spread from one host to another, and they accumulate in the environment even after the death of the organisms [39]. Immunity to biotoxins does not develop in the same way as immunity to pathogens. Based on their interaction with target sites, the toxicity of biotoxins can range from low to acute to extremely harmful in humans. The level of toxicity and information on health hazards is described by the median lethal dose (LD50). Biotoxins are of major concern as food contaminants, since they are toxic at very low doses [40]. Food contamination with biotoxins is a serious public health concern, and is often linked to cancer incidences and mortality. Therefore, timely screening of these toxins in food stuffs is highly desirable.

Marine biotoxins produced by diverse microalgal species are a result of a natural phenomenon called algal boom. Of several thousands of known marine algal species, more than 100 species are known to produce toxins. Examples of marine biotoxins include saxitoxins, tetrodotoxins, azaspiracid, okadaic acid, ovatoxins, and palytoxins. These toxins have been responsible for incidents of large-scale death of aquatic organisms. Several marine organisms are able to accumulate these biotoxins without any apparent adverse effects and thus are significant sources of contaminated seafood [41]. Human exposure to marine biotoxins can cause intoxication with a wide range of symptoms including diarrhea, vomiting, nausea, abdominal cramps, respiratory distress, and paralysis. The monitoring of marine biotoxins in aquatic organisms is crucial for seafood control, since these toxins may cause serious illness in humans. Mouse bioassay (MBA) is frequently used for detection of new or unknown marine biotoxins in seafood production areas [42]. Different analytical methods applied for identification and quantification of multiple marine biotoxins are high-performance liquid chromatography (HPLC-UV), liquid chromatography coupled with mass spectrometry (LC-MS), and liquid chromatography coupled with fluorescence (LC-FL). Immunoassays, cell-based assays, and receptor-binding assays (RBAs) have also been employed in both screening and quantification of marine biotoxins [43,44].

Plant biotoxins are phytochemicals or secondary metabolites that confer protection against herbivores, insects, and predators, as well as resistance against pathogens. Some of the biotoxins produced by plants include ricin, abrin, opium alkaloids, tropane alkaloids, and pyrrolizidine alkaloids. These biotoxins are present in commonly consumed human foods such as fruit and vegetables. Human exposure to food contaminated with even trace amounts of plant biotoxins may result in acute or chronic effects, development of intoxication symptoms, and adverse effects on the liver due to formation of harmful metabolites. Different approaches such as liquid chromatography-mass spectrometry, immunoassays, surface plasma resonance (SPR), immunosensors, and aptasensors have been employed for detection of plant-derived toxins in different food analytes [45–47]. Detecting plant toxins in food and tracing them to their source is a primary responsibility of food manufacturers, retailers, processors, and distributors which would, in turn, lessen the food safety issues.

Fungi such as yeasts and molds are among the major contributors of food spoilage and contamination. Mycotoxins produced by fungi that routinely contaminate food products can cause serious health hazards in humans when ingested. Ingestion of mycotoxin-contaminated food has been linked to food poisoning (mycotoxicosis) and illness, particularly hepatic, gastrointestinal, and carcinogenic diseases [48,49]. Fungal infestation of agricultural crops is one of the ways in which entry of mycotoxins into the food chain is facilitated. The International Agency for Research on Cancer (IARC) has classified aflatoxins and ochratoxin A as potent human carcinogens among the several mycotoxins identified [50]. Aflatoxins are secondary metabolites produced by produced by several species of *Aspergillus*. Aflatoxin metabolites are generally found in cereals, nuts, oilseeds, spices, animal milk, and dairy products. Among the different types of aflatoxin produced by fungal species, the most toxic and strong carcinogenic toxin is aflatoxin B1 (AFB1). Ochratoxins are produced by *Aspergillus spp.* and *Penicillium spp.* and generally exist in three forms: ochratoxin A (OTA), ochratoxin B (OTB), and ochratoxin C (OTC). High levels of ochratoxin contamination are often detected in dried fruits, seeds, cereals, coffee, beverages, and spices. Food contamination by ochratoxins has been directly linked to nephrotoxicity, liver toxicity, and carcinogenicity in humans and animals [51]. Although aflatoxin and ochratoxin remain among the most toxic mycotoxins identified, deoxynivalenol, fumonisins, zearalenone, patulin, citrinin, ergot alkaloids, and nivalenol are also known to exhibit potent toxicities to different degrees [50]. Several experimental studies demonstrated oxidative stress-mediated DNA damage, DNA/RNA mutations, formation of DNA adducts, and gene modifications in cells on exposure to high levels of mycotoxins [49,52–54]. Thin-layer chromatography, high-performance liquid chromatography, gas chromatography-mass spectrometry, liquid chromatography-mass spectrometry, enzyme-linked immunosorbent assay (ELISA), fluorescence polarization immunoassay, molecular imprinting, and microarray technology are some of the analytical techniques employed for detection of mycotoxins [55].

5.2.6 Chemical Contaminants

Foods may become contaminated by a wide variety of chemical contaminants introduced into the environment by humans, either intentionally or accidentally. Examples of chemical contaminates are dioxins, polychlorinated biphenyls (PCBs), polybrominated biphenyls (PBBs), polycyclic aromatic hydrocarbons (PAH), nitrites, nitrates, and heavy metals [56,57]. These contaminants can be incorporated during food processing, transportation, and storage, or they can already be present in raw materials. The presence of these chemical contaminants in foods can have harmful effects on the gastrointestinal, respiratory, reproductive, immune, and central nervous systems [58]. Many chemical contaminants (such as dioxins, polycyclic aromatic hydrocarbons, acrylamide, nitrosamines, and acetaldehydes) have been confirmed as carcinogenic to humans [59,60]. Antibiotics, drugs, and hormones used during food production are also chemical contaminants that can threaten food safety [61]. Several reports suggest that antibiotic residues in foods are likely to induce growth of antibiotic-resistant bacteria, and promote development of severe pathologies such as cancers, bone marrow impairment, and reproductive disorders [62]. The growth of industry increases the use of agrochemicals, and certain natural and anthropogenic activities can make a significant contribution to the presence of chemical contaminants in foods and food products [63]. Radioactive compounds, pollutants, and chemicals generated by industry are sources of food contamination and causes of food spoilage. The widespread use or distribution of chemical contaminations in foods may result in increased levels along the food chain and potentially have adverse health effects in humans. For instance, bisphenol A (BPA), used in the packaging of foods and beverages, has the propensity to cause endocrine disruption effects and developmental toxicity in humans [64,65]. Concerns have also been raised over the safety of hormones in foods and their potential health impacts in humans [66,67].

Exposure to chemicals is associated with a range of health effects that can have a significant impact on the global burden of food-borne disease. However, information on regional estimates of food-borne illness attributed to chemical exposures and methods of estimation are limited. The Chemical and Toxins Disease Task Force (CTTF) established by the WHO identifies methyl mercury, lead, arsenic, cadmium, dioxin, and dioxin-like compounds as high-priority chemicals for estimation of the burden of food-borne disease [68]. The chemical associated with the majority of illnesses is dioxin, which makes the largest contribution in countries in South-East Asia. A surveillance-based study on food-borne disease outbreaks in India from the period 2008–2018 revealed a rise in food-borne diseases due to chemical contamination of food [69].

Controlling the level of chemical contaminants present in the environment built into foods and their elimination from the process of food production is a challenging task [70]. The chemical-activated luciferase expression (CALUX) bioassay is a rapid and highly sensitive analytical assay that offers possibilities for detection of dioxins and dioxin-like compounds in food samples [71]. Several authors also reported efficient use of high-resolution gas chromatography combined with mass spectrometry (GC-MS) and liquid chromatography-mass spectrometry (LC-MS) for qualitative and quantitative detection of chemical compounds, agrochemicals, chemical residues, carcinogenic components, hormones, and antibiotics in various foods and food products [72–80]. Other routinely used analytical techniques for detection of chemical contaminants are Fourier transform infrared (FT-IR), atomic absorption spectroscopy (AAS), and near-infrared spectroscopy (NIRS) [81–83]. These analytical methods are costly, involve sophisticated instruments, require well-trained staff, and use expensive chemicals. Therefore, the need for low-cost, continuous, specific, real-time, and rapid contaminant detection methods is urgent. In this regard, sensing technology has the potential to overcome the limitations of sophisticated analytical tools [84–86]. Developments in electrosensing strategies have proved beneficial in electrochemical detection of several chemical contaminants in foodstuffs [87].

5.3 SENSORS AND SENSING TECHNOLOGY

5.3.1 Introduction to Sensors

The basis of sensor-based technology is the specific capture of an analyte by an affinity element and transduction of functionally related output into a measurable signal. A sensor is a device that

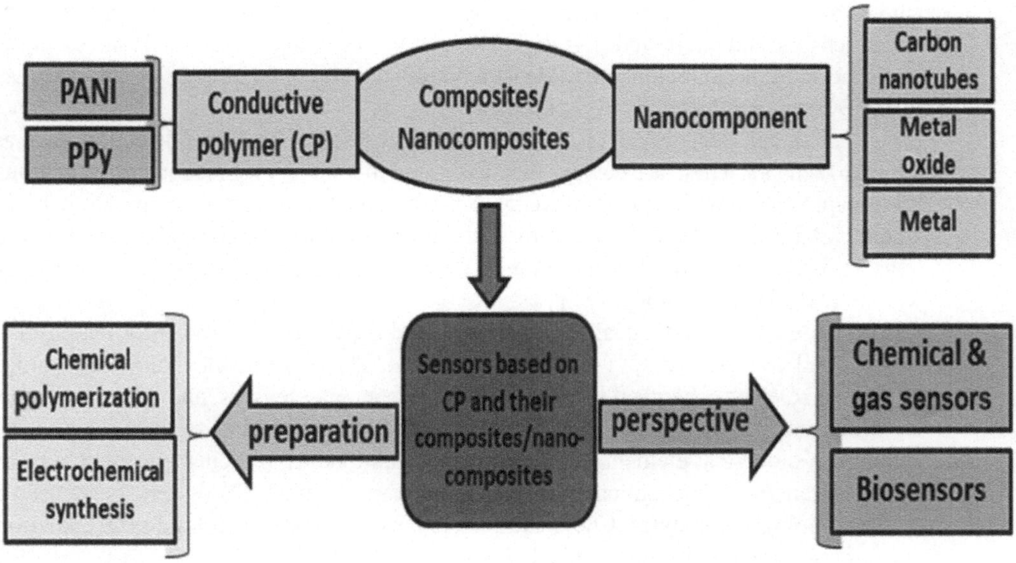

FIGURE 5.2 Sensors based on conductive polymers and their composites/nanocomposites.

converts a physical phenomenon into an electrical signal. Sensor-based devices are used for detection and measurement of various physical and chemical properties of chemical compounds such as temperature, pH, chemicals, pressure, flow, position, light intensity. Sensors offer advantages of low detection limits, good stability, and reproducibility. Figure 5.2 represents conductive polymers based sensors and their composite/nanocomposites.

Based on the physical quantity or target to be detected, sensors can be categorized into different categories [88].

(1) Active and passive sensors: An active sensor is a device that requires an external energy source to operate. Some examples of active sensor-based technologies are scanning electron microscopes, radar, LiDAR, sonar, and X-ray. However, passive sensors generate signals without using external energy. Some examples are infrared sensors, visible light cameras, and radio telescopes. Active sensors are usually used for accurate measurements like positioning or mapping, and passive sensors are mostly used for remote sensing applications.
(2) Contact and noncontact sensors: Contact sensors require physical contact with the target to measure various physical quantities. There are many examples of contact sensors, like thermocouples, potentiometers, and resistance temperature detectors. Noncontact sensors measure or detect physical quantities without any physical contact with the target. They are basically used to measure properties like thickness, distance, and proximity. Examples of noncontact sensors are infrared temperature sensors, magnetic sensors, and optical sensors.
(3) Absolute and relative sensors: An absolute sensor provides measurements with respect to an absolute scale, like an absolute pressure sensor, which measures pressure based on vacuum or absolute zero pressure, whatever may be the atmospheric pressure. A relative sensor gives information about measurements with respect to a fixed or variable reference, like a relative humidity sensor, which measures the change in humidity relative to previous information.
(4) Analog and digital sensor: An analog sensor is used to measure physical quantities and convert them into an electrical signal that is proportional to the quantity to be measured. Some

examples of analog sensors are thermocouples, resistance temperature detectors, capacitive pressure sensors, and photodiodes. A digital sensor produces an output signal that is a digital representation of measured quantity. Measurements such as conductivity, pH, and dissolved oxygen (DO) are conducted by digital sensors.

(5) Detection of signal: Sensors can also be classified according to the signal detection, such as:
   - Physical sensors: These are devices that detect and measure a physical quantity and transform it into an electrical or optical signal. A thermocouple, converting temperature into an output voltage that is readable by a voltmeter, is an example of a physical sensor. Physical sensors are basically used in the biomedical field, along with micro-electromechanical system (MEMS) technology.
   - Chemical sensors: A chemical sensor detects and measures chemical information in an analyte and converts it into a useful output signal. There are various kinds of chemical sensors, like electrochemical sensors, optical sensors, gas sensors, and biosensors, each of which has different principles for detection of chemicals. For example, biosensors combine a biological element (enzymes, antigen/antibodies, proteins, etc.) with a transducer to transform biochemical reactions into electrical signal. These sensors are used to monitor biological analytes. Chemical sensors have applications in food safety, environmental pollution monitoring, clinical diagnostics, etc.
   - Thermal sensors: Thermal sensors or temperature sensors are devices that measure and detect temperature variations of a system. They transform temperature into an electrical output that can be read by other devices. Thermal sensors are of different types, each with specific operating principles and applications. Some common examples are thermocouples, infrared sensors, resistance temperature detectors, and semiconductor thermal sensors.
   - Biological sensors: A biological sensor, also referred to as a biosensor, mainly detects and measures biological information, like biochemical, physiological, or environmental activities. They use the interaction between biological components (antibodies/antigens, enzymes, or living cells) and a transducer to convert a biological signal into a measurable output signal.

In the field of emerging technologies for point-of-care devices, biosensors are gaining popularity for their numerous advantages such as being rapid, sensitive, specific, cost-effective, and user-friendly. Biosensors are effective and innovative analytical tools for fast and efficient biological sensing functions. Biosensors essentially use a biological component (such as receptors, enzymes, nucleic acids, cells, or antibodies) and a signal transducer that converts the biochemical signal into physically detectable signals. The signal displayed is a direct measurement of biochemical reactions occurring on a transducer surface. The development of biosensors is considered a significant breakthrough in the field of materials science and biochemistry. In 1956, Leland Charles Clark Jr. was the first to develop a biosensor device, which was used for measuring the oxygen levels in blood. Later, he demonstrated an amperometric electrode-based enzyme for glucose detection. After more than a decade, Guilbault and Montalvo developed a potentiometric biosensor for urea detection. Since then, rapid development has been made in the biosensor field with growing research interest in a variety of disciplines.

5.3.2 Different Features of Sensors

- Transfer function: This gives a functional relationship between physical input signal and electrical output.
- Sensitivity: Sensitivity refers to the ability of a sensor to detect small changes in the quantity to be measured. A highly sensitive sensor can respond to even minute changes.

- Span or dynamic range: This refers to the range of input physical signals that the sensor can measure or detect accurately [89]. The dynamic range is mostly expressed by the ratio between the maximum value to be measured and the minimum detectable value. It is an important characteristic of sensors due to its ability to detect signals within a wide range of magnitudes.
- Accuracy: Accuracy refers to how closely the measurement of a sensor matches the expected value. It represents the sensor's ability to give reliable and precise data without any deviation or error.
- Hysteresis: This refers to a situation when the outputs of any sensor do not follow a consistent path with respect to the change in input signal in cycle up or down. It can be observed when there is a delay in response from the sensor.
- Linearity: This is the ability of a sensor to follow a linear relationship between the output and input signal. However, various sensors behave nonlinearly for reasons such as environmental conditions and manufacturing defects, which causes error in the final data.
- Noise: This is attributed to unwanted disturbances or fluctuations in the output signal of a sensor which ultimately limit the sensor's performance in terms of accuracy or reliability. Sensors may have noise from various sources such as thermal noise, electrical noise, sensor drift, and environmental noise. To overcome the effects of noise, different methods are used, such as ground shielding, amplification, and filtering, to enhance the overall sensor performance.
- Resolution: This is defined as the smallest fluctuation or change that a sensor can detect. It refers to the extent of precision with which a sensor can work. Resolution depends on many factors like the design of a sensor, measurement range, and technological aspects.
- Bandwidth: A sensor has a finite range of frequencies at which it can respond to changes in various physical parameters. A sensor with wider bandwidth is able to detect high-frequency changes in any measured quantity. The bandwidth of a sensor depends on various physical factors like design or technological aspects, and external factors like noise or any fluctuations in the sensor.

5.3.3 Application of Sensors

Although conventional methods are highly effective and sensitive, they are also time-consuming and labor-intensive, which makes them less applicable in field settings. The use of sensor technology has gained considerable attention over the years since these devices are highly sensitive, rapid, reliable, cost-effective, and suitable for in situ analysis. Sensors and biosensors have found extensive applications in diverse sectors as point-of care monitoring devices. These analytical devices have found a wide variety of applications in food safety, monitoring of environmental pollutants, medicine, and drug development. In particular, detection methods using sensors and biosensors are becoming more common in the food safety industry compared to conventional analytical methods. Biosensor-based methods have been useful for onsite detection of food-borne organisms. The optical transducers employed in design of biosensors include total internal reflection, surface plasmon resonance (SPR), fluorescence resonance energy transfer (FRET), and chemi- and bio-luminescence. Depending on the form of the transduction element, sensors and biosensors can be optical, electrochemical, or mechanical. Several authors have reported fabrication of electrochemical biosensors for rapid detection of pathogenic bacteria, including *Escherichia coli*, *Staphylococcus aureus*, and *Clostridium perfringens*. Highly sensitive and rapid electrochemical biosensors for selective capture of norovirus, hepatitis viruses, human enterovirus, and human rotaviruses have also been developed, with prospects for their use in food contaminant detection. Electrochemical and optical biosensor systems are highly convenient for detection of mycotoxins and have frequently been used in food analyte detection. Recently, fluorescence-based optical biosensors that use different fluorophores for monitoring of mycotoxins have received special interest in the food processing industries. Recent advances in engineered materials and research progress in electrochemical sensors have also shown significant advantages in detection of heavy metals and pesticides in foods.

5.4 CONDUCTING POLYMER NANOCOMPOSITE (CPN)-BASED SENSORS

Conducting polymers (CPs), referred to as synthetic metals [90], possess π-bonds on their backbone, facilitating the migration of electrons in the polymeric chain. Alan J. Heeger, Alan G. MacDiarmid, and Hideki Shirakawa were honored with the Nobel Prize in Chemistry in 2000 for their significant contribution in discovering CPs. CPs exhibit unexpected electronic properties like larger surface area, have excellent mechanical properties, and have a strong tendency to accept additional electrons due to the presence of delocalized π-electrons. They are also capable of conducting electricity and absorbing light at low energies showing optical transitions [91]. Some common conducting polymers are polyacetylene (PA), polyaniline (PANI), polypyrrole (PPy), polythiophene (PTh), poly(para-phenylene) (PPP), poly(phenylenevinylene) (PPV), polyfuran (PF), and their derivatives. Figure 5.3 shows the chemical structure of some of these polymers. Conjugated polymers in their undoped state exhibit low electrical conductivity ranging between 10^{-10}and10^{-5} S/cm. However, their conductivity can be increased significantly by doping. Various dopants, like inorganic salts, peptides, or biomolecules and organic dopants, can be utilized to increase the conductivity of neutral CPs [92]. For example, PANI in an undoped state does not conduct electricity; however, conductivity significantly improved by 10^2–10^3 S/cm or higher by doping. Due to high conductivity, CP is considered a suitable candidate for a wide range of applications like fabrication of electro-optic devices, superconductors, nanoelectronic devices, and biosensors. However, CPs need further chemical modification or functionalization with other materials to address their inherent drawbacks, such as inadequate mechanical strength, limited solubility in general organic solvents, high production costs, and restricted biocompatibility [93]. Nanostructured conducting polymers have attracted significant attention in recent years because of their unique characteristics. They offer various advantages over their bulk counterparts, like large surface area, enhanced electrical conductivity, flexibility, and tunable properties. CP is copolymerized with various nanomaterials and dopants, like

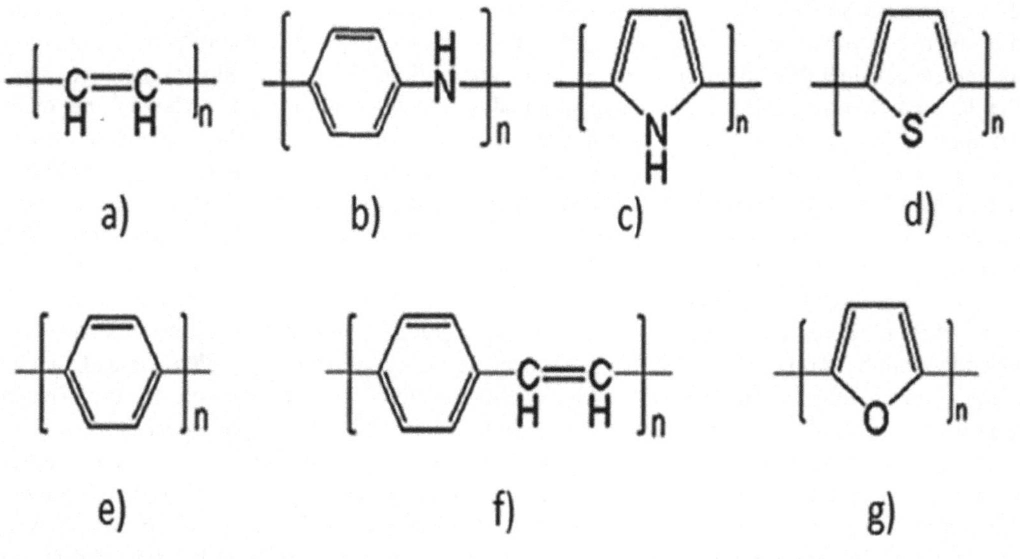

FIGURE 5.3 Chemical structures of (a) polyacetylene (PA), (b) polyaniline (PANI), (c) polypyrrole (PPy), (d) polythiophene (PTh), (e) poly(p-phenylene) (PPP), (f) poly(phenylenevinylene) (PPV), and (g) polyfuran (PF).

graphene, metal nanoparticles, clays, chitosan, and proteins, to form nanocomposites for biosensing and electrochemical applications [94].

5.4.1 Synthesis of Conducting Polymer Nanocomposites

Table 5.2 summarizes some main differences between chemical and electrochemical polymerization.

5.4.1.1 Chemical Polymerization

Chemical polymerization is one of the widely used techniques for synthesizing nanocomposites based on CP. Two common types of polymerization methods are used in this process: addition polymerization and step-growth polymerization, which is initiated by oxidizing agents like ammonium per sulfate or hydrogen per oxides [95]. Anions are mostly used as doping agents for oxidative CP. They can accept anions as counter ions to maintain charge neutrality. For example, polypyrrole, which is a conducting polymer, exhibits electrical conductivity when doped with some ions/molecules (like lithium perchlorate ($LiClO_4$), para-toluene sulfonate (p-TS), and naphthalene sulfonic acid) [96]. Composite films of polypyrrole and polyvinylidine fluoride can also be prepared through chemical oxidation. In this process, ammonium per sulfate is used to facilitate the polymerization reaction [97]. In another study, Armour et al. chemically synthesized polyfuran, a conducting polymer, using trichloroacetic acid as a catalyst and pyridinium chlorochromate (PCC) as an oxidizing agent [98]. Further characterization showed that the prepared material possesses proton acceptor properties and electrical conductivity. The synthesis of polythiophene grafted poly(methacrylate) (PTh-g-PMA)/ZnO hybrid through in situ chemical polymerization was reported [99]. The (PTh-g-PMA)/ZnO hybrid material has many applications in fields like electronics, sensors, and photovoltaic devices. Mostafaei et al. reported the synthesis of polyaniline/ZnO hybrid nanocomposite [100], which can be used for anticorrosive coating. Hybrid materials like Pt/PANI/grapheme nanosheets prepared through interfacial polymerization show exceptional electrocatalytic properties[101]. The studied material also shows promising results for various applications, especially in the construction of biosensors or electrochemical sensors. Baghayeri et al. prepared poly(para-phenylenediamine) (PpPDA)-Fe_3O_4 nanocomposite through emulsion polymerization. The resulting method has been useful for hydrogen peroxide biosensors with a detection limit of 0.21 μM [102].

TABLE 5.2
Differences between electrochemical and chemical polymerization

Chemical polymerization	Electrochemical polymerization
This process depends on initiator, catalyst, or heat to commence the reaction by chain growth or crosslinking method.	This process is initiated by utilizing electric potential to a monomeric solution which further generates ions to undergo reaction.
In this process, polymers of various forms can be prepared, including powdered form, thick film, etc., depending on the requirements.	In terms of electrochemical setup, this method is categorized as potentiostatic electrochemical polymerization and galvanostatic electrochemical polymerization.
This type of reaction is mostly used for bulk production of polymers, occurring in a homogeneous mixture of monomers in the absence of solvent.	In this process, composition of the polymer can be modified by changing the conditions like monomer concentration, adding dopants, etc.
In general, conductivity of polymers prepared through this process is lower compared to electrochemical process.	**This process gives precise control over the surface morphology of the resulting material by adjusting parameters like potential applied, electrolytes composition, etc.**

5.4.1.2 Electrochemical Synthesis

Electrochemical polymerization is another technique employed to synthesize CP-based nanocomposites. It involves a very simple and versatile technique in which a working electrode, a counter electrode, and a reference electrode are used to monitor the potential. A solution containing a monomer along with suitable solvent and electrolyte is also required [97,103]. The working electrode is made of platinum or gold, for example. In this process, monomers with low oxidation potential capable of eletrophilic substitution are favored. It is important to select a solvent and electrolyte that is stable at the oxidation potential of the monomer, to prevent unwanted reactions and maintain ionic conductivity. Conductive polymer film can be prepared by electrochemical reactions using galvanostatic or potentiostatic methods [104]. Each method has its advantages, and the choice can be made according to the requirement of the polymerization reactions. Diaz and Kanazaw introduced electrochemical polymerization of pyrrole on a platinum electrode. The monomer undergoes oxidation at the surface of the electrode, forming soluble oligomers surrounding the electrode in the diffusion layers. Polypyrrole films fabricated by a precise electrochemical technique show enhanced conductivity and excellent stability. Various literature surveys have shown that CPNs are being synthesized by researchers for use in biosensors due to their great electrocatalytic activities. One of the important features of CPNs is their capability to facilitate charge transfer in the electrode and immobilized biomolecules. Zhang et al. reported a composite material based on polycarbazole (PCZ) and nitrogen-doped graphene (N-GE) for detecting 4-nitrophenol (4-NP) [105]. The resulting material showed great catalytic properties for reduction of 4-NP. It demonstrated a very low detection limit (0.062 μM), suggesting high sensitivity of the sensor. Hybrid film based on graphene oxide (GO) and PANI can be prepared successfully. It exhibits better electrical conductivity and electroactivity in a wide range of pH values, making it a promising material for sensor-based applications [106].

5.4.2 Characterization and Analysis of Conducting Polymers

Some characterization techniques used for conducting polymers are discussed below.

5.4.2.1 Fourier Transform Infrared (FT-IR) Spectroscopy

FT-IR spectroscopy is employed to detect various functional groups present in the nanocomposite. Olad and Azhar studied the FT-IR spectra of biopolymer/clay/conducting polymer nanocomposite to inspect the probable interactions among the nanocomposite components [107]. The spectra of starch (St), montmorillonite (MMT), polyaniline (PANI), St-MMT, and St-MMT/PANI have been reported. The absorption peaks of starch (3419 cm^{-1}), MMT (1031 cm^{-1}), and PANI (1550, 1460, 1286 and 790 cm^{-1}) were found to be present in the St-MMT/PANI spectrum with slight shifting. The stretching vibration band of –OH in the nanocomposites was found to be weak compared to the native starch. According to the findings, it was confirmed that the nanocomposite was successfully prepared.

5.4.2.2 Differential Scanning Calorimeter (DSC) and Thermo-Gravimetric Analysis (TGA)

From these analytical techniques, thermal properties like thermal stability, glass transition temperature (Tg), decomposition temperature, and melting point can be investigated. A TGA thermograph usually interprets the variation of weight of material with temperature. Pati et al. developed conductive thermosetting resin using polyaniline (PANI) and 2-Methacryloyloxyethyl acid phosphate, the latter being used as a dopant. A DSC study was conducted to estimate the activation energy and curing behavior in the resin. It was reported that 5wt% of PANI resin exhibited longer post-curing time, while 20 wt% PANI resin needed a shorter period of time to yield the cured material [108].

5.4.2.3 X-ray Diffraction (XRD)

X-ray diffraction is mostly used to investigate the amorphous and crystalline behavior of any material. XRD analysis for PANI-ZnO nanocomposites was carried out by Mostafaei and Zolriasatein. They tried to compare the effects of ZnO nanorods at different percentages on PANI [109]. Diffraction angle for PANI was found to be at $2\theta = 19.31°$ and $2\theta = 25.72°$. They reported the d-spacing at 4.593 Å and 3.461 Å, respectively, and the crystal size of PANI was found to be 4.4 nm. The crystal size of nanocomposites was enhanced to 5.3 nm, 5.5 nm, and 8.2 nm when ZnO nanorods were loaded at 1 wt%, 2 wt%, and 4 wt%, respectively. There was a slight shift in significant peaks of PANI, and intensity was also enhanced with an increase in the percentage of ZnO nanorods in PANI-ZnO.

5.4.2.4 Morphological Study

Scanning electron microscopy (SEM) studies the surface morphology, indicating the roughness or smoothness in the material. Anderson et al. reported the scanning electron micrographs for polypyrrole (PPy)/indium tin oxide (ITO) films synthesized in supercritical carbon dioxide ($scCO_2$) and acetonitrile [110]. The $scCO_2$-synthesized PPy was composed of small, raised granular PPy nodules between 0.5 and 3.0 μm on a flat, continuous PPy surface. Maeda et al. synthesized stable PPy-SiO_2 colloid by employing 20 nm SiO_2 particles as dispersant [111]. The TEM image revealed a characteristic "raspberry" morphology. A separation of 5–10 nm among the SiO_2 particles was reported, suggesting the existence of a thinner layer, probably a monolayer of PPy on SiO_2 particles.

5.4.2.5 Electrical Conductivity

Modak et al. reported the electrical conductivity of a grapheme nanosheet (GN) covered with PANI through an in situ chemical oxidative polymerization. They found that the electrical conductivity of PANI-GN composites was enhanced by increasing the amount of grapheme nanosheet in the PANI matrix [112].

5.4.3 Conducting Polymer Nanocomposite-Based Sensors

Recently, CPNs have arisen as highly favorable materials for sensor application with great selectivity, stability, and cyclability. For sensing techniques, they enable fast response times and quick detection of target analytes. Incorporating fillers like carbon nanotubes, clay, TiO_2, and graphene into the conducting polymer can boost the overall properties, making them suitable for highly selective, stable, and sensitive sensors [113]. In addition, on adding the requisite amount of nano-sized particles, gas sensors, chemical sensors, or biosensors with improved performance can be designed.

The sensors based on CPNs have shown great potential as new point-of-care sensing systems for detection of heavy metals, gases, pesticides, hormones, enzymes, and small biomolecules with high specificity and sensitivity. Menegazzo et al. prepared PANI nanofibers in the presence of a dopant (camphor sulfonic acid; CSA) through interfacial polymerization. The doping process improved the responsive behavior of PANI towards ammonia detection [114]. A report suggested that PANI-In_2O_3composite prepared by chemical polymerization exhibits high sensitivity towards H_2 detection compared to CO and NO_2 gas [115]. Virji et al. prepared PANI/$CuCl_2$ composite for the detection of H_2S gas and reported that the composite material showed greater efficiency for detecting H_2S gas[116], due to the greater surface area and smaller dimension of PANI allowing faster diffusion of gas into the film. Another report indicates that a PANI/Poly vinyl pyrrolidone (PVP) blend has better sensitivity for $CHCl_3$than CH_2Cl_2 [117]. Huang et al. also developed reduced graphene oxide (RGO)/PANI and reported the responsive behavior for NH_3 gas [118], which was found to be higher compared to PANI and graphene sensors. A Pt/PANI/graphene nanosheet (GN) composite shows electrocatalytic activity towards the detection of H_2O_2 and glucose due to various characteristics including higher conductivity, surface area, and enzyme absorption capability [119]. There were also reports by Kaushik et al. of preparing PANI-WO_3 nanocomposite via vacuum thermal evaporation, and responsive behavior for NOx gas was studied. The resulting system has a quick response time, indicating interaction of NOx with π-electrons of PANI, which leads to improvement in conductivity and sensitivity [120]. Zampetti et al. fabricated a TiO_2/poly(3,4-ethylenedioxythiophene)/poly(4-styrenesulfonate) system through electro-spinning technique. The hybrid system has exceptional reversible and reproducible behavior for NO_2 gas. The hybrid also showed fast response, with detection limit up to 5 ppb for the toxic gas [121]. A polypyrrole (PPy)-reduced graphene oxide (rGO) (PPy/rGO) hybrid developed through in situ polymerization exhibited a high response that was 2.5 times greater than PPy. The exceptional gas-sensing behavior of the material was attributed to high electron mobilization in graphene that helped in fast charge transport. It might also be due to the large surface area in the system that increases the adsorption of NH_3 molecules [122].

In recent years, conducting polymers and their composites/nanocomposites have attracted researchers in the fabrication of biosensors because of their excellent properties, namely immobilization of biomolecules and fast electron transfer. It has been reported that PANI/polystyrene (PS) composite nanofibers prepared via an electro-spinning technique were used for sensing hydrogen peroxide (H_2O_2) and glucose [123]. The greater surface area and exceptional electrochemical behavior of PANI are the reasons behind such properties. In another report, Fiorito et al. prepared the cupper hexacyanoferrate/PPy composite as mediator for hydrogen peroxide detection. They found that the resulting system was exceptionally sensitive, stable, and had great electrocatalytic properties for H_2O_2 reduction around sodium/potassium ions [124]. Likewise, electromagnetic poly(p-phenylenediamine) (P*p*PDA)-Fe_3O_4 nanocomposite was synthesized via emulsion polymerization by Baghayeri et al., which showed good reproducible results with high responsive behavior, with a detection limit of 0.21 μM for the reduction of H_2O_2. The Fe_3O_4 nanoparticle behaved as an enzyme-mobilizing agent that improves the sensing application of the nanocomposite [125]. Poly(aminobenzenesulfonic acid)-modified single-walled carbon nanotubes (PABS–SWNTs) were fabricated using a simple layering method. The PANI/PABS–SWNTs multilayer film obtained was found to be stable and showed high electrocatalytic performance to identify β-nicotinamide adenine dinucleotide (NADH) at lower potential, and it was very selective towards NADH [126]. Polyaniline/Au composite hollow spheres were developed based on polystyrene/sulfonated polystyrene core/shell gel particle templates [127]. The catalytic behavior was examined on glassy carbon electrode (GCE). The electrical conductivity of the resulting system was three times greater than the pure form. The composite was also immobilized on GCE surface for fabricating a sensor, which had exceptional catalytic properties for the oxidation of dopamine. Polypyrrole/reduced graphene oxide hybrid sheets were formed by Qian et al. with an Au nanoparticle for detecting dopamine. The sensor showed excellent sensitivity, with a detection limit of 18.29 pM [128]. The authors also indicated the detection of enzyme, nucleic acid, and protein with the resulting material. Polyaniline (PA)-coated Fe_3O_4 nanoparticle-carbon-nanotube (CNTs) nanocomposite shows remarkable glucose-sensing behavior [129]. It was suggested that CNT be used to enhance the conductivity and surface area of the material for immobilization of glucose oxidase. The nanocomposite also exhibited great sensitivity and reproducibility. Yan et al. assembled gold nanoparticles (AuNPs) on a silver chloride@polyaniline (PANI) core–shell (AgCl@PANI) surface via adsorbing glucose oxidase enzyme on Au/AgCl/PANI electrode [130]. At neutral pH, the resulting AuNP–AgCl@PANI system revealed superior electroactivity. It also provided a biocompatible environment for better loading of enzymes. The AuNP was advantageous as a catalyst for oxidation and reduction of H_2O_2. The biosensor proved to be very stable and sensitive towards the analyte, having a detection limit of 4 pM. Pt nanoparticle/PANI hydrogel showed excellent sensitivity towards glucose enzymes, with a rapid response time, and was able to identify quite low amounts of gas [131]. A nanocomposite based on graphene/polyaniline/gold nanoparticles for biosensing of glucose was developed, and it was found that the material was stable and favorable for glucose detection at low concentrations (0.6 μM) [132]. A nanocomposite film composed of polyaniline (PANI) and multiwalled carbon nanotubes (MWCNT) via electro-polymerization technique is an excellent choice for cholesterol biosensor, having 6800 nA mM^{-1} sensitivity [133]. It was found that the bioelectrode responded very fast and could identify cholesterol ranging between 1.29 and12.93 mM. Radhapyari et al. developed a graphene–polyaniline nanocomposite for identification of the antimalarial drug artesunate in biological fluids and pharmaceutical setup [134]. The biosensor showed excellent properties to identify lower percentages of the drug, and it exhibited high repeatability and reproducibility.

5.5 APPLICATION OF CP NC-BASED SENSORS FOR FOOD SAFETY ANALYSIS

Sensors utilizing conductive polymers and their composites have attracted great interest in the field of food safety analysis due to several advantages, including high sensitivity, rapid response time, and stability. They have the ability to modify both physical and chemical properties by employing numerous substituents. Another important feature is that the sensors can be operated at ambient temperature [135]. CPs have attracted significant attention for their unique capability to serve as an immobilization matrix for bioenzymes and bioactive species. They also function as a transducer for converting any bioanalytical signal to an electrical signal [136]. Despite the fact that conducting polymers are useful for chemical and biological sensing, they have limitations in terms of sensitivity and selectivity. To address these limitations, conducting polymer nanostructures have been utilized, which can offer fast response times owing to their porous structure and large surface area [137]. For example, a nanofiber thin film of PANI fabricated through interfacial polymerization had a much greater response time compared to a thin film of pure PANI. Adding a nanocomponent like CNT, SiO_2, or bentonite into the matrix of a conducting polymer might improve the mobility of chain in the conducting polymers. It can also alter the affinity of the composite system and even behave as a catalyst [138]. The interaction between electron donor and acceptor enables the development of several nanocomposites, showcasing synergetic properties that find applications in biosensors or electrochemical sensors.

Due to metabolic processes, food-borne pathogens emit gases and odors which are known to cause food degradation. Arshak et al. developed a conducting polymer nanocomposite-based sensor for quantification of odors released by bacterial pathogens with a potential applications in food safety [139]. Immobilization of DNA into CPs was studied broadly for identification of several DNA target sequences. A study was conducted on an electrochemical impedance-based DNA biosensor based on polypyrrole (PPy) and multiwalled carbon nanotubes to a functionalized glassy carbon electrode [140]. The electrode showcased a large surface area and excellent electrochemical activities, suitable for sensing DNA. Electrochemical polymerization was used by Khan and Dhayal to fabricate chitosan (CS)-polyaniline conducting biopolymer film on indium tin oxide electrode for identification of toxic constituents such as ochratoxin A [141]. The resulting film showed enhanced performance for identification of ochratoxin A due to numerous carboxylic and hydroxyl groups on the PANI/CS matrix that helped immobilization of the rabbit antibody (IgGs).

ACKNOWLEDGMENTS

Archana Deka acknowledges the Indian Council of Medical Research, Govt. of India for ICMR-Research Associate Fellowship (Sanction No. 2021-8051/PROTEOMICS-BMS).

CONFLICT OF INTEREST

The authors declare no conflict of interest.

REFERENCES

1. World Health Organization. *WHO initiative to estimate the global burden of foodborne diseases. Fourth formal meeting of the Foodborne Disease Burden Epidemiology Reference Group (FERG): Sharing new results, making future plans, and preparing ground for the countries*. WHO Department of Food Safety, Geneva, 2014, https://iris.who.int/bitstream/handle/10665/159844/9789241507950_eng.pdf
2. World Health Organization. *Global health observatory data repository: Road traffic deaths data by country*, 2021, www.who.int/gho/road_safety/mortality/traffic_deaths_number/en/. Retrieved on January 24, 2023
3. Devleesschauwer, B.; Haagsma, J. A.; Mangen, M. J.; Lake, R. J.; Havelaar, A. H. The global burden of foodborne disease. In: *Food safety economics*. Springer, Cham., 2018, pp. 107–122.
4. Scallan, E.; Hoekstra, R. M.; Angulo, F. J.; Tauxe, R. V.; Widdowson, M. A.; Roy, S. L.; Jones, J. L.; Griffin, P. M. Foodborne illness acquired in the United States—major pathogens. *Emerg. Infect. Dis.* 2011, *17*, 7.
5. Pires, S. M.; Desta, B. N.; Mughini-Gras, L.; Mmbaga, B. T.; Fayemi, O. E.; Salvador, E. M.; Gobena, T.; Majowicz, S. E.; Hald, T.; Hoejskov, P. S.; Minato, Y. Burden of foodborne diseases: think global, act local. *Curr. Opin. Food Sci.* 2021, 39, 152–9.
6. Grace, D. Food safety in low and middle income countries. *Int. J. Environ. Res. Public Health.* 2015, 12, 10490–507.
7. Käferstein, F. Foodborne diseases in developing countries: aetiology, epidemiology and strategies for prevention. *Int. J. Environ. Health Res.* 2003, *13 Suppl 1*, S161–8. doi: 10.1080/0960312031000102949
8. Radovanovic, R. Food safety: the global problem as a challenge for future initiatives and activities. In Advances in food protection: Focus on food safety and defense. Magdy Hefnawy (Ed.). Springer Netherlands., 2011, pp. 27–48.
9. Uyttendaele, M.; Franz, E.; Schlüter, O. Food safety, a global challenge. *Int. J. Environ. Res. Public Health.* 2016, *13*, 67.
10. Gram, L.; Ravn, L.; Rasch, M.; Bruhn, J. B.; Christensen, A. B.; Givskov, M. Food spoilage—interactions between food spoilage bacteria. *Int. J. Food Microbiol.* 2002, *78*, 79–97.
11. World Health Organization [WHO] www.who.int/mediacentre/news/releases/2015/foodborne-disease-estimates/en/
12. Havelaar, A. H.; Kirk, M. D.; Torgerson, P. R.; Gibb, H. J.; Hald, T.; Lake, R. J. et al. World Health Organization global estimates and regional comparisons of the burden of foodborne disease in 2010. *PLoS Med.* 2015, *12*, e1001923.
13. Bintsis, T. Foodborne pathogens. *AIMS Microbiol.* 2017, *3*, 529–563.
14. Capita, R.; Alonso-Calleja, C. Antibiotic-resistant bacteria: a challenge for the food industry. *Crit. Rev. Food Sci. Nutr.* 2013, *53*, 11–48.
15. Olaimat, A. N.; Al-Holy, M. A.; Shahbaz, H. M.; Al-Nabulsi, A. A.; Abu Ghoush, M. H.; Osaili, T. M.; Ayyash, M. M.; Holley, R. A. Emergence of antibiotic resistance in listeria monocytogenes isolated from food products: a comprehensive review. *Compr. Rev. Food Sci. Food Saf.* 2018, *17*, 1277–1292.
16. Verraes, C.; Van Boxstael, S.; Van Meervenne, E.; Van Coillie, E.; Butaye, P.; Catry, B.; de Schaetzen, M. A.; Van Huffel, X.; Imberechts, H.; Dierick, K.; Daube, G.; Saegerman, C.; De Block, J.; Dewulf, J.; Herman, L. Antimicrobial resistance in the food chain: a review. *Int. J. Environ. Res. Public Health.* 2013, 10 , 2643–69.
17. Founou, L. L.; Founou, R. C.; Essack, S. Y. Antibiotic resistance in the food chain: a developing country-perspective. *Front Microbiol.* 2016, *7*, 1881.
18. Shrestha, A.; Uzal, F. A.; McClane, B. A.. Enterotoxic clostridia: *Clostridium perfringens* enteric diseases. *Microbiol. Spectr.* 2018, *6*. doi:10.1128/microbiolspec.GPP3-0003-2017
19. Mayer, C. L.; Leibowitz, C. S.; Kurosawa, S.; Stearns-Kurosawa, D. J. Shiga toxins and the pathophysiology of hemolytic uremic syndrome in humans and animals. *Toxins (Basel)*. 2012, *4*, 1261–1287.
20. Wassenaar, T. M.; Zimmermann, K. Lipopolysaccharides in food, food supplements, and probiotics: should we be worried?. Eur. *J. Immunol.* 2018, *8*, 63–69.
21. Wu, H.; Wang, Y.; Li, H.; Meng, L.; Zheng, N.; Wang, J. Effect of food endotoxin on infant health. *Toxins*. 2021, *13*, 298.
22. Law, J. W. F.; Ab Mutalib, N. S.; Chan, K. G.; Lee, L. H. Rapid methods for the detection of foodborne bacterial pathogens: principles, applications, advantages and limitations. *Front Microbiol.* 2015, *5*, 770.

23. Foddai, A. C.; Grant, I. R. Methods for detection of viable foodborne pathogens: current state-of-art and future prospects. *Appl. Microbiol. Biotechnol.* 2020, 104, 4281–4288.
24. Ding, J. L.; Ho, B. Endotoxin detection–from limulus amebocyte lysate to recombinant factor C. *SubcellBiochem.* 2010, *53*, 187–208.
25. Thorne, P. S.; Perry, S. S.; Saito, R.; O'Shaughnessy, P. T.; Mehaffy, J.; Metwali, N.; Keefe, T.; Donham, K. J.; Reynolds, S. J. Evaluation of the Limulus amebocyte lysate and recombinant factor C assays for assessment of airborne endotoxin. *Appl. Environ. Microbiol.* 2010, *76*, 4988–4995.
26. Josić, D.; Rešetar, D.; Peršurić, Ž.; Martinović, T.; Pavelić, S. K. Detection of microbial toxins by-omics methods: a growing role of proteomics. In Proteomics in *food science*. Academic Press, 2017, pp. 485–506. https://doi.org/10.1016/B978-0-12-804007-2.00029-1
27. Alahi, M. E. E.; Mukhopadhyay, S. C. Detection methodologies for pathogen and toxins: a review. *Sensors*. 2017, *17*, 1885.
28. D'Souza, D. H. Update on foodborne viruses: types, concentration and sampling methods. In Advances in microbial food safety. Woodhead Publishing, 2015, pp. 102–116. https://doi.org/10.1533/9781782421153.1.102
29. Rzeżutka, A.; Cook, N. Review of currently applied methodologies used for detection and typing of foodborne viruses. In Global issues in food science and technology. Academic Press, 2009, pp. 229–246. https://doi.org/10.1016/B978-0-12-374124-0.00015-6
30. Bosch, A.; Gkogka, E.; Le Guyader, F. S.; Loisy-Hamon, F.; Lee, A.; Van Lieshout, L.; Marthi, B.; Myrmel, M.; Sansom, A.; Schultz, A. C.; Winkler, A.; Zuber, S.; Phister, T. Foodborne viruses: Detection, risk assessment, and control options in food processing. *Int. J. Food Microbiol.* 2018, 285, 110–128.
31. Dorny, P.; Praet, N.; Deckers, N.; Gabriël, S. Emerging food-borne parasites. *Vet Parasitol.* 2009, *163*, 196–206.
32. Gabriël, S.; Dorny, P.; Saelens, G.; Dermauw, V. Foodborne parasites and their complex life cycles challenging food safety in different food chains. *Foods*. 2023, *12*, 142.
33. Stryiński, R.; Łopieńska-Biernat, E.; Carrera, M. Proteomic insights into the biology of the most important foodborne parasites in Europe. *Foods*. 2020, *9*, 1403.
34. Budu-Amoako, E.; Greenwood, S. J.; Dixon, B. R.; Barkema, H. W.; McClure, J. T. Foodborne illness associated with Cryptosporidium and Giardia from livestock. *J. Food Prot.* 2011, *74*, 1944–55.
35. Almeria, S.; Dubey, J. P. Foodborne transmission of Toxoplasma gondii infection in the last decade. An overview. *Res Vet Sci.* 2021, *135*, 371–385.
36. Weinstock, J. V.; Leung, J. Parasitic diseases: helminths. In Yamada's textbook of gastroenterology. Wiley, 2022, pp. 3039–3078. https://doi.org/10.1002/9781119600206.ch147
37. Emameh, R. Z.; Purmonen, S.; Sukura, A.; Parkkila, S. Surveillance and diagnosis of zoonotic foodborne parasites. *Food Sci. Nutr.* 2018, *6*, 3–17.
38. Chalmers, R.; Robertson, L. J.; Dorny, P.; Jordan, S. Parasite detection in food: Current status and future needs for validation. *Trends Food Sci. Technol.* 2020, *99*, 337–350.
39. Chen, S.; Chu, L. T.; Chen, T. H. Colorimetric detection of active botulinum neurotoxin using Cu^{2+} mediated gold nanoparticles agglomeration. *Sens. Actuators B Chem.* 2016, *235*, 563–567.
40. Rajput, J. K. Nanomaterial-based sensors as potential remedy for detection of biotoxins. *Food Control.* 2022, 135, 108686.
41. Visciano, P.; Schirone, M.; Berti, M.; Milandri, A.; Tofalo, R.; Suzzi, G. Marine biotoxins: occurrence, toxicity, regulatory limits and reference methods. *FrontMicrobiol.* 2016, *7*, 1051.
42. Christian, B.; Luckas, B. Determination of marine biotoxins relevant for regulations: from the mouse bioassay to coupled LC-MS methods. *Anal. Bioanal. Chem.* 2008, *391*, 117–134.
43. Reverté, L.; Soliño, L.; Carnicer, O.; Diogène, J.; Campàs, M. Alternative methods for the detection of emerging marine toxins: biosensors, biochemical assays and cell-based assays. *Mar. Drugs*. 2014, *12*, 5719–5763.
44. Rossini, G. P. Functional assays in marine biotoxin detection. *Toxicology*. 2005, *207*, 451–462.
45. Mol, H. G. J.; Van Dam, R. C. J.; Zomer, P.; Mulder, P. P. Screening of plant toxins in food, feed and botanicals using full-scan high-resolution (Orbitrap) mass spectrometry. *Food Addit. Contam. Part A Chem.* 2011, *28*, 1405–1423.
46. Rubina, A. Y.; Dyukova, V. I.; Dementieva, E. I.; Stomakhin, A. A.; Nesmeyanov, V. A.; Grishin, E. V.; Zasedatelev, A. S. Quantitative immunoassay of biotoxins on hydrogel-based protein microchips. *Anal. Biochem.* 2005, *340*, 317–329.

47. Hodnik, V.; Anderluh, G. Toxin detection by surface plasmon resonance. *Sensors*. 2009, *9*, 1339–1354.
48. da Rocha, M. E. B.; Freire, F. D. C. O.; Maia, F. E. F.; Guedes, M. I. F.; Rondina, D. Mycotoxins and their effects on human and animal health. *Food Control*. 2014, *36*, 159–165.
49. Awuchi, C. G.; Ondari, E. N.; Nwozo, S.; Odongo, G. A.; Eseoghene, I. J.; Twinomuhwezi, H.;Ogbonna, C. U.; Upadhyay, A. K.; Adeleye, A. O.; Okpala, C. O. R. Mycotoxins' toxicological mechanisms involving humans, livestock and their associated health concerns: a review. *Toxins*. 2022, *14*, 167.
50. Abnet, C. C. Carcinogenic food contaminants. *Cancer Invest*. 2007, *25*, 189–196.
51. Reddy, L.; Bhoola, K. Ochratoxins—food contaminants: impact on human health. *Toxins*. 2010, *2*, 771–779.
52. Liu, B. H.; Yu, F. Y.; Wu, T. S.; Li, S. Y.; Su, M. C.; Wang, M. C.; Shih, S. M. Evaluation of genotoxic risk and oxidative DNA damage in mammalian cells exposed to mycotoxins, patulin and citrinin. *Toxicol. Appl. Pharmacol*. 2003, *191*, 255–263.
53. Palma, N.; Cinelli, S.; Sapora, O.; Wilson, S. H.; Dogliotti, E. Ochratoxin A-induced mutagenesis in mammalian cells is consistent with the production of oxidative stress. *Chem. Res. Toxicol*. 2007, *20*, 1031–1037
54. Pfohl-Leszkowicz, A.; Grosse, Y.; Kane, A.; Creppy, E. E.; Dirheimer, G. Differential DNA adduct formation and disappearance in three mouse tissues after treatment with the mycotoxin ochratoxin A. *Mutat. Res. Fundam. Mol. Mech. Mutagen*. 1993, *289*, 265–273.
55. Singh, J.; Mehta, A. Rapid and sensitive detection of mycotoxins by advanced and emerging analytical methods: a review. *Food Sci. Nutr*. 2020, *8*, 2183–2204.
56. Rather, I. A.; Koh, W. Y.; Paek, W. K.; Lim, J. The sources of chemical contaminants in food and their health implications. *Front Pharmacol*. 2017, *8*, 830.
57. da Costa Silva, L. M.; dos Santos, V. P. S.; Salgado, A. M.; Pereira, K. S. Biosensors for contaminants monitoring in food and environment for human and environmental health. In *State of the art in biosensors,* Croatia: InTech., 2013, pp. 152–168. DOI: 10.5772/55617
58. Choudhury, S.; Medina-Lara, A.; Smith, R.; Daniel, N. Research on health impacts of chemical contaminants in food. *Bull. World Health Organ*. 2022, *100*, 180.
59. Abnet, C. C. Carcinogenic food contaminants. *Cancer Invest*. 2007, *25*, 189–196.
60. Saini, R. V.; Vaid, P.; Saini, N. K.; Siwal, S. S.; Gupta, V. K.; Thakur, V. K.; Saini, A. K. Recent advancements in the technologies detecting food spoiling agents. *J. Funct. Biomater*. 2021, *12*, 67.
61. Nerin, C.; Aznar, M.; Carrizo, D. Food contamination during food process. *Trends Food Sci. Technol*. 2016, *48*, 63–68.
62. Arsène, M. M. J.; Davares, A. K. L.; Viktorovna, P. I.; Andreevna, S. L.; Sarra, S.; Khelifi, I.; Sergueïevna, D. M. The public health issue of antibiotic residues in food and feed: causes, consequences, and potential solutions. *Vet. World*. 2022, *15*, 662–671
63. Thomson, B.; Rose, M. Environmental contaminants in foods and feeds in the light of climate change. *Qual. Assur. Saf. Crops Foods*. 2011, *3*, 2–11.
64. Hafezi, S. A.; Abdel-Rahman, W. M. The endocrine disruptor Bisphenol A (BPA) exerts a wide range of effects in carcinogenesis and response to therapy. *Curr. Mol. Pharmacol*. 2019, *12*, 230–238.
65. Konieczna, A.; Rutkowska, A.; Rachoń, D. Health risk of exposure to Bisphenol A (BPA). *Rocz PanstwZaklHig*. 2015, *66*, 5–11.
66. Malekinejad, H.; Rezabakhsh, A. Hormones in dairy foods and their impact on public health – a narrative review article. *Iran. J. Public Health*. 2015, *44*, 742–758.
67. Palacios, O. M.; Cortes, H. N.; Jenks, B. H.; Maki, K. C. Naturally occurring hormones in foods and potential health effects. *Toxicol. Res. Appl*. 2020, *4*, 1–12. 10.1177/2397847320936281
68. Gibb, H.; Devleesschauwer, B.; Bolger, P. M.; Wu, F.; Ezendam, J.; Cliff, J.; Zeilmaker, M.; Verger, P.; Pitt, J.; Baines, J.; Adegoke, G.; Afshari, R.; Liu, Y.; Bokkers, B.; van Loveren, H.; Mengelers, M.; Brandon, E.; Havelaar, A. H.; Bellinger, D. World Health Organization estimates of the global and regional disease burden of four foodborne chemical toxins, 2010: a data synthesis. *F1000Res*. 2015, *4*, 1393.
69. Bisht, A.; Kamble, M. P.; Choudhary, P.; Chaturvedi, K.; Kohli, G.; Juneja, V. K.; Sehgal, S.; Taneja, N. K. A surveillance of food borne disease outbreaks in India: 2009–2018. *Food Control*. 2021, *121*, 107630.
70. Hoogenboom, L. P.; Hamers, A. M.; Bovee, T. H. Bioassays for the detection of growth-promoting agents, veterinary drugs and environmental contaminants in food. *Analyst. 1999*, *124*, 79–85.

71. Windal, I.; Denison, M. S.; Birnbaum, L. S.; Van Wouwe, N.; Baeyens, W.; Goeyens, L. Chemically activated luciferase gene expression (CALUX) cell bioassay analysis for the estimation of dioxin-like activity: critical parameters of the CALUX procedure that impact assay results. *Environ. Sci. Technol.* 2005, *39*, 7357–7364.
72. Xu, M. L.; Gao, Y.; Wang, X.; Han, X. X.; Zhao, B. Comprehensive strategy for sample preparation for the analysis of food contaminants and residues by GC–MS/MS: a review of recent research trends. *Foods*. 2021, *10*, 2473.
73. Hird, S. J.; Lau, B. P. Y.; Schuhmacher, R.; Krska, R. Liquid chromatography-mass spectrometry for the determination of chemical contaminants in food. *TrAC.* 2014, *59*, 59–72.
74. Lutter, P.; Savoy-Perroud, M. C.; Campos-Gimenez, E.; Meyer, L.; Goldmann, T.; Bertholet, M. C.; Mottier, P.; Desmarchelier, A.; Monard, F.; Perrin, C.; Robert, F.; Delatour, T. Screening and confirmatory methods for the determination of melamine in cow's milk and milk-based powdered infant formula: validation and proficiency-tests of ELISA, HPLC-UV, GC-MS and LC-MS/MS. *Food Control.* 2011, *22*, 903–913.
75. Paradis, D.; Bérail, G.; Bonmatin, J. M.; Belzunces, L. P. Sensitive analytical methods for 22 relevant insecticides of 3 chemical families in honey by GC-MS/MS and LC-MS/MS. *Anal. Bioanal. Chem.* 2014, *406*, 621–633.
76. Amulen, D. R.; Spanoghe, P.; Houbraken, M.; Tamale, A.; de Graaf, D. C.; Cross, P.; Smagghe, G. Environmental contaminants of honeybee products in Uganda detected using LC-MS/MS and GC-ECD. *PLoS one*. 2017, *12*, e0178546.
77. Rivera-Pérez, A.; López-Ruiz, R.; Romero-González, R.; Frenich, A. G. A new strategy based on gas chromatography–high resolution mass spectrometry (GC–HRMS-Q-Orbitrap) for the determination of alkenylbenzenes in pepper and its varieties. *Food Chem.* 2020, *321*, 126727.
78. Ono, H.; Chuda, Y.; Ohnishi-Kameyama, M.; Yada, H.; Ishizaka, M.; Kobayashi, H.; Yoshida, M. Analysis of acrylamide by LC-MS/MS and GC-MS in processed Japanese foods. *Food Addit. Contam.* 2003, *20*, 215–220.
79. Chafi, S.; Ballesteros, E. A simple, efficient, eco-friendly sample preparation procedure for the simultaneous determination of hormones in meat and fish products by gas chromatography—mass spectrometry. *Foods*. 2022, *11*, 3095.
80. Xiao, Y.; Liu, S.; Gao, Y.; Zhang, Y.; Zhang, Q.; Li, X. Determination of antibiotic residues in aquaculture products by liquid chromatography tandem mass spectrometry: recent trends and developments from 2010 to 2020. *Separations*. 2022, *9*, 35.
81. Rodriguez-Saona, L. E.; Allendorf, M. E. Use of FTIR for rapid authentication and detection of adulteration of food. *Annu. Rev. Food Sci. Technol.* 2011, *2*, 467–483
82. Adu, J. K.; Fafanyo, D.; Orman, E.; Ayensu, I.; Amengor, C. D.; Kwofie, S. Assessing metal contaminants in milled maize products available on the Ghanaian market with atomic absorption spectrometry and instrumental neutron activation analyser techniques. *Food Control*. 2020, *109*, 106912.
83. Teye, E.; Huang, X. Y.; Afoakwa, N. Review on the potential use of near infrared spectroscopy (NIRS) for the measurement of chemical residues in food. *Am. J. Food Sci. Technol.* 2013, *1*, 1–8.
84. Bhavadharini, B.; Kavimughil, M.; Malini, B.; Vallath, A.; Prajapati, H. K.; Sunil, C. K. Recent advances in biosensors for detection of chemical contaminants in food—a review. *Food Anal. Methods.* 2022, *15*, 1545–1564.
85. Chobtang, J.; De Boer, I. J.; Hoogenboom, R. L.; Haasnoot, W.; Kijlstra, A.; Meerburg, B. G. The need and potential of biosensors to detect dioxins and dioxin-like polychlorinated biphenyls along the milk, eggs and meat food chain. *Sensors*. 2011, *11*, 11692–11716.
86. McGrath, T. F.; Elliott, C. T.; Fodey, T. L. Biosensors for the analysis of microbiological and chemical contaminants in food. *Anal. Bioanal. Chem.* 2012, *403*, 75–92.
87. Moro, G.; De Wael, K.; Moretto, L. M. Challenges in the electrochemical (bio) sensing of nonelectroactive food and environmental contaminants. *Curr. Opin. Electrochem.* 2019, *16*, 57–65.
88. Naresh, V.; Lee, N. A review on biosensors and recent development of nanostructured materials-enabled biosensors. *Sensors*. 2021, *21*, 1109.
89. Galar, D.; Kumar, U. Sensors and data acquisition. In *eMaintenance: essential electronic tools for efficiency*, Academic Press, 2017, pp. 1–72. https://doi.org/10.1016/B978-0-12-811153-6.00001-4.

90. Malhotra, B. D.; Ali, A. Biopolymeric nanostructures: biosensors and bioimaging. In *Nanomaterials for biosensors, fundamentals and applications micro and nano technologies*, Elsevier, 2018, pp. 127–144. 10.1016/B978-0-323-44923-6.00001-7
91. Delory, E.; Toma, D.M.; del Rio Fernandez, J.; Cervantes, P.; Ruiz, P.; Memè, S. Ocean in situ sensors crosscutting innovations. In *Challenges and innovations in ocean in situ sensors measuring inner ocean processes and health in the digital age*, Elsevier, 2019, pp. 117–171. https://doi.org/10.1016/C2015-0-06798-3
92. Balint, R.; Cassidy, N. J.; Cartmell, S. H. Conductive polymers: towards a smart biomaterial for tissue engineering. *Acta Biomater.* 2014, *10*, 2341–2353.
93. Namsheer, K.; Rout, C. S. Conducting polymers: a comprehensive review on recent advances in synthesis, properties and applications. *RSC Adv.* 2021, *11*, 5659–5697.
94. Thadathil, A.; Pradeep, H.; Joshy, D.; Ismail, Y. A.; Periyat, P. Polyindole and polypyrrole as a sustainable platform for environmental remediation and sensor applications. *Mater. Adv.* 2022, *3*, 2990–3022.
95. Kumar, D.; Sharma, R. C. Advances in conductive polymers. *Eur. Polym. J.* 1998, *34*, 1053–1060.
96. Chitte, H. K.; Bhat, N. V.; Walunj, V. E.; Shinde, G. N. Synthesis of polypyrrole using ferric chloride ($FeCl_3$) as oxidant together with some dopants for use in gas sensors. *J. Sens. Technol.* 2011, *1*, 47–56.
97. Sharma, S.; Sudhakara, P.; Omran, A. A. B.; Singh, J.; Ilyas, R. A. Recent trends and developments in conducting polymer nanocomposites for multifunctional applications. *Polymers.* 2021, *13*, 2898.
98. Armour, M.; Davies, A.G.; Upadhyay, J.; Wassermann, A. Coloured electrically conducting polymers from furan, pyrrole and thiophene. *J. Polym. Sci.* 1967, *5*, 1527–1538.
99. Peng, X.; Zhang, L.; Chen, Y.; Li, F.; Zhou, W. In situ preparation and fluorescence quenching properties of polythiophene/ZnO nanocrystals hybrids through atom-transfer radical polymerization and hydrolysis. *Appl. Surf. Sci.* 2010, *256*, 2948–2955.
100. Mostafaei, A.; Nasirpouri, F. Epoxy/polyaniline–ZnO nanorods hybrid nanocomposite coatings: synthesis, characterization and corrosion protection performance of conducting paints. Prog. *Org. Coat.* 2014, *77*, 146–159.
101. Qiu, J. D.; Shi, L.; Liang, R. P.; Wang, G. C.; Xia, X. H. Controllable deposition of a platinum nanoparticle ensemble on a polyaniline/graphene hybrid as a novel electrode material for electrochemical sensing. *Eur. J. Chem.* 2012, *18*, 7950–7959.
102. Baghayeri, M.; Zare, E. N.; Lakouraj, M. M. A simple hydrogen peroxide biosensor based on a novel electro-magnetic poly(p-phenylenediamine)@Fe_3O_4 nanocomposite. Biosens. *Bioelectron.* 2014, *55*, 259–265.
103. Gangopadhyay, R.; De, A. Conducting polymer nanocomposites: a brief overview. Chem. *Mater.* 2000, *12*, 608.
104. Rawat, N. K.; Ghosh, R. Conducting polymer-based nanobiosensors, nanosensors for smart cities. In B. Han, V.K. Tomer, T. A. Nguyen, A. Farmani & P. K. Singh (Eds.), *Nanosensors for Smart Cities.* Elsevier, 2020, pp. 129–142. https://doi.org/10.1016/B978-0-12-819870-4.00008-6
105. Zhang, Y.; Wu, L.; Lei, W.; Xia, X.; Xia, M.; Hao, Q. Electrochemical determination of 4-nitrophenol at polycarbazole/N-doped graphene modified glassy carbon electrode. *Electrochim. Acta.* 2014, *146*, 568–576.
106. Tang, Y.; Wu, N.; Luo, S.; Liu, C.; Wang, K.; Chen, L. One-Step electrodeposition to layer-by-layer graphene–conducting-polymer hybrid films. Macromol. *Rapid Commun.* 2012, *33*, 1780–1786.
107. Olad, A.; Azhar, F. F. Eco-friendly biopolymer/clay/conducting polymer nanocomposite: characterization and its application in reactive dye removal. *Fibers Polym.* 2014, *15*, 1321–1329.
108. Pati, S.; Kumar, V.; Goto, T.; Takahashi, T.; Yokozeki, T. Synthesis and characterization of PANI/P-2M conductive composites: thermal, rheological, mechanical, and electrical properties. *Polym. Compos.* 2019, *40*, 4321–4328.
109. Mostafaei, A.; Zolriasatein, A. Synthesis and characterization of conducting polyaniline nanocomposites containing ZnO nanorods. *Prog. Nat. Sci.: Mater. Int.* 2012, *22*, 273–280.
110. Anderson, P. E.; Badlani, R. N.; Mayer, J.; Mabrouk, P. A. Electrochemical synthesis and characterization of conducting polymers in supercritical carbon dioxide. J. *Am. Chem. Soc.* 2002, *124*, 10284–10285.
111. Maeda, S.; Armes, S. P. Surface area measurements on conducting polymer-inorganic oxide nanocomposites. *Synth. Met.* 1995, *73*, 151–155.

112. Modak, P.; Kondawar, S. B.; Nandanwar, D. V. Synthesis and characterization of conducting polyaniline/grapheme nanocomposites for electromagnetic interference shielding. *Proc. Mater. Sci.* 2015, *10*, 588–594.
113. Park, H.-W.; Kim, T.; Huh, J.; Kang, M.; Lee, J. E.; Yoon, H. Anisotropic growth control of polyaniline nanostructures and their morphology-dependent electrochemical characteristics. *ACS Nano.* 2012, *6*, 7624–7633.
114. Menegazzo, N.; Herbert, B.; Banerji, S.; Booksh, K. S. Discourse on the utilization of polyaniline coatings for surface plasmon resonance sensing of ammonia vapour. *Talanta.* 2011, *85*, 1369–1375.
115. Sadek, A. Z.; Wlodarski, W.; Shin, K.; Kaner, B. R.; Kalantar-zadeh, K. A layered surface acoustic wave gas sensor based on a polyaniline/In_2O_3nanofibre composite. *Nanotechnol.* 2006, *17*, 4488.
116. Virji, S.; Fowler, J. D.; Baker, C. O.;Huang, J.; Kaner, R. B.; Weiller, B. H. Polyaniline nanofiber composites with metal salts: chemical sensors for hydrogen sulfide. *Small.* 2005, *1*, 624–627.
117. Vijayakumar, N.; Subramanian, E.; Padiyan, D. P. Conducting polyaniline blends with the soft template poly(vinyl pyrrolidone) and their chemosensor application. *Int. J. Polym. Mater. Polym. Biomater.* 2012, *61*. https://doi.org/10.1080/00914037.2011.610054
118. Huang, X.; Hu, N.; Gao, R.; Yu, Y.; Wang, Y.; Yang, Z.; Kong, E. S.-W.; Weia, H.; Zhang, Y. Reduced graphene oxide–polyaniline hybrid: preparation, characterization and its applications for ammonia gas sensing. *J. Mater. Chem.* 2012, *22*, 22488–22495.
119. Qiu, J.-D.; Shi, L.; Liang, R.-P.; Wang, G.-C.; Xia, X.-H. Controllable deposition of a platinum nanoparticle ensemble on a polyaniline/graphene hybrid as a novel electrode material for electrochemical sensing. *Chem. Eur. J.* 2012, *18*, 7950–7959.
120. Kaushik, A.; Khan, R.; Gupta, V.; Malhotra, B. D.; Ahmad, S.; Singh, S. P. Hybrid cross-linked polyaniline-WO_3nanocomposite thin film for NOx gas sensing. *J. Nanosci. Nanotechnol.* 2009, *9*, 1792–1796.
121. Zampetti, E.; Pantalei, S.; Muzyczuk, A.; Bearzotti, A.; De Cesare, F.; Spinella, C.; Macagnano, A. A high sensitive NO2 gas sensor based on PEDOT–PSS/TiO2 nanofibres. Sens. *Actuators B Chem.* 2013, *176*, 390–398.
122. Sun, J.; Shu, X.; Tian, Y.; Tong, Z.; Bai, S.; Luo, R. Facile preparation of polypyrrole-reduced graphene oxide hybrid for enhancing NH_3 sensing at room temperature. *Sens. Actuators B Chem.* 2017, *241*, 658–64.
123. Aussawasathien, D.; Dong, J. H.; Dai, L. Electrospun polymer nanofiber sensors. *Synth. Met.* 2005, *154*, 37–40.
124. Fiorito, P. A.; de Torresi, S. I. C. Hybrid nickel hexacyanoferrate/polypyrrole composite as mediator for hydrogen peroxide detection and its application in oxidase-based biosensors. *J. Electroanal. Chem.* 2005, *581*, 31–7.
125. Baghayeri, M.; Zare, E. N.; Lakouraj, M. M. A simple hydrogen peroxide biosensor based on a novel electro-magnetic poly(p-phenylenediamine)@Fe_3O_4 nanocomposite. *Biosens. Bioelectron.* 2014, *55*, 259–65.
126. Liu, J.; Tian, S.; Knoll, W. Properties of polyaniline/carbon nanotube multilayer films in neutral solution and their application for stable low-potential detection of reduced β-nicotinamide adenine dinucleotide. *Langmuir.* 2005, *21*, 5596–9.
127. Feng, X.; Mao, C.; Yang, G.; Hou, W.; Zhu, J.-J. Polyaniline/Au composite hollow spheres: synthesis, characterization, and application to the detection of dopamine. *Langmuir.* 2006, *22*, 4384–9.
128. Qian, T.; Yu, C.; Zhou, X.; Wu, S.; Shen, J. Au nanoparticles decorated polypyrrole/reduced graphene oxide hybrid sheets for ultrasensitive dopamine detection. *Sens. Actuators B Chem.* 2014, *193*, 759–63.
129. Liu, Z.; Wang, J.;Xie, D.; Chen, G. Polyaniline-coated Fe_3O_4nanoparticle–carbon-nanotube composite and its application in electrochemical biosensing. *Small.* 2008, *4*, 462–6.
130. Yan, W.; Feng, X.; Chen, X.; Hou, W.; Zhu, J. J. A super highly sensitive glucose biosensor based on Au nanoparticles–AgCl@polyaniline hybrid material. *Biosens. Bioelectron.* 2008, *23*, 925–31.
131. Zhai, D.; Liu, B.; Shi, Y.; Pan, L.; Wang, Y.; Li, W.; Zhang, R.; Yu, G. Highly sensitive glucose sensor based on Pt nanoparticle/polyaniline hydrogel heterostructures. *ACS Nano.* 2013, *7*, 3540–3546.
132. Xu, Q.; Gu, S.-X.; Jin, L.; Zhou, Y.; Yang, Z.; Wang, W.; Hu, X. Graphene/polyaniline/gold nanoparticles nanocomposite for the direct electron transfer of glucose oxidase and glucose biosensing. *Sens. Actuators B Chem.* 2014, *190*, 562–9.

133. Dhand, C.; Arya, S. K.; Datta, M.; Malhotra, B. D. Polyaniline–carbon nanotube composite film for cholesterol biosensor. *Anal. Biochem*. 2008, *383*, 194–9.
134. Radhapyari, K.; Kotoky, P.; Das, M. R.; Khan, R. Graphene–polyaniline nanocomposite based biosensor for detection of antimalarial drug artesunate in pharmaceutical formulation and biological fluids. *Talanta*. 2013, *111*, 47–53.
135. Lin, C-H.; Lin, J-H.; Chen, C-F.; Ito, Y.; Luo, S-C. Conducting polymer-based sensors for food and drug analysis. *J Food Drug Anal*. 2021, *29*, https://doi.org/10.38212/2224-6614.3374
136. Nguyen, H. H.; Lee, S. H.; Lee, U. J.; Fermin, C, D.; Kim, M. Immobilized enzymes in biosensor applications. *Materials (Basel)*. 2019, *12*, 121.
137. Samtham, M.; Singh, D.; Hareesh, K.; Devan, R. S. Perspectives of conducting polymer nanostructures for high-performance electrochemical capacitors. *J. Energy Storage*. 2022, *51*, 104418.
138. Dong, Y. Z.; Choi, K.; Kwon, S. H.; Nam, J-D.; Choi, H. J. Nanoparticles functionalized by conducting polymers and their electrorheological and magnetorheological applications. *Polymers (Basel)*. 2020, *12*, 204.
139. Arshak, K.; Adley, C.; Moore, E.; Cunniffe, C.; Campion, M.; Harris, J. Characterisation of polymer nanocomposite sensors for quantification of bacterial cultures. *Sens. Actuators B Chem*. 2007, *126*, 226–31.
140. Xu, Y.; Ye, X.; Yang, L.; He, P.; Fang, Y. Impedance DNA biosensor using electropolymerized polypyrrole/multiwalled carbon nanotubes modified electrode. *Electroanalysis*. 2006, *18*, 1471–1478.
141. Khan, R.; Dhayal, M. Chitosan/polyaniline hybrid conducting biopolymer base impedimetric immunosensor to detect Ochratoxin-A. *Biosens. Bioelectron*. 2009, *24*, 1700–5.

6 Cellulose-Based Nanomaterials Solving the Food Crisis

Ajaz Ahmad Dar, Akhtar Hussain Malik, Arif Hassan Dar, Mohd Azhardin Ganayee, and Adil Majeed Rather
Corresponding author: Adil Majeed Rather

6.1 INTRODUCTION

The application of green, renewable, and sustainable materials has become progressively more important for producing various high-value products owing to their low or negligible impact on environment [1,2]. This area of research has attracted the interest of large numbers of people across industries and academics because these materials provide an alternative solution to the ever-depleting nonrenewable sources, environmental pollution, global warming, energy, and food crisis [3]. In this context, cellulose, starch, alginate, chitin, chitosan, and gelatin have been revealed to be promising candidates with regard to their abundant availability from various resources [4]. Among all of them, cellulose is by far the most abundant renewable material obtained from the biosphere, and it can be found in plants, algae, tunicates, and some bacteria [5].

Cellulose has several unique properties that make it desirable for sustainable product development. It is a strong, lightweight, and versatile material with excellent mechanical properties. It has a high tensile strength, making it suitable for reinforcing composite materials. Cellulose fibers can be processed into various forms, including films, fibers, aerogels, and nanoparticles, enabling their use in different industries [6]. Cellulose is one of the most important, abundant, environmentally friendly, bioresorbable, and sustainable biopolymers in nature [7,8]. A linear syndiotactic homopolymer consists of a linear chain from several hundreds to several thousands of repeating units of β-D-glucopyranose (n), covalently linked by 1–4-glycosidic bonds. (Figure 6.1). The number of repeating units n is highly dependent on the source of the original cellulose (e.g., 10,000 in natural wood, 15,000 in cotton, and 44,000 in the genus *Valonia*). Cellulose is generally known to be fibrillary and semicrystalline [9] and is obtained directly from natural plants such as wood, cotton, and hemp, or from microbes.

One of the key advantages of cellulose is its biocompatibility and nontoxic nature. It can be used in biomedical applications, such as drug delivery systems, tissue engineering scaffolds, and wound dressings. Cellulose-based materials can also be modified to improve their properties or to incorporate additional functionalities, further expanding their potential applications. In the past, cellulose-based materials have been widely used in numerous fields, including oil/water separation [10–16], sustained drug delivery [17,18], CO_2 capture [19–21], and energy applications [22–24]. Rapid advances in the field of nanotechnology have opened up prospects for cellulose. The nanoscale structures of cellulose (e.g., nanofibrils or nanocrystals) offer a unique combination of properties including flexible surface chemistry, transparency, low thermal expansion, high elasticity, renewability, sustainability, chemical stability, and economy [25,26]. Cellulose has gained

DOI: 10.1201/9781003389408-6

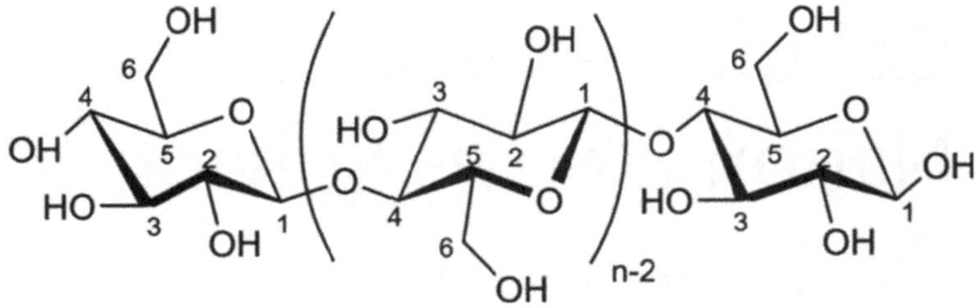

FIGURE 6.1 Chemical structure of cellulose.

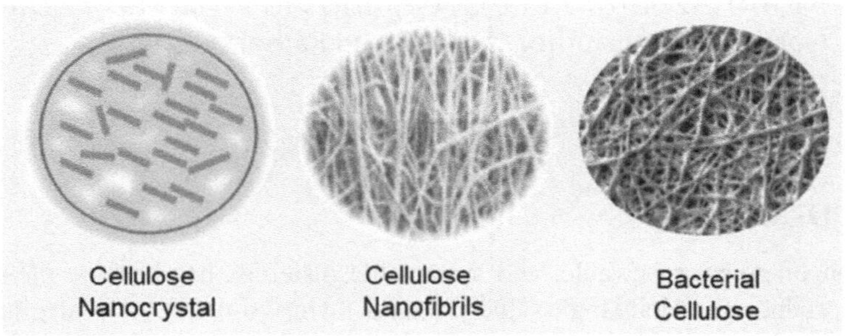

FIGURE 6.2 Different types of cellulose-based nanomaterials.

prominence in the form of nanocellulose that can be used as advanced and novel biomaterials owing to its unusual qualities. Cellulosic nanomaterials (CNMs) are the prime examples of green, nontoxic nanomaterials that are obtained from renewable resources. These nanomaterials exhibit exceptional mechanical strength, light weight, and biocompatibility. The terminology is currently lacking standardization; however, nanocellulose is typically classified into three types based on the method of preparation and morphological characteristics (Figure 6.2). These types are referred to as CNC (cellulose nanocrystals), CNF (cellulose nanofibrils), and BC (bacterial cellulose) [27]. CNC are rod-like in shape, having a width of 3–5 nm and a length of 50–500 nm [28]. CNCs are commonly produced by a strong acid treatment (e.g., with hydrochloric, sulfuric, or phosphoric acids) of fibrils [29]. CNF typically has a width of 4–20 nm and a length between 500 nm and few micrometers [30]. When comparing CNCs to CNFs, CNCs can be likened to nanoscale rice-like structures, while CNFs resemble either uncooked spaghetti (with a low aspect ratio of less than 100) or cooked spaghetti (with a high aspect ratio exceeding 300).

Nanostructured cellulose fibrils of another type similar to CNFs are known as cellulose microfibrils (CMFs). CMFs are cellulose fibrils that measure 10 to 100 nanometers in width and can be several tens of micrometers long [31,32]. BC is biosynthesized by specific bacteria with many characteristics differing from CNC and CNF. BC is composed of ribbon-like fibers with a width in the range 0.01–0.10 m and a length between 100 nm to the micron level, and the fibers interweave to form a 3D network [33]. The distinctive characteristics of BC, including its uniform structure, high crystallinity, and purity, along with the absence of other polysaccharides commonly found in other types of nanocelluloses, have generated considerable interest in the field of sustainable development. These unique properties of BC, such as its remarkable mechanical strength (e.g., modulus > 100 GPa, a value comparable to steel), have motivated researchers to explore its potential applications. The characteristics and sources of various types of nanocellulose are presented in Table 6.1.

Cellulose-based nanomaterials have the potential to contribute to the solution of the global food crisis in a number of ways. One possibility is the use of cellulose nanomaterials as food additives to

TABLE 6.1
Types and sources of nanocellulose

Type	Source	Average size
Cellulose nanocrystals (CNCs)	Wood, cotton, straw	Width: 50–70 nm Length: 100 nm–10 μm
Cellulose nanofibrils (CNFs)	Straw, wood, cotton, tubers	Width: 50–60 nm Length: 1–10 μm
Cellulose nanowhiskers (CNWs)	Cotton, wood, hemp	Width: 2–60 nm Length: 100–500 nm
Bacterial nanocellulose (BC)	Sugars, alcohols	Width: 5–70 nm Length: 1–10 μm

improve the nutritional value of foods. For example, cellulose nanocrystals have been shown to have a high water-holding capacity, which can be used to improve the moisture content and texture of food products [34]. They can also be used as a replacement for fat in food products, which can help to reduce caloric content and improve the nutritional profile of the food [35]. Another potential use of cellulose nanomaterials in the food industry is the development of food packaging materials that are more sustainable and biodegradable [34]. Traditional food packaging materials, such as plastics, can take hundreds of years to decompose and contribute significantly to the problem of plastic pollution. Cellulose nanomaterials, on the other hand, can be used to create packaging materials that are biodegradable and have a much lower environmental impact. In addition, cellulose-based nanomaterials have the potential to be used in food safety applications, such as the development of antimicrobial films and coatings for food packaging. These films and coatings could help to prevent the growth of harmful bacteria on the surface of food products [36,37], reducing the risk of food-borne illness. Overall, the use of cellulose nanomaterials in the food industry has the potential to improve the nutritional value of foods, reduce the environmental impact of food production and packaging, and help to address the global food crisis.

6.2 SOURCE AND STRUCTURE OF CELLULOSE AND ITS DERIVED NANOCELLULOSE

Cellulose can be isolated from a wide variety of sources from the plant and animal kingdoms. The properties and size of the cellulose thus obtained depend directly on the source isolated. Broadly, there are four different polymorphs of cellulose, I, II, III, and IV, which can be transformed from one to another by chemical and thermal treatments. Cellulose I is a native cellulose, which is found into two allomorphs, I_α and I_β.

Numerous plants and plant-based materials have been recognized as highly promising sources for extracting cellulose and nanocellulose. These sources encompass a wide range of options, including both hardwood and softwood species, as well as various agricultural crops and residues. For instance, hardwood and softwood trees, such as oak and pine, are known to contain abundant cellulose that can be utilized for extraction. Additionally, agricultural crops like sugar beet, cotton, and potato tubers are potential sources of cellulose. Agricultural residues, including soybean stalks, banana rachis (stalk), sisal, rice husk, hemp, flax, and coconut husk, have also been investigated for their cellulose content. Researchers have explored different extraction methods and processing techniques to obtain cellulose and nanocellulose from these materials. By harnessing the cellulose-rich components of these plants, the aim is to develop sustainable and ecofriendly alternatives to conventional materials. The utilization of these abundant plant resources not only offers a renewable and environmentally friendly solution but also contributes to reducing dependency on nonrenewable resources and to mitigating environmental challenges [38]. Wood is a natural composite material abundant in cellulose, hemicelluloses, nanocellulose, and lignin. Its porous and anisotropic structure combines low density, high strength, toughness, and stiffness. Cellulose provides strength, while hemicelluloses offer flexibility and adhesion, nanocellulose enhances properties at the nanoscale, and lignin acts as a binder and increases durability. Wood's porosity contributes to its low density and efficient resistance to forces. Its anisotropic nature adds strength and toughness. These properties make wood versatile and ecofriendly in construction, furniture, paper production, and more [39]. A high-quality source material of cellulose and nanocellulose are cotton fibers because of their relatively low noncellulosic components [40]. The isolation of nanocellulose involves vigorous chemical and/or mechanical processes, through multistep process from the wood.

The animal sources of nanocellulose include some tunicates (marine invertebrates of a group that includes sea squirts), which produce cellulose in the outer tissue, named the tunic, from which a purified cellulose portion called tunicin can be obtained. Tunicin, a natural polysaccharide found in tunicates, is known for its high crystallinity and composition of pure CI_β allomorph cellulose. The cellulose in tunicin forms nanofibrils with a large aspect ratio, providing mechanical strength and structural integrity. These nanofibrils of tunicate cellulose possess a remarkably high specific surface area of 150–170 m^2g^{-1} [41,42]. The combination of high crystallinity, CI_β allomorph type, large aspect ratio (60–70), and high specific surface area makes tunicin cellulose a unique and potentially valuable material. Tunicin cellulose holds promise for various technological and biomedical applications due to its exceptional properties. Various species of algae have also been considered as sources of cellulose and nanocellulose. Microfibrils of cellulose obtained from algae cell wall with high aspect ratio (>40) can be processed through acid hydrolysis and mechanical refining [43]. The structures of microfibrils of cellulose isolated from *Valonia* algae are primarily of I_α crystalline type with a square cross section (20 × 20 nm). Also, *M. denticulate* microfibrils are primarily of the CI_β crystalline type with a rectangular cross section (5 × 20–30 nm) [44]. Another source of cellulose is certain types of bacteria in which cellulose is a byproduct of the primary metabolic processes. The extensively used bacterial species *Gluconacetobacter xylinus* produces a thick gel composed of microfibrils of cellulose and 97–99% water. The bacterial cellulose crystallites have a degree of polymerization typically between 2000 and 6000 and are primarily of the CI_α crystalline type. In comparison to plant cellulose, bacterial cellulose is of high chemical purity, devoid of hemicelluloses and lignin. The microfibril formation and crystallization can be altered by adjusting the culturing conditions, acid concentration, temperature, sonication duration, and reaction time of the bacteria. However, celluloses obtained from both sources (bacteria and various plants) have similar molecular structures [45]. Cellulose is a semicrystalline polycarbohydrate composed of a repeated unit of two anhydroglucose rings combined through oxygen, which is covalently bonded with one glucose ring C1 and C4 of the adjacent glucose ring (β-1,4-glycosidic bonds). The degree of polymerization *n* can be up to 20,000, where *n* is dependent on the source of cellulose material. The presence of hydrogen bonding between the hydrogen atom of the C-3 hydroxyl group of one unit and oxygen atoms of the adjacent ring molecules stabilizes the linkage between them, causing the linear conformation of the cellulose chain.

The combination of intramolecular and intermolecular hydrogen bonds in cellulose crystallites creates a rigid structure. These bonds strengthen the cellulose constituents and enhance their stability, resisting water and certain chemicals. This property maintains cellulose's structural integrity and protects against degradation, and the combination of hydrogen bonds in cellulose results in a rigid, water-repellent nature. For example, acid hydrolysis removes the noncrystalline part, thereby yielding cellulose fragments with improved crystallinity [46]. Additionally, exposure to concentrated alkali or other solvents, and even rigorous mechanical grinding, results in a decrease in the crystallite size and crystallinity. Currently, a two-phase model is used to describe the structural organization of cellulose, which consists of crystalline and noncrystalline domains [47]. However, investigations have revealed that the presence of paracrystalline portions on the crystalline surfaces provides an improved model of cellulose structure. The fundamental constituents of long and thin elementary nanofibrils and their microfibrils are alternated nanocrystallites, along with the nanoscale noncrystalline domains.

The lateral sizes of elementary nanofibrils vary based on the source of cellulose. For natural cellulose from plants and woods, sizes range from 3 to 4 nm, while *Valonia algae* cellulose can be 10–15 nm. *Valonia algae* cellulose nanofibrils can reach lengths of several microns. Factors like extraction method and processing conditions influence the exact dimensions. These variations impact the properties of cellulose-based materials [48]. Recently, a more detailed model of the supermolecular structure of natural cellulose includes the elementary nanofibril constructed from orientated nanocrystallites and noncrystalline nanodomains arranged along the fibril. A thin paracrystalline layer is also located on the surface of the crystalline core, and the crystallites can contain local defects caused by the ends of the chains. The proposed model simplifies explanation to various chemical, biochemical, and physicomechanical properties of natural cellulose. For instance, it has been observed that the noncrystalline domains are weak and constitute available points on the elementary fibrils. Therefore, in processes like acidic and enzymatic hydrolysis, alcoholysis, and acetolysis, glycosidic bonds are cleaved, leading to longitudinal fibril splitting and the release of crystalline particles. These released nanocrystallites share lateral dimensions with elementary nanofibrils but vary in length (50–200 nm). The study of nanocrystallites is crucial for understanding cellulose behavior and potential applications due to their unique characteristics, including high mechanical strength, stiffness, and optical transparency. These properties make nanocrystallites appealing for diverse fields such as materials science, nanotechnology, and biomedicine.

6.3 PREPARATION AND CHARACTERISTICS OF NANOCELLULOSE

The strong mechanical disintegration of the plant cell wall degrades the original structure of cellulose fiber to nanofibrils or their microfibril bundles within the 10–100 nm diameter range, depending on the disintegration power. Different techniques are to be used for extraction of nanofibrils or microfibril from various materials and substances, for example, homogenization, microfluidization, grinding, cryocrushing, and ultrasonication, as shown in Figure 6.3.

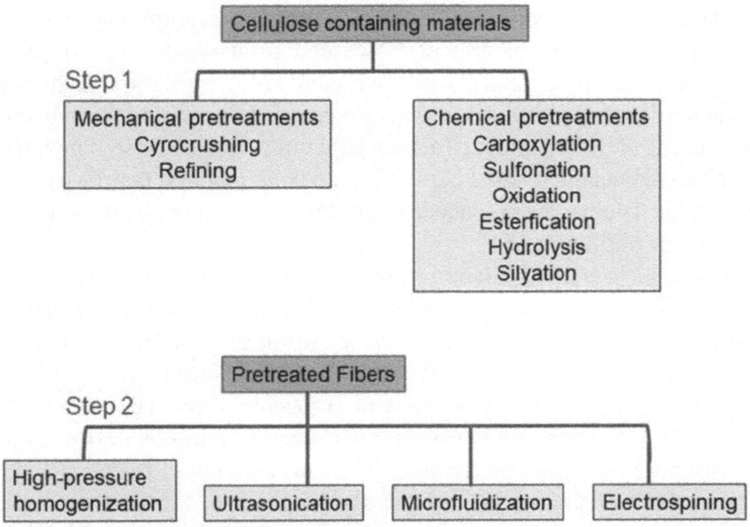

FIGURE 6.3 Steps involved in the preparation of nanocellulose.

The homogenization method is widely used for both large-scale and laboratory-scale fabrication of nanofibrils. Homgenization is a mechanical process that involves reducing the size via various mechanisms such as pressure drop, shear fore, turbulent flow, and interparticle collisions. During homogenization, a sample containing cellulose fibers is passed through a homogenizer machine, which subjects the material to high pressure and shear force. These forces cause the fibers to undergo extensive mechanical disruotion and fibrillation, resulting in the formation of nanofibrils. The extent of cellulose fibrillation achived through homogenization depends mainly on two factors: the number of homogenization cycles and the applied pressure. Increasing the number of cycles allows for further disruption and fibrillation of the fibers, resulting in finer nanofibrils. The higher applied pressure enhances the efficiency of disruption pass through the machine, and leads to intense shear force and increased fibrillation [49].

However, some issues may occur during the production of nanofibrillated cellulose from pulp, such as insufficient disintegration and clogging issues in the homogenizer when pumping pulp through a small orifice. In order to ensure smoother processing in the homogenizer, various mechanical pretreatments are commonly employed before homogenization such as grinding, milling, refining, cryocrushing, or ultrasonication [50]. By employing these mechanical pretreatments before homogenization, the pulp fibers can be adequately fabricated for processing through a small orifice. These methods facilitate better disintegration, reduce clogging, and improve the overall efficiency of the homogenization process. However, the high energy consumption, prior chemical treatments using acid hydrolysis, oxidation, enzymatic hydrolysis, and certain other pretreatment methods cause excessive damage to the crystalline structure of the cellulose nanofibrils [51].

Another gadget that uses a constant shear rate is a microfluidizer. Pumping the liquid slurry through a Z-shaped chamber causes it to experience strong shear forces. Lee et al. examined the impact of how many times a microfluidizer passed a slurry of microcrystalline cellulose on the morphology of the resulting nanofibrils [52]. They discovered that after 10–15 passing cycles, the aspect ratio of the nanofibrillar bundles rose; however, an additional 20 passes caused the nanofibrils to aggregate due to an increase in surface area and surface hydroxyl group concentration. To create nanofibrils, three different kinds of cellulose pulp from empty palm fruit bunches were refined and microfluidized [53]. Morphological characterization showed that microfluidization could produce nanofibers with a more uniform size distribution. Another technique for separating cellulose fibers into nanosized fibrils is grinding. The fibrillation mechanism in the grinder utilizes shear forces to degrade the cell wall structure and individualize the nanoscale fibrils [32]. The extent of fibrillation is dependent on the distance between the disks, the morphology of the disk channels, and the number of passes through the grinder. However, a primary advantage of grinder processing is that additional mechanical pretreatments are not required [49]. Furthermore, the heat produced by friction during the fibrillation process led to water evaporation, increasing the pulp consistency from 2% to 3.2%. As a result of the grinding, two main structures were disclosed: first, untwisted fibrils, and second, twisted and entangled nanofibers. Iwamoto et al. reported that 14 passes were required for sufficiently fibrillated pulp to be produced, while extra cycles of up to 30 passes for the pulp fiber did not promote further fibrillation [54].

Cryocrushing is a mechanical fibrillation technique that is used to break down frozen cellulose. In this method, the cellulose material is frozen at cryogenic temperatures before being mechanically divided into smaller fibrils. The resulting cryocrushed fibrils are typically between 0.1 and 1.0 μm in diameter [55]. Cryocrushing starts with freezing the cellulose material to very low temperatures, frequently using liquid nitrogen or another type of cryogenic agent. The cellulose becomes more fragile and prone to breaking when it is frozen. The frozen cellulose is next treated mechanically, usually with a crushing or grinding mechanism. The frozen cellulose fractures and fragments into smaller fibrils as a result of the mechanical pressure applied to it [56]. The application of high-pressure forces to the frozen cellulosic fibers leads to rupturing of the cell walls due to the pressure exerted by the ice crystals and liberates the nanofibers [32]. The cryocrushed fibers then disperse uniformly in water using a disintegrator. This procedure is applicable to various cellulose materials before homogenization. Wang and Sain used cryocrushing and high-pressure fibrillation to create nanofibers from soybean stock [57]. In comparison to water, the nanofibers thus produced had a better ability to disperse in acrylic emulsion. However, due to its significant energy consumption, the cryocrushing method is expensive and has low productivity.

High-intensity ultrasonication has also been used to isolate nanofibers from a variety of cellulosic materials, including pure cellulose, pulp, banana peels, rice debris, and microfibrillated cellulose [58–60]. The findings demonstrate that ultrasonication of the cellulose samples can produce a mixture of microscale and nanoscale fibrils. The diameters of the fibrils thus obtained range widely from 20 nm to several microns, indicating that some nanofibrils are peeled from the fibers while others remain on the fiber surface [61]. As a result, this approach produces aggregated fibrils with a wide variation of widths. Additionally, it has been discovered that ultrasonic processing can change the crystalline structure of some cellulose fibers. Wang and Cheng examined how various cellulose fibers' degree of fibrillation was affected by temperature, concentration, power, size, time, and distance from the probe tip [62]. Longer fibers were shown to be less defibrillated, and higher power and temperature were found to generate superior fibrillation. Larger distances between the probe and the beaker and higher pulp concentrations were not beneficial for fibrillation. For the manufacture of nanocellulose, it was discovered that blending and high-intensity ultrasonication treatments worked best. All ultrasound techniques employ a lot of energy and can significantly reduce the amount of nanofibrils produced and the length of the fibrils.

6.4 STRATEGIES TO OVERCOME THE GLOBAL FOOD CRISIS

Many scholars and decision-makers with an interest in global food security are worried that humanity will experience serious problems in the coming years. Global food consumption is changing because of population increase and economic disparity, and the production of food could become more challenging and expensive due to climate change, fluctuating energy prices, soil degradation, and water scarcity. To address the problems of food security, it is necessary to manage growing food shortages for a population that is always growing globally. Between now and 2050, the world's population is expected to increase by 1.7 billion people, putting increasing strain on the limited resources utilized to produce our food. The current model of an expanding population dependent on limited resources is obviously unsustainable, which highlights how crucial it is to strive for resource efficiency within a circular economy. A number of recent studies reveal that anywhere between 30 and 50% of food produced is never consumed and inevitably goes to waste [63]. Food waste is a serious problem with significant economic ramifications. US Americans squander an astounding 400 pounds (180 kg) of food yearly, which costs the country USD 218 billion (1.3% of GDP) on average. Similar to this, each European person wastes 173 kg of food annually, costing an estimated EUR 143 billion in 2012. Food waste is a problem in Canada as well; the country loses about CAD 25 billion in food annually, or around 2% of its GDP. These numbers demonstrate the substantial financial losses brought about by food waste. Comprehensive approaches that address challenges of food security while also delivering possible benefits for the economy and the environment are essential to solving this problem. These approaches should center on production, distribution, and consumption patterns [64,65]. There are several strategies that can be implemented to help overcome global food crises:

- Improving agricultural productivity: This can be done through the use of modern technologies such as precision farming, genetically modified crops, and drip irrigation to increase crop yields and reduce waste [66].
- Investing in infrastructure: Improving transportation and storage systems can help to reduce food waste and make it easier to distribute food to areas in need [67].

- Promoting sustainable agriculture: Using sustainable farming practices can help to preserve natural resources and reduce the negative environmental impacts of agriculture [68].
- Reducing food waste: It is estimated that up to one third of all food produced in the world is wasted. Reducing food waste can help to ensure that more food is available to feed the growing global population [69].
- Increasing international aid: Providing financial and logistical support to countries facing food insecurity can help to alleviate the effects of food crises [70].
- Strengthening food systems: This can be done through the development of more efficient supply chains, the promotion of local agriculture, and the establishment of safety nets and other programs to protect the most vulnerable members of society [71].
- International cooperation: Addressing global food insecurity requires cooperation among different countries and organizations. This can include sharing information and resources, as well as coordinating efforts to increase food production, reduce food waste, and improve food distribution [72].
- Supporting smallholder farmers: Supporting smallholder farmers with better access to resources like seeds, fertilizer, and credit can help them to increase their crop yields and improve their livelihoods [73].

6.5 NANOCELLULOSE FOR OVERCOMING THE FOOD CRISIS

The mass production of nanocellulose from natural resources has been established as a primary protocol to design new efficient organic molecules [74–76]. A large number of organic molecules have been reported and explored in biosystems to overcome overall food crises [77]. However, organic chemists have long pursued different stimuli-specific strategies for activating molecules and enabled their applications in controlled drug release, sensing, imaging, and signal amplification. Among the applications, developing systems by consuming nanocellulose using small organic molecules has a long history and has been extensively implemented and explored over time. Nanocellulose displays pseudoplastic behavior, which is akin to some gels or fluids with viscosity under typical circumstances. A feature known as pseudoplasticity describes a material's ability to flow more easily as shear stress or strain rate increases. When nanocellulose is subjected to shear forces, such as churning or applying tension, it might act like a gel or fluid with increased viscosity at rest but become less viscous [78]. This property is also known as thixotropy. As a noncaloric stabilizer and gellant in food applications, nanocellulose is extremely intriguing due to its high viscosity at low nanocellulose concentrations [79]. In a variety of culinary products, nanocellulose can be utilized as a thickener, flavor carrier, and suspension stabilizer [80], and is useful for producing filings, crushes, chips, soups, wafers, and puddings [78,81,82]. The rheological behavior of nanocellulose gel has been recognized as a particularly interesting application for nanocellulose in the food industry [80,83–85]. Overall, the use of cellulose nanomaterials in the food industry has the potential to improve the nutritional value of foods, reduce the environmental impact of food production and packaging, and help to address the global food crisis.

6.6 CONCLUSIONS AND FUTURE PROSPECTS

Academics and policymakers alike are paying attention to the issues surrounding global food security because they understand the potentially serious problems facing humanity in the near future. Population growth and economic inequality have increased food demand globally, putting further pressure on agricultural systems. However, a number of problems, such as climate change, fluctuating energy prices, soil erosion, and water scarcity, further endanger our capacity to meet that demand. Together, these difficulties make food production more difficult and more costly. This chapter focuses on possible approaches to the world food issue, with a particular emphasis on the benefits of nanocellulose in the field of food technology. The potential for innovation in nanocellulose, a cellulose substance made at the nanoscale from plant sources, is discussed. It is a fascinating contender for a variety of uses in the food sector due to its special qualities and adaptability. The chapter goes into the origins, makeup, and fabrication processes of nanocellulose, giving readers an understanding of the variety of raw materials from which this substance can be made. Readers acquire a clearer comprehension of the processes involved in producing nanocellulose by learning more about these characteristics. The chapter also looks at how nanocellulose might be used in the field of food technology. Due to its pseudoplastic characteristic, which causes viscosity to drop with rising shear stress, it can be used as a thickening agent, coating, or gel. The production, preservation, and general quality of food could all be improved by these characteristics. The chapter not only discusses the benefits of nanocellulose but also expands on those benefits by exploring more general tactics and solutions needed to address the world food crisis. It acknowledges the significance of managing growing food shortages due to an expanding global population. The chapter suggests that nanocellulose can help overcome the difficulties in achieving global food security by incorporating it and its potential applications in creative ways. The goal of this chapter is to reach a wider audience of people who are interested in learning about and realizing the promise of nanocellulose for resolving the world food crisis. This includes academics, decision-makers in government, researchers, and members of the business community who are looking for information on new technologies and approaches to creating resilient and sustainable food systems.

REFERENCES

[1] J.K.Pandey,K.R.Reddy,A.K.Mohanty,andM.Misra,*HandbookofPolymernanocomposites:Processing, Performance and Application*. Springer, Berlin, 2013.

[2] H. Kargarzadeh *et al.*, "Recent developments in nanocellulose-based biodegradable polymers, thermoplastic polymers, and porous nanocomposites," *Progress in Polymer Science*, vol. 87, pp. 197–227, 2018.

[3] D. Trache *et al.*, "Nanocellulose: from fundamentals to advanced applications," *Frontiers in Chemistry*, vol. 8, pp. 392, 2020.

[4] D. Trache, A. F. Tarchoun, M. Derradji, O. Mehelli, M. H. Hussin, and W. Bessa, "Cellulose fibers and nanocrystals: preparation, characterization, and surface modification," in V. Kumar, P. Guleria, N. Dasgupta, and S. Ranjan (eds), *Functionalized Nanomaterials I*. CRC Press, Boca Raton, 2020, pp. 171–190.

[5] D. Trache, K. Khimeche, A. Mezroua, and M. Benziane, "Physicochemical properties of microcrystalline nitrocellulose from Alfa grass fibres and its thermal stability," *Journal of Thermal Analysis and Calorimetry*, vol. 124, no. 3, pp. 1485–1496, 2016.

[6] B. Seantier, D. Bendahou, A. Bendahou, Y. Grohens, and H. Kaddami, "Multi-scale cellulose based new bio-aerogel composites with thermal super-insulating and tunable mechanical properties," *Carbohydrate Polymers*, vol. 138, pp. 335–348, 2016.

[7] A. J. Sayyed, L. V. Mohite, N. A. Deshmukh, and D. V. Pinjari, "Swelling kinetic study with mathematical modeling of cellulose pulp in aqueous N-methyl-morpholine-N-oxide solution," *Reaction Kinetics, Mechanisms and Catalysis*, vol. 133, no. 1, pp. 101–115, 2021.

[8] A. J. Sayyed, N. A. Deshmukh, and D. V. Pinjari, "A critical review of manufacturing processes used in regenerated cellulosic fibres: viscose, cellulose acetate, cuprammonium, LiCl/DMAc, ionic liquids, and NMMO based lyocell," *Cellulose,* vol. 26, no. 5, pp. 2913–2940, 2019.

[9] I. M. Saxena, and R. M. Brown Jr, "Cellulose biosynthesis: current views and evolving concepts," *Annals of Botany,* vol. 96, no. 1, pp. 9–21, 2005.

[10] A. M. Rather, N. Jana, P. Hazarika, and U. Manna, "Sustainable polymeric material for the facile and repetitive removal of oil-spills through the complementary use of both selective-absorption and active-filtration processes," *Journal of Materials Chemistry A*, vol. 5, no. 44, pp. 23339–23348, 2017.

[11] A. Shome, A. M. Rather, and U. Manna, "Chemically reactive protein nanoparticles for synthesis of a durable and deformable superhydrophobic material," *Nanoscale Advances,* vol. 1, no. 5, pp. 1746–1753, 2019.

[12] A. Shome *et al.*, "Design of a waste paper-derived chemically 'reactive' and durable functional material with tailorable mechanical property following an ambient and sustainable chemical approach," *Chemistry–An Asian Journal,* vol. 16, no. 14, pp. 1988–2001, 2021.
[13] Z. Li, L. Zhong, T. Zhang, F. Qiu, X. Yue, and D. Yang, "Sustainable, flexible, and superhydrophobic functionalized cellulose aerogel for selective and versatile oil/water separation," *ACS Sustainable Chemistry & Engineering*, vol. 7, no. 11, pp. 9984–9994, 2019.
[14] F.-P. Wang *et al.*, "Sustainable, superhydrophobic membranes based on bacterial cellulose for gravity-driven oil/water separation," *Carbohydrate Polymers*, vol. 253, 117220, 2021.
[15] G. Wang *et al.*, "A cellulose sponge with robust superhydrophilicity and under-water superoleophobicity for highly effective oil/water separation," *Green Chemistry*, vol. 17, no. 5, pp. 3093–3099, 2015.
[16] S. Zhou, P. Liu, M. Wang, H. Zhao, J. Yang, and F. Xu, "Sustainable, reusable, and superhydrophobic aerogels from microfibrillated cellulose for highly effective oil/water separation," *ACS Sustainable Chemistry & Engineering*, vol. 4, no. 12, pp. 6409–6416, 2016.
[17] A. M. Rather, S. Mahato, K. Maji, N. Gogoi, and U. Manna, "'Reactive' nano-complex coated medical cotton: a facile avenue for tailored release of small molecules," *Nanoscale,* vol. 9, no. 42, pp. 16154–16165, 2017.
[18] A. Shome, A. M. Rather, A. Ghosal, B. K. Bhunia, B. B. Mandal, and U. Manna, "Rational chemical engineering in natural protein derived functional interface," *ACS Sustainable Chemistry & Engineering,* vol. 7, no. 8, pp. 7502–7509, 2019.
[19] A. M. Rather, P. Srikrishnarka, A. Baidya, A. Shome, T. Pradeep, and U. Manna, "Evaluating the impact of tailored water wettability on performance of CO_2 capture," *ACS Applied Energy Materials,* vol. 3, no. 11, pp. 10541–10549, 2020.
[20] S. Geng, J. Wei, S. Jonasson, J. Hedlund, and K. Oksman, "Multifunctional carbon aerogels with hierarchical anisotropic structure derived from lignin and cellulose nanofibers for CO_2 capture and energy storage," *ACS Applied Materials & Interfaces,* vol. 12, no. 6, pp. 7432–7441, 2020.
[21] Y. Miao *et al.*, "A facile method for in situ fabrication of silica/cellulose aerogels and their application in CO_2 capture," *Carbohydrate Polymers,* vol. 236, pp. 116079, 2020.
[22] X. Wang, C. Yao, F. Wang, and Z. Li, "Cellulose-based nanomaterials for energy applications," *Small,* vol. 13, no. 42, pp. 1702240, 2017.
[23] H. Zhu *et al.*, "Wood-derived materials for green electronics, biological devices, and energy applications," *Chemical Reviews,* vol. 116, no. 16, pp. 9305–9374, 2016.
[24] R. Sabo, A. Yermakov, C. T. Law, and R. Elhajjar, "Nanocellulose-enabled electronics, energy harvesting devices, smart materials and sensors: a review," *Journal of Renewable Materials,* vol. 4, no. 5, pp. 297–312, 2016.
[25] M. Jonoobi, J. Harun, A. P. Mathew, M. Z. B. Hussein, and K. Oksman, "Preparation of cellulose nanofibers with hydrophobic surface characteristics," *Cellulose,* vol. 17, no. 2, pp. 299–307, 2010.
[26] E. Espino-Pérez, S. Domenek, N. Belgacem, C. Sillard, and J. Bras, "Green process for chemical functionalization of nanocellulose with carboxylic acids," *Biomacromolecules,* vol. 15, no. 12, pp. 4551–4560, 2014.
[27] A. Blanco, M. C. Monte, C. Campano, A. Balea, N. Merayo, and C. Negro, "Nanocellulose for industrial use: cellulose nanofibers (CNF), cellulose nanocrystals (CNC), and bacterial cellulose (BC)," in *Handbook of Nanomaterials for Industrial Applications*: Elsevier, 2018, pp. 74–126.
[28] O. M. Vanderfleet, and E. D. Cranston, "Production routes to tailor the performance of cellulose nanocrystals," *Nature Reviews Materials,* vol. 6, no. 2, pp. 124–144, 2021.
[29] Y. Habibi, L. A. Lucia, and O. J. Rojas, "Cellulose nanocrystals: chemistry, self-assembly, and applications," *Chemical Reviews,* vol. 110, no. 6, pp. 3479–3500, 2010.
[30] J. Li *et al.*, "Emerging food packaging applications of cellulose nanocomposites: a review," *Polymers,* vol. 14, no. 19, pp. 4025, 2022.
[31] S. H. Osong, S. Norgren, and P. Engstrand, "Processing of wood-based microfibrillated cellulose and nanofibrillated cellulose, and applications relating to papermaking: a review," *Cellulose*, vol. 23, no. 1, pp. 93–123, 2016. doi: 10.1007/s10570-015-0798-5
[32] I. Siró, and D. Plackett, "Microfibrillated cellulose and new nanocomposite materials: a review," *Cellulose,* vol. 17, no. 3, pp. 459–494, 2010. doi: 10.1007/s10570-010-9405-y
[33] F. G. Blanco Parte *et al.*, "Current progress on the production, modification, and applications of bacterial cellulose," *Critical Reviews in Biotechnology,* vol. 40, no. 3, pp. 397–414, 2020.

[34] F. Hemmati, S. M. Jafari, and R. A. Taheri, "Optimization of homogenization-sonication technique for the production of cellulose nanocrystals from cotton linter," *International Journal of Biological Macromolecules,* vol. 137, pp. 374–381, 2019.
[35] A. Serpa, J. Velásquez-Cock, P. Gañán, C. Castro, L. Vélez, and R. Zuluaga, "Vegetable nanocellulose in food science: a review," *Food Hydrocolloids,* vol. 57, pp. 178–186, 2016.
[36] V. Mihalca *et al.*, "Protein-based films and coatings for food industry applications," *Polymers,* vol. 13, no. 5, pp. 769, 2021.
[37] A. B. Perumal, R. B. Nambiar, J. Moses, and C. Anandharamakrishnan, "Nanocellulose: recent trends and applications in the food industry," *Food Hydrocolloids,* vol. 127, pp. 107484, 2022.
[38] E. Abraham *et al.*, "Environmental friendly method for the extraction of coir fibre and isolation of nanofibre," *Carbohydrate Polymers,* vol. 92, no. 2, pp. 1477–1483, 2013.
[39] E. M. Fernandes, R. A. Pires, J. F. Mano, and R. L. Reis, "Bionanocomposites from lignocellulosic resources: properties, applications and future trends for their use in the biomedical field," *Progress in Polymer Science,* vol. 38, no. 10–11, pp. 1415–1441, 2013.
[40] K. Oksman *et al.*, "Review of the recent developments in cellulose nanocomposite processing," *Composites Part A: Applied Science and Manufacturing,* vol. 83, pp. 2–18, 2016.
[41] B. L. Peng, N. Dhar, H. Liu, and K. Tam, "Chemistry and applications of nanocrystalline cellulose and its derivatives: a nanotechnology perspective," *Canadian Journal of Chemical Engineering,* vol. 89, no. 5, pp. 1191–1206, 2011.
[42] Y. Zhao, Y. Zhang, M. E. Lindström, and J. Li, "Tunicate cellulose nanocrystals: preparation, neat films and nanocomposite films with glucomannans," *Carbohydrate Polymers,* vol. 117, pp. 286–296, 2015.
[43] R. J. Moon, A. Martini, J. Nairn, J. Simonsen, and J. Youngblood, "Cellulose nanomaterials review: structure, properties and nanocomposites," *Chemical Society Reviews,* vol. 40, no. 7, pp. 3941–3994, 2011.
[44] S. J. Hanley, J.-F. Revol, L. Godbout, and D. G. Gray, "Atomic force microscopy and transmission electron microscopy of cellulose from Micrasterias denticulata; evidence for a chiral helical microfibril twist," *Cellulose,* vol. 4, no. 3, pp. 209–220, 1997.
[45] A. Ashjaran, M. E. Yazdanshenas, A. Rashidi, R. Khajavi, and A. Rezaee, "Overview of bio nanofabric from bacterial cellulose," *Journal of the Textile Institute,* vol. 104, no. 2, pp. 121–131, 2013.
[46] I. Reiniati, A. N. Hrymak, and A. Margaritis, "Recent developments in the production and applications of bacterial cellulose fibers and nanocrystals," *Critical Reviews in Biotechnology,* vol. 37, no. 4, pp. 510–524, 2017.
[47] D. Ciolacu, and V. I. Popa, "Cellulose allomorphs: structure, accessibility and reactivity," *Environmental Engineering and Management Journal,* vol. 10, no. 3, pp. 467–468, 2011.
[48] D. Klemm, B. Heublein, H. P. Fink, and A. Bohn, "Cellulose: fascinating biopolymer and sustainable raw material," *Angewandte Chemie International Edition,* vol. 44, no. 22, pp. 3358–3393, 2005.
[49] S. Kalia, S. Boufi, A. Celli, and S. Kango, "Nanofibrillated cellulose: surface modification and potential applications," *Colloid and Polymer Science,* vol. 292, no. 1, pp. 5–31, 2014.
[50] M. Jonoobi, J. Harun, P. M. Tahir, A. Shakeri, S. SaifulAzry, and M. D. Makinejad, "Physicochemical characterization of pulp and nanofibers from kenaf stem," *Materials Letters,* vol. 65, no. 7, pp. 1098–1100, 2011.
[51] S. Rebouillat and F. Pla, "State of the art manufacturing and engineering of nanocellulose: a review of available data and industrial applications," *Journal of Biomaterials and Nanobiotechnology*, vol. 4, pp. 165–188, 2013.
[52] S.-Y. Lee, S.-J. Chun, I.-A. Kang, and J.-Y. Park, "Preparation of cellulose nanofibrils by high-pressure homogenizer and cellulose-based composite films," *Journal of Industrial and Engineering Chemistry,* vol. 15, no. 1, pp. 50–55, 2009.
[53] A. Ferrer, I. Filpponen, A. Rodríguez, J. Laine, and O. J. Rojas, "Valorization of residual Empty Palm Fruit Bunch Fibers (EPFBF) by microfluidization: production of nanofibrillated cellulose and EPFBF nanopaper," *Bioresource Technology,* vol. 125, pp. 249–255, 2012.
[54] S. Iwamoto, A. N. Nakagaito, H. Yano, and M. Nogi, "Optically transparent composites reinforced with plant fiber-based nanofibers," *Applied Physics A,* vol. 81, no. 6, pp. 1109–1112, 2005.
[55] A. Alemdar, and M. Sain, "Isolation and characterization of nanofibers from agricultural residues–Wheat straw and soy hulls," *Bioresource Technology,* vol. 99, no. 6, pp. 1664–1671, 2008.

[56] A. N. Frone, D. M. Panaitescu, and D. Donescu, "Some aspects concerning the isolation of cellulose micro-and nano-fibers," *UPB Scientific Bulletin, Series B: Chemistry and Materials Science,* vol. 73, no. 2, pp. 133–152, 2011.
[57] B. Wang, and M. Sain, "Isolation of nanofibers from soybean source and their reinforcing capability on synthetic polymers," *Composites Science and Technology,* vol. 67, no. 11-12, pp. 2521–2527, 2007.
[58] A. Farooq *et al.*, "Cellulose from sources to nanocellulose and an overview of synthesis and properties of nanocellulose/zinc oxide nanocomposite materials," *International Journal of Biological Macromolecules,* vol. 154, pp. 1050–1073, 2020.
[59] P. Khawas, and S. C. Deka, "Isolation and characterization of cellulose nanofibers from culinary banana peel using high-intensity ultrasonication combined with chemical treatment," *Carbohydrate Polymers,* vol. 137, pp. 608–616, 2016.
[60] S. Rezanezhad, N. Nazanezhad, and G. Asadpur, "Isolation of nanocellulose from rice waste via ultrasonication," *Lignocellulose,* 2015.
[61] Q. Cheng, S. Wang, and Q. Han, "Novel process for isolating fibrils from cellulose fibers by high-intensity ultrasonication. II. Fibril characterization," *Journal of Applied Polymer Science,* vol. 115, no. 5, pp. 2756–2762, 2010.
[62] S. Wang, and Q. Cheng, "A novel process to isolate fibrils from cellulose fibers by high-intensity ultrasonication, Part 1: Process optimization," *Journal of Applied Polymer Science,* vol. 113, no. 2, pp. 1270–1275, 2009.
[63] D. Gunders, and J. Bloom, *Wasted: How America is Losing up to 40 percent of its Food from Farm to Fork to Landfill.* NRDC Food Waste Report, 2nd edition. NRDC, New York, 2017.
[64] J. Bloom, *American Wasteland: How America throws away Nearly Half of its Food (and what we can do about it).* Da Capo Press, London, 2010.
[65] S. Mercier, M. Mondor, S. Villeneuve, and B. Marcos, "The Canadian food cold chain: a legislative, scientific, and prospective overview," *International Journal of Refrigeration,* vol. 88, pp. 637–645, 2018.
[66] T. Talaviya, D. Shah, N. Patel, H. Yagnik, and M. Shah, "Implementation of artificial intelligence in agriculture for optimisation of irrigation and application of pesticides and herbicides," *Artificial Intelligence in Agriculture,* vol. 4, pp. 58–73, 2020.
[67] L. Chabada, H. C. Dreyer, A. Romsdal, and D. J. Powell, "Sustainable food supply chains: towards a framework for waste identification," in *Advances in Production Management Systems. Competitive Manufacturing for Innovative Products and Services: IFIP WG 5.7 International Conference, APMS 2012, Rhodes, Greece, September 24-26, 2012, Revised Selected Papers, Part I,* 2013: Springer, pp. 208–215.
[68] T. Gomiero, M. G. Paoletti, and D. Pimentel, "Energy and environmental issues in organic and conventional agriculture," *Critical Reviews in Plant Sciences,* vol. 27, no. 4, pp. 239–254, 2008.
[69] K. L. Thyberg, and D. J. Tonjes, "Drivers of food waste and their implications for sustainable policy development," *Resources, Conservation and Recycling,* vol. 106, pp. 110–123, 2016.
[70] M. Haile, "Weather patterns, food security and humanitarian response in sub-Saharan Africa," *Philosophical Transactions of the Royal Society B: Biological Sciences,* vol. 360, no. 1463, pp. 2169–2182, 2005.
[71] R. Kanter, and S. Boza, "Strengthening local food systems in times of concomitant global crises: reflections from Chile," *American Journal of Public Health,* vol. 110, no. 7, pp. 971–973, 2020.
[72] P. Shetty, "Incorporating nutritional considerations when addressing food insecurity," *Food Security,* vol. 1, pp. 431–440, 2009.
[73] S. Fan, and C. Rue, "The role of smallholder farms in a changing world," *The role of smallholder farms in food and nutrition security,* Springer, pp. 13–28, 2020.
[74] J. R. Pires, V. G. Souza, and A. L. Fernando, "Valorization of energy crops as a source for nanocellulose production–current knowledge and future prospects," *Industrial Crops and Products,* vol. 140, pp. 111642, 2019.
[75] M. Noremylia, M. Z. Hassan, and Z. Ismail, "Recent advancement in isolation, processing, characterization and applications of emerging nanocellulose: a review," *International Journal of Biological Macromolecules,* vol. 226, pp. 954–976, 2022.
[76] Y. Shi *et al.*, "Functionalization of nanocellulose applied with biological molecules for biomedical application: a review," *Carbohydrate Polymers,* vol. 285, pp. 119208, 2022.

[77] A. Zaghloul, M. Saber, S. Gadow, and F. Awad, "Biological indicators for pollution detection in terrestrial and aquatic ecosystems," *Bulletin of the National Research Centre,* vol. 44, no. 1, pp. 1–11, 2020.
[78] S. Şılbır, and Y. Göksungur, "Nanocellulose production and its food applications," *2nd Congress on Food Structure Design, Antalya, October 26–28,* 2016.
[79] P. Huang, C. Wang, Y. Huang, and M. Wu, "Structure and properties of cellulose nanofibrils," *Nanocellulose: From Fundamentals to Advanced Materials,* pp. 53–80, 2019.
[80] R. Mu *et al.*, "Recent trends and applications of cellulose nanocrystals in food industry," *Trends in Food Science & Technology,* vol. 93, pp. 136–144, 2019.
[81] D. R. Bogati, *Cellulose based Biochemicals and their Applications*, Bachelor's thesis, Saimaa University of Applied Sciences, 2011.
[82] V. S. More *et al.*, "Isolation and purification of microbial exopolysaccharides and their industrial application," *Microbial Polymers: Applications and Ecological Perspectives,* vol. 6, pp. 69–86, 2021.
[83] H. Biao, T. Li-rong, D. Da-song, O. Wen, L. Tao, and C. Xue-rong, "Preparation of nanocellulose with cation–exchange resin catalysed hydrolysis," *Biomaterials Science and Engineering,* vol. 6, pp. 139–152, 2011.
[84] D. Klemm *et al.*, "Nanocellulose as a natural source for groundbreaking applications in materials science: today's state," *Materials Today,* vol. 21, no. 7, pp. 720–748, 2018.
[85] D. Klemm *et al.*, "Nanocelluloses: a new family of nature-based materials," *Angewandte Chemie International Edition*, vol. 50, no. 24, pp. 5438–5466, 2011.

7 Occurrence, Application, and Acceptance of Nanomaterials in Food

Barasa Malakar, Sweety Das, and Mousham Talukdar
Corresponding author: Barasa Malakar

7.1 INTRODUCTION

Nanotechnology has been considered as an appealing technology that has transformed the food sector over the past few decades. It is a technology on the nanometer scale and deals with atoms, molecules, or macromolecules with a size of approximately 1–100 nm to generate and utilize materials that have novel properties (Singh et al., 2017). Nanotechnology is the science of the mixture, pattern, and characterization use of small-sized materials. It is the fine art of particle manipulation. It is a growing field in science research and development in nanoscience (Tsagkaris et al., 2018). The demand for nanoparticle-based substances has been enhanced in the food industry, as many of them hold significant elements and have been found to be nontoxic (Roselli et al., 2003). They have also been found to be stable at elevated pressure and temperature (Sawai, 2003). Nanomaterials bring about a substantial variance in food quality and safety, and also the health benefits that food delivers. Many industries and researchers are coming up with novel approaches, strategies, and products that involve a direct administration of nanotechnology in food science (Dasgupta et al., 2015).

Nanomaterials are ecofriendly and cost-effective to be employed in food packaging. Nanotechnology can enhance the barrier, mechanical, and antimicrobial properties that are desirable in the food packaging materials (Figure 7.1). Nanosensors can also notify the state of the food throughout transport and storage, which will be greatly favorable (Chadha et al., 2022).

The food market demands new technologies, which are crucial to keep market leadership in the food processing industry by producing authentic, fresh, convenient, and flavorful food products, extending the product's freshness and shelf life with enhanced food quality (Alfadul and Elneshwy, 2010). According to a definition in a report by the European Nanotechnology Gateway, a food is a nanofood when nanoparticles, or nanotechnology techniques, are utilized during cultivation, production, processing, or packaging of the food product (Ameta et al., 2020). Nanoencapsulates are comprised of a core containing the bioactive ingredient and an organic shell. Nanoencapsulation serves to enhance the stability, delivery, and bioavailability of nutrients like minerals and vitamins, agrochemicals, or pharmaceuticals (Anu Puri et al., 2009; Peters et al., 2016).

Nanotechnology offers appealing opportunities in the food industry, such as for food safety and quality control, as well as the fabrication of new food additives and supplements (Dimitrijevic et al., 2015). In the food industry, nanotechnology can also be used for the manufacture of packages with improved thermal and mechanical properties, and safety. Nanosensors immersed in food packaging systems are utilized to warn consumers when foods have expired. Nanotechnology can also be used to make healthier foods (Gokularaman et al., 2017). Various nanomaterials have enormous applications (Figure 7.2) in the creation of food products and the enhancement of nutritional value (Shafiq et al., 2020). In the food industry, nanotechnology can be employed to boost food

 DOI: 10.1201/9781003389408-7

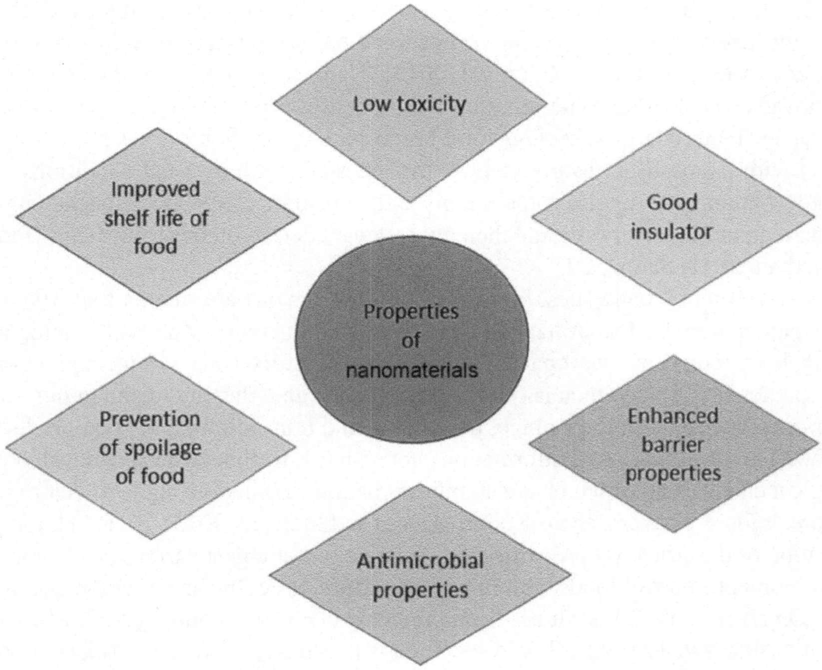

FIGURE 7.1 Properties of nanomaterials.

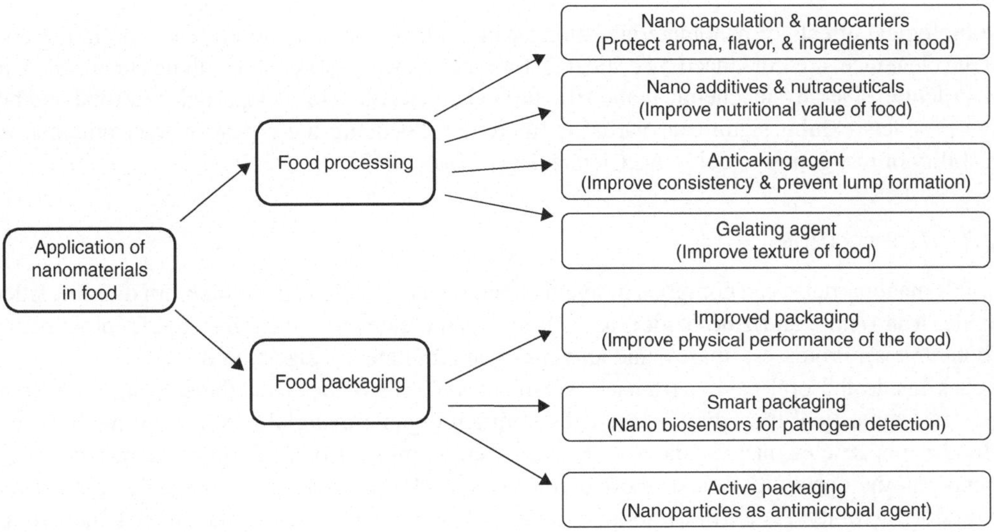

FIGURE 7.2 Applications of nanomaterials in various aspects of the food industry.

quality, shelf life, cost, safety, and nutritional benefits (Sozer and Kokini, 2009; McClements and Hang, 2017).

Nanotechnology is the field which deals with materials on the nanoscale. In the food sector, nanoparticles are used as additives, such as preservatives, flavoring agents, antimicrobial agents, and packaging substances. Nanoparticles of titanium dioxide, silver, zinc, zinc oxide, silicon dioxide, platinum, and gold are used widely in the food industry in various forms (Alfadul and

Elneshwy, 2010). In human food processing, nanocapsules have been employed as nanosized ingredients, nutritional supplements, and in functional foods, and as nanoadditives to furnish protective barriers, taste, and flavor (Mahmoud, 2015). Nanotechnology contributes to the development of innovative packaging materials that can boost the safety and shelf life of products by providing barrier materials or detecting food-borne pathogens. Some food packaging materials are equipped with nanosensors to track either the internal or the external conditions of the food products and containers throughout the supply and transport chain. Such packaging can track humidity and temperature over time and then give relevant details on these attributes (Prasanna and Shanmuganathan, 2011; Samal, 2017).

Materials consisting of aggregates, fragments, or filaments that are smaller than 100 nm are now classified as nanomaterials. The utilization of nanosensors in conjunction with intelligent systems of packaging, both active and smart packaging, is a very sensitive and swift technique for detecting food-borne pathogens, heavy metals, and contaminants and thereby maintaining food safety. Nanosensors are functional in the storage, packaging, and transportation of food products because they are allowed to sense and signal information about quality, freshness, and chemical, physical, and microbiological changes, all of which are significant parameters in averting food deterioration from dust, moisture, light, off-flavors, and off-odors (Duran and Marcato, 2013). Nanotechnology permits the modification of the structure, properties, and interactions amongst various food materials, enabling the development of novel foods with improved texture, taste, freshness, color, and stability (Yu et al., 2018; Onyeaka et al., 2022). It has been predicted that nanotechnology will revolutionize the entire food industry, transforming all the ways food is produced, processed, packaged, transported, and consumed.

7.2 CHARACTERISTICS OF FOOD NANOPARTICLES

The biological effects of nanomaterials, along with their imposition and dispensation in the body, and the clearance, are influenced by critical factors such as size, shape, and both physical and chemical structure. The physicochemical and structural characteristics of nanoparticles in food and beverage products exhibit significant variation, thereby influencing the digestive tract outcome and possibility of instigating toxicity (McClements and Xiao, 2017).

7.2.1 Composition

In foods, nanoparticles can comprise inorganic components (such as silver, titanium dioxide, silicon dioxide, iron oxide, and zinc oxide), as well as organic elements (including lipids, proteins, and carbohydrates). Proteases, lipases, and amylases can facilitate the digestion of proteins, lipids, and starches inside the oral cavity, stomach, small intestine, or colon. Nonetheless, digestion in the upper gastrointestinal tract may be evaded by certain organic materials utilized in the fabrication of food nanoparticles, namely dietary fibers as well as mineral oils. Chemical tendency to react depends on the constitution, and is often connected with the capacity of inorganic nanoparticles for produce toxicity. As an illustration, certain inorganic nanoparticles, like silver nanoparticles, undergo dissolution, releasing ionic particles which trigger unfavorable chemical or biochemical reactions, while others, such as titanium dioxide nanoparticles, remain comparatively inert (Pradeep and Anshup, 2009).

7.2.2 Dimensions

The dimensions of food nanoparticles exhibit significant variability, spanning from the nanolevel (such as surfactant micelles) to numerous nanometers (as seen in lipid, protein, or carbohydrate nanoclusters). This range is contingent upon the materials employed and the manufacturing methods employed in their production (Yada et al., 2014). Faster dissolution or digestion in gastrointestinal fluids is typically experienced by smaller nanoparticles when compared to their larger counterparts with similar compositions.

7.2.3 Aggregation State

Nanoparticles classified as food-grade can either occur like solitary singular molecules or aggregate into bunches, exhibiting variations in size, shape, and cohesion. Significant influences on the gastrointestinal fate of nanoparticles, including the capacity they possess to navigate in the gastrointestinal tract (GIT) secretion, the mucous membrane, are exerted when the size of nanoparticle clusters significantly exceeds that of individual nanoparticles. Therefore, determining the genuine productive dimensions of nanoparticles at their area of action is crucial, instead of focusing on the magnitude of initial nanomaterials introduced as components of diet.

7.2.4 Interphase Attributes

Interfacial properties of edible nanomaterials frequently influence gastrointestinal fate and, consequently, their potential to cause negative well-being outcomes (Weir et al., 2012; Magnuson et al., 2011; Powell et al., 2010).

7.3 TYPES OF FOOD NANOPARTICLES

Usually, nanoparticles found in food can be classified based on their composition, that is, as organic or inorganic.

7.3.1 Inorganic Nanoparticles

Various kinds of nanoparticles utilized in food products predominantly consist of inorganic substances, including silver, iron oxide, titanium dioxide, silicon dioxide, and zinc oxide (Pietroiusti et al., 2016). These molecules can exist in crystalline or amorphous states at room temperature, and their shapes might vary from spherical to nonspherical. They exhibit diverse surface properties and sizes contingent upon the primary materials and preparation methods employed during production.

7.3.1.1 Silver Nanoparticles

In the food industry, silver (Ag) nanoparticles find application in numerous roles. They have been utilized as antimicrobial agents in both food products and materials used for food packaging (Gaillet and Rouanet 2015; Hajipour et al., 2012; Pulit-Prociak et al., 2015). Estimates suggest that adults might ingest approximately 20 to 80 micrograms of silver per day, and just a small portion of that amount is in nanoparticle form (Frohlich and Frohlich, 2016). Currently, there are limited data available regarding the possible harm caused by silver nanoparticles consumed in food. Some studies indicate no harm, while others suggest significant harm.

7.3.1.2 Titanium Dioxide Nanoparticles

Titanium dioxide particles are utilized as functional components in specific food products to impart distinctive visual qualities like heightened luminosity and brilliance (Weir et al., 2012). However, these constituents encompass various particle dimensions, including a notable portion with diameters less than 100 nanometers, thus qualifying as nanoparticles (Warheit, 2015). As an example, food-safe titanium dioxide (E171) fine particles procured from various production methods were found to have an average particle diameter of approximately 110 nm, and over 36% of the particles measured were smaller than 100 nm (Weir, 2012). Consuming a single piece of chewing gum may lead to the ingestion of 1.5–5.1 milligrams of titanium dioxide (TiO_2) nanoparticles (Chen, 2013). It is important to highlight that children's intake of titanium dioxide nanoparticles was 2–4 times greater than that of adults. This difference can be attributed to the fact that products popular among children, such as candy bars, chewing gum, and soft drinks, contained some of the highest TiO_2 nanoparticle levels. TiO_2 nanoparticles can exhibit variations in size, shape, crystal structure, surface characteristics, and aggregation status, all of which can influence their fate and toxicity within the GIT.

7.3.1.3 Zinc Oxide Nanoparticles

Nanoscale zinc oxide can serve as a root of zinc in nutraceutical and enhanced foods because zinc is a crucial element in maintaining human health and overall well-being (Wang, 2014). Zinc oxide nanoparticles can also find application in packaging of foods as antimicrobial agents to safeguard against contamination by harmful bacteria (Sirelkhatim, 2015).

7.3.1.4 Silicon Dioxide Nanoparticles

Silicon dioxide nanoparticles are incorporated into specific powdered food products as powder flow enhancer to improve their movement characteristics, like salts, frosting sugar, flavor enhancer, powdered milk, and dehydrated mixtures (Dekkers, 2011; Peters, 2012). Most of the particles in food-grade silicon dioxide ingredients typically fall within the size range of 100 to 1000 nm, although there might also exist a notable fraction of smaller particles. Estimates suggest that individuals typically consume approximately 20 to 50 milligrams of SiO_2 per day (Frohlich and Frohlich, 2016).

7.3.1.5 Iron Oxide Nanoparticles

Iron oxide nanoparticles can find application in food products as color additives or as sources of readily absorbable iron (Raspopov, 2011; Zimmermann and Hilty, 2011; Hilty, 2010). Use of FeO_2 in the role of coloring agent in the United States is severely restricted, specifically limited to a maximum of 0.1% of weight in sausages, primarily as a component of casings (WHO in Safety Evaluation of Certain Food Additives and Contaminants, 2000). Estimates indicated that individuals who consume these products typically ingest approximately 450 micrograms (μg) of iron oxide per day (WHO in Safety Evaluation of Certain Food Additives and Contaminants, 2000). It was suggested that the capacity of FeO_2 nanomaterials to produce reactive oxygen species is a more probable operation behind possible toxicity (Wu, 2014).

7.3.2 Organic Nanoparticles

These types of nanomaterials consists predominantly of natural compounds, including lipids, proteins, and carbohydrates. They tend to exist in a liquefied, gel-like, or rigid state (either crystal-like or noncrystalline) at typical room temperatures, contingent on the structure and handling situations. The majority of natural nanomaterials frequently employed for food production are spheroid in shape, although under certain conditions, such as nanofiber formation, they may assume a nonspherical configuration. Organic substances exhibit diverse behaviors as they traverse different sections of the human GIT. For instance, they can undergo dissolution, precipitation, aggregation, or digestion in the oral cavity, gut, intestinal tract, and large intestine, contingent upon the formations and configuration (McClements and Xiao, 2017).

7.3.2.1 Lipid Nanoparticles

Lipid nanoparticles are commonly found in numerous commercially available food items and are under investigation for potential use in additional products (McClements, 2013). A significant portion of these globules of oils in products exist within the ultrafine scale, with a diameter (d) of less than 100 nanometers. Various forms of lipid nanoparticles can be found in food products, encompassing micelles, vesicles, fat crystals, and oil droplets, each differing in configuration, arrangement, and size. The size of liposomal nanomaterials can range from several nm (as seen in surface-active agent micelles) to numerous nanometers (found in globules of oils or rigid lipid nanomaterials). Lipid nanoparticles intended for use in food products typically consist of nonpolar fats (like triacylglycerols (TAGs), diacylglycerols (DAGs), monoacylglycerols (MAGs), hydrocarbons, and terpenes) and polar fats (comprising free fatty acids (FFAs), surface-active agents, and phosphatides). Some categories of lipid nanoparticles, those presently employed for food products and those under consideration for future use, resist digestion by gastrointestinal enzymes (McClements and Rao, 2011; McClements and Xiao, 2012).

7.3.2.2 Protein Nanoparticles

The prevalent polypeptide nanomaterials encountered in food products are milk protein aggregates present in cow's milk and various milk items. These micelles consist of small groupings of casein molecules along with phosphocalcium ions (Livney, 2010; Holt, 2003). The nanoarchitecture of casein micelles likely evolved as a natural solution to efficiently deliver essential nutraceutical (polypeptide and minerals) for babies (Oftedal, 2012). Protein nanoparticles typically comprise a group of protein molecules that have aggregated due to physical interactions (such as hydrophobic forces, interaction of hydrogen, van der Waals forces, or electrostatic attraction) and molecular bonds (for instance, sulfur–sulfur bond). Protein nanoparticles can exhibit a wide size range, spanning from several nm (singular globulin) to several hundred nm (examples include milk protein micelles, maize protein, wheat prolamin, milk serum, and protein of soybean molecules). While polypeptide nanomaterials are commonly circular in shape, this is feasible for engineering fiber-based formations and various geometries. Polypeptide nanomaterials which undergo digestion towards the upper GIT are improbable for induced toxicity, as they yield comparable absorption byproducts to conventional protein forms, namely amino acids and peptides.

7.3.2.3 Carbohydrate Nanoparticles

Nanoparticles composed of carbohydrates are commonly constructed using polysaccharides, which can be either absorbable or nondigestible. These polymers of sugars include plant fiber, carrageen, plant-based thickeners, and xanthan (Myrick, 2014; Joye, 2014). Carbohydrate nanoparticles can be formed by disassembling larger naturally occurring structures, like starch particles, chitosan filaments, and plant fibers. These nanoparticles can assume either circular or noncircular shapes contingent upon the source, and their digestibility within the upper GIT may vary, ranging from digestible to indigestible. The extent to which dietary fibers can be digested significantly influences the potential gastrointestinal outcome of carbohydrate nanomaterials.

7.3.2.4 Complex Nanoparticles

Numerous varieties of nanoparticles employed in food products are created through the amalgamation of diverse components, including lipids, proteins, and carbohydrates (Joye, 2014; Jones and McClements, 2010; McClements, 2012). For instance, liquid droplets are usually generated with the electrically charged binding of the polypeptide and carbohydrate polymers with opposing electrical charges. Nanomaterials of lipids enveloped with nanoscale laminates could be produced through sequential electrostatic precipitation of organic polymers with opposing charges towards the surfaces of fat droplets. This type of nanoscale laminate could significantly affect the gastrointestinal destiny of the particles ingested.

7.4 VARIOUS ASPECTS AND ROLES OF NANOTECHNOLOGY IN THE FOOD SECTOR

7.4.1 Nanotechnology in the Food Industry

Diverse utilizations have been noted in multiple areas, encompassing the precise delivery of nutrients and/or bioactive molecules via nanoencapsulation, the use of biological sensors for pathogen detection and quantification, and the preservation of fruits and vegetables through edible films (Durán et al., 2013). Nanotechnology enhances production operation in the food industry, yielding products with improved quality and the introduction of new functionalities. Consideration related to nanotechnologies begins with the production phase, enabling technical innovations in precision agriculture for the improvement of plant growth and the enhancement of capabilities in detecting and resisting pests and allelopathy. A transformation in the food industry has been brought about by nanotechnology, which involves the creation of creative transportation platform for producing nanoparticle-based pesticides, the enhancement of dietary components, and the generation of recent additions via a bioactive encapsulating system. Food products are safeguarded against the impacts of temperature and water content through nanoencapsulations, simultaneously managing the controlled dispensing of edible components and their relationship within the culinary framework. For example, water-fearing molecules find delivery carriers in liposomes, and oral bioavailability is improved by nanoemulsions (Chellaram et al., 2014). Recently, in the realm of food nanotechnology, industrial applications have ranged from the development of smart packaging to the formulation of interactive food products.

7.4.2 Nanotechnology in Food Packaging

Modern technology in packaging, referred to as active packaging, involves altering the condition of packaged food items to enhance both the sensory attributes and safety of food products. This, in turn, extends the shelf life by preserving the quality of the food items (Singh et al., 2017). Food processing and preservation also make use of nanotechnology. A protective barrier is provided by this application to hide various tastes and flavors. Controlled release and enhanced dispersal are facilitated by nanotechnology for food ingredients and water-insoluble dietary supplements (McClements et al., 2016). The field of biopolymer nanocomposites is garnering interest as it holds promise for improved mechanical and barrier properties, rendering them better suited for a broader array of food packaging applications. Anticipated for the near future, the foremost nanotechnology utilization in the food industry involves the combination of nanomaterials into packaging materials or storage containers. The primary objective is to extend product shelf life while maintaining freshness. An added advantage of nanoclays lies in their ability to enhance tensile strength and thermal properties, leading to the emergence of polymer-clay nanocomposites as novel materials for food packaging. The hybridization of nanoclays in low-density polyethylene films played a role in retarding the outward diffusion, resulting in the preservation of elevated levels of antibiotic and fungicidal extracted oils (Shemesh et al., 2015). In food packaging, the utilization of carbon nanotubes can enhance physical characteristics and introduce anti-infective properties (Rungraeng et al., 2012). As a result of its decomposability, bio-friendliness, anti-infective attributes, lack of toxicity, and adaptable compounds and substance attributes, chitosan is regarded as a highly promising choice for antimicrobial food packaging with broad-ranging potential applications (Aresta et al., 2013; Dutta et al., 2009). In the realm of active nanoenabled packaging, the Asia-Pacific region, particularly Japan, holds the dominant position, accounting for 45% of the existing outlet, which was priced at USD 1.86 billion in 2008. This sector was expected to expand to USD 3.43 billion by 2014, a yearly growth rate of 13%. In the USA, Japan, and Australia, enhanced and smart packaging is already effectively employed to prolong product shelf life, preserve nutrient profile, and ensure bacterial safety.

7.4.3 Nanotechnology in Agricultural Production

Nanoformulated fungicides and other agripesticides in nano form are utilized during agricultural production to improve power when compared with standard pesticides. Research has been conducted on nanoencapsulates utilizing chitosan, a polysaccharide derived from chitin, for a variety of agripesticides (Kashyap et al., 2015). Novel fungicide formulations have been investigated using solid triglycerides and nanoencapsulated polymers that contain carbendazim and tebuconazole.

7.4.4 Nanotechnology in Feed

Development and efficacy testing of nanomaterials (NMs) are underway with the aim of replacing antibiotics. Their function includes the absorption of microbes and harmful elements for enhanced nutrient digestibility. The worldwide issue of mycotoxin contamination in animal feed, which can result in serious illnesses in farm animals, has led to investigations into the use of nanoparticles, including nanoclays such as smectite-based clay nanoparticles, and other nanomaterials like thermoplastic-based fragments. These studies have demonstrated their capacity to absorb fungal toxins within the GIT of animals, providing protection against mycotoxin-related illnesses (Kuzma et al., 2008; Shi et al., 2006). The reduction of material size to the nano range has also exhibited the ability to enhance nutrient absorption, as evidenced in the study of iron and selenium.

7.4.5 Nanotechnology in Food

Food additives, primarily used to improve food consistency throughout handling and storing, enhance the quality of items, and enrich nutrition power and bioaccessibility within the food, constitute the primary focus of most nanoscale technology practice in food formation. SAS, which is frequently employed, is manufactured from either sparked or fumigated silica and comprises aggregates of main molecules of nm in size range. In addition to SAS, various commercially produced powder flow enhancers are employed in food items. These include calcium silicate, sodium aluminosilicate, dicalcium phosphate, sodium ferrocyanide, and microcrystalline cellulose. However, there is a lack of detail in the reviewed composition regarding whether these substances exist in a (partial) nanoscale form. There is a growing interest in assessing the dietary consumption of titanium dioxide in food. A recent study revealed that a range of 5% to 36% of the TiO_2 present in conventional edible goods falls within the nanoscale range (Weir et al., 2012). Antimicrobial substances including nanosized TiO_2 are employed in conjunction with other substances or elements like nickel oxide and cobalt to deactivate food-borne pathogens (Amna et al., 2013). Researchers are studying Fe^{2+} nanoparticles as a food additive with the aim of promoting health, as iron deficiency stands as one of the top widespread microelement imbalances across the world. Improving the solubility and bioavailability of iron compounds that have low solubility in acidic environments can be accomplished by decreasing their primary particle size, thereby increasing their specific surface area. Iron compounds with a nanostructured configuration could potentially be beneficial for fortifying certain foods with iron (Zimmermann and Hilty, 2011). A contrast was made between the antimicrobial capabilities of nisin, a polycyclic antibacterial peptide, when encapsulated and when in its unencapsulated state, in skimmed milk containing *Listeria* and in cheese. Better inhibitory properties were observed in milk when free nisin was used at higher temperatures. Conversely, at lower temperatures, such as refrigeration, improved results were seen in both milk and cheese when encapsulated nisin was utilized (da Silva Malheiros et al., 2010; Imran et al., 2012).

7.4.6 Environmental Impact of Nanotechnology

When released into the environment, nanoparticles' unique characteristics can lead to adverse environmental effects. Furthermore, in addition to possessing inherent harmful characteristics, nanoparticles, owing to their distinct shape, exterior properties, or charge, can potentially engage in undesirable interactions with chemicals or attach themselves to nutrients. This applies equally to nanoparticles employed in food packaging. Comparing the available toxicity data for lethal and sublethal effects, it was noted that these concentrations were considerably below the threshold for causing biological effects, suggesting a minimal level of risk. However, it should be noted that with the development of new particles and applications, as well as the accumulation of additional information about their fate, behavior, and entry routes into the environment, this estimation might undergo alterations. Additionally, it should be recognized that environmental organisms have the potential to accumulate nanomaterials once these substances have entered their surroundings. In line with the disclosure originating from the manufacturing, handling, and utilization of nanoscale substances, one must monitor the fate of the starting materials and their transformation products within the target compartments through activities such as life-cycle analyses and exposure scenarios. Regarding the formulation of potential scenarios for the fate of nanoparticles in the environment, information is progressively becoming accessible. Within soil, due to their extensive and active surfaces, nanoparticles have the capacity to bind and transport pollutants such as nonferrous metals or organic compounds, consequently posing a warning to groundwater. It seems undeniable that the nanoparticles utilized in food packaging make an extremely negligible contribution to the total environmental concentration. An additional captivating aspect of the environmental impact of nanoparticles involves their application as so-called nanoadditives to serve two contrasting purposes: enhancing the degradation and stabilization of polymers in different environmental scenarios, as well as bolstering their resilience under various environmental conditions (Shafiq et al., 2020).

7.5 APPLICATIONS OF NANOTECHNOLOGY IN FOOD PROCESSING

Consumable food items are produced by converting agricultural and animal products through food processing. Significant improvements can be made in food processing in terms of achieving intelligent nutrient delivery, implementing nanoscale encapsulation of nutraceuticals, utilizing biotechnology for protein separation, rapidly sampling biological and chemical impurities, improving solubility and delivery, and addressing color-related perspectives within food systems. Various developing applications of nanoscale technology in the field of food science are being represented. The potential for nanoscale technology which supports the creation of functional or integrative foods that fulfill the requisites of the human body and improve the efficient delivery of nutrients is recognized. The idea is that numerous nanocapsules, which contain essence or color amplifier, or additional nutraceutical components (like vitamins), will stay inactive within meals until they are activated through the customer (Amin et al., 2015). Notable success has been achieved by a bakery situated in Western Australia through the incorporation of nanocapsules of oil extracted from tuna fish (a good source of linolenic fatty acid) into the in-demand commodity Tip-Top UP bread. Efforts are underway at Unilever to decrease the size of emulsion particles, a factor influencing texture, to develop low-fat ice creams. The aspiration is to reduce emulsion usage by up to 90%, resulting in a decrease in fat percentage ranging from 16% to approximately 1% (Verma and Gupta, 2017).

Controlled release and improved dispersibility for food ingredients and water-insoluble dietary supplements are provided by nanotechnology as well (McClements et al., 2016). Incorporating nanotechnology in food processing entails achieving exhaustive textures for food components, encapsulating food additives, pioneering novel essence and impression, and managing fragrance discharge to boost absorption rates in nutritional enhancer (Cho et al., 2019). In the food industry, nanotechnology has brought about a transformation through the invention of creative delivery systems for producing agrochemicals in nano form, the improvement of nutritional content, and the generation of new products via the encapsulation of bioactives. There is no doubt that nanoencapsulation enhances the nutritional characteristics of the product (Yoo et al., 2004). Certainly, on the nanoscale, the manipulation of food components enables the enhancement of fragrance and consistency, or the creation of novel flavors, while simultaneously elevating nutritional content. Food products are shielded from heat or wetness, and controlled discharge of food components, as well as their communications alongside substances within the nutrient composition, is overseen through the encapsulation of nanomaterials. As an illustration, liposomes serve as carriers for nonpolar particles, and nanodispersion enhances the absorption rate when taken orally (Chellaram et al., 2014). The efficient delivery of nutrient cells excluding altering the palate or pigmentation of food is achieved through the utilization of nanocochleates (Farooq et al., 2019).

Recently, there has been a range of manufacturing purposes of food nanoscale technology, starting with the use of smart packaging in the development of engaging food products. Food wastage due to various types of infectious diseases is minimized by advanced food packaging and systems relying on nanotechnology. Significant effectiveness as antimicrobial agents has been demonstrated by various nanoparticles including TiO_2, MgO_2, ZnO_2, nanosilvers, and CNTs (carbon nanotubes) (Salgado et al., 2019). Moreover, there have been recent developments in utilizing specific nanomaterials for the detection or removal of hazardous substances and infectious agents in food products. Within intelligent packaging, a unified e-taste sensor comprises an array of nanosized sensors highly attuned to vapor emitted by food disposal. It is a system that provides a distinct and observable indicator of a meal's freshness by means of a sensor strip that undergoes a color change (Neethirajan et al., 2011). Ferromagnetic oxide nanomaterials are employed in the food industry for the swift determination of food-related diseases. Nanoscale sensors possess a high detection capacity and are capable of identifying microscopic organisms, hazardous materials, and impurities in various food items, contributing to the safety of food products (He et al., 2016). Advances across various magnitudes of food products, consisting of surface, flavor, organoleptic properties, product pigmentation intensity, and the consistency and expiry date of these items, are propelled by the utilization of nanoscale technology in food production (Rodrigues, 2017). Boost abilities, like enhanced discharge effectiveness and envelopment attributes, are exhibited by nanoparticles when compared to conventional encapsulation techniques. Nanoencapsulation involves addressing the flavor and odor management aspects of potent ingredients within the food matrix. It also includes the controlled discharge of effective elements, ensuring the timely accessibility of the desired elements, and safeguarding against various origins of toxification, including heat, moisture, and biological and chemical deterioration, throughout storage, manufacturing, and application (Drusch et al., 2009).

New potential for improving the efficiency of food product packaging is held by the utilization of nanotechnology in packaging of foods. Their ability to encapsulate key elements and enhance functionality, stability, and bioavailability has led to the production and industrial application of numerous nanoparticles (Abraham, 2016). Food packaging utilizes nanofillers for their desired functions and uses in food packing. In biometric sensors, these nanoparticles serve a dual role by safeguarding food from ecological components and integrating attributes into wrapping supplies, thus presenting various opportunities within the food packaging industry (Arroyo, 2019). In the food industry, ZnO_2, a nanostructured composite, finds use in advance packaging of foods because of its crucial antioxidant effect. In the packaging of food, nanoparticles play a role not only in antimicrobial packaging but also in the extensive use of nanocomposites and nanolaminates. The production of biodegradable packaging through nanotechnology reduces environmental pollution. In the food industry, the packaging of food items utilizes a nanostructure known as a silicate layer. Advances in sensor technology within smart food packaging offer insights into material quality, safety, and half-life (Schaefer et al., 2018). The packaging of food products also incorporates the use of nanocomposites, which possess improved characteristics for withstanding thermal stress during food processing, transportation, and storage. Presently, nanocomposites are employed in beer bottles, extending their shelf life by as much as six months. In the food industry, there is potential for the utilization of carbon-based graphene nanoplates in food product packaging due to their heat-resistant properties. In material packaging, silver nanoparticles hold significance and find long-term use. For instance, these particles protect substances they cover from contamination. In the realm of food research and the meal preservation industry, silver nanoparticles have been underscored as important by many researchers. Nevertheless, it should be noted that only certain approaches utilizing silver nanoparticles have received certification from the European Food Safety Authority (EFSA) for potential recycling in food preservation and packaging (Dinh et al., 2015).

Some antiseptics with new attributes, like ZnO_2, MgO_2, NiO, and nanosilvers, are produced through nanotechnology. At the nanoscale, these nanoparticles have demonstrated possible utilization and disinfectant attributes. Polymer matrices incorporate these nanoparticles to bestow beneficial qualities, like disinfectant actions, and to improve packaging properties (Momin et al., 2013). The utilization of nanotechnology can extend to creating materials of food packaging alongside heightened thermal and/or automated properties in the food industry, thereby bolstering safety. Nanotechnology-enabled "intelligent" and "dynamic" food packaging offers numerous benefits compared to traditional packaging approaches, including the provision of superior packaging materials with enhanced mechanical durability, resistance characteristics, disinfectant coatings, and nanosensors for identification of infectious agents, thereby notifying customers of the food's safety status (Mihindukulasuriya and Lim, 2014). Potent antibacterial effects at low concentrations and greater stability under harsh environmental conditions can be achieved through the use of inorganic nanoparticles. The utility of nanoparticles extends beyond antimicrobial food packaging; the active utilization of nanocomposites and nanolaminates in food packaging creates barriers against extreme thermal and mechanical stressors, resulting in extension of food item expiry dates. As a result, the integration of nanomaterials into packing materials enhances the standard of food items, leading to the prolongation of their storage duration. Effective transport mechanisms with all aforementioned standards are provided through the application of nanotechnology in the creation of encasing, suspensions, biodegradable polymer substrate, normal solutions, and association suspension.

7.6 SAFETY CONCERNS IN RELATION TO NANOMATERIALS

Although nanoparticles have found substantial application in the field of food nanotechnology, their potential toxicity remains inadequately understood. When using nanoparticles, two major safety concerns revolve around allergen presence and the release of heavy metals. The accelerated application of nanoparticles in food products, in the absence of essential knowledge and regulatory measures, is presenting environmental and human health risks beyond the preferred pace (Ranjan et al., 2014). As per research conducted by the British Royal Society, there is a potential future threat of a catastrophe related to nanotoxicity that could impact humanity (Amini et al., 2014). In order to determine which food products are safe and offer benefits, it is essential to attain a comprehensive grasp of the attributes of nanomaterials, encompassing solubility, dimensions, makeup, and surface properties. While certain distinctive traits of nanomaterials render them attractive for various applications, their suitability for use in the context of food applications, where there is a potentially high risk to human health, is a subject open to debate (Ameta et al., 2020).

7.6.1 Size

The unique properties of nanoparticles are greatly influenced by their size, as the nanoparticles' surface region is intricately linked with their size. Respiration's connection with surface area recognition has been established, with reports indicating that pulmonary inflammation can be induced by certain nanoparticles (Qiao et al., 2015). The toxicity of these nanomaterials is influenced not just by their chemical makeup but also by the amount and the deposition site. In vitro dermal cell cytotoxicity observation can be influenced by particle size, making it an important factor to consider. An immune system response can be initiated by nanoparticles that are absorbed through different absorption routes. The diminished dimensions of these nanoparticles enable them to pass through various biological obstacles. Subsequently, they have the capacity to accumulate in organs such as central nervous system. Therefore, the safety suitability of these nanomaterials for utilization in the food and food-related sectors is contingent upon nanoparticle size.

7.6.2 Chemical Composition

The reagents employed during nanoparticle production might possess toxicity. A portion of these reagents may persist in the end product, potentially leading to contact with the harmful substances which are not associated with the nanoparticles themselves. For example, certain documented harmful effects associated with carbon nanotubes and semiconductor nanomaterials can be assigned with the existence of leftover reagents from the synthesis process. Our comprehension of possible adverse effects of carbon nanotubes could be obstructed by the residual reagents and contaminants. Iron ions and contaminants can expedite oxidative stress in cells. The chemical composition also encompasses the crucial factor of crystallinity. Three different levels of crystallinity in titanium oxide result in distinct cytotoxic effects (Suker and Albadran, 2013).

7.6.3 Solubility

The toxicity of nanoparticles is influenced by their solubility as well, making it another critical factor to consider. Comparatively, hydrophilic titanium oxide nanoparticles that are soluble exhibit higher toxicity than their insoluble counterparts (Oberdörster, 2001). The literature has also documented the correlation between nanoparticle oxide solubility and toxicity (Brunner et al., 2006).

7.6.4 Surface Structure

The surfaces of nanomaterials can also effect cytotoxicity. The toxicologic properties of nanoparticles absorbed within the human body can be influenced by major factors such as roughness, charge, hydrophobicity, and surface composition (Kirchner et al., 2005). Covering nanomaterials with water-attracting polymers such as polyethylene glycol (PEG) can offer a degree of control over the adverse impact of nanomaterials. Study has shown that nanoparticles with a positive charge exhibit greater toxicity when contrast to their unfavorable or impartial counterparts.

7.6.5 Routes of Nanomaterials Exposure

Exposure to nanoparticles can occur through various entry routes, including dermal, respiratory, and digestive pathways. Upon absorption, these nanomaterials have the potential to enter the bloodstream and accumulate in various tissues, including the brain, and may initiate immune responses. Airborne nanomaterials are likely emitted into the atmosphere during the manufacturing of nanomaterials utilized in the food and associated sectors. It is essential to bear this in mind, and safeguarding the respiratory health of employees from nanoparticle inhalation should be a priority. The gastrointestinal pathway represents another significant avenue for nanoparticle absorption. Through mucociliary clearance, nanoparticles can also find entry into the respiratory tract, ultimately leading them into the digestive system. An alternative primary avenue of contact between humans and nanomaterials is through the skin.

7.7 NANOMATERIAL CONTAMINANTS IN FOOD

Origins from inadvertent nanoparticle impurities in food can arise from environmental factors, nanomaterials employed during the cultivation and rearing of plants and animals, unintended discharge from food wrapping substances containing nanomaterials, and residual matters in food originating from the utilization of nanoparticles as food enhancer and coating layer in food-related equipment (Magnuson, 2009). For example, researchers are exploring nanoemulsion and nanoencapsulation technologies to enhance the absorption and effectiveness of fertilizers, herbicides, and insecticides. Consequently, alterations in the levels of these substances in food may occur, potentially resulting in an increase or decrease in their presence (Perez-de-Luque and Rubiales, 2009). The potential presence of nanoparticles in food can also be attributed to environmental contamination stemming from various industries. This can occur when this set of substances are found in the surroundings where food is being cultivated or in the H_2O utilized during food processing. The existence of numerous established industrial contaminants discovered in foods has already shown the possibility of environmental pollution caused by these various industries impacting the food supply (Magnuson et al., 2011).

While nanotechnology offers numerous benefits to the food industry in food packaging, it is essential not to overlook the safety concerns linked to nanomaterials. Safety concerns regarding nanoparticles are deliberated upon by numerous professionals, with particular emphasis placed on the potential migration of nanomaterials from wrapping materials into the nutrition and its influence on consumer well-being (Bradley et al., 2011; Jain et al., 2016). Despite considering a material as GRAS, it is essential to acquire additional studies to evaluate the risks posed by its nano equivalents, because the physicochemical characteristics differ significantly when in the nanoscale state compared to the macrostate. The diminutive dimensions of these nanoparticles could increase the potential for accumulation within the organs and tissues of the body (Savolainen et al., 2010). As an illustration, when exposed, silica nanoparticles employed as anticaking agents may provoke cytotoxic effects in human lung cells (Athinarayanan et al., 2014). Dissolution is influenced by a multitude of factors, with relation to the surface structure of particles, concentration, surface tension, clustering, and adhesion, all playing crucial roles. A framework for investigating the migration of particles from food packaging has been developed by Cushen et al. (2014), who conducted an examination of the leaching of silver and copper from nanocomposites. The proportion of nanofiller in the nanocomposites emerged as one of the most crucial factors controlling migration, exerting more influence than particle size, temperature, or contact time. Given the unique properties of each nanomaterial, toxicity assessments are likely to be established on a case-by-case basis (Mahler et al., 2012).

7.8 DETECTION OF NANOMATERIALS IN FOOD

Nanoscience's application in the food sector includes, to some extent, reducing the size of typical food components like fats and proteins. This often leads to challenges in distinguishing the nanoscale version from the larger, bulk form of these materials. Furthermore, the existence of naturally existing nanoparticles in food presents a difficulty in differentiating among the indigenous and purposely crafted nanoparticles (Tiede et al., 2008; Onyeaka et al., 2022). Prominent approaches for identifying nanoparticles in food matrices include microscopy techniques like scanning electron microscopy (SEM), transmission electron microscopy (TEM), scanning transmission electron microscopy (STEM), X-ray microscopy (XRM), scanning probe microscopy (SPM) and atomic force microscopy (AFM), chromatographic techniques like high-performance liquid chromatography (HPLC) and ultra-performance liquid chromatography (UPLC), field flow fractionation (FFF), spectroscopy techniques, Fourier transform infrared spectroscopy (FT-IR), dynamic light scattering (DLS), capillary electrophoresis, and centrifugation and filtration techniques as shown in Figure 7.3.

7.9 CONSUMER PERCEPTION

The perception of new technologies such as nanotechnology is significantly influenced by people's emotions, while reactions to information about nanotechnology are determined by their values. The latest surveys reveal that food nanotechnology is perceived differently by consumers in different

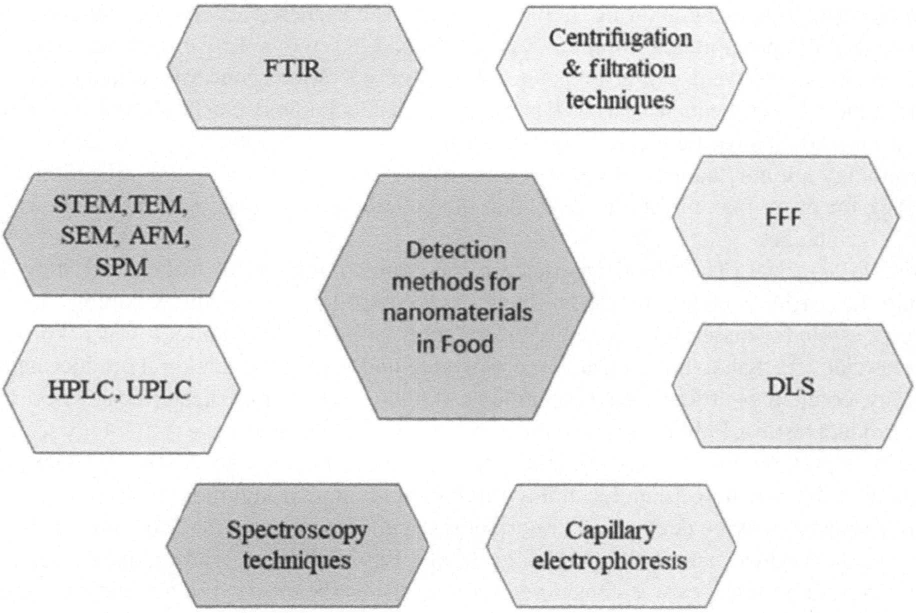

FIGURE 7.3 Different methods of detection of nanoparticles in food.

countries around the globe. While European consumers generally hold a positive outlook on nanotechnology's potential in various applications, they harbor skepticism concerning the utilization of nanoparticles in food. In the USA, the situation and consumer perception vary notably. According to a recent survey on nanotechnology, a substantial 80% of the participants indicated that they had little or no prior exposure to information about nanotechnology. Nevertheless, they anticipate numerous advantages from nanotechnology in terms of improving food safety and quality. Nonetheless, a survey supported by the National Science Foundation in 2006, which focused on the audience perspective on nanoscale technology items, revealed that US American customers are open to using certain items that incorporate nanomaterials, including in cases where wellness as well as safety matters exist, as long as the potential benefits are substantial. Nanomaterials alongside diverse physical characteristics for utilization in the food can be generated through different preparation technologies. Nevertheless, the public's perception of this novel technology remains uncertain. Various regulatory bodies, including the US Environmental Protection Agency (USEPA), the International Organization for Standardization (ISO), the Organization for Economic Cooperation and Development (OECD), the National Institute for Occupational Safety and Health (NIOSH), the Health and Consumer Protection Directorate of the European Commission (HCPDEC), as well as the Food and Drug Administration (FDA), have published multiple instructions addressing the possible danger associated with nanoparticles. Reports indicate that nanoparticles have the potential to improve the safety of food by improving the effectiveness of food packaging, extending storage life, and enhancing the nutraceutical content of food, all while preserving the palate and physical attributes of food items.

Food products can incorporate nanotechnology, be included in food packaging, or be employed in food processing. Nanotechnology's application to food as well as food products appears to receive higher acceptance from experts compared to consumers. This discrepancy may arise because many experts in the field of nanotechnology possess asymmetrical information, giving them a greater understanding of risk and benefit assessments than consumers have access to. Considered by experts were the benefits in terms of food freshness and safety, along with the broader advantages for the environment and food manufacturing. In particular, it was considered acceptable if nanotechnology had the capacity to prohibit deterioration of food, prolong the storage life of food, and decrease the requirement for excessive packing. Consumer studies raised various concerns, which encompassed possible unintended consequences as well as the belief that technology can be abused. The unacceptability of agrifood nanoscale technology even stems from the perception that products containing this technology are not "natural." Concerns were also expressed regarding the utilization of nanotechnology for profit maximization rather than the creation of improved food products with clear consumer advantages.

Increased approval of agrifood nanoscale technologies might result from establishing clarity regarding the parties accountable for producing and controlling nanoscale technology items, and those responsible for disseminating safety information to the general public. It was acknowledged by commercializers that, to boost customer approval of and belief in agricultural produce nanoscale technology, companies utilizing nanotechnology in their products may need to carry out comprehensive product testing. Public receptivity to agricultural produce nanoscale technology seems to be influenced, in part, by fundamental principles driving nanotechnology goods, the attributes of these products, and the pricing of nanotechnology items. Studying the significance of marketing, particularly branding, may be deemed an important research topic to stimulate customer acquisitions of agricultural produce nanoscale technology items. Emphasizing the advantages to consumers through marketing messages was considered crucial, just as the creation of a "reliable brand" was deemed significant. Some evidence suggests that agricultural food nanoscale technology might be better accepted by certain population groups than others. Individuals with a traditional mindset might recognize higher danger associated with the utilization of agricultural produce nanoscale technologies, as opposed to those who have been exposed to embracing novel technologies (Giles et al., 2015).

7.10 CONCLUSIONS

In recent years, the popularity of the utilization of structures on the nanometer scale in the food sector is increasing and, hence, interest and activities in this aspect have been immensely focused. The transparency of safety issues and environmental impact should be the prime concern when dealing with the fabrication of nanotechnology in food systems, and compulsory testing of nanofoods is needed before they are delivered in the market for the use of the public. Nanotechnology is considered to be a nonharmful and ecofriendly "green chemistry" approach. Nanomaterials have exhibited a range of exceptional elements considered unique to their bulk materials. The abundant availability of nanomaterials is a perfect choice for large-scale manufacture for various applications. Active, edible packaging, and biodegradable materials are considered among the food industry's main objectives due to the growing need for alternative packaging materials that are recyclable, renewable, easily degradable, and require little or almost no disposal. Nanomaterials offer a wide range of economic benefits, and are ecofriendly, decomposable, and renewable. These nanomaterials play a significant role in the food industry, which includes packaging, processing, and preservation of food. Nanomaterials also increase the shelf life of foodstuffs by protecting them from moisture and various climatic conditions.

Although advances in nanobiotechnology are made daily, the field is not without its challenges. The environmental effects and safety issues must be a first priority in research to cope with the hazards associated with the use of nanomaterials and nanodevices in the food industry. Nanotechnology still has immense potential, and new applications are being investigated in different arenas of the food industry. It is clear that security and safety issues are surfacing and must be carefully considered and addressed in the future. Nanoparticles are capturing a great deal of attention in the food arena, since both organic and inorganic nanoparticles have the potential to boost food nutritional attributes, safety, and quality.

REFERENCES

Abraham, A. 2016. *Understanding the effect of phytochemical coated silver nanoparticles on mammalian cells and the protein interactions with the surface corona of these nanoparticles.* Ph.D. Thesis, RMIT University, Melbourne, Australia.

Alfadul, S.M., Elneshwy A.A. 2010. Use of nanotechnology in food processing, packaging and safety — review. *Afr J Food Agric Nutr Dev* 10(6):2719–2739.

Amin, A.A., Salama, M.F., Seliem, E.I. Mahmoud K.F. 2015. Encapsulation of nano carotenoids; evaluation of stability and safety. *Int J Curr Microbiol App Sci* 4(2):1017–1029.

Amini, S.M., Gilaki, M., Karchani, M. 2014. Safety of nanotechnology in food industries. *Electron Physician* 6:962–968.

Ameta, S.K., Rai, A.K., Hiran, D., Ameta, R., Ameta, S.C. 2020. Use of nanomaterials in food science. In: *Biogenic Nanoparticles and Their Use in Agro-Ecosystems*; Springer: Singapore, 457–488.

Amna, T., Hassan, M.S., Yousef, A., Mishra, A., Barakat, N.A.M., Khil, M.S., et al. 2013. Inactivation of foodborne pathogens by NiO/TiO_2 composite nanofibers: A novel biomaterial system. *Food Bioprocess Technol* 6:988–996.

Anu Puri, K., Brandon Smith, L., Lee, J.H., Yavlovich, A., Heldman, E., Blumenthal, R. 2009. Lipid-based nanoparticles as pharmaceutical drug carriers: From concepts to clinic. *Crit Rev Ther Drug Carrier Syst* 26:523–580.

Aresta, A., Calvano, C.D., Trapani, A., Cellamare, S., Zambonin, C.G., De Giglio, E. 2013. Development and analytical characterization of vitamin(s)-loaded chitosan nanoparticles for potential food packaging applications. *J Nanoparticle Res* 15(4):523–580.

Arroyo, B.J., Santos, A.P., de Melo, E.d.A., Campos, A., Lins, L., Boyano-Orozco, L.C. 2019. Bioactive compounds and their potential use as ingredients for food and its application in food packaging. In *Bioactive Compounds*; Elsevier: Amsterdam, The Netherlands, 143–156.

Athinarayanan, J., Periasamy, V.S., Alsaif, M.A., Al-Warthan, A.A., Alshatwi, A.A. 2014. Presence of nanosilica (E551) in commercial food products: TNF-mediated oxidative stress and altered cell cycle progression in human lung fibroblast cells. *Cell Biol Toxicol* 30:89–100.

Bradley, E.L., Castle, L., Chaudhry, Q. 2011. Applications of nanomaterials in food packaging with a consideration of opportunities for developing countries. *Trends Food Sci Technol* 22(11):604–610.

Brunner, T.J., Wick, P., Manser, P., Spohn, P., Grass, R.N., Limbach, L.K., Bruinink, A. et al. 2006. In vitro cytotoxicity of oxide nanoparticles: Comparison to asbestos, silica, and the effect of particle solubility. *Environ Sci Technol* 40(14):4374–4381.

Chadha, U., Bhardwaj P., Selvaraj S.K., Arasu K., Praveen S., Pavan A. et al. 2022. Current trends and future perspectives of nanomaterials in food packaging application. *J Nanomaterials* 2022:2745416.

Chellaram, C., Murugaboopathi, G., John, A., Sivakumar, R., Ganesan, S., Krithika, S. et al. 2014. Significance of nanotechnology in food industry. *APCBEE Procedia* 2014, 8:109–113.

Chen, X.X. Cheng, B., Yang Y.X., Cao, A., Liu, J.H., Du, L.J. et al. 2013. Characterization and preliminary toxicity assay of nano- titanium dioxide additive in sugar-coated chewing gum. *Small* 9:1765–1774.

Cho, Y.H., Jones, O.G. 2019. Assembled protein nanoparticles in food or nutrition applications. *Adv Food Nutr Res* 2019, 88:47–84.

Cushen, M., Kerry, J., Morris, M., Cruz-Romero, M., Cummins, E. 2014. Evaluasian and simulation of silver and copper nanoparticle migration from polyethylene nanocomposites to food and an associated exposure assessment. *J Agric Food Chem* 62(6):1403–1411.

da Silva Malheiros, P., Daroit, D.J., da Silveira, N.P., Brandelli, A. 2010. Effect of nanovesicle-encapsulated nisin on growth of Listeria monocytogenes in milk. *Food Microbiol* 27:175–178.

Dasgupta, N., Ranjan, S., Mundekkad, D., Ramalingam, C., Shanker, R., Kumar, A. 2015. Nanotechnology in agrofood: From field to plate. *Food Res Int* 69:381–400.

Dekkers, S. Krystek, P., Peters, R.J.B., Lankveld, D.P.K., Bokkers, B.G.H., van Hoeven-Arentzen, P.H. et al. 2011. Presence and risks of nanosilica in food products. *Nanotoxicol* 5:393–405.

Dimitrijevic, M., Karabasil, N., Boskovic, M., Teodorovic, V., Vasilev, D., Djordjevic, V et al. 2015. Safety aspects of nanotechnology applications in food packaging. *Procedia Food Sci* 5:57–60.

Dinh, N.X., Quy, N.V., Huy, T.Q., Le, A.T. 2015. Decoration of silver nanoparticles on multiwalled carbon nanotubes: Antibacterial mechanism and ultrastructural analysis. *J Nanomater* 16:63.

Drusch, S., Mannino, S. 2009. Patent-Based review on industrial approaches for the microencapsulation of oils rich in polyunsaturated fatty acids. *Trends Food Sci Technol* 20:237–244.

Duran, ´ N., Marcato, P.D. 2013. Nanobiotechnology perspectives. Role of nanotechnology in the food industry: A review. *Int J Food Sci Tech* 48:1127–1134.

Dutta, P.K., Tripathi, S., Mehrotra, G.K., Dutta, J. 2009. Perspectives for chitosan based antimicrobial films in food applications. *Food Chem* 114:1173–1182.

Gaillet, S., Rouanet, J.M. 2015. Silver nanoparticles: Their potential toxic effects after oral exposure and underlying mechanisms–a review. *Food Chem Toxicol* 77:58–63.

Giles E.L., Kuznesof S., Clark B., Hubbard C., Frewer L.J. 2015. Consumer acceptance of and willingness to pay for food nanotechnology: A systematic review. *J Nanopart Res* 17:467.

Gokularaman, S., Cruz, S.A., Pragalyaashree, M., Nishadh, A. 2017. Nanotechnology approach in food packaging-review. *J Pharm Sci Res* 9:1743–1749.

Farooq, S. Gull, A. Ahmad, T. Masoodi, F. Rather, S.A. Dar, A.H. 2019. Emerging role of nanotechnology in food industry. In *Proceedings of the 5th International Conference on Nanotechnology for Better Living (NBL-2019)*, April 7–11 2019, Kashmir, India.

Frohlich, E.E., Frohlich, E. 2016. Cytotoxicity of nanoparticles contained in food on intestinal cells and the gut microbiota. *Int J Mol Sci* 17:1–22.

Hajipour, M.J. et al. 2012. Antibacterial properties of nanoparticles. *Trends Biotechnol* 30:499–511.

He, X., Deng, H., Hwang, H.M. 2016. The current application of nanotechnology in food and agriculture. *J Food Drug Anal* 24:277–289.

Hilty, F.M. et al. 2010. Iron from nanocompounds containing iron and zinc is highly bioavailable in rats without tissue accumulation. *Nat Nanotech* 5:374–380.

Holt, C., de Kruif, C.G., Tuinier, R., Timmins, P.A. 2003. Substructure of bovine casein micelles by small-angle X-ray and neutron scattering. *Colloid Surf A* 213:275–284.

Imran, M., Revol-Junelles, A.M., Rene, N., Jamshidian, M., Akhtar, M.J., Arab-Tehrany, E., et al. 2012. Microstructure and physico-chemical evaluation of nano-emulsion-based antimicrobial peptides embedded in bioactive packaging films. *Food Hydrocolloids* 29:407–419.

Jain, A., Shivendu, R., Nandita, D., and Chidambaram, R. 2016. Nanomaterials in food and agriculture: An overview on their safety concerns and regulatory issues. *Crit Rev Food Sci Nutr* 58(2):297–317.

Joye, I.J., Davidov-Pardo, G., McClements, D.J. 2014. Nanotechnology for increased micronutrient bioavailability. *Trends Food Sci Tech* 40:168–182.

Jones, O.G., McClements, D.J. 2010. Functional biopolymer particles: Design, fab- rication, and applications. *Comp Rev Food Sci Food Saf* 9:374–397.

Kashyap, P.L., Xiang, X., Heiden, P. 2015. Chitosan nanoparticle-based delivery systems for sustainable agriculture. *Int J Biol Macromol* 77:36–51.

Kirchner C, Liedl T, Kudera S, Pellegrino T, Muñoz JA, Gaub HE, Stölzle S et al. 2005. Cytotoxicity of colloidal CdSe and CdSe/ZnS nanoparticles. *Nano Lett* 5(2):331–338.

Kuzma, J., Romanchek, J., Kokotovich, A. 2008. Upstream oversight assessment for agrifood nanotechnology: A case studies approach. *Risk Analysis* 28,1081–1098.

Livney, Y.D. 2010. Milk proteins as vehicles for bioactives. *Curr Opin Colloid Interface Sci* 15:73–83.

Magnuson, B.A., Jonaitis, T.S., Card, J.W. 2011. A brief review of the occurrence, use, and safety of food-related nanomaterials. *J Food Sci* 76:R126–R133.

Magnuson, B.A. 2009. Nanoscale materials in foods: Existing and potential sources. In: Al- Taher F, Jackson L, DeVries JW, editors. *Intentional and unintentional contaminants in food and feed*; American Cancer Society (ACS) Symposium Series. Washington, DC. 1020:47–55.

Mahler, G.J., Esch, M.B., Tako, E., Southard, T.L., Archer, S.D., Glahn, R.P., et al. 2012. Oral exposure to polystyrene nanoparticles affects iron absorption. *Nat Nanotech* 7:264–271.

Mahmoud, M.B. 2015. Nanotechnology in food industry; advances in food processing, packaging and food safety, *Int J Curr Microbiol App Sci* 4(5):345–357.

McClements, D.J. 2013. Edible lipid nanoparticles: Digestion, absorption, and potential toxicity. *Prog Lipid Res* 52:409–423.

McClements, D.J., Rao, J. 2011. Food-grade nanoemulsions: Formulation, fabrication, properties, performance, biological fate, and potential toxicity. *Crit Rev Food Sci Nutr* 51:285–330.

McClements, D.J., Xiao, H. 2012. Potential biological fate of ingested nanoemulsions: Influence of particle characteristics. *Food Funct* 3:202–220.

McClements, D.J., DeLoid, G., Pyrgiotakis, G., Shatkin, J.A., Xiao, H., Demokritou, P. 2016. The role of the food matrix and gastrointestinal tract in the assessment of biological properties of ingested engineered nanomaterials (iENMs): State of the science and knowledge gaps. *NanoImpact* 3:47–57.

McClements D.H., Xiao H. 2017. Is nano safe in foods? Establishing the factors impacting the gastrointestinal fate and toxicity of organic and inorganic food-grade nanoparticles. *NPJ Science of Food* 1:6.

Mihindukulasuriya, S.D.F., Lim, L.T. 2014. Nanotechnology development in food packaging: A review. *Trends Food Sci Technol* 40:149–167.

Myrick, J.M., Vendra, V.K., Krishnan, S. 2014. Self-assembled polysaccharide nanostructures for controlled-release applications. *Nanotech Rev* 3:319–346.

Neethirajan, S., Jayas, D.S. 2011. Nanotechnology for the food and bioprocessing industries. *Food Bioprocess Technol* 4:39–47.

Oberdörster, G. 2001. Pulmonary effects of inhaled ultrafine particles. *Int Arch Occup Environ Health* 74:1–8.

Oftedal, O.T. 2012. The evolution of milk secretion and its ancient origins. *Animal* 6:355–368.

Onyeaka, H., Passaretti, P., Miri, T., Al-Sharify, Z.T. 2022. The safety of nanomaterials in food production and packaging. *Curr Res Food Sci* 5:763–774.

Perez-de-Luque, A., Rubiales, D. 2009. Nanotechnology for parasitic plant control. *Pest Manag Sci* 65(5):540–5.

Peters, R. et al. 2012. Presence of nano-sized silica during in vitro digestion of foods containing silica as a food additive. *ACS Nano* 6:2441–2451.

Peters, R.J.B., Bouwmeester, H., Gottardo, S., Amenta, V., Arena, M., Brandhoff, P., et al. 2016. Nanomaterials for products and application in agriculture, feed and food. *Trends Food Sci Technol* 54:155–164.

Pietroiusti, A., Magrini, A., Campagnolo, L. 2016. New frontiers in nanotoxicology: Gut microbiota/microbiome-mediated effects of engineered nanomaterials. *Toxicol Appl Pharma* 299:90–95.

Powell, J.J., Faria, N., Thomas-McKay, E., Pele, L.C. 2010. Origin and fate of dietary nanoparticles and microparticles in the gastrointestinal tract. *J Autoimmun* 34:J226–J233.

Pradeep, T., Anshup. 2009. Noble metal nanoparticles for water purification: A critical review. *Thin Solid Films* 517:6441–6478.

Prasanna, P., Shanmuganathan, S. 2011. Nanotechnology in food industry. *Bionano Frontier* 4(1):6–11.

Pulit-Prociak, J., Stoklosa, K., Banach, M. 2015. Nanosilver products and toxicity. *Environ Chem Lett* 13:59–68.

Qiao, H., Liu, W., Gu, H., Wang, D., Wang, Y. 2015. The transport and deposition of nanoparticles in respiratory system by inhalation. *J Nanomater* 1–8.

Ranjan, S., Dasgupta, N., Chakraborty, A.R., Melvin Samuel, S., Ramalingam, C., Shanker, R., Kumar, A., 2014. Nanoscience and nanotechnologies in food industries: Opportunities and research trends. *J Nanoparticle Res* 16:2464.

Raspopov, R.V., Trushina, E.N., Gmoshinsky, I.V., Khotimchenko, S.A. 2011. Bioavailability of nanoparticles of ferric oxide when used in nutrition. Experimental results in rats. *Vopr Pitan* 80:25–30.

Rodrigues, S.M., Demokritou, P., Dokoozlian, N., Hendren, C.O., Karn, B., Mauter, M.S. 2017. Nanotechnology for sustainable food production: Promising opportunities and scientific challenges. *Environ Sci Nano* 4:767–781.

Roselli, M., Finamore, A., Garaguso, I., Britti, M.S., Mengheri, E. 2003. Zinc oxide protects cultured enterocytes from the damage induced by Escherichia coli. *J Nutri* 133:4077–4082.

Rungraeng, N., Cho, Y.C., Yoon, S.H., Jun, S. 2012. Carbon nanotube polytetrafluoroethylene nanocomposite coating for milk fouling reduction in plate heat exchanger. *J Food Eng* 111:218–224.

Salgado, P.R., Di Giorgio, L., Musso, Y.S., Mauri, A.N. 2019. Bioactive packaging: Combining nanotechnologies with packaging for improved food functionality. In Nanomaterials for Food Applications; Elsevier: Amsterdam, The Netherlands, 233–270.

Samal, D. 2017. International use of nanotechnology in food industry: A review. *J Environ Agri Biotechnol* 2(4):2270–2278.

Savolainen, K., Pylkkänen, L., Norppa, H., Falck, G., Lindberg, H., Tuomi, T., et al. 2010. Nanotechnologies, engineered nanomaterials and occupational health and safety — a review. *Saf Sci* 6:1–7.

Sawai, J. 2003. Quantitative evaluation of antibacterial activities of metallic oxide powders (ZnO, MgO and CaO) by conductimetric assay. *J Microbiol Method* 54:177–182.

Schaefer, D., Cheung, W.M. 2018. Smart packaging: Opportunities and challenges. *Procedia CIRP* 72:1022–1027.

Shafiq, M., Anjum, S., Hano, C., Anjum, I., Abbasi, B.H. 2020. An overview of the applications of nanomaterials and nanodevices in the food industry. *Foods* 9:148.

Shemesh, R., Krepker, M., Goldman, D., Danin-Poleg, Y., Kashi, Y., Nitzan, N., et al. 2015. Antibacterial and antifungal LDPE films for active packaging. *Polymers Advan Technol* 26:110–116.

Shi, Y.H., Xu, Z.R., Feng, J.L., Wang, C.Z. 2006. Efficacy of modified montmorillonite nanocomposite to reduce the toxicity of aflatoxin in broiler chicks. *Animal Feed Sci Technol* 129:138–148.

Sirelkhatim, A. et al. 2015. Review on zinc oxide nanoparticles: Antibacterial activity and toxicity mechanism. *Nano Micro Lett* 7:219–242.

Singh, T., Shukla, S., Kumar, P., Wahla, V., Bajpai, V.K., 2017. Application of nanotechnology in food science: Perception and overview. *Front Microbiol* 8:1501.

Sozer, N., Kokini, J.L. 2009. Nanotechnology and its applications in the food sector. *Trends Biotechnol* 27:82–89.

Suker, D.K., Albadran, R.M. 2013. Cytotoxic effects of titanium dioxide nanoparticles on rat embryo fibroblast REF-3 cell line in vitro. *Eur J Exp Biol* 3(1):354–363.

Tiede, K., Boxall, A.B.A., Tear, S.P., Lewis, J., David, H., Hassellov, M. 2008. Detection and char- acterization of engineered nanoparticles in food and the environment. *Food Add Contam* 25(7):795–821.

Tsagkaris, A.S., Tzegkas, S.G., Danezis, G.P. 2018. Nanomaterials in food packaging: State of the art and analysis. *J Food Sci Technol* 55(8):2862–2870.

Verma, S., Gupta, S. 2017. Nanotechnology and food science: Tomorrow design the food. *Int J Curr Microbiol App Sci* 6(9):3553–3561.

Wang, Y.L. et al. 2014. A combined toxicity study of zinc oxide nanoparticles and vitamin C in food additives. *Nanoscale* 6:15333–15342.

Warheit, D.B., Brown, S.C., Donner, E.M. 2015. Acute and subchronic oral toxicity studies in rats with nanoscale and pigment grade titanium dioxide particles. *Food Chem Toxicol* 84:208–224.

Weir, A., Westerhoff, P., Fabricius, L., Hristovski, K., von Goetz, N. 2012. Titanium dioxide nanoparticles in food and personal care products. *Environ Sci Technol* 46:2242–2250.

Wu, H.H., Yin, J.J., Wamer, W.G., Zeng, M.Y., Lo, Y.M. 2014. Reactive oxygen species- related activities of nano-iron metal and nano-iron oxides. *J Food Drug Anal* 22:86–94.

WHO. 2000. Safety evaluation of certain food additives and contaminants. World Health Organization Food Additives Series 44; WHO: Geneva.

Yada, R.Y. et al. 2014. Engineered nanoscale food ingredients: Evaluation of current knowledge on material characteristics relevant to uptake from the gastro-intestinal tract. *Comp Rev Food Sci Food Saf* 13:730–744.

Yoo, S.M., Keum, K.C., Yoo, S.Y., Choi, J.Y., Chang, K.H., Yoo, N.C. et al. 2004. Development of DNA microarray for pathogen detection. *Biotechnol Bioprocess Eng* 9(2):93–99.

Yu, H., Park, J.Y., Kwon, C.W., Hong, S.C., Park, K.M., Chang, P.S. 2018. An overview of nanotechnology in food science: Preparative methods, practical applications, and safety. *J Chem* 1–10.

Zimmermann, M.B., Hilty, F.M. 2011. Nanocompounds of iron and zinc: Their potential in nutrition. *Nanoscale* 3:2390–2398.

8 Global Hunger

The Need for Smart and Sustainable Agriculture

Tarun Sharma, Rachid Mrabet, Akashdeep Singh, Neeraj Kumar, Devanshi Baghla, Sachin Kumar, Bharat Bhushan Rana, and Garima Chauhan
Corresponding author: Akashdeep Singh

8.1 INTRODUCTION

The productivity of agricultural systems has undergone major transformations (agricultural revolutions) which are influenced by the development of civilization, technology, and general human advances. However, exceptionally large increases in population, especially in African and south Asian countries in the last century, have had many unforeseen consequences, mainly on changes in ecological conditions that severely impacted the security of food supply. With such a rise in world population, it has been estimated that global agricultural production will need to be doubled to meet increasing food demand (Tilman et al., 2011). In this scenario, hunger and malnutrition are the most common challenges faced by mankind. Although the terms are very closely related, they have a different meaning. Hunger, which is an unscientific term, means a condition in which people lack the basic food intake to provide them with the energy and nutrients for fully productive lives. Insufficient nutrient intake and impaired absorption, also called malabsorption, are two main reasons for the occurrence of hunger. Malnutrition is a physical condition in which either nutrients are deficient, termed undernutrition, or there is an excess of certain nutrients, termed overnutrition (WFP, 2007). In developing countries, nearly one third of children are underweight or stunted, and nutrition is a cause of more than one third of deaths among children under 5 years of age (Horton et al., 2009). As of 2015, almost 800 million people were chronically undernourished, and at the same time, 500 million people were obese.

Global hunger is not an intractable issue, because 2,100 kcals per person per day supplies enough energy for the majority of daily activities, and the average amount of food produced globally is 2,796 kcals per person per day, significantly above the amount needed (Hoddinott and Rosegrant, 2013). The causes of hunger and malnutrition range from limited food supplies and low purchasing power to intricate relationships between nutrition, sanitation, and infectious diseases that result in poor health. For reducing hunger and malnutrition, especially undernutrition, food security is of the foremost importance. Food security exists when people have adequate physical, social, or economic access to food (FAO, 2010). Food security has generally had four dimensions: food availability; stability of supply; access to adequate quantities and varieties of safe, good-quality food; and the utilization of that food by the body. It is clear from history that the green revolution resulted in increased agricultural production and food supply per person (FAOSTAT, 2011), significantly reduced undernourishment, drastically changed agricultural markets, and exacerbated food-related health problems.

DOI: 10.1201/9781003389408-8

Agriculture is an "open-air factory" where economic activity depends heavily on climate to produce food and many other goods necessary to sustain human needs. Changes in climatic patterns affect characteristics of all agricultural sciences including agronomy, plant physiology, crop protection, genetics, and plant breeding (Mueller, 2013). Climate change influences surface and air temperatures and alters precipitation patterns and CO_2 concentration, posing a severe threat to the global food production system (Shrestha, 2019). Global warming, one of the determinants of increased GHGs, is a serious threat today. According to the Intergovernmental Panel on Climate Change (IPCC) (Field et al., 2014), temperatures have risen steadily over the past three decades, with the 2000s being the warmest decade. The Earth may warm by 1.4 to 5.8 °C over the next century, according to a variety of global climate models and development scenarios (Pachauari and Reisinger, 2007). Among factors such as floods, drought, high-intensity rainfall, erratic rain, and increased temperatures, changes in precipitation patterns may have a more substantial role in affecting food productivity than rising temperatures (Parry, 1990). The temperature rise is likely to reduce the yield of many crops, as the degree days required from sowing to maturity of the crop will be achieved more rapidly, reducing the crop duration (Mahato, 2014). If both the temperate and tropical regions experience an increase in temperature of 2 °C, aggregate production of cereal crops is expected to decrease. Climate change impact is more prevalent in tropical regions, as crops of this region are closer to high-temperature optima and thereby experience high-temperature stress during elevated levels of temperature. Along with temperature and rainfall, other weather factors like humidity and wind speed also affect crop yields, and without these factors, there has been a risk of overestimating the cost of climate change (Malhi et al., 2021).

Agriculture sits at the nexus of climate change and food security issues (Rani and Reddy, 2023). The United Nations Food and Agriculture Organization (FAO) has projected that food production will need to be increased by a minimum of 60% to meet the dietary needs of the projected global population of 9 billion by the year 2050. However, most recent literature shows that key food commodities such as rice, corn, and wheat are projected to decrease globally in quantity and quality through reductions in the amount of protein and mineral concentrations (Challinor et al., 2014). According to Rani and Reddy (2023), climate change will reduce rice production and livestock productivity under climatic variability. Yields of staple cereals in industrialized and less-industrialized regions appear to have stagnated, approaching a "yield plateau," a level beyond which production can no longer rise, under conventional or corporate agriculture. According to Grassini et al. (2013), yield plateaus have already been reached in around one third of the world's rice and wheat farms.

Plant physiological changes are particularly responsive to changes in temperature and precipitation, but dramatic alterations are more likely to have an influence than changes in the typical climate (Rammig et al., 2012). As there is a species-specific threshold for various plants, the response to climate change varies with plant species and the stage of the plant's growth. Plants respond to changes in temperature and precipitation by growing longer roots and undergoing physiological changes, but drastic changes are more likely to have an impact than fluctuations in the usual environment. There are numerous types of plants that experience changes in the angle at which their roots grow and reductions in yield (Gray and Brady, 2016). Besides changes in crops, agriculture fauna are also expected to change. Agricultural pests are directly or indirectly affected by climatic changes. Direct impact accounts for pests' reproduction, development, survival, and dispersal. Indirectly, climate change affects the relationship between pests and their environment and other insect pests such as natural enemies, competitors, vectors, and mutualists (Prakash et al., 2014). As insects are poikilothermic organisms, their body temperature depends on the temperature of their surroundings, which makes temperature the most important factor affecting insect behavior, distribution, development, and reproduction (Kocmánková et al., 2009). The complexity of physiological effects exerted by rising temperature and CO_2 can significantly affect the interaction between crops and insect pests and the interaction of insect pests and their ecosystem, which include their natural enemies too (Hare, 1992; Caulfield and Bunce, 1994; Roth and Lindroth, 1995).

Supply chains for the movement of agricultural products, as well as the flow of money and information about the products, are known as agriculture value chains or agriculture supply chains. The supply of farm inputs, processing of those inputs, and subsequent steps until the finished product is ready for consumption are all included in the agriculture value chains (Aramyan et al., 2006). To boost operational efficacy and efficiency, these supply chains can be expanded to incorporate more cooperating members (businesses) (Crandall et al., 2010; Mentzer et al., 2001). There are four pillars of food security, namely food availability, food access, food utilization, and food consumption. The agricultural value chain can contribute to food security in almost all parts except food consumption, as the latter depends more on people's decisions on consuming food, and their decisions are influenced by their perception of the food need, regardless of food availability and access. These value chains can increase access to food in a number of ways, including through the production of food using efficient and effective technology, the provision of harvest and post-harvest handling technologies that minimize losses, and the dissemination of knowledge regarding appropriate products and specifications, thereby reducing food waste. By lowering food losses and preventing food wastage, food availability or supply could be increased. It is estimated that around one third of the world's total food output is lost or squandered, including in both wealthy and developing nations. FAO (2010) report shows that food loss and wastage have not been reduced to the extent needed. So, it is of immense importance to implement agricultural value chains in appropriate integrated ways, commencing with obtaining sufficient and accurate information on where food loss and waste occur in the value chain and across the region.

8.2 CLIMATE-SMART AGRICULTURE: DRIVING TRANSFORMATIONS FOR RESTORING FOOD SECURITY AND REDUCING THE HUNGER GAP

Virtually all continental and marine regions of the world are impacted by recent anthropogenic changes in the environment. Climate change, land degradation, and biodiversity loss are three different faces of the same central challenge (IPBES, 2018). According to this report, more than 75% of Earth's land areas are substantially degraded, and if the trend continues, 95% of the Earth's land could become degraded by 2050, triggering cascading breakdowns through time and scales. That is why it is said that reversing the desertification process is like climbing back up again; the world is in a dramatically different position, and agricultural (food, farming) systems, infrastructures, policies, and institutions were designed for a climate that no longer exists. Consequently, tipping the Earth's system into a new settled, or balanced, state is like falling off a void or a cliff. Globally, about 84% of farms are smallholders having less than 2 ha of farm area and producing one third of the world's food (Lowder et al., 2016; Ricciardi et al., 2018). Smallholder farming is a critical fulcrum for sustainable development, food security, and the hunger gap paradox (Cohn et al., 2017; Ricciardi et al., 2021). However, most of the world's hunger is in South Asia and sub-Saharan Africa. These subregions are also known for widespread poverty and extensive areas of low agricultural productivity. In fact, a meta-analysis by Ricciardi et al. (2021) concluded that smaller farms, on average, have higher yields and comprise greater biodiversity at the farm and landscape scales than do larger corporate farms. However, their sustainability is questioned with rising land challenges (Pretty, 2018).

Recent multiple-scale crises and globalized fragile food systems have brought into sharp focus the shift to new narratives such as regenerative, just, and climate-smart agriculture (Anderson and Rivera-Ferre, 2021; Tittonell et al., 2022). In addition, the world is increasingly affected by desertification, aridity, and progressive drought even in regions where the phenomena were judged not critical, such as Europe. From a meta-analysis by van Dijk et al. (2021), if climate change is taken into account, the total food demand is expected to increase by +30% to +62%, and the population at risk of hunger from –91% to +30% between 2010 and 2050. While the human population growth rate has dropped to below 1% per year, the growth rate of the livestock population has risen to 2.4%

a year. Hence, an additional crisis will come from livestock-accelerated growth to feed the changing diets of the world's population. By 2050, according to the UN, world meat consumption is likely to be 120% greater than it was in 2000. In other terms, the world will not (or is not on track to) achieve zero hunger and degradation neutrality if it does not embrace pre-emptive systematic change based on the climate-land-nature nexus to swerve to inclusive and efficient agri-food systems.

With the dominance of corporate agriculture, the world has shifted to a so-called Global Standard Diet; however, this convergence has induced high nutrition gaps and increased hunger (Kammlade et al., 2017). In addition, monocropping systems are increasingly implemented and receive policy support to influence such shifts. Most of our food comes from a small number of species. Four crops (wheat, rice, maize, and soybeans) account for almost 60% of the calories grown by farmers (Khoury et al., 2014). In fact, a large amount of the world's food is grown in vast, lightly habited lands, such as the Canadian prairies, the US plains, the Russian steppes, the Ukrainian chernozem, and the Brazilian interior, and shipped to tight, densely populated places or urban areas. In other terms, a recent study by Kinnunen et al. (2020) found that a quarter of the global population that consumes crops such as wheat, rice, barley, rye, beans, millet, and sorghum requires food grown at least 5,200 kilometers away. In the face of such trends in globalization and food trade, local food movements have strengthened, and food sovereignty communications and treatises are preconizing and valuing small-scale farming and local markets.

Anderson and Rivera-Ferre (2021) distinguished extractive and regenerative food systems under which different labels fit. One such label, climate-smart agriculture (CSA), was first introduced by the FAO as a global development goal to guide the transformation of agricultural systems through the integration of climate adaptation, mitigation, and food and nutrition security (FAO, 2010). According to the FAO, CSA is an open concept and approach to developing the technical, policy, and investment conditions for achieving three main objectives (see Box 8.1). In other terms, CSA options possess emergent properties allowing high resilience, but also have thresholds if not implemented correctly. Hence, CSA should not be perceived as only shifts in paradigms but also as the promotion of "old" traditional practices. All these types of agriculture embrace the spirit of Agenda 2030 (Bernard and Lux, 2017). With the introduction of CSA, the FAO and other international organizations are showing that achieving sustainability requires multiobjective and multicriteria decision-making models. Hence, international organizations and research institutions (such as CGIAR) that played major roles in ensuring the Green Generation have shifted in paradigm and now support CSA (Vermeulen et al., 2010).

BOX 8.1 PILLARS OF CLIMATE-SMART AGRICULTURE

According to FAO (2013) and Lipper et al. (2014), the three core interlinked principles of CSA, with a focus on the most vulnerable, including smallholder producers, women, the poor, and marginalized groups, are as follows:

- Increase agricultural productivity by improved farm sustainability and economic growth in support of equitable increases in incomes;
- Build climate resilience into agricultural and farming systems while reducing risks through robust actions in the face of uncertainty and enhancing desirable ecosystem services;
- Reduce agricultural greenhouse gas emissions (produce more healthy food with lower emissions through low-emission growth strategies and by reducing the conversion of pastures, forests, and wetlands) and enhance carbon storage and sequestrations in soils and lands (through increasing resource-use efficiency).

The terminology is still vague, including a large array of options (policies and technologies from local to global scales) and hence open to interpretation and framing, as discussed by Struik and Kuyper (2017). Henceforth, a universal definition is unlikely to be agreed upon due to cross-cutting issues related to the three pillars, and strategic thinking is advised at a local scale while implementing the pillars. In this regard, the CSA approach has allowed academics and other resource practitioners to express the interdependencies and constraints between the pillars, and to offer solutions that bridge them. Hence, for unpacking the complexity, CSA was promoted by governments, NGOs, and international organizations as a systemic, solutions-focused, and impact-driven approach that can contribute functionally to reducing vulnerability and ensuring sustainable progress in capacities of farming systems while halting land systems changes.

The CSA pillars operate at different geo-economic, institutional, and spatial scales and hence suppose diverse perspectives which entail a transformative agenda, integrative global governance, and knowledge co-production and sharing. There is no "one size fits all," and CSA involves strong flexibility and social participation to empower farmers and upscaling. In addressing emerging challenges, CSA supports exploring new knowledge, developing innovative solutions, appropriating frontier technologies, and piloting effective and holistic policies and incentives. In this regard, reliable, timely, and equitable access to institutions, inputs, and finance to support resource-use efficiency is of prime importance for small-scale farming systems without undermining their capacity to continue producing food in the future.

Currently, several uncoupled competing agricultural or farming systems coexist, and the challenge is to move to integrated sustainable models to fully unlock land potentials and optimize the delivery of benefits and services. There is a trend towards global recognition of disaggregated but not rebranded CSA options which include conservation agriculture (CA), organic farming/agro-ecology, agroforestry, sustainable land-based management systems, soil-improving cropping systems, ecosystem-based management, small-scale low-cost irrigation, soil and water conservation, and grazing land/livestock management (Smith et al., 2020; Boix-Fayos and de Vente, 2023) (Figure 8.1). All these options emphasize the wise use of natural functionalities of (agro)ecosystems. Countless farmers have used and are using these options. A growing number of countries are recognizing CSA options in their national adaptation programs (NAPs), nationally appropriate mitigation actions (NAMAs), climate change investment plans, national determined contribution (NDC) pledges, and land degradation neutrality (LDN) targets to accelerate the transition towards sustainability in food and agriculture (Mrabet & Moussadek, 2022; Goddard et al., 2022). In this sense, bold and rigorous commitments as well as greater coherence across multilateral policy processes are needed to appropriately downscale the CSA options with long-lasting impacts, reducing gaps and unnecessary overlaps.

Kassam et al. (2022) tracked the worldwide adoption of CA to an estimated 205 million hectares, which is a testimony to stepping beyond conventional agriculture and high-level shifts in paradigms and related policies and motivations. In other words, threshold effectiveness is achieved when all three CA principles are used and CA technologies are applicable in diverse biophysical conditions ranging from low to high rainfall areas with varied edaphic conditions. CA has become the cornerstone of dryland systems in certain regions of the USA, Canada, Australia, Spain, South Africa, and other countries. In these countries, CA is gaining ground as a climate-proof approach

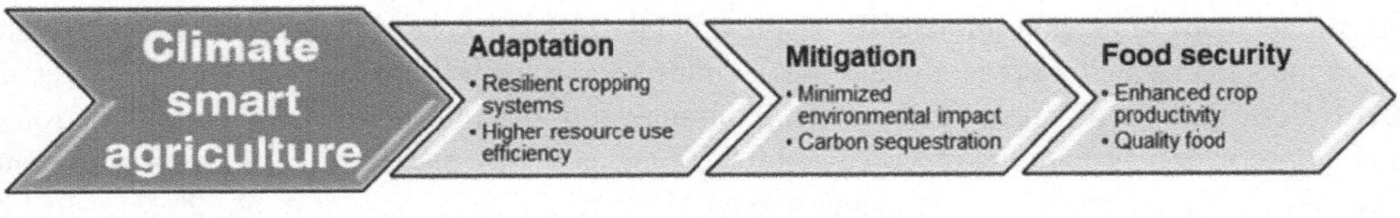

FIGURE 8.1 Pillars of climate-smart agriculture and their potential benefits.

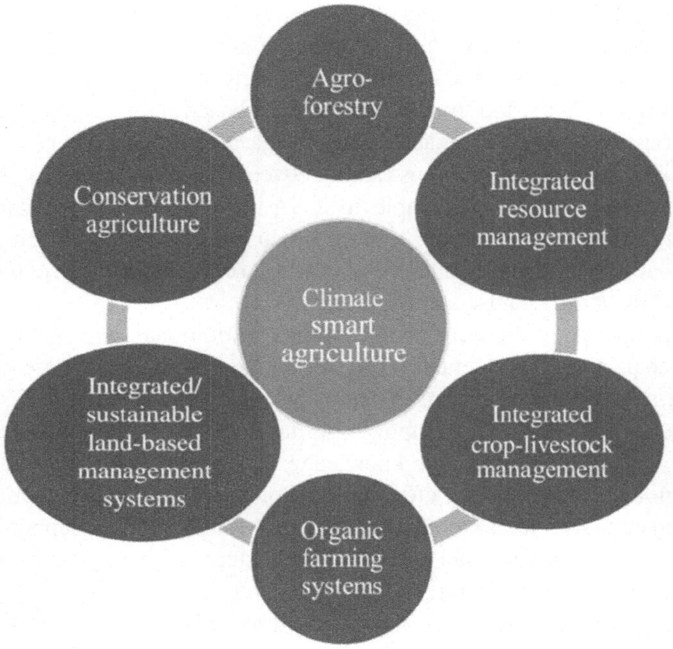

FIGURE 8.2 Climate-smart agriculture options.

or as a negative emission option, or both. Enhancing water-use efficiency and hence offsetting climate change impacts through CA principles has been recognized by several authors (Khatri-Chhetri et al., 2023). Later authors reported that CSA technologies and practices for wheat and rice production systems in India helped to generate multiple benefits: enhancing farm productivity, increasing technical efficiency, and reducing GHG emissions in many smallholder farming systems (Figure 8.2).

It is worth noticing that mitigation is leveraged to support food security and adaptation, rather than hampering or harnessing them (Lipper et al., 2014; Zhao et al., 2023). CSA aims to implement soil-based strategies and long-term soil fertility dynamics that restore soil functions and health (Paustian et al., 2016). The potential for carbon sequestration in soils (especially on desertified and degraded lands) is large (Amelung et al., 2020), and progress can be made using proven biodiversity, crop, range, and forest management practices under CSA. In comparison to a baseline without soil carbon sequestration, Frank et al. (2017) estimate that soil carbon sequestration might reduce calorie loss associated with agricultural mitigation efforts by 65%, preventing 60–225 million people from becoming hungry.

Systematic documentation of the impact of CSA options started in the early 1930s and continued and intensified over time (Kassam and Kassam, 2020; Bhattacharjee et al., 2023). It is well established that by converting to CSA options, food availability and security increase due to higher and stable resource-use efficiency and crop and animal productivity, as well as to higher resilience to climate change and variability (Lipper et al., 2014). In other words, it is ascertained that healthy, nutrient-rich soils are developed with CSA options and hence global agricultural and food systems are made more environmentally friendly. In fact, efficient crop yields due to CSA strengthen global food chains and improve global food and nutrition security (Lipper et al., 2014; Smith et al., 2020). Considerable progress has been made in analyzing the impacts of CSA options (agroforestry, conservation agriculture, resource-conserving technologies, etc.) in different regions in terms of increasing food security and reducing hunger (Dev et al., 2023; Kumar et al., 2023). Most of these CSA options have a positive impact on households' food security, measured in terms of food consumption and dietary diversity. Food security is increased by raising crop productivity and reducing the risk of crop failures by lessening the adverse impacts of climate change (Smith et al., 2020). Lal and Stewart (2013) reported that crop yields in Africa, South Asia, and the Caribbean could be doubled or tripled through the adoption of technologies based on laws of sustainable soil management or CSA. However, integrated land use planning is indispensable to heighten the delivery of ecosystem services with trifling environmental and social externalities.

Several meta-analyses concluded that CA systems are better performing than conventional tillage systems (CT) under a range of crop species and climate regimes (Das et al., 2021; Su et al., 2021a, 2021b; Sun et al., 2020; Jat et al., 2020; Li et al., 2018; Laik et al., 2014; Pittelkow et al., 2015). Several authors (Li et al., 2019; Page et al., 2019; Ogle et al., 2019) found that CA systems improved water balance and dynamics, soil health and conservation, and hence crop responses to changes and variability of climate events. In contrast to 16 sub-Saharan African countries, Corbeels et al. (2020) showed that average yields under CA systems are 3.7% higher for six major crop species and 4.0 to 8.4% higher for maize than CT systems. A meta-analysis using data from 9,686 paired site–year comparisons across South Asia in a variety of cropping systems found that CA systems provided 5.8% higher mean yield than conventional agricultural practices (Jat et al., 2020). In a study by Laik et al. (2014), under CA systems, yields of wheat and rice increased by 46–54% and 10–24%, respectively, over conventional tillage. In a review by Das et al. (2021), the CA systems increased yields of crops from 2 to 200% depending on crop rotations and years of implementation.

Around 1,000 million hectares of agroforestry systems are present worldwide (Zomer et al., 2014). As CA systems, they support improved soil quality and carbon sequestration, preserve soil fertility and nutrient cycling, and manage soil erosion, while giving local residents access to food and a source of income. They improve the preservation of biodiversity, ecological harmony and restoration, and climate change resilience (Mbow et al., 2014; Newaj et al., 2016; Altieri et al., 2015; Lorenz and Lal, 2014; Rosenstock et al., 2019). Increased food production, diverse lifestyles, economic opportunities, healthier diets, the rehabilitation of damaged lands, fewer and/or milder dust storms, and more favorable regional climates and shade for people and their livestock are all benefits of this integration (Ellison et al., 2017; Kuyah et al., 2021; Rosenstock et al., 2019).

Approximately 2.8 million farmers, mostly in India, manage 71.5 million hectares of organically farmed land, accounting for 1.5% of the world's total agricultural land. Organic farming is performed in 186 countries (Willer, 2020). The top three nations, which together account for 62% of all organic farmed land, are Australia (37.5 million hectares), Argentina (3.6 million hectares), and China (3.1 million hectares). Despite obstacles, organic farming has grown significantly over the past 10 years, but an analysis of this expansion shows that the industry has not yet reached maturity (Willer, 2020). A study by Pimentel et al. (2005) found that organic farming produced the same yields of corn and soybeans as conventional farming, but used 30% less energy, less water, and no pesticides. The authors also found that the regenerative system conserved more water in the soil, induced less erosion, maintained better soil quality, and conserved more biological resources than conventional agriculture. Pimentel et al. (2005) found that crop yields were improving over time due to soil health improvement, and that the crop produced higher yields under drought conditions. They also concluded that the net economic return per hectare under organic farming is often equal to or higher than that of conventionally produced crops. Reganold and Wachter (2016) added that even though crop yields were higher under conventional systems, organic agricultural systems deliver healthy products, greater ecosystem services, and social benefits. They reported that organic farming permitted better water balance due to higher moisture storage and infiltration in soils. Such soil improvements can protect farmers from the consequences of lower crop yields if shifted from conventional agriculture.

No single soil management system or approach will safely feed the planet, and a blend of innovative farming systems is needed to ensure sustainability for farmers and their communities (Paustian et al., 2016). If CA systems are systematically integrated with trees (i.e., agroforestry) and livestock, and embrace organic farming attributes, the adoption of such CSA integrations and hence benefits will be higher and more widespread (Goddard et al., 2022; Boix-Fayos and de Vente, 2023). It was also reported that CA systems mitigate deforestation due to improved agricultural soil health and productivity (Mulimbi et al., 2022). Such farming successes are certainly linked to complementary benefits with new land use planning, including water and nutrient cycling and soil fertility management and regeneration at the landscape level, and farmers have embarked on integrated pest/disease management at the farm gate to control and alter predator dynamics to guarantee crop production and diversity. Such integration of CSA options allows new efficiencies and creative disruptions while using fewer external inputs and expending less energy. These CSA pathways will help to broaden the audience of support while increasing land productivity and smallholder innovativeness, resilience, and profitability and hence paving ways to zero hunger.

All CSA options are scale-dependent. Constraints and enabling drivers have been extensively discussed in the CSA literature (Lipper et al., 2014; Mrabet and Moussadek, 2022). It is doubtful that national and global food security and climate goals, embedded in the SDGs, can be achieved sustainably without CSA, especially in the low-income world. CSA adoption will help to unpack dynamics, benefits, and synergies among the triple land challenges which will certainly allay hunger and malnutrition. In fact, during the last UN food system summit, there seems to have been a political, environmental, and social momentum to promote a transition toward sustainable and greener agriculture.

8.3 AN ECOSYSTEM-BASED APPROACH FOR SUSTAINABLE AGRICULTURE

The term ecosystem-based approach refers to a human-centered, nature-based strategy that takes into account the integration of soil, water, and life resources that encourages conservation and sustainable use to combat climate change's effects while achieving environmental conservation goals and socioeconomic development objectives, with a view to restoring or enhancing the ability of human-natural linkages to sustain themselves (Richter et al., 2015). Simply put, it emphasizes the importance of working in sync with natural systems to achieve productivity while maintaining environmental health. The main focal areas of the ecosystem-based strategy for sustainable agriculture are the use of environmentally friendly pesticides and organic manures, development of agrobiodiversity, agroforestry, and reforestation, as well as multilayer farming with crops regionally adapted to climate changes (Figure 8.3). With an estimated 9.8 billion people on the planet by 2050, increased food and housing costs will have a detrimental knock-on effect on land degradation, biodiversity, and waste production (Seto et al., 2012). Besides this, 795 million people suffered from hunger in 2017, despite the fact that more food was produced globally than was needed to satisfy the caloric needs of the world's population (FAO, 2015). This issue was prevalent in developing nations with high rates of poverty and underdeveloped large-scale agriculture. Therefore, to feed the world's growing population, global food output must increase by 70–110% by 2050 (Wu et al., 2018).

Crop and animal productivity, financial stability, and farm household prosperity are directly affected by extreme weather events, such as temperature fluctuations, frequency, and intensity of rainfall duration, as smallholder farmers are widely regarded as being disproportionately vulnerable to climate change. Small landholders in developing countries are likely to experience significant productivity reductions due to increased climate variability and climate change in the coming decades. Therefore, farmer adaptation to climate change in the long term (by assuring the ongoing provision of ecosystem services both onsite and offsite) and in the short term (through resilience enrichment of production units) is likely to benefit from ecosystem-based management practices that emphasize biodiversity conservation, restoration, and farm sustainability (Altieri and Nicholls, 2008).

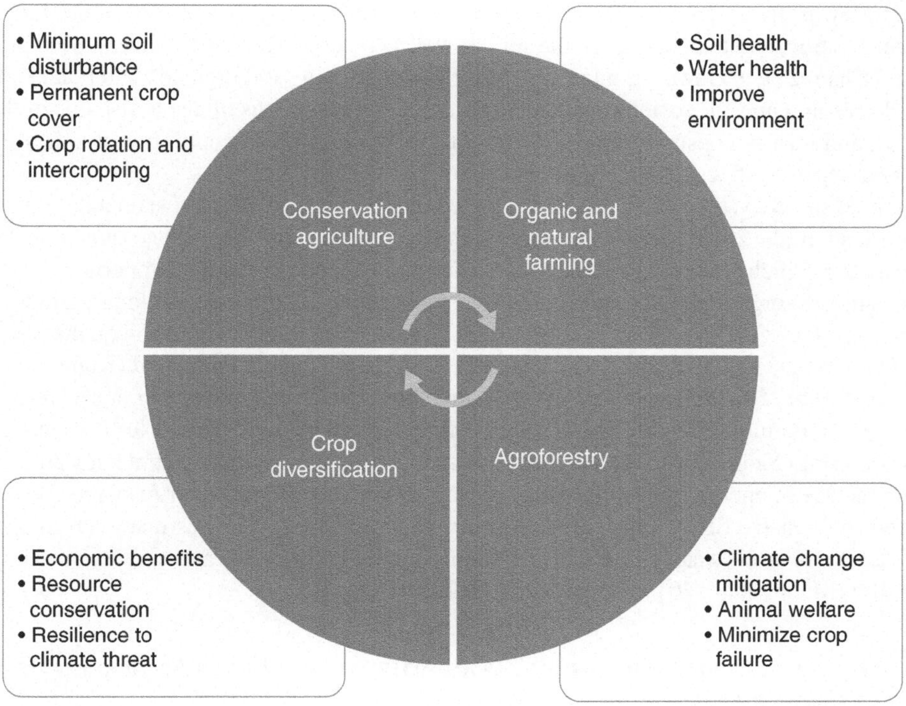

FIGURE 8.3 Ecosystem-based approaches for sustainable agriculture.

8.3.1 Soil and Environmental Health through Conservation Agriculture-Based Practices

In farm land, nonsustainability may be caused by soil erosion, reduced organic matter, salinization, and acidity due mainly to continuous deep conventional tillage, which imparts soil structure, reduces water infiltration, and causes soil compaction and hardness. Among conventional practices, intensive tillage practices, particularly repeated plowing and harrowing, have resulted in significant depletion of the soil organic carbon pool to a level of 60–90 Pg in recent times (Lal, 1999; Hussain et al., 2021). Farm machinery-based operations have been responsible for loss of 66–90 Pg of soil organic carbon since 1750 (Lal, 2004). Therefore, to ensure farming productivity and sustainability of available resources, significant changes in agricultural production systems are necessary. Three interdependent principles underpin conservation agriculture (CA): minimum or no mechanical soil disturbance, permanent soil cover, and diversification of cropping systems either by sequences or rotations; along with good agronomic practices, this is a concept that appeared as a concern of sustainability of agriculture globally. The first principle, minimum or no mechanical soil disturbance, is based on reduced or least tillage practices such as minimum or zero tillage. Zero-till or no-till enhances the soil health with reduced soil aggregate disintegration and enhanced soil nutrient availability status and soil organic carbon sequestration. Zero-till practices have been found to elevate the soil organic carbon pool at a rate of 57±14 g carbon m^{-2} yr^{-1} (West and Post, 2002; Hussain et al., 2021). This potential is further boosted with inclusion of cover crops which have the ability to increase soil organic carbon by between 10.9 and 14.3 Mg ha^{-1} (de Torres et al., 2021). Agricultural sustainability implies climate-smart agriculture which increases crop productivity, reduces greenhouse gas emissions, and enhances the achievement of food (Nyanga et al., 2020) and nutritional security to meet the present requirements without imperiling future potential. The practice of tillage eases soil-related constraints to crop production, populations (macro- and micro-flora and fauna), and crop growth. In addition to the preparation of seed beds, tillage has also been used to control moisture, aeration, and temperature in the soil, as well as incorporation of fertilizers, pesticides, and manures (Khan, 2019). However, intensive tillage was destroying soil organic matter and soil structure, and speeding up soil erosion (Schneider et al., 2012). Continuous tillage is also one of the most expensive and energy-consuming farm operations, increasing the susceptibility of soil to erosion and having negative effects on soil health.

A number of ecosystems have recently adopted and promoted CA for sustainable crop intensification (Saad et al., 2016). Soil inversion by repeated continuous deep tillage quickens decay of organic matter, which affects soil quality (physical, chemical, and biological properties). In addition to increasing soil organic matter and soil carbon sequestration, no-tillage for long periods of time also retains soil water (Bhattacharyya et al., 2009; 2015) and improves overall soil quality and health and soil structure (Jat et al., 2018), which results in more sustainable crop yield (Kumar and Rana, 2021; Kumar et al., 2022a) than intensive tillage practices. However, CA has minimal adverse effects on ecosystems, minimizes losses due to soil runoff and erosion, improves soil fertility, and reduces carbon footprint (Yadav et al., 2020). Zero-/low-till conservation farming strategies are aimed at reducing the use of energy inputs like diesel for higher net returns and better resource conservation compared to intensive tillage practices that consume more energy. CA has many advantages over CT, not least enhanced water-use efficiency, higher microbial diversity and activity, and improved soil fertility (Muoni et al., 2019; Kumar et al., 2022b).

8.3.2 Natural and Organic Farming as an Alternate to Uneven Application of Chemical Fertilizers

In late nineteenth century, the green revolution with the use of chemical fertilizers ensured food security. However, to feed the ever-growing global population and to meet our standard of living needs, agricultural production needs to be enhanced further with maintenance of soil quality and quantity of arable land (Krasilnikov et al., 2022). However, recent reports of degrading soil quality or soil health with overuse or imbalanced use of chemical fertilizers have made the agricultural fraternity reconsider the use of chemical fertilizers. Uneven application of fertilizers can be controlled with the adoption of nature-based farming practices such as bio-based fertilizers and organic or natural farming, which ultimately protect our soil, soil microorganisms, water sources, and related ecosystem. The surplus food grain production due to intensive chemical application, causing a reduction in the productivity of the soil, has started to stagnate and even decline.

The potential to supply plant nutrients and enhance soil health has highlighted the ability of bio-based fertilizers to replace chemical fertilizers. Easy availability of raw material, nondependence on nonrenewable energy sources, and minimized environmental impacts further strengthen the candidature of bio-based fertilizers in the food production chain (Kurniawati et al., 2023). Eighty-six million tonnes of biowaste was generated in 2017 by the European Union, and its unmanaged disposal has had serious environmental consequences (European Environment Agency, 2020). Bio-based fertilizers are being supported by the European Commission to support the implementation of a circular economy-based approach in enhancing the sustenance of agricultural production practices (Kurniawati et al. 2023). Several studies are being carried out to generate cheap and nutrient-rich bio-based fertilizers. Nitrogen-rich fertilizers are being synthesized from biowaste using scrubbing technology that can match the nutrient-supplying capacity of chemical fertilizers (Sigurnjak et al., 2019). Several studies have supported the positive influence of bio-based fertilizers on soil health indicators such as physical, chemical, and biological properties and other ecosystem services. Elevated bacterial populations, better organic matter levels, and improvements in soil texture were reported by Cheong et al. (2020) when food waste and anaerobic digest were applied as a soil amendment. Biochar-based application was reported to enhance the bacterial population and, consequently, the mineralization of organic carbon (Ibrahim et al., 2022).

Natural farming is becoming increasingly popular among farmers who are looking for environmentally friendly and farmer-friendly alternatives to conventional farming. Natural farming is a holistic alternative to high-cost chemical input-based agriculture and is very effective in addressing the uncertainties of climate change. Keeping such a vast potential in consideration, natural farming has covered over 500,000 hectares of arable land in India, where this production system is being used as a primary practice (Kumar et al., 2023). In natural farming, natural fertilizers and local seeds are used to reduce farming expenses and promote local food production without purchasing costly fertilizers and pesticides externally. A microbial inoculum (*Beejamrit* and *Jeevamrit*) is added to soil to promote microbial populations and their enzymatic activities, thereby improving soil fertility and productivity. Several homemade formulations such as *Agniastra*, *Brahmastra*, and *Neemastra* are being utilized as biopesticides to manage crop pests (Mishra, 2018). Increase in soil organic matter in conjunction with elevated microbial biodiversity can act as a core practice to boost soil health (Saharan et al., 2023). Enhanced organic matter and microbial populations enhance nutrient availability, strengthening the resistance as well as the stability of agro-ecosystems (Maougal et al., 2020; Nisrina et al., 2021). Here, intercropping is a primary component which compensates production cost by its income and makes farming a close to zero-budget activity. Intercropping avoids the exploitation of specific nutrients as in monoculture or based cropping practices and thus sustains the optimum nutrient cycling in natural ecosystems. Apart from such practices, *Achhadana* (biomulching) is also a primary practice under natural farming-based systems. *Bos indicus* (humped cow) dung and local species of soil earthworms increase organic matter most effectively and have the highest concentrations of microorganisms.

In a scientific study, it was emphasized that bio-based nitrogen sources can substitute chemical sources of nitrogen and act as sustainable alternatives to meet our crop nutrient requirements (Luo et al., 2021). Apart from this, production and processing of agricultural crops generate a significant amount of biowaste which requires special attention for safe disposal. Nevertheless, farm-based wastes can be utilized as raw material for production of manures and composts which result in yield levels comparable to those of their nutrient-supplying chemical counterparts (Orden et al., 2021). Organic farming-based inputs, such as farmyard manure, compost based on municipal waste, and sewage and sludge generated from cities, have been reported to enhance soil properties such as cation exchange capacity, organic matter content, available soil, and plant nutrient status as compared to soils where no such inputs were applied (Oueriemmi et al., 2021).

In contrast to intensive agriculture and organic farming, crops in natural farming only need 10% water and 10% power. Therefore, natural farming has the potential to produce more food at a lower cost, which moves towards food sustainability, reduces risks, and offers protection from climate change uncertainties.

8.3.3 Agrobiodiversity Promotion and Sustainable Agriculture

In order to increase the productivity of crop types and livestock breeds, scientific breeding efforts have mostly relied on the qualities found in agricultural biodiversity, which is an important component for the sustainability of the food supply. Agriculture biodiversity is already a crucial part of breeding and farming practices throughout the world. A greater understanding of it may enable nations to make the most of their current resources for joint nutrition and environmental results. It includes a variety of crops and their wild relatives, trees, and cattle as a source of nutrient-dense foods that are often tailored to regional and low-input agricultural systems and are socially and culturally acceptable. Agricultural biodiversity can boost farming systems' production in a variety of growing environments by increasing soil fertility, mitigating the incidence of insect pests and diseases (which are typically more resilient to adverse conditions), and improving food security. Better nutrition and improved health are delivered by diversified diets, which are based on diverse farming methods for better socioeconomics and livelihoods of small and marginal farmers in

TABLE 8.1
Ecosystem-based practices for enhanced soil health

Location	Crop/cropping system	Duration (years)	Soil organic carbon (t/ha/yr)	Reference
Mediterranean regions	Cereal and legume rotations	6–72	+0.3	Francaviglia et al., 2017
China	Cereals (maize, wheat, and rice) and oilseeds (soybean)	6.5	+0.25	Du et al., 2017
India (Indo-Gangetic Plains)	Rice-wheat	2–26	+0.54	Powlson et al., 2016
Worldwide	Various	3–43	+0.86	Mangalassery et al., 2015
Mediterranean region	Agricultural and forestry crops	10.6	+0.27	Aguilera et al., 2013
Worldwide	Agricultural crops (cover crops)	8.5	+0.56	Jian et al., 2020
Mediterranean region	Horticultural crops and annual cover crops	5.6	+0.43	Morugán-Coronado et al., 2020
Mediterranean region	Horticultural crops and perennial cover crops	6.7	+1.01	Morugán-Coronado et al., 2020
Worldwide	Agricultural crops (cover crops)	11.9	+0.32	Poeplau and Don, 2015

developed nations. However, agricultural biodiversity will also be important to combat the ill effects of climate change.

Agrobiodiversity with a greater number of species or more genetic diversity within species leads to higher sustainability and productivity of the system (Hector and Hopper, 2002). However, in agricultural systems, improved resistance to pest and disease outbreaks via wise use of both inter- and intra-specific variety serves as the primary mechanism for more production and yield stability (Table 8.1).

8.4 SMART AGRICULTURAL PRACTICES FOR ENHANCING AGRICULTURAL PRODUCTIVITY

Agriculture has always been the backbone of any successful civilization since the earliest historical records. Land was cultivated to grow food, and cities were established near water bodies for sustainable farming practices. This era in agriculture was dependent on manpower and animal-drawn implements. Farmers generally used traditional implements such as sickles and hoes. This era can be regarded as the agricultural era 1.0. The industrial revolution of the eighteenth century saw a boom in the agriculture sector. Many machines driven by either steam or diesel were invented and brought to the field, which would help in preparing land for agricultural activities and reduce the dependence on human labor. Besides the industrial revolution, during the late eighteenth century the green revolution increased dependence on the use of synthetic fertilizers to increase agricultural productivity (Rose and Chilvers, 2018). The dependence on chemicals to increase crop productivity came at the cost of loss of biodiversity, land degradation, loss of soil fertility, salinization, and other environmental implications. This era can be regarded as the agricultural era 2.0.

During the twentieth century, the rapid development in science and technology led to further growth in the agricultural sector. Robots, precision agricultural tools like global positioning systems (GPS), geographical information systems (GIS), conservation agriculture, and site-specific nutrient management helped in overcoming some of the damage done by the excessive use of chemical fertilizers. This was the agricultural era 3.0. The current agricultural era, 4.0, involves recent technologies such as nanotechnology, artificial intelligence (AI) and machine learning (ML), sensors, big data analytics, the Internet of Things (IoTs), drone technology, robots, and remote sensing. Innovation in any sector is crucial for the development of society (Wolfert et al., 2017). The agriculture sector, which supports the global population, needs input from such advanced technologies in order to accomplish the task of achieving food security and enhancing productivity. Smart agricultural tools are technological devices that are designed to improve the productivity of the system, cropping efficiency, and sustainability (Javaid et al., 2022). They enable farmers to remotely access fields and monitor crop plants at any moment. Automation technologies have made management-related work efficient and reduced dependence on human labor, which is time-consuming and economically straining. Real-time data collection, and precise soil and crop assessment for nutrient and water stresses are some of the benefits of smart farming.

8.4.1 Technologies in Smart Agriculture

8.4.1.1 Internet of Things (IoT)

IoT is a technology that enables devices to be connected remotely and still be accessed all times about the agricultural farm. Through IoT, farmers will be able to continuously monitor crops and livestock and retrieve information remotely using their smartphones (Dhanaraju et al., 2022). Integration of IoT and monitoring technologies such as sensors has enabled farmers to access weather uncertainties and forecast possible crop yield levels. The process of connected technology can be achieved by utilizing technologies such as sensors, GIS, GPS, remote sensing, drones, satellites, AI, and decision support systems (DSS). These proficiencies have reformed the agriculture sector in terms of resource conservation, combating climate effects, and improving crop productivity. Real-time monitoring and on-the-go application of inputs have substantially enhanced the input use efficiencies with site-specific application of finite inputs available.

Shrinking resources such as water require special attention in present times. The impact of digital technologies on hydrological aspects such as water harvesting, flow rates, application efficiencies, and need-based applications significantly influences the resource and system productivity (Yong et al., 2018). Sensors and monitoring technologies can accurately assess the soil and crop moisture so that need-based applications can be used (Mekala and Viswanathan, 2017).

Climate change poses a huge threat to the food security of the world. According to estimates, 40% of crop yields are threatened by pests and diseases. The setback to the agriculture system is primarily due to the lack of strong technologies to curb the losses. Amidst the losses, the use of sensors accompanied by other smart systems, such as automated irrigation systems for water management and drone technology for the spraying of fertilizers and pesticides, helps to monitor crop health and determine the exact amounts of pesticides to be remitted to reduce losses (Mittal and Singh, 2007; Mekala and Viswanathan, 2017; Yong et al., 2018).

8.4.1.2 Agricultural Drones

Drone technology is not new, but its use in the agriculture sector is refreshing. Recent scientific developments, especially in enhancing the payload capacity, stability, and flight duration of unmanned aerial vehicles or drones, have significantly boosted the adoption of this emerging technology in the field of agriculture (Arza-García and Burgess, 2022). The military around the globe has been using drones to patrol difficult terrain. Similarly, drones in agriculture can be used to monitor the growth and development of crop plants. It is difficult for a single farmer to cover an entire farm and focus on each crop. Monitoring of dynamic changes in agro-ecosystems is a cumbersome task for farmers on the ground, even with the use of satellite-based imagery, which has less spatial resolution with limited monitoring operations available (Liu and Li, 2023). To overcome this struggle, drones are fitted with various sensors (optical, thermal, radar, multispectral, and hyperspectral) which monitor and derive heavy loads of data with key observations of agro-ecosystems and ongoing cycles. Drone-based data are interpreted as field maps and used to derive indices such as the normalized differential vegetation index (NDVI) and the crop water stress index (CWSI), which helps farmers assess the crop health status and availability of resources such as water in agricultural fields. Based on the inputs received via sensors, the drones can decide to irrigate/fertilize the individual crop plants (Dhanaraju et al., 2022).

Drone technology finds other applications in monitoring water resources for flood and drought management, studying climate change, monitoring soil degradation, conducting vegetation surveys, and cloud seeding (Natarajan et al., 2023). The only problem with drone technologies is that, because of the drift, they cannot be used when the wind speed is high. Under such conditions, unmanned terrestrial vehicles play a vital role in carrying out field operations and reducing dependence on human labor. They are more efficient and precise in doing tasks than humans themselves. Such devices are capable of planting crops at the desired depth and removing particular weeds from the field by spraying herbicides, flaming, or uprooting. The identification of weeds is the result of feeding artificial intelligence or machine learning-based software or applications with multiple sets of photographs of weeds which are to be targeted (Hassanein and El-Sheimy, 2018; Crimaldi et al., 2020). The AI is capable of further analyzing the data feed and responding accordingly. Although the initial step of training the AI system is time-consuming, the results are promising. On-the-go weed uprooting/destroying robots based on artificial intelligence have shown great potential in the field of weed management.

8.4.1.3 Sensors

Sensors are an integral part of precision farming, and they help in the successful automation of agricultural farms. They are capable of assessing soil moisture status, fertility status, temperature, and physical and chemical properties of soil, such as pH, in real time. Soil nutrients such as nitrogen, phosphorus, potassium, and organic matter are key inputs in crop production. Nitrogen variation can be assessed based on near-infrared radiation (NIR) sensors. Soil organic matter determination is based on visible and infrared spectral reflectance regions. Nutrient predictions for nitrogen and phosphorus are based on NIR spectrophotometry technologies (Maleki et al., 2006; Kuang and Mouazen, 2013; Lvova and Nadporozhskaya, 2017). The data collected are used to feed an artificial neural network (ANN), which in turn helps in making relevant decisions pertaining to increasing crop yields while combating any abiotic or biotic stresses (Chen et al., 2004; Muhammad et al., 2020). The networks are capable of providing data in real time, thereby enabling farmers to make informed decisions pertaining to fertilizer application, irrigation scheduling, and pest management. Besides management in production, resource-use efficiency is also enhanced by reduction in wastage of applied nutrients and water. Regular monitoring of standing crop plants allows farmers to prepare for yield changes due to disease or other stresses. The drawback to the use of this technology is the skill set required by the stakeholders. Additionally, the initial investment cost is high for marginal farmers.

8.4.1.4 Precision Systems

Fertilizers are vital to the growth and productivity of intensive cropping systems. Shrinking land resources put stress on the agriculture production systems, resulting in the excessive use of chemical fertilizers to narrow the production gaps. SSNM using smart agriculture technologies helps in the precise application of nutrients (Sarkar et al., 2017). The IoT-based system employs GPS, remote sensing, hyperspectral imaging tools, and the normalized difference vegetation index (NDVI) in order to analyze crop health, vigor, and nutrient deficiencies, if any (Lu et al., 2022). Based on the NDVI values, it becomes clear whether a crop is suffering from biotic stresses or abiotic stresses. The use of variable rate applicators through unmanned aerial vehicles (UAVs) or unmanned terrestrial vehicles (UTVs) can be considered an effective management strategy to improve the efficiency of the pesticides and fertilizers applied.

8.4.1.5 Livestock Monitoring Systems

Livestock monitoring systems use sensors and cameras to track the health and behavior of livestock, providing farmers with information about feed consumption, water intake, and disease outbreaks. This information can help farmers manage their livestock more effectively and prevent losses.

8.4.1.6 Automated Irrigation Systems

According to the United Nations, around 2 billion people live in countries that are experiencing high water stress. As per reports from the FAO, agriculture accounts for 70% of global water withdrawal. Since the water demand is increasing in every sector, the first cut in its supply is from the agriculture sector. As a result, the brunt of dwindling resources in the wake of a burgeoning population is borne by the farming sector (Kansara et al., 2015). Under such circumstances, the use of IoT devices becomes essential, as they enable the DSS to take appropriate decisions based on the type of soil, the moisture content in the soil, the type of crop, and other factors, to efficiently operate the efficient irrigation systems such as drip and sprinkler systems. Drip and sprinkler-based pressurized irrigation systems can be controlled using sensor-based prescriptions, GPS-based application controllers, and variable rate applicators. The flow and pressure can be controlled wirelessly for need-based inputs (Dhanaraju et al., 2022). This can help farmers optimize water use, reduce water waste, and improve crop yields (Hassan et al., 2020).

8.4.1.7 Crop Management Software

Crop management software can be used to track crop growth, predict yields, and optimize crop inputs such as fertilizers and pesticides. Crop or farm management software helps store, analyze, and provide recommendations for the farm operations. The basic aim of such computer-based tools is to curtail production costs, enhance crop quality parameters, and reduce the environmental footprint of agricultural inputs. Planning vital production operations, timely implementation in a proper manner, and automation in real time has been a key character of such software tools. This can help farmers make informed decisions about their crops and optimize their farming operations. Examples of important crop management software are the farm-level geographic information system (FGIS), the decision support system for agro-technology transfer (DSSAT), and the agriculture production systems simulator (APSIM) (Saiz-Rubio and Rovira-Más, 2020).

Smart agricultural tools no doubt help in improving the agriculture sector and make it more sustainable and efficient, but the problems at the stakeholders' end make the adoption of technology sluggish. The agriculture sector has long been dependent on human labor for carrying out various field practices. The technologies may replace the human workforce and prove to be more economical, but the initial cost of investment is too high for marginal farmers. With no support from government agencies, it becomes difficult for farmers to step ahead of others, hence the huge yield gaps. The majority of the farming community in developing nations lack the basic education to understand the working of complex modern technologies, which hinders the adoption of these technologies (Kimiti et al., 2009). Hence, farmers in developing nations tend to stick to traditional methods of crop production (Khan et al., 2007).

8.5 FUTURE PROSPECTS

The issue of global hunger needs the immediate attention of world leaders and innovative solutions from the scientific community. A paradigm shift towards smart and sustainable agriculture practices is critical to address the hunger crisis. Smart agriculture encompasses the IoT, sensors, drones, and other advanced technologies to strengthen the productivity of agriculture systems. They are tools not only for optimizing resource-use efficiencies (water, nutrients, and pesticides) but also for minimizing the production of waste. AI and ML also help farming communities to prepare for challenges in crop production and make informed decisions. For sustaining crop production and minimizing threats to food security, there is a need to embrace diversification of existing systems, conserving agriculture and following integrated approaches. Climate-resilient agriculture practices, which may include carbon sequestration techniques such as regenerative agriculture or conservation agriculture, use of biochar, and bioenergy crop production, contribute to food security as well as help to mitigate climate change. It is impossible to improve food production without strengthening the stakeholders with the required knowledge and access to resources. Collaboration from governments with regard to policy formulation and institutional support in respect to transfer of technologies (smart and sustainable practices) is pivotal. For the world to see the achievement of the UN's "Zero Hunger" sustainable development goal (SDG) it is necessary to share knowledge, harness innovations, and advocate unbiased policies.

8.6 CONCLUSIONS

Achieving the SDGs related to hunger and food security requires the implementation of smart and sustainable agriculture practices. These practices encompass climate-smart agriculture, which involves using modern farming tools and techniques that reduce greenhouse gas emissions (GHGs), preserve natural resources, and increase resilience to climate change. The use of smart farming technologies, such as precision agriculture and data-driven decision-making, can also increase efficiency and productivity in food production, leading to improved food security. To combat global hunger and improve food security, it is crucial to address the underlying causes of poverty and inequality, which often contribute to limited access to food. Sustainable agriculture can help reduce poverty and promote social equity by supporting smallholder farmers and investing in rural development. Moreover, by improving soil health, sustainable agriculture can increase the nutritional value of crops, ultimately leading to improved health outcomes. By doing so, we can ensure that everyone has access to nutritious food while safeguarding the planet's resources for future generations.

REFERENCES

Aguilera, E., Lassaletta, L., Gattinger, A., and Gimeno, B.S. 2013. Managing soil carbon for climate change mitigation and adaptation in Mediterranean cropping systems: A meta-analysis. *Agriculture, Ecosystem and Environment* 168: 25–36.

Altieri, M., and Nicholls, C. 2008. Scaling up agroecological approaches for food sovereignty in Latin America. *Development* 51: 472–480. https://doi.org/10.1057/dev.2008.68

Altieri, M.A., Nicholls, C.I., Henao, A., and Lana, M.A. 2015. Agroecology and the design of climate change-resilient farming systems. *Agronomy for Sustainable Development* 35: 869–890.

Amelung, W., Bossio, D., de Vries, W., Kogel-Knabner, I., Lehmann, J., Amundson, R., Bol, R., Collins, C., Lal, R., and Leifeld, J. 2020. Towards a global-scale soil climate mitigation strategy. *Nature Communication* 11 (1): 1–10. https://doi.org/10.1038/s41467- 020-18887-7

Anderson, M.D., and Rivera-Ferre, M. 2021. Food system narratives to end hunger: Extractive versus regenerative. *Current Opinion in Environmental Sustainability* 49: 18–25. https://doi.org/10.1016/j.cosust.2020.12.002

Aramyan, L., Ondersteijn, C., Ov, K., and Lansink, A.O. 2006. Performance indicators in Agri-food production chains. In: Ondersteijn C.J.M., Wijnands J.H.M., Huirne R.B.M., Kooten O. (eds), Quantifying the agri-food supply chain. Springer, Wageningen, pp. 47–64.

Arza-García, M., and Burgess, A.J. 2022. Drones in the sky: Towards a more sustainable agriculture. *Agriculture* 13(1): 84.

Bernard, B. and Lux, A. 2017. How to feed the world sustainably: An overview of the discourse on agroecology and sustainable intensification. *Regional Environmental Change* 17: 1279–1290. https://doi.org/10.1007/s10113-016-1027-y

Bhattacharjee, S., Panja, A., Panda, M., Dutta, S., Dutta, S., Kumar, R., Kumar, D., Yadav, M.R., Minkina, T., Kalinitchenko, V.P. et al. 2023. How did research on conservation agriculture evolve over the years? A bibliometric analysis. *Sustainability* 15: 2040. https://doi.org/10.3390/su15032040

Bhattacharyya, R., Das, T.K., Sudhishri, S., Dudwal, B., Sharma, A.R., Bhatia, A., and Singh, G. 2015. Conservation agriculture effects on soil organic carbon accumulation and crop productivity under a rice-wheat cropping system in the western Indo-Gangetic Plains. *European Journal of Agronomy* 70: 11–21.

Bhattacharyya, R., Prakash, V., Kundu, S., Srivastva, A.K., and Gupta, H.S. 2009. Soil aggregation and organic matter in a sandy clay loam soil of the Indian Himalayas under different tillage and crop regimes. *Agriculture, Ecosystem and Environment* 132: 126–134.

Boix-Fayos, C., and de Vente, J. 2023. Challenges and potential pathways towards sustainable agriculture within the European Green Deal. *Agricultural Systems* 207: 103634. https://doi.org/10.1016/j.agsy.2023.103634

Caulfield, F., and Bunce, J.A. 1994. Elevated atmospheric carbon dioxide concentration affects interactions between *Spodoptera exigua* (Lepidoptera: Noctuidae) larvae and two host plant species outdoors. *Environmental Entomology* 23(4): 999–1005.

Challinor, A.J., Watson, J., Lobell, D.B., Howden, S.M., Smith, D.R., and Chhetri, N. 2014. A meta-analysis of crop yield under climate change and adaptation. *Nature Climate Change* 4(4): 287–291.

Chen, F., Kissel, D.E., West, L.T., Adkin, W., Clark, R., Rickman, D., and Luvall, J.C. 2004. Field scale mapping of surface soil clay concentration. *Precision Agriculture* 5: 7–26.

Cheong, J.C., Lee, J.T.E., Lim, J.W., Song, S., Tan, J.K.N., Chiam, Z.Y., Yap, K.Y., Lim, E.Y., Zhang, J., Tan, H.T.W., and Tong, Y.W. 2020. Closing the food waste loop: Food waste anaerobic digestate as fertilizer for the cultivation of the leafy vegetable, xiao bai cai (*Brassica rapa*). *Science of Total Environment* 715: 136789. https://doi.org/10.1016/j.scitotenv.2020.136789

Cohn, A.S., Newton, P., Gil, J.D.B., Kuhl, L., Samberg, L., Ricciardi, V., Manly, J.R., and Northrop, S. 2017. Smallholder agriculture and climate change. *Annual Review of Environment and Resources* 42 (1): 347–375. https://doi.org/10.1146/annurev-environ-102016- 060946

Corbeels, M., Naudin, K., Whitbread, A.M., Kuhne, R., and Letourmy, P. 2020. Limits of conservation agriculture to overcome low crop yields in sub-Saharan Africa. *Nature Food* 1: 447–454.

Crandall, R.E., Crandall, W.R., and Chen, C.C. 2010. *Principles of supply chain management*. CRC Press, Boca Raton.

Crimaldi, M., Cristiano, V., De Vivo, A., Isernia, M., Ivanov, P., and Sarghini, F. 2020. Neural network algorithms for real time plant diseases detection using UAVs. In *Innovative biosystems engineering for sustainable agriculture, forestry and food production*, 827–35. Springer Nature, Switzerland AG, 2020. https://doi.org/10.1007/978-3-030-39299-4

Das, T.K., Ghosh, S., Das, A., Sen, S., Datta, D., Ghosh, S., Raj, R., Behera, B., Roy, A., Vyas, A., and Rana, D. 2021. Conservation agriculture impacts on productivity, resource-use efficiency and environmental sustainability: A holistic review. *Indian Journal of Agronomy* 66: 111–127.

de Torres, M.A.R.R., Carbonell-Bojollo, R.M., Moreno-García, M., Ordóñez-Fernández, R., and Rodríquez-Lizana, A. 2021. Soil organic matter and nutrient improvement through cover crops in a Mediterranean olive orchard. *Soil Tillage Research* 210: 104977.

Dev, P., Khandelwal, S., Yadav, S.C., Arya, V., Mali, H.R., and Poonam. 2023. Climate based smart agriculture: Need for food security and sustainability. *International Journal of Environment and Climate Change* 13(3): 224–231.

Dhanaraju, M., Chenniappan, P., Ramalingam, K., Pazhanivelan, S., and Kaliaperumal, R. 2022. Smart farming: Internet of Things (IoT)-based sustainable agriculture. *Agriculture* 12(10): 1745.

Du, Z., Angers, D.A., Ren, T., Zhang, Q., and Li, G. 2017. The effect of no-till on organic C storage in Chinese soils should not be overemphasized: A metaanalysis. *Agriculture, Ecosystem and Environment* 236: 1–11. doi: 10.1016/j.agee.2016.11.0

Ellison, D., Cindy, E., Morri, C.E., Locatelli, B., Sheil, D., Cohen, J., Murdiyarsoi, D., Gutierrez, V., van Noordwijk, M., Creed, I.F., Pokorny, J., Gaveau, D., Spracklen, D.V., Tobella, A.B., Ilstedt, U., Teuling, A.J., Gebrehiwot, S. G., Sands, D.C., Muys, B., Verbist, B., Springgay, E., Sugandi, Y., and Sullivan, C.A. 2017. Trees, forests and water: Cool insights for a hot world. *Global Environmental Change* 43:56–61.

European Environment Agency. 2020. Bio-waste in Europe turning challenges into opportunities. https://doi.org/10. 2800/630938

FAO. 2010. Climate-smart agriculture: Policies, practices and financing for food security, adaptation and mitigation. Food and Agriculture Organization of the United Nations, Rome. www.fao.org/3/i1881e/i1881e00.htm

FAO. 2013. Climate smart agriculture sourcebook. Food and Agriculture Organization of the United Nations, Rome. www.fao.org/3/i3325e/i3325e.pdf

FAO. 2015. L'état de l'insécurité alimentaire dans le monde. Objectifs internationaux 2015 de réduction de la faim: des progrès inégaux. Food and Agriculture Organization of the United Nations, Rome.

Field, C.B., Barros, V.R., Mastrandrea, M.D., Mach, K.J., Abdrabo, M.K., Adger, N., and Yohe, G.W. 2014. Summary for policymakers. In *Climate change 2014: impacts, adaptation, and vulnerability. Part A: global and sectoral aspects. contribution of Working Group II to the Fifth Assessment Report of the Intergovernmental Panel on Climate Change* pp. 1–32. Cambridge University Press.

Food Agriculture Organization 2011. FAOSTAT data. Food Agriculture Organization, Rome, Italy. http://faostat.fao.org/

Francaviglia, R., Di Bene, C., Farina, R., and Salvati, L. 2017. Soil organic carbon sequestration and tillage systems in the Mediterranean Basin: A data mining approach. *Nutrient Cycling in Agroecosystem* 107: 125–137. doi: 10.1007/s10705-016-9820-z

Frank, S., Havlík, P., Soussana, J.F., Levesque, A., Valin, H., Wollenberg, E., and Obersteiner, M. 2017. Reducing greenhouse gas emissions in agriculture without compromising food security?. *Environmental Research Letters* 12(10): 105004.

Goddard, T., Kassam A., and Mkomwa, S. 2022. Moving Paradigms — Conservation Agriculture with Alternative Agronomics to Minimize Inputs. In: S. Mkomwa and A. Kassam (eds). Conservation Agriculture in Africa: Climate Smart Agricultural Development. CABI publisher. ISBN: 9781789245745. pp. 189–205. DOI: 10.1079/9781789245745.0010

Grassini, P., Eskridge, K.M., and Cassman, K.G. 2013. Distinguishing between yield advances and yield plateaus in historical crop production trends. *Nature Communications* 4: 2918. https://doi.org/10.1038/ncomms3918

Gray, S.B., and Brady, S.M. 2016. Plant developmental responses to climate change. *Developmental Biology* 419(1): 64–77.

Hare, J.D. 1992. Herbivore-Natural Enemy. *Plant resistance to herbivores and pathogens: Ecology, evolution, and genetics* pp. 278.

Hassan, A., Aitazaz, A.F., Farhat, A., Bishnu, A., and Travis, E. 2020. Precision irrigation strategies for sustainable water budgeting of potato crop in Prince Edward Island. *Sustainability* 12: 2419.

Hassanein, M., and El-Sheimy, N. (2018). An effcient weed detection procedure using low-cost UAV imagery system for precision agriculture applications. *International Archives of the Photogrammetry, Remote Sensing and Spatial Information Sciences* 42 : 181–7.

Hector, A., and Hooper, R. 2002. Darwin and the first ecological experiment. *Science* 295: 639–640.

Hoddinott, J.F., and Rosegrant, M.W. 2013. *Investments to reduce hunger and undernutrition.*

Horton, S., Shekar, M., McDonald, C., Mahal, A., and Brooks, J.K. 2009. *Scaling up nutrition: What will it cost.* World Bank Publications.

Hussain, S., Hussain, S., Guo, R., Sarwar, M., Ren, X., Krstic, D., and El-Esawi, M. A. 2021. Carbon sequestration to avoid soil degradation: A review on the role of conservation tillage. *Plants* 10(10): 2001.

Ibrahim, M.M., Guo, L., Wu, F., Liu, D., Zhang, H., Zou, S., Xing, S., and Mao, Y. 2022. Field-applied biochar-based MgO and sepiolite composites possess CO_2 capture potential and alter organic C mineralization and C-cycling bacterial structure in fertilized soils. *Science of Total Environment* 813: 152495. https://doi.org/10.1016/j.scitotenv.2021.152495

IPBES. 2018. The IPBES assessment report on land degradation and restoration. Montanarella, L., Scholes, R., and Brainich, A. (eds.), *Secretariat of the Intergovernmental Science-Policy Platform on Biodiversity and Ecosystem Services*, Bonn, Germany. pp. 744.

Jat, H.S., Choudhary, K.M., Nandal, D.P., Yadav, A.K., Poonia, T., Singh, Y., Sharma, P.C., and Jat, M.L. 2020. Conservation agriculture–based sustainable intensification of cereal systems leads to energy conservation, higher productivity and farm profitability. *Environmental Management* 65(6): 774–786.

Jat, H.S., Datta, A., Sharma, P.C., Kumar, V., Yadav, A.K., and Choudhary, M. 2018. Assessing soil properties and nutrient availability under conservation agriculture practices in a reclaimed sodic soil in cereal-based systems of North West India. *Achieves of Agronomy and Soil Science* 64: 531–545.

Javaid, M., Haleem, A., Singh, R.P., and Suman, R. 2022. Enhancing smart farming through the applications of agriculture 4.0 technologies. *International Journal of Intelligent Networks* 3: 150–164.

Jian, J., Du, X., Reiter, M.S., and Stewart, R.D. 2020. A meta-analysis of global cropland soil carbon changes due to cover cropping. *Soil Biology and Biochemistry* 143:107735.

Kammlade, S., Khoury, C.K., Sotelo, S., Achicanoy, H., Bjorkman, A., and Navarro-Racines, C. 2017. The changing global diet. International Center for Tropical Agriculture (CIAT). https://ciat.cgiar.org/the-changing-global-diet/

Kansara, K., Zaveri, V., Shah, S., Delwadkar, S., and Jani, K. 2015. Sensor based automated irrigation system with IOT: A technical review. *International Journal of Computer Science and Information Technologies* 6(6): 5331–5333.

Kassam, A., and L. Kassam. 2020. *Rethinking food and agriculture: New ways forward.* Elsevier. ISBN 978-0-12-816410-5

Kassam, A., Friedrich, T., and Derpsch, R. 2022. Successful experiences and lessons from conservation agriculture worldwide. *Agronomy* 12(4): 769.

Khan, A. 2019. Tillage and crop production. Agronomic Crops, volume 1: Production Technologies, Springer Nature Singapore Pvt. Ltd. M. Hasanuzzaman (ed.), *Agronomic Crops* 115.

Khan, A.R., Dubey, M.K., Bisen, P.K., and Saxena, K.K. 2007. Constraints faced by farmers of Narsing Kheda village of Sihore district. *Young* 8: 16.

Khatri-Chhetri, A., Sapkota, T.B., Maharjan, S., Konath, N.C., and Shirsath, P. 2023. Agricultural emissions reduction potential by improving technical efficiency in crop production. *Agricultural Systems* 207:103620. https://doi.org/10.1016/j.agsy.2023.103620

Khoury, C.K., Bjorkman, A.D., Dempewolf, H., Ramirez-Villegas, J., Guarino, L., Jarvis, A., and Struik, P.C. 2014. Increasing homogeneity in global food supplies and the implications for food security. *Proceedings of the National Academy of Sciences* 111(11), 4001–4006.

Kimiti, J.M., Odee, D.W., and Vanlauwe, B. 2009. Area under grain legumes cultivation and problems faced by smallholder farmers in legume production in the semi-arid Eastern Kenya. *Journal of Sustainable Development in Africa* 11(4): 305–315.

Kinnunen, P., Guillaume, J.H.A., Taka, M., D'Odorico, P., Siebert, S., Puma, M.J., Jalava M., and Kummu, M. 2020. Local food crop production can fulfil demand for less than one-third of the population. *Nature Food* 1: 229–237. https://doi.org/10.1038/s43016-020-0060-7

Kocmánková, E., Trnka, M., Juroch, J., Dubrovský, M., Semerádová, D., Možný, M., and Žalud, Z. 2009. Impact of climate change on the occurrence and activity of harmful organisms. *Plant Protection Science, 45*(Special Issue).

Krasilnikov, P., Taboada, M.A., and Amanullah. 2022. Fertilizer use, soil health and agricultural sustainability. *Agriculture* 12(4): 462.

Kuang, B., and Mouazen, A.M. 2013. Non-biased prediction of soil organic carbon and total nitrogen with vis-NIR spectroscopy, as affected by soil moisture content and texture. *Biosystems Engineering* 114: 249–258.

Kumar, R., Kumar, S., Yashavanth, B.S., Venu, N., Meena, P.C., Dhandapani, A., and Kumar, A. 2023. Natural farming practices for chemical-free agriculture: implications for crop yield and profitability. *Agriculture* 13(3): 647.

Kumar, S., and Rana, S.S. 2021. Conservation agriculture and weed management effects on growth indices and yield attributes of maize-wheat cropping system. *Himachal Journal of Agricultural Res*earch 47(2): 187–197.

Kumar, S., Rana, S.S., and Sharma, N. 2022a. Long-term tillage and weed management effects on weed shifts, phytosociology and crops productivity. *Indian Journal of Weed Sciences* 54(2): 165–173. http://dx.doi.org/10.5958/0974-8164.2022.00031.4

Kumar, S., Rana, S.S., Kumar, R., and Sharma, N. 2022b. Effects of conservation tillage and weed management on soil microbial community and enzymatic activity. *Bangladesh Journal of Botany* 51(3): 425–431.

Kurniawati, A., Toth, G., Ylivainio, K., and Toth, Z. 2023. Opportunities and challenges of bio-based fertilizers utilization for improving soil health. *Organic Agriculture* 1–16.

Kuyah, S., Sileshi, G.W., Nkurunziza, L., Chirinda, N., Ndayisaba, P.C., Dimobe, K., and Öborn, I. 2021. Innovative agronomic practices for sustainable intensification in sub-Saharan Africa. A review. *Agronomy for Sustainable Development* 41: 16. https://doi.org/10.1007/s13593-021-00673-4

Laik, R., Sharma, S., Idris, M., Singh, A.K., Singh, S.S., Bhatt, B.P., Saharawat, Y., Humphreys E., and Ladha, J.K. 2014. Integration of conservation agriculture with best management practices for improving system performance of the rice–wheat rotation in the Eastern Indo–Gangetic Plains of India. *Agriculture, Ecosystems and Environment* 195: 68–82.

Lal, R. 1999. Soil management and restoration for C sequestration to mitigate the accelerated greenhouse effect. *Progress in Environmental Science* 1: 307–326.

Lal, R. 2004. Carbon sequestration in dryland ecosystems. *Environmental Management* 33: 528–544.

Lal, R., and Stewart, B.A. 2013. Principles of Sustainable Soil Management in Agroecosystems. Advanced in soil sciences. CRC Press. pp. 552.

Li, Q., Li, H., Zhang, L., Zhang, S., and Chen, Y. 2018. Mulching improves yield and water-use efficiency of potato cropping in China: A meta-analysis. *Field Crops Research* 221: 50–60. doi: 10.1016/j.fcr.2018.02.017

Li, Y., Li, Z., Cui, S., Jagadamma, S., and Zhang, Q.P. 2019. Residue retention and minimum tillage improve physical environment of the soil in croplands: A global meta-analysis. *Soil Tillage Research* 194: 104292.

Lipper, L., Thornton, P., Campbell, B.M., Baedeker, T., Braimoh, A., Bwalya, M., and Torquebiau, E.F. 2014. Climate-smart agriculture for food security. *Nature Climate Change* 4(12): 1068–1072.

Liu, Z., and Li, J. 2023. Application of Unmanned Aerial Vehicles in Precision Agriculture. *Agriculture* 13(7): 1375.

Lorenz, K., and Lal, R. 2014. Soil organic carbon sequestration in agroforestry systems. A review. *Agronomy for Sustainable Development* 34(2): 443–445.

Lowder, S.K., Skoet, J., and Raney, T. 2016. The number, size, and distribution of farms, smallholder farms, and family farms worldwide. *World Development* 87:16–29. https://doi. org/10.1016/j.worlddev.2015.10.041

Lu, Y., Liu, M., Li, C., Liu, X., Cao, C., Li, X., and Kan, Z. 2022. Precision fertilization and irrigation: Progress and applications. *AgriEngineering* 4(3): 626–655.

Luo, H., Robles-Aguilar, A.A., Sigurnjak, I., Michels, E., and Meers, E. 2021. Assessing nitrogen availability in biobased fertilizers: Effect of vegetation on mineralization patterns. *Agriculture* 11: 870.

Lvova, L., and Nadporozhskaya, M. 2017. Chemical sensors for soil analysis: Principles and applications. In Series Nanotechnology in the Agri-Food Industry; New Pesticides and Soil Sensors; Grumezescu, A.M., Ed.; Elsevier: Amsterdam, The Netherlands, 2017; Volume 10, pp. 637–678.

Mahato, A. 2014. Climate change and its impact on agriculture. *International Journal of Scientific and Research Publications* 4(4): 1–6.

Maleki, M.R., Van Holm, L., Ramon, H., Merckx, R., De Baerdemaeker, J., and Mouazen, A.M. 2006. Phosphorus Sensing for Fresh Soils using Visible and Near Infrared Spectroscopy. *Biosystems Engineering* 95: 425–436.

Malhi, G. S., Kaur, M., and Kaushik, P. 2021. Impact of climate change on agriculture and its mitigation strategies: A review. *Sustainability* 13(3): 1318.

Mangalassery, S., SjÖGersten, S., Sparkes, D.L., and Mooney, S.J. 2015. Examining the potential for climate change mitigation from zero tillage. *Journal of Agricultural Sciences* 153: 1151–1173. doi: 10.1017/S0021859614001002

Maougal, R.T., Kechid, M., Ladjabi, C., and Djekoun, A. 2020. PGPR Characteristics of Rhizospheric bacteria to understand the mechanisms of Faba bean growth. *Proceedings of the National Academy of Sciences USA* 66: 27.

Mbow, C., Van Noordwijk, M., Luedeling, E., Neufeldt, H., Minang, P.A. and Kowero, G. 2014. Agroforestry solutions to address food security and climate change challenges in Africa. *Current Opinion in Environmental Sustainability* 6: 61–67.

Mekala, M.S., and Viswanathan, P.A. 2017. Survey: Smart agriculture IoT with cloud computing. In *Proceedings of the 2017 International Conference on Microelectronic Devices, Circuits and Systems (ICMDCS)*, Vellore, India, 10–12 August 2017; IEEE: Vellore, India, 2017; pp. 1–7.

Mentzer, J.T., DeWitt, W., Keebler, J.S., Min, S., Nix, N.W., Smith, C.D., and Zacharia, Z.G. 2001. Defining supply chain management. *Journal of Business logistics* 22(2): 1–25.

Mishra, S. 2018. Zero Budget Natural Farming: Are This and Similar Practices the Answers; Nabakrushna Choudhury Centre for Development Studies (NCDS): Odisha, India, 2018.

Mittal, A., and Singh, A. 2007. Microcontroller based pest management system. In *Proceedings of the Second International Conference on Systems (ICONS'07), Martinique*, France, 22–28 April 2007; IEEE: Martinique, France, 2007; p. 43.

Morugán-Coronado, A., Linares, C., Gómez-López, M.D., Faz, Á., and Zornoza, R. 2020. The impact of intercropping, tillage and fertilizer type on soil and crop yield in fruit orchards under Mediterranean conditions: A meta-analysis of field studies. *Agricultural Systems* 178: 102736.

Mrabet, R., and Moussadek, R. 2022. Climate smart agriculture development in Africa. In: S. Mkomwa and A. Kassam (eds). *Conservation Agriculture in Africa: Climate Smart Agricultural Development.* CABI publisher. ISBN: 9781789245745. pp. 17–65. DOI: 10.1079/9781789245745.0002

Mueller, C. 2013. African lessons on climate change risks for agriculture. *Annual Review of Nutrition* 33: 395–411.

Muhammad, S.F., Shamyla, R., Adnan, A., Tariq, U., and Yousaf, B.Z. 2020. Role of IoT technology in agriculture: A systematic literature review. *Electronics* 9: 319.

Mulimbi, W., Nalley, L., Nayga, R.M. Jr., and Gaduh, A. 2022. Are consumers willing to pay for conservation agriculture? The case of white maize in the Democratic Republic of the Congo. *Natural Resources Forum* 1–20. https://doi.org/10.1111/1477-8947.12268

Muoni, T., Mhlanga, B., Forkman, J., Sitali, M. and Thierfelder, C. 2019. Tillage and crop rotations enhance populations of earthworms, termites, dung beetles and centipedes: evidence from a long-term trial in Zambia. *Journal of Agricultural Science* 1–11.

Natarajan, K., Karthikeyan, R., and Rajalingam, S. 2023. Importance of drone technology in agriculture. *Drone Technology: Future Trends and Practical Applications* 351–374.

Newaj, R., Chaturvedi, O.P., and Handa, A.K. 2016. Recent development in agroforestry research and its role in climate change adaptation and mitigation. *Indian Journal of Agroforestry* 18(1): 1–9.

Nisrina, L., Effendi, Y., and Pancoro, A. 2021. Revealing the role of Plant Growth Promoting Rhizobacteria in suppressive soils against Fusarium oxysporum f.sp. cubense based on metagenomic analysis. *Heliyon* 7: e07636.

Nyanga, P.H., Umar, B.B., Chibamba, D., Mubanga, K.H., Kunda-Wamuwi, C.F., and Mushili, B.M. 2020. Reinforcing ecosystem services through conservation agriculture in sustainable food systems. L. Rusinamhodzi (ed.). *The Role of Ecosystem Services in Sustainable Food Systems*. Academic Press, pp. 119–133. https://doi.org/10.1016/B978-0-12-816436-5.00006-8

Ogle, S., Alsaker, C., Baldock, J., Bernoux, M., Breidt, F., McConkey, B.G., Regina, K., and Vazquez-Amabile, G.G. 2019. Climate and soil characteristics determine where No-Till management can store carbon in soils and mitigate greenhouse gas emissions. *Scientific Rep*orts 9:11665. doi: 10.1038/s41598-019-47861-7

Orden, L., Ferreiro, N., Satti, P., Navas-Gracia, L.M., Chico-Santamarta, L., and Rodríguez, R.A. 2021. Effects of onion residue, bovine manure compost and compost tea on soils and on the agroecological production of onions. *Agriculture* 11: 962.

Oueriemmi, H., Kidd, P.S., Trasar-Cepeda, C., Rodríguez-Garrido, B., Zoghlami, R.I., Ardhaoui, K., Prieto-Fernández, Á., and Moussa, M. 2021. Evaluation of composted organic wastes and farmyard manure for improving fertility of poor sandy soils in arid regions. *Agriculture* 11: 415.

Pachauari, R.K., and Reisinger, A. 2007. *Climate Change 2007: Synthesis Report*. Contribution of Working Groups I, II and III to the Fourth Assessment Report on Intergovernmental Panel on Climate Change; Intergovernmental Panel on Climate Change (IPCC): Geneva, Switzerland, 2007; ISBN 92-9169-122-4

Page, K.L., Dang, Y.P., Dalal, R.C., Reeves, S., Thomas, G., and Wang, W. 2019. Changes in soil water storage with no-tillage and crop residue retention on a Vertisol: Impact on productivity and profitability over a 50-year period. *Soil Tillage Research* 194: 104319.

Parry, M. 1990. The potential impact on agriculture of the greenhouse effect. *Land Use Policy* 7(2): 109–123.

Paustian, K., Lehmann, L., Ogle, S., Reay, D., Robertson, G.P., and Smith, P. 2016. Climate-smart soils. *Nature* 532: 49–57.

Pimentel, D., Hepperly, P., Hanson, J., Douds, D., and Siedel, R. 2005. Environmental, energetic, and economic comparisons of organic and conventional farming systems. *BioScience* 55(7): 573–582.

Pittelkow, C.M., Linquist, B.A., Lundy, M.E., Liang, X., Van Groenigen, K.J., Lee, J., and Van Kessel, C. 2015. When does no-till yield more? A global meta-analysis. *Field Crops Research* 183: 156–168.

Poeplau, C., and Don, A. 2015. Carbon sequestration in agricultural soils via cultivation of cover crops—A meta-analysis. *Agriculture, Ecosystem and Environment* 200: 33–41.

Powlson, D.S., Stirling, C.M., Thierfelder, C., White, R.P., and Jat, M.L. 2016. Does conservation agriculture deliver climate change mitigation through soil carbon sequestration in tropical agro-ecosystems? *Agriculture, Ecosystem and Environment* 220: 164–174. doi: 10.1016/j.agee.2016.01.005

Prakash, A., Rao, J., Mukherjee, A.K., Berliner, J., Pokhare, S.S., Adak, T., Munda, S., and Shashank, P.R. 2014. *Climate Change: Impact on Crop Pests*. Odisha, India: Applied Zoologists Research Association (AZRA), Central Rice Research Institute. ISBN 81-900947-2-7.

Pretty, J. 2018 Intensification for redesigned and sustainable agricultural systems. *Science* 362: 908. https://doi.org/10.1126/science. aav0294

Rammig, A.W., Bartholomeus, R.P., Bonfante, A., De Lorenzi, F., Dury, M., Gloning, P., and Pereira, M. A. 2012. Plant's perspective of extremes: Terrestrial plant responses to changing climatic variability.

Rani, P., and Reddy, R.G. 2023. Climate change and its impact on food security. *International Journal of Environment and Climate Change* 13(3): 104–108.

Reganold, J., and Wachter, J. 2016. Organic agriculture in the twenty-first century. *Nature Plants* 2: 15221. https://doi.org/10.1038/nplants.2015.221.

Ricciardi, V., Mehrabi, Z., Wittman, H., James, D., and Ramankutty, N. 2021. Higher yields and more biodiversity on smaller farms. *Nature Sustainability* 4(7): 651–657. https://doi. org/10.1038/s41893-021-00699-2

Ricciardi, V., Ramankutty, N., Mehrabi, Z., Jarvis, L., and Chookolingo, B. 2018. How much of the world's food do smallholders produce? *Global Food Security* 17: 64–72. https:// doi.org/10.1016/j.gfs.2018.05.002

Richter, C.H., Xu, J., and Wilcox, M.A. 2015. Opportunities and challenges of the ecosystem approach. *Futures* 67: 40–51. https://doi.org/10.1016/j.futures.2014.12.002

Rose, D.C., and Chilvers, J. 2018. Agriculture 4.0: Broadening responsible innovation in an era of smart farming. *Frontiers in Sustainable Food Systems* 2: 87.

Rosenstock, T.S., Dawson, I.K., Aynekulu, E., Chomba, S., Degrande, A., Fornace, K., Jamnadass, R., Kimaro, A., Kindt, R., Lamanna, C., and Malesu, M. 2019. A planetary health perspective on agroforestry in Sub-Saharan Africa. *One Earth* 1(3): 330–344.

Roth, S. K., and Lindroth, R. L. 1995. Elevated atmospheric CO_2: effects on phytochemistry, insect performance and insect-parasitoid interactions. *Global Change Biology* 1(3): 173–182.

Saad, A.A., Das, T.K., Rana, D.S., Sharma, A.R., Bhattacharyya, R., and Lal, K. 2016. Energy auditing of a maize-wheat-green gram cropping system under conventional and conservation agriculture in irrigated north-western Indo-Gangetic plains. *Energy* 116: 293–305.

Saharan, B.S., Tyagi, S., Kumar, R., Vijay, Om, H., Mandal, B.S., and Duhan, J.S. 2023. Application of jeevamrit improves soil properties in zero budget natural farming fields. *Agriculture* 13: 196.

Saiz-Rubio, V., and Rovira-Más, F. 2020. From smart farming towards agriculture 5.0: A review on crop data management. *Agronomy* 10(2): 207.

Sarkar, D., Meena, V.S., Haldar, A., and Rakshit, A. 2017. Site-Specific Nutrient Management (SSNM): A unique approach towards maintaining Soil Health. *Adaptive Soil Management: From Theory to Practices* 69–88.

Schneider, F., Steiger, D., Ledermann, T., Fry, P., and Rist, S. 2012. No-tillage farming: Co-creation of innovation through network building. *Land Degradation and Development* 23: 242–255.

Seto, K.C., Güneralp, B., and Hutyra, L.R. 2012. Global forecasts of urban expansion to 2030 and direct impacts on biodiversity and carbon pools. *Proceedings of National Academy of Sciences of USA* 109(40): 16083–16088. https://doi.org/10.1073/pnas.1211658109

Shrestha, S. 2019. Effects of climate change in agricultural insect pest. *Acta Scientific Agriculture* 3(12): 74–80.

Sigurnjak, I., Brienza, C., Snauwaert, E., De Dobbelaere, A., De Mey, J., Vaneeckhaute, C., Michels, E., Schoumans, O., Adani, F., and Meers, E. 2019. Production and performance of bio-based mineral fertilizers from agricultural waste using ammonia (stripping-)scrubbing technology. *Waste Management* 89: 265–274. https://doi.org/10.1016/j.wasman.2019.03.043

Smith, P., Calvin, K., and Nkem, J. 2020. Which practices co-deliver food security, climate change mitigation and adaptation, and combat land degradation and desertification? *Global Change Biology* 26: 1532–1575. https://doi.org/10.1111/gcb.14878

Struik, P. C., and Kuyper, T. W. 2017. Sustainable intensification in agriculture: The richer shade of green. A review. *Agronomy for Sustainable Development* 37: 39. doi: 10.1007/s13593-017-0445-7

Su, Y., Gabrielle, B., and Makowski, D. 2021a. The impact of climate change on the productivity of conservation agriculture. *Nature Climate Change* 11: 628–633. https://doi.org/10.1038/s41558-021-01075-w

Su, Y., Gabrielle, B., Beillouin, D., and Makowski, D. 2021b. High probability of yield gain through conservation agriculture in dry regions for major staple crops. *Scientific Reports* 11: 3344. https://doi.org/10.1038/s41598-021-82375-1

Sun, W.J., Canadell, J.G., Yu, L.J., Yu, L.F., Zhang, W., Smith, P., Fischer, T., and Huang, Y. 2020. Climate drives global soil carbon sequestration and crop yield changes under conservation agriculture. *Global Change Biology* 26(6): 3325–3335. https://doi.org/ 10.1111/gcb.15001

Tilman, D., Balzer, C., Hill, J., and Befort, B. L. 2011. Global food demand and the sustainable intensification of agriculture. *Proceedings of the national academy of sciences USA* 108(50): 20260–20264.

Tittonell, P., El Mujtar, V., Felix, G., Kebede, Y., Laborda, L., Lujan, Soto R., and De Vente, J. 2022. Regenerative agriculture—agroecology without politics? *Frontiers in Sustainable Food Systems* 6: 84426. https://doi.org/10.3389/fsufs.2022.844261

van Dijk, M., Morley, T., and Rau, M.L. 2021. A meta-analysis of projected global food demand and population at risk of hunger for the period 2010–2050. *Nature Food* 2:494–501. https://doi.org/10.1038/s43016-021-00322-9

Vermeulen, S.J., Aggarwal, P.K., Ainslie, A., Angelone, C., Campbell, B.M., Challinor, A.J., Hansen, J., Ingram, J.S.I., Jarvis, A., Kristjanson, P., Lau, C., Thornton, P.K., and Wollenberg, E. 2010. Agriculture, food security and climate change: Outlook for knowledge, tools and action. CCAFS Report 3. Copenhagen, Denmark: CGIAR-ESSP Program on Climate Change, Agriculture and Food Security.

West, T.O., and Post, W.M. 2002. Soil organic carbon sequestration rates by tillage and crop rotation: A global data analysis. *Soil Science Society of America Journal* 66: 1930–1946.

WFP. 2007. *World hunger series 2007: Hunger and health*. WFP, United Nations, Rome. www.wfp.org/sites/default/ files/World Hunger Series 2007 Hunger and Health EN.pdf.

Willer, H. 2020. The world of organic agriculture: Statistics and emerging trends 2020. Switzerland: Research Institute of Organic Agriculture (FiBL) & Bonn, Germany: IFOAM-Organics International: Frick.

Wolfert, S., Ge, L., Verdouw, C., and Bogaardt, M.J. 2017. Big data in smart farming — A review. *Agricultural Systems* 153: 69–80. doi: 10.1016/j.agsy.2017.01.023

Wu, W., Yu, Q., You, L., Chen, K., Tanga, H., and Liu, J. 2018. Global cropping intensity gaps: Increasing food production without cropland expansion. *Land Use Policy* 76: 515. https://doi.org/10.1016/j.landusepol.2018.02.032

Yadav, G.S., Babu, S., Das, A., Mohapatra, K.P., Singh, R., Avasthe, R.K., and Roy, S. 2020. No-till and mulching enhances energy use efficiency and reduce carbon footprint of a direct-seeded upland rice production system. *Journal of Cleaner Production*: 122700.

Yong, W., Shuaishuai, L., Li, L., Minzan, L., Ming, L., Arvanitis, K.G., Grorgieva, C., and Sigrimis, N. 2018. Smart sensors from ground to cloud and web intelligence. *IFAC Papers OnLine* 51: 31–38.

Zhao, M., Shi, R., and Du, R. 2023. The pathway to China's carbon–Neutral agriculture: Measures, potential and future strategies. *Chinese Political Science Review*. https://doi.org/10.1007/s41111-023-00237-9

Zomer, R.J., Trabucco, A., Coe, R., Place, F., Van Noordwijk, M., and Xu, J.C. 2014. Trees on farms: an update and reanalysis of agroforestry's global extent and socio-ecological characteristics. *Working Paper* 179. World Agroforestry Centre (ICRAF) Southeast Asia Regional Program, Bogor.

9 Cultivation and Conservation of Crops

Perspectives for Agriculture through Nanotechnology

Lakshmayya NSV, Chandan Kumar Gautam, Kiran Yellappa Vajanthri, B Sumithra[1], Sanjeeb Kumar Mandal, Amilia Nongbet, Jibanjyoti Panda, and Bishwambhar Mishra
Corresponding author: Bishwambhar Mishra

9.1 INTRODUCTION

Innovative farming or growing and managing crops sustainably is critical for accomplishing Zero Hunger, one of the United Nations' 17 sustainable development goals (SDGs). Food cultivation and delivery around the world are under enormous strain as a result of population growth, shifting climate, pollution in the environment, and increased energy and water needs. Agriculture today consumes a startling quantity of resources. Worldwide yearly yield of crops, for example, breaches 3 billion metric tons, requiring 187 million tons of fertilizer, 4 million tons of pesticides, 2.7 trillion cubic meters of water (roughly 70% of all freshwater utilized for consumption throughout the world), and more than two quadrillion British thermal units (BTU) of energy [1]. By the estimates of the Food and Agriculture Organization (FAO) [2], the global population is expected to exceed 10 billion by 2050, increasing food demand by 50%, particularly in emerging nations. Furthermore, roughly 815 million people are already nutritionally deficient, with a further 2 billion anticipated to fall into this demographic group by 2050. This predicament necessitates significant adjustments in the world's agricultural infrastructures. Contemporary studies have revealed that nanotechnology holds the possibility to strengthen the crop production industry by boosting the efficacy of crop contributions and providing remedies for agro and environmental challenges in order to optimize the production of food and its safety. As a result, potential ways for nanotechnology to benefit agriculture have received a lot of attention [3]. However, despite the large number of study findings accessible in this discipline, the published literature still lacks a complete overview of the subject that addresses the practical implications of nanotechnology in many agricultural sectors. Here, we aggregate empirical information from multiple articles that address various elements of agricultural nanotechnology. The various applications of nanomaterials in agriculture, including nanofertilizers, nanopesticides, nanobiosensors, and nanomaterials for soil management, are analyzed in depth. The destiny of nanomaterials in soil-plant systems is addressed, along with an extensive assessment of possible hazards to soil and its ecology (Figure 9.1 illustrates the multiple climatic barriers and their mitigation by nanotechnology, resulting in improved crop quality and quantity).

 DOI: 10.1201/9781003389408-9

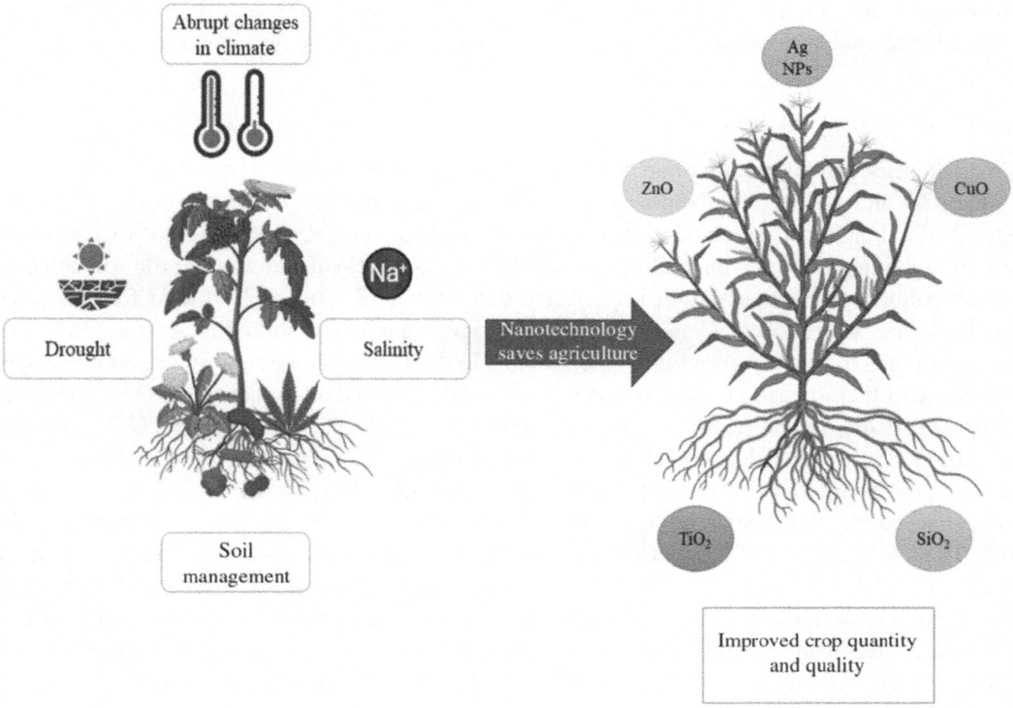

FIGURE 9.1 Application of nanoparticle aids in combating harsh environmental conditions.

9.2 CHALLENGES IN EXISTING AGRICULTURAL PRACTICES

Nowadays, the rise in agricultural productivity is primarily attributable to the exploitation of new regions, which is not ecologically sound for the foreseeable future. As a result, diverse regions must be included in the farming industry. The agricultural industry must develop more quickly than the growth of populations, and this goal must be met in an efficient way that does not harm the environment even more. Green intensive cultivation, which includes more efficient use of arable assets such as water and land, is the impetus underlying advancement [4]. Urbanization is removing a growing percentage of agricultural land while also putting strain on present-day land usage and ecosystems. Furthermore, the growing need for agricultural products will result in an enormous shift in the utilization of land. Harvesters who support the theory of global warming and its anthropomorphic (manmade) provenance are extremely inclined towards lowering emissions of greenhouse gases; nonetheless, cultivators have a tendency to be better equipped to adjust to evolving conditions than to minimize dangerous element pollution. Climate change and adverse weather patterns are major contributors to worldwide hunger and food insecurity. They have a detrimental effect on subsistence and all areas of food availability (accessibility, stability, and so on), as well as contributing to various forms of deficiencies concerning childcare and nourishment [5].

9.3 NANOTECHNOLOGY FOR AGRICULTURAL PRODUCTIVITY

The number of discoveries and articles in the areas of containerization, surveillance of quality, nutraceutical delivery, and novel advances in nanotechnology is expanding in the agricultural industry and business [6]. Nanotechnologies have the potential to be used throughout the production of goods. The use of nanotechnology for agricultural purposes also benefits farmers by boosting the quality of the soil, increasing crop yields, reducing water requirements, and so on. It entails using nanofertilizers, nanosensors to ensure safety and quality, and even nanopackaging. [7].

9.3.1 Crop Improvement

In a broad sense, supplementing nutrients (element fertilization) is unavoidable for increasing crop productivity and soil fertility. Nonetheless, proper fertilizer management is seen as a critical prerequisite for long-term agricultural success [8]. To address the growing hunger crisis, extensive farming is being practiced, which ultimately contributes to a cyclical pattern of fertile soil depletion and agricultural production loss. It is projected that around 40% of the land used for agriculture worldwide has severely deteriorated, resulting in a significant loss of soil richness as a result of the aforementioned extensive agricultural practices [9]. In the same vein, numerous investigations have demonstrated that the use of nanomaterials has a good influence on sprouting and also the development of plants. For instance, the use of multiwalled carbon nanotubes (MWCNTs) improves the germination of seeds in a variety of agricultural crops such as tomato, corn, soybean, barley, wheat, maize, peanut, and garlic [10]. Furthermore, nano SiO_2, TiO_2, and zeolite applications favorably enhance crop plant seed establishment [11]. In conjunction with sprouting, nanomaterials like ZnO, TiO^2, MWCNTs, FeO, ZnFeCu-oxide, and hydroxyfullerenes have been shown to promote the development and growth of crops while improving nutritional value in a variety of agricultural crops including peanut, soybean, mung bean, wheat, onion, spinach, tomato, potato, and mustard. Fullerols, as OH-functionalized fullerenes, for example, have been shown to have beneficial impacts on vegetative development [12] (Table 9.1 addresses the effects of various nanoparticles on different plant species).

TABLE 9.1
Effects of various nanoparticles on different plant species

Nanomaterial	Crop species	Mode of application	Duration of treatment	Response recorded	Reference
ZnO	*Coffea arabica*	Foliar spray	45 days	Enhanced growth, biomass accumulation, and net photosynthesis	[10]
ZnO	*Triticum aestivum*	Mixed with growth substrate	Growth cycle	Increased grain yield and biomass accumulation	[26]
CuO	*Spinacia oleracea*	Mixed with soils	60 days	Improved photosynthesis and biomass production	[27]
ZnO	*Nicotiana tabacum*	Hydroponics	21 days	Positive effect on growth physiology, increased metabolites, enzymatic activities, and anatomical properties of plants	[28]
AgNPs	*Triticum aestivum*	Mixed with pot soils	Trifoliate stage	Improved growth and tolerance to heat stress	[29]
TiO_2 and SiO_2	*Oryza sativa*	Foliar application	55 days	Mitigated Cd toxicity and improved growth by stimulating antioxidant potential and inhibiting Cd translocation	[30]
ZnO, CuO, and AgNPs	*Prunus domestica* fruits	Fruit spray	4 days	Suppressed grey mold symptoms caused by *B. cinerea* and soil-borne diseases	[31]
AgNPs	*Vigna unguiculata*	Foliar application	7 days	Showed no phytotoxicity, but could inhibit growth of *Xanthomonas axonopodis pv. malvacearum* and *Xanthomonas campestris pv. campestris* in vitro	[32]

9.3.2 Smart Nanofertilizers

It is predicted that around 40% of the land used for agriculture worldwide has severely deteriorated, resulting in an enormous decline in soil vitality because of so-called extreme farming procedures. As a result, enormous amounts of fertilizer are needed to boost soil health and crop yields. It has also been unambiguously noted that fertilizers account for one third of yields from crops, with the remainder dependent on the efficiency with which additional inputs to agriculture are used. However, standard fertilizer nutrient utilization performances are only 30–40%. For instance, the nutrient utilization effectiveness of traditional fertilizers has remained steady for a few decades, with nitrogen (N) at 30–35%, phosphorus (P) at 18–20%, and potassium (K) at 35–40% [13]. Nanofertilizers may be the most promising option for alleviating macro- and micronutrient deficiencies by means of improved nutrient usage efficacy and overcoming the ongoing issue of eutrophication. Nanofertilizers, which have the goal of controlling the flow of nutrients according to the needs of the harvests whilst minimizing disproportionate expenses, offer tremendous possibilities. Traditional nitrogen fertilizers, for example, have huge deductions from the soil due to leakage, evaporation, or even deterioration of up to 50–70%, which diminishes fertilizer performance and raises manufacturing expenses [14]. Several nanofertilizer variations have been developed using various materials, including zeolite, copper, silica, aluminum, carbon, zinc, and nitrogen [15]. Recent advances in nanotechnology have included the application of certain significant nanomaterials as carrier molecules for delivering a variety of nutrients, like macro- and micronutrients, as well as plant growth-promoting compounds, including nanoclays, mesoporous silica nanoparticles (MSN), hydroxyapatite nanoparticles, and polymeric nanoparticles [16]. A group of scientists has also investigated the influence of Zn/B nanofertilizers on coffee shoots [17]. Furthermore, nanofertilizers are already employed in a variety of agricultural systems, including hydroponic cultivation, aeroponic techniques, and irrigation [18].

9.3.3 Effective Nanopesticides

Using pesticides is common practice in agriculture in order to increase productivity. Pesticides with unusual and effective properties are constantly being produced for targeted goals. As a result, a variety of pesticides are hidden in abundance year after year. However, substantially less insecticide (0.1%) reaches the intended insect infestations. The remaining 99.99% of insecticides used are tainted. Polluted pesticides in the soil have serious repercussions for people as well as the food chain [19]. Pesticide tolerance develops as a consequence of the widespread distribution of pesticides in terms of diseases, invasive plants, and bugs. All of these negative consequences can be avoided by utilizing biopesticides instead of chemical pesticides. However, the development is gradual, and the effectiveness of biopesticides against insects was primarily reliant on the natural world.

All such challenges can be readily solved by substituting artificial pesticides for nanopesticides. Nanopesticides can provide long-term control of pests by diligently and precisely dispersing the active ingredients [20]. Nanopesticides behave very differently than traditional pesticides. Because the particles are so small, they travel in dispersion and dissipate variations. Given the tiny dimensions, the functional elements will be more soluble. As a result, use of nanopesticides rather than traditional pesticides has a substantially lower environmental impact. Overall, when nanopesticides are used in smaller amounts, they save a lot of water and energy [21]. The release of effective ingredients towards various crop-related pests will be an important strategy in years to come for pest management in the agricultural sector. The antibacterial behavior of nanoparticles has been demonstrated to be effective against microbes such as viruses, bacteria, and fungi. Silver (Ag), copper (Cu), and aluminum (Al) are examples of inorganic nanomaterials with pesticidal characteristics [14]. The elimination of infectious agents in infested leaves is detected when silver-silica nanoparticles are sprayed on them, and the procedure only takes three days to complete.

Copper (Cu) nanoparticles exhibit excellent antimicrobial and antifungal effects [22]. For example, to improve biological suitability, accessibility, and physiochemical characteristics, atrazine is enclosed by a biopolymer [23]. It has been demonstrated that employing polymeric material-encased atrazine over native atrazine improves the stability of the soil and increases herbicidal activity while decreasing soil dispersion [24]. Crude atrazine typically pollutes crops and water sources in the ground. Atrazine in nanoformulations was more efficient than traditional atrazine for combating designated weeds. Foliar association of an atrazine nanoformulation with mustard (India) demonstrates that medicinal efficacy proceeds directly via the vascular tissues of the leaves [25]. This demonstrates that element nanoformulation has the potential to retain a small amount of herbal effectiveness.

9.3.4 Crop Protection

Nanotechnology has a wide range of possibilities in protecting crops, but it is most frequently employed to create slow-release encapsulating agrochemicals. Innovative nanomaterials based on metals, polymers, and inorganic nanoparticles have been designed for enhancing smart nanosystems capable of capturing and immobilizing micro- and macronutrients while allowing their graded dissolution in the soil to boost fertilizer efficacy. Furthermore, the creation of nanosensors enables the identification of diseases and pests in plants [33]. Numerous varieties of nanoparticles made of metals, including Ag, Fe, Cu, and Zn, can be utilized in two ways: as nanofertilizers to improve the germination of seeds and plant growth, and as pesticides or fungicides combating pathogenic microbes and pests. Ag nanoparticles were the first to be used in agriculture to combat disease in crops. It is the most researched, well-characterized, and popular antimicrobial agent. Many soil-borne diseases have been inhibited by the use of silver nanoparticles, including *Phytophthora parasitica*, *Meloidogyne spp.*, and *Fusarium spp.* [34]. Recently conducted research found that using chitosan nanoparticles improved resistance to diseases in plants. Because of its ease of production, it is an excellent medium for medication delivery and transfer of genes. The use of chitosan nanoparticles increased disease resistance in the tea plant by increasing defense-related proteins like polyphenol oxidase, phenylalanine ammonia-lyase, and -1,3-glucanase [35].

9.3.4.1 Precision Agriculture Management Systems

To address the multiple issues faced in farms, the agriculture sector's digital evolution is currently seen as essential. The use of ecological surveillance and automation in agriculture is fast expanding in order to create more efficient and profitable cropping systems and instruments. Precision agriculture and smart farming have the potential to pave the way in this trend. The combination of sophisticated technologies with traditional agricultural practices to accomplish fine-grid management of crops is referred to by both of these terms. Smart farming systems can give farmers useful environmental information from their field cultivation, allowing them to increase their marketability and revenue. This sort of technical innovation is beneficial to practically all aspects of agriculture, from sowing and feeding operations to crop defense and processing techniques [36]. Precision agriculture's destiny depends on contemporary technical breakthroughs and remote sensing techniques utilizing unmanned aerial vehicles (UAV) and various types of smart sensor technology. UAVs allow earth observation (EO) to improve precision, conduct more frequent and more thorough site surveillance, and explore enormous (inaccessible) regions. Sensors and UAVs are components of the Internet of Things (IoT). A substantial number of initiatives have been published in recent years concerning the most appropriate network topology and technology for smart agricultural operations. Farmers can gather and evaluate valuable data using IoT agriculture apps. Major landowners and small-scale farmers can embrace the promise of the IoT marketplace for agricultural land by implementing smart technologies to boost their production's profitability and durability. Farmers will have to accommodate the need for agro-IoT alternatives due to the increase of human needs [37].

9.3.4.2 Nanoencapsulation

Agrochemicals are essential in the farming process. According to current estimates, up to 7.5 million tonnes of agrochemicals, comprising fertilizers, insecticides, herbicides, fungicides, and hormonal substances, have been deployed worldwide in 2022 [38]. Nonetheless, more than 90% of these compounds are not target-specific. The results certainly waste money, as well as posing a severe risk to the natural world, damaging both terrestrial and aquatic environments. Nanoemulsions have been among the most basic and efficient nutritional supplement and medicinal administration methods in the encapsulation sector, and they can be made from a wide range of organic and chemical-based constituents such as oily substances, detergents, cosurfactants, weighing intermediaries, maturation antagonists, emulsifier, and gel-forming substances [39]. For a variety of reasons, nanoemulsions have been found to boost the absorption and utilization of different kinds of lipophilic compounds enclosed within them, which could prove beneficial in enhancing the biological efficacy of agrochemicals. The primary goal of fertilizer encasing is to maximize the nutrient utilization rate while reducing mineral content losses. Pesticide encapsulation is a more prevalent application of nanoencapsulation than fertilizer encapsulation. Inorganic nanoparticles of zinc (Zn), copper (Cu), Fe, cerium (Ce), and titanium (Ti) have been thoroughly investigated under lab and pilot conditions when contrasted with enclosure of these nutrients into nanoparticulate structures [40]. Chitosan is a naturally generated polymer with numerous agricultural applications. Chitosan has the potential to be used in cultivation as an independently operated or an encapsulating matrix for nutrients because of its intrinsic growth promotion and antibacterial capabilities [41]. A team of researchers used this cyclodextrin characteristic to create complex mixtures featuring two plant-derived chemicals, carvacrol and linalool, which have insect-killing and repellent effects. To simplify the application to crops, the cyclodextrins were employed to modify chitosan prior to creating chitosan nanoparticles. Given the extremely volatile nature of natural substances, nanocompounding extended their lifespan in the atmosphere [42]. A further investigation from the same team employed zein, a naturally occurring protein, to create nanoencapsulates of plant-derived pesticides such as citronella oil, geraniol, eugenol, and cinnamaldehyde [43].

9.3.4.3 Nanosensors

Utilizing nanotechnology breakthroughs, nanobiosensors, or nanosensors paired with so-called immobilized bioreceptor testers sensitive to specific analyte molecules, were first produced in the twenty-first century. Many nanoscale incorporations are now being observed in agriculture, with a variety of nanocapsules and nanodevices being used for the identification and treatment of plant illnesses, as nanobiosensors may be focused on a particular region of value. Many nanomaterials are utilized for sewage treatment and evaluation of soil, and some are employed to give nutrients as well as pesticides to plants at certain locations [14]. Salicylic acid (a phenolic phytohormone) promotes growth in plants, water loss, the process of photosynthesis and other properties. It is an important plant ingredient, and there is a need to determine salicylic acid content in plants. For this purpose, electrical-based nanobiosensors with electrodes made of Au enhanced using Cu nanoparticles (copper nanoparticles) have been employed [44]. An electrochemical nanobiosensor and a C-nanotube (carbon nanotube) electrode customized with Au nanoparticle coating were utilized to detect triazophos (an insecticide) in after-harvest vegetables and crops [45].

9.4 SOIL FERTILITY MANAGEMENT

Given human-induced activities in addition to shifts in daily routine, the agricultural yield and profitability of agricultural land throughout the globe has been adversely impacted. These developments could turn disastrous in the not-too-distant future, resulting in drought and starvation, particularly in emerging nations. As a result, significant focus needs to be drawn so as to deal with these anticipated disasters, eventually paving the way for ecologically sound agriculture. In today's world, it is critical to create smart materials that can slowly and steadily distribute nutrients to selected locations, thereby leading to a more sustainable ecosystem. Several forms of nanoparticles have been invented, and they have either functioned as nourishment for plants or as carriers for the same [46].

9.4.1 Effect on Soil Organic Matter

Organic and inorganic minerals within the soil can react with nanomaterials, resulting in a separation of the soil structure's liquid and solid states. The destiny and behavior of nanomaterials in aquatic environments are mostly inferred from literature, but information regarding the outcome and behavior of nanomaterials in ecosystems of soil is scarce. Soil organic matter has an instantaneous effect on topsoil chemical, physical, and biological attributes, and its buildup in soil is an important challenge for improved crop productivity and reducing the impact of global warming. A recent study found inconsistent impacts of nanomaterials on soil organic matter interactions due to differences in laboratory settings, soil characteristics, and nanomaterial dose [47]. In broader terms, the TiO_2 nanoparticles have no effect on the fluctuations of microbial-derived soil organic matter; however, they can reduce its degradation by enhancing its durability. Conversely, the use of ZnO nanoparticles reduced the efficacy of litter-derived organic carbon breakdown by as much as 130% owing to lowered microbial activity [48]. A further investigation discovered that using Fe_2O_3 nanoparticles decreased greenhouse gas emissions by up to 30%, suggesting minimal soil organic matter degradation with nanomaterial utilization in topsoil. These nanomaterials can trap organic carbon in soil by minimizing the release of carbon dioxide, which may serve to reduce the impact of global warming [49].

9.4.2 Effect on Soil Microflora

Soil microbes are excellent gauges of the state of the soil because they have an immediate bearing on the quality of the soil, particularly soil organic matter interactions and cycling of nutrients. As 40–70% of the overall soil bacteria are found in these microaggregates, nanomaterials are localized in soil microaggregates and are in proximity with the majority of resident microbial populations. The detrimental effect of these nanoparticles on microbial activity and functionality greatly relies on the kind of nanoparticle, with inorganic nanoparticles (metal and metal oxide) being more hazardous than organic nanoparticles (fullerenes and carbon nanotubes). Given their distinct cell wall architectures, gram-negative microorganisms are frequently less reactive to nanomaterials rather than gram-positive microbes [50]. Nevertheless, there is no agreement in the scientific community as to whether silver nanoparticles are more harmful to gram-negative microorganisms. Carbon-based nanoparticles had a major effect on the gene expression and mechanisms of soil microbial populations (*Archaea*, *Bacteria*, and *Eukarya*) engaged in the C and N cycles, whereas the S and P phases appeared comparatively unaffected. Metals and metal oxide nanoparticles, on the other hand, may operate as antibacterial compounds [51]. Furthermore, in isolated cultures, these nanomaterials reduced the development of plant growth-promoting rhizobacteria (PGPRs) along with other species of bacteria participating in the N cycle [50]. The differing impacts of nanoparticles on the biomass of microbes and activity were mostly determined by the amount of nanoparticles introduced to the soil, as it was discovered that Ag nanoparticles had only an undesirable effect on microbial functionality once an excessive amount of Ag was given to the substrate. The majority of nanoparticles (Ag, Al_2O_3, TiO_2, CuO, ZnO, etc.) exhibited detrimental effects on the microbial communities of soil, but experiments using Si, Fe, Au, Pd, and Ag_2S had hardly any impact [50]. Recently, it was discovered that biological nanomaterials are less damaging to soil microbial life than artificially synthesized analogues, and hence their use has been promoted as a method to reduce nanotoxicity in soils [52].

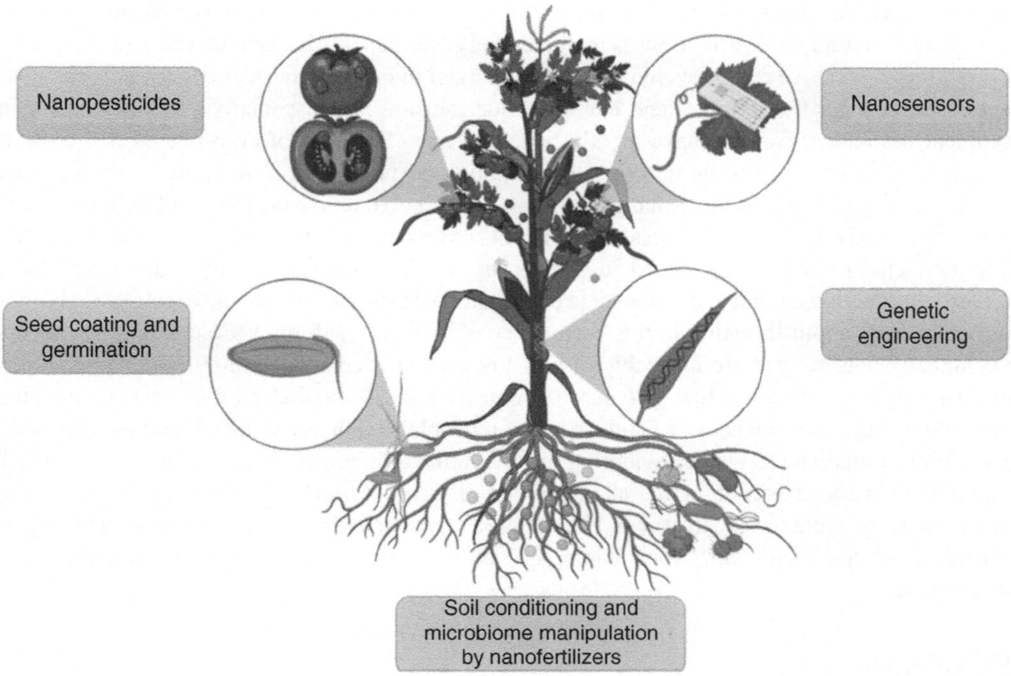

FIGURE 9.2 Overview of applications of nanotechnology in the agricultural sector.

9.5 CAN NANOTECHNOLOGY PROVIDE EFFECTIVE SOLUTIONS?

Through precision farming, nanotechnology is making its way through the door of the agricultural industry, since it may be used sensibly in accordance with the general organization of ecologically responsible farming, in particular by offering environmentally benign and innovative solutions. The recent rise of nanotechnology in medications and pharmacology has created new potential for integrating nanotechnology basics into the agricultural industry as well. The combination of nanotechnology and other technological domains has cleared the path for the creation of revolutionary agricultural goods, technological advances, and applications [53]. Nanotechnology has enormous opportunities to enhance today's agricultural system through plant input, cattle, animal husbandry, fishing, and seafood conservation. With constant surveillance infrastructure, the final objective of agri-nanotechnology is to maximize crop output (yields) whilst minimizing crop inputs (e.g., agrochemicals, fertilizers, and herbicides) [54]. Nanotechnology has the capacity to boost the yield and quality of crops at all levels of the farming process (from storing seeds to preparing, sprouting, fertigation, and post-harvest). Nanotechnology is significant in enhancing agricultural output through nitrogen regulation and quality of water oversight for long-term growth [55]. Figure 9.2 gives an overview of various approaches through nanotechnology for crop protection.

9.6 CONCLUSIONS AND FUTURE PERSPECTIVES

Growing crops sustainably has become critical for ensuring crop security and safeguarding the environment. Farming requires modernization and invention to cope with the mounting demand for food of the world's growing population while also ensuring environmental sustainability. As a fundamental premise for ecological farming, minimal agrochemicals with inexpensive manufacturing costs but increased yields must be employed. The manufacture and deployment of nanoagrochemicals in agriculture has increased investigators' and scientists' belief in the effectiveness of nanotechnology in agribusiness. The most sensible instruments for contemporary farming are nanosized devices for delivery and continuous distribution. Nanotechnology can benefit the agricultural and food sector by providing distinctive resources like nanofertilizer and micronutrient distribution, agrochemical-encased nanocarrier frameworks, and nanopesticide transmission, especially green pesticides and biological pesticides. Farmers have greatly benefited from handling of crop-related debris, water management, soil treatment, agroproduct preparation and preservation, identification of diseases of crops, and tracking growth of plants with nanosensors. Nonetheless, prior to being used, such nanoproducts must be thoroughly tested. The results of this discipline's study tend to be confined to laboratory settings, and the associated guidelines and standards are likewise little chronicled. Given the costly production capital and inadequate economic rewards, cultivators and commercial individuals are hesitant about investing in this sector. The aforementioned innovations might not become economically feasible and are incapable of meeting current needs concerning food manufacturing and distribution. There are a few instances of industrial goods on shelves that require immediate intervention. The consequences of nanoparticles on people's wellness, comfort, and environmental ramifications remain unknown. Considering the utilization of nanoparticles to mitigate crop pitfalls, an unambiguous investigation of the chemical and physical features of the ground in each of the cultivation areas ought to be performed. Humanitarian and ecotoxicological study can help us comprehend the complicated connections among agriculture, nanoscale agrochemicals, dosage phases, and humans.

REFERENCES

[1] M. Kah, N. Tufenkji, J.C. White, Nano-enabled strategies to enhance crop nutrition and protection, *Nat Nanotechnol.* 14 (2019) 532–540. https://doi.org/10.1038/s41565-019-0439-5

[2] Food and Agriculture Organization of the United Nations, (n.d.). www.fao.org/home/en

[3] X. Gao, S.M. Rodrigues, E. Spielman-Sun, S. Lopes, S. Rodrigues, Y. Zhang, A. Avellan, R.M.B.O. Duarte, A. Duarte, E.A. Casman, G.V. Lowry, Effect of soil organic matter, soil pH, and moisture content on solubility and dissolution rate of CuO NPs in soil, *Environ Sci Technol.* 53 (2019) 4959–4967. https://doi.org/10.1021/acs.est.8b07243.

[4] N. Ramankutty, Z. Mehrabi, K. Waha, L. Jarvis, C. Kremen, M. Herrero, L.H. Rieseberg, Trends in global agricultural land use: Implications for environmental health and food security, *Annu Rev Plant Biol.* 69 (2018) 789–815. https://doi.org/10.1146/annurev-arplant-042817-040256.

[5] S. Knapp, M.G.A. van der Heijden, A global meta-analysis of yield stability in organic and conservation agriculture, *Nat Commun.* 9 (2018) 3632. https://doi.org/10.1038/s41467-018-05956-1

[6] D. Thangadurai, J. Sangeetha, R. Prasad, *Nanotechnology for Food, Agriculture, and Environment*, Springer International Publishing, Cham, 2020. https://doi.org/10.1007/978-3-030-31938-0

[7] M. Thiruvengadam, G. Rajakumar, I.-M. Chung, Nanotechnology: Current uses and future applications in the food industry, *3 Biotech.* 8 (2018) 74. https://doi.org/10.1007/s13205-018-1104-7

[8] C. Li, Y. Li, Y. Li, G. Fu, Cultivation techniques and nutrient management strategies to improve productivity of rain-fed maize in semi-arid regions, *Agric Water Manag.* 210 (2018) 149–157. https://doi.org/10.1016/j.agwat.2018.08.014.

[9] X. He, H. Deng, H. Hwang, The current application of nanotechnology in food and agriculture, *J Food Drug Anal.* 27 (2019) 1–21. https://doi.org/10.1016/j.jfda.2018.12.002

[10] L. Rossi, L.N. Fedenia, H. Sharifan, X. Ma, L. Lombardini, Effects of foliar application of zinc sulfate and zinc nanoparticles in coffee (Coffea arabica L.) plants, *Plant Physiol Biochem.* 135 (2019) 160–166. https://doi.org/10.1016/j.plaphy.2018.12.005

[11] K.M. Manjaiah, R. Mukhopadhyay, R. Paul, S.C. Datta, P. Kumararaja, B. Sarkar, Clay minerals and zeolites for environmentally sustainable agriculture, in: *Modified Clay and Zeolite Nanocomposite Materials*, Elsevier, 2019: pp. 309–329. https://doi.org/10.1016/B978-0-12-814617-0.00008-6

[12] T.R. Shojaei, M.A.M. Salleh, M. Tabatabaei, H. Mobli, M. Aghbashlo, S.A. Rashid, T. Tan, Applications of nanotechnology and carbon nanoparticles in agriculture, in: *Synthesis, Technology and Applications of Carbon Nanomaterials*, Elsevier, 2019: pp. 247–277. https://doi.org/10.1016/B978-0-12-815757-2.00011-5

[13] K.S. Subramanian, A. Manikandan, M. Thirunavukkarasu, C.S. Rahale, Nano-fertilizers for balanced crop nutrition, in: *Nanotechnologies in Food and Agriculture*, Springer International Publishing, Cham, 2015: pp. 69–80. https://doi.org/10.1007/978-3-319-14024-7_3
[14] P. Shukla, P. Chaurasia, K. Younis, O.S. Qadri, S.A. Faridi, G. Srivastava, Nanotechnology in sustainable agriculture: Studies from seed priming to post-harvest management, *Nanotechnol Environ Eng*. 4 (2019) 11. https://doi.org/10.1007/s41204-019-0058-2
[15] S. León-Silva, R. Arrieta-Cortes, F. Fernández-Luqueño, F. López-Valdez, Design and production of nanofertilizers, in: *Agricultural Nanobiotechnology*, Springer International Publishing, Cham, 2018: pp. 17–31. https://doi.org/10.1007/978-3-319-96719-6_2
[16] R. Kumar, M. Ashfaq, N. Verma, Synthesis of novel PVA–starch formulation-supported Cu–Zn nanoparticle carrying carbon nanofibers as a nanofertilizer: Controlled release of micronutrients, *J Mater Sci*. 53 (2018) 7150–7164. https://doi.org/10.1007/s10853-018-2107-9
[17] S.-L. Wang, A.D. Nguyen, Effects of Zn/B nanofertilizer on biophysical characteristics and growth of coffee seedlings in a greenhouse, *Res Chem Intermed*. 44 (2018) 4889–4901. https://doi.org/10.1007/s11164-018-3342-z
[18] A. Jain, S. Ranjan, N. Dasgupta, C. Ramalingam, Nanomaterials in food and agriculture: An overview on their safety concerns and regulatory issues, *Crit Rev Food Sci Nutr*. 58 (2018) 297–317. https://doi.org/10.1080/10408398.2016.1160363
[19] L. Natrayan, S. Yogeshwaran, L. Yuvaraj, M.S. Kumar, Effect of graphene reinforcement on mechanical and microstructure behavior of AA8030/graphene composites fabricated by stir casting technique, 2019: pp. 020012. https://doi.org/10.1063/1.5131599
[20] A.S. Hashem, S.S. Awadalla, G.M. Zayed, F. Maggi, G. Benelli, Pimpinella anisum essential oil nanoemulsions against Tribolium castaneum—insecticidal activity and mode of action, *Environ Sci Poll Res*. 25 (2018) 18802–18812. https://doi.org/10.1007/s11356-018-2068-1
[21] K. Seeniappan, B. Venkatesan, N.N. Krishnan, T. Kandhasamy, S. Arunachalam, R.K. Seeta, M.V. Depoures, A comparative assessment of performance and emission characteristics of a DI diesel engine fuelled with ternary blends of two higher alcohols with lemongrass oil biodiesel and diesel fuel, *Energy Environ*. 33 (2022) 1134–1159. https://doi.org/10.1177/0958305X211051323
[22] K. Khalid, X. Tan, H.F. Mohd Zaid, Y. Tao, C. Lye Chew, D.-T. Chu, M.K. Lam, Y.-C. Ho, J.W. Lim, L. Chin Wei, Advanced in developmental organic and inorganic nanomaterial: A review, *Bioengineered*. 11 (2020) 328–355. https://doi.org/10.1080/21655979.2020.1736240
[23] T. Hofmann, G.V. Lowry, S. Ghoshal, N. Tufenkji, D. Brambilla, J.R. Dutcher, L.M. Gilbertson, J.P. Giraldo, J.M. Kinsella, M.P. Landry, W. Lovell, R. Naccache, M. Paret, J.A. Pedersen, J.M. Unrine, J.C. White, K.J. Wilkinson, Technology readiness and overcoming barriers to sustainably implement nanotechnology-enabled plant agriculture, *Nat Food*. 1 (2020) 416–425. https://doi.org/10.1038/s43016-020-0110-1
[24] P. Zhang, Z. Guo, S. Ullah, G. Melagraki, A. Afantitis, I. Lynch, Nanotechnology and artificial intelligence to enable sustainable and precision agriculture, *Nat Plants*. 7 (2021) 864–876. https://doi.org/10.1038/s41477-021-00946-6
[25] S. Ahmar, T. Mahmood, S. Fiaz, F. Mora-Poblete, M.S. Shafique, M.S. Chattha, K.-H. Jung, Advantage of nanotechnology-based genome editing system and its application in crop improvement, *Front Plant Sci*. 12 (2021). https://doi.org/10.3389/fpls.2021.663849
[26] W. Du, J. Yang, Q. Peng, X. Liang, H. Mao, Comparison study of zinc nanoparticles and zinc sulphate on wheat growth: From toxicity and zinc biofortification, *Chemosphere*. 227 (2019) 109–116. https://doi.org/10.1016/j.chemosphere.2019.03.168
[27] Y. Wang, Y. Lin, Y. Xu, Y. Yin, H. Guo, W. Du, Divergence in response of lettuce (var. ramosa Hort .) to copper oxide nanoparticles/microparticles as potential agricultural fertilizer, *Environ Poll Bioavailability*. 31 (2019) 80–84. https://doi.org/10.1080/26395940.2019.1578187
[28] M. Mazaheri Tirani, M. Madadkar Haghjou, A. Ismaili, Hydroponic grown tobacco plants respond to zinc oxide nanoparticles and bulk exposures by morphological, physiological and anatomical adjustments, *Funct Plant Biol*. 46 (2019) 360. https://doi.org/10.1071/FP18076
[29] M. Iqbal, N.I. Raja, Z.-U.-R. Mashwani, M. Hussain, M. Ejaz, F. Yasmeen, Effect of silver nanoparticles on growth of wheat under heat stress, *Iran J Sci Technol Trans A: Sci*. 43 (2019) 387–395. https://doi.org/10.1007/s40995-017-0417-4

[30] M. Rizwan, S. Ali, M.Z. ur Rehman, S. Malik, M. Adrees, M.F. Qayyum, S.A. Alamri, M.N. Alyemeni, P. Ahmad, Effect of foliar applications of silicon and titanium dioxide nanoparticles on growth, oxidative stress, and cadmium accumulation by rice (Oryza sativa), *Acta Physiol Plant.* 41 (2019) 35. https://doi.org/10.1007/s11738-019-2828-7

[31] A.A. Malandrakis, N. Kavroulakis, C.V. Chrysikopoulos, Use of copper, silver and zinc nanoparticles against foliar and soil-borne plant pathogens, *Sci Total Environ.* 670 (2019) 292–299. https://doi.org/10.1016/j.scitotenv.2019.03.210

[32] G.L. Vanti, V.B. Nargund, B.K.N, R. Vanarchi, M. Kurjogi, S.I. Mulla, S. Tubaki, R.R. Patil, Synthesis of *Gossypium hirsutum* -derived silver nanoparticles and their antibacterial efficacy against plant pathogens, *Appl Organomet Chem.* 33 (2019). https://doi.org/10.1002/aoc.4630

[33] G. Medina-Pérez, F. Fernández-Luqueño, E. Vazquez-Nuñez, F. López-Valdez, J. Prieto-Mendez, A. Madariaga-Navarrete, M. Miranda-Arámbula, Remediating polluted soils using nanotechnologies: Environmental benefits and risks, *Pol J Environ Stud.* 28 (2019) 1013–1030. https://doi.org/10.15244/pjoes/87099

[34] S.V.& A.V. Mohammadhosein Keshvadi, F. Karimi, Comparative study of antibacterial inhibitory effect of silver nanoparticles and garlic oil nanoemulsion with their combination, *Biointerface Res Appl Chem.* 9 (2019) 4560–4566. https://doi.org/10.33263/BRIAC96.560566.

[35] R.M. Bentolhoda Amanzadi, E. Mirzaei, G. Hassanzadeh, P. Mahdaviani, S. Boroumand, M. Abdollahi, A. Abdolghaffari, Chitosan-based layered nanofibers loaded with herbal extract as wound-dressing materials on wound model studies, *Biointerface Res Appl Chem.* 9 (2019) 3979–3986. https://doi.org/10.33263/BRIAC94.979986

[36] S.K.Y. Donzia, H.-K. Kim, H.J. Hwang, A software model for precision agriculture framework based on smart farming system and application of IoT gateway, 2019: pp. 49–58. https://doi.org/10.1007/978-3-319-96806-3_4

[37] J.C. Guillermo, A. García-Cedeño, D. Rivas-Lalaleo, M. Huerta, R. Clotet, IoT architecture based on wireless sensor network applied to agricultural monitoring: A case of study of Cacao crops in Ecuador, 2019: pp. 42–57. https://doi.org/10.1007/978-3-030-04447-3_3

[38] FAOSTAT, (n.d.). www.fao.org/faostat/en/#data/RP

[39] S. Akhavan, E. Assadpour, I. Katouzian, S.M. Jafari, Lipid nano scale cargos for the protection and delivery of food bioactive ingredients and nutraceuticals, *Trends Food Sci Technol.* 74 (2018) 132–146. https://doi.org/10.1016/j.tifs.2018.02.001

[40] R. Raliya, V. Saharan, C. Dimkpa, P. Biswas, Nanofertilizer for precision and sustainable agriculture: Current state and future perspectives, *J Agric Food Chem.* 66 (2018) 6487–6503. https://doi.org/10.1021/acs.jafc.7b02178

[41] R.V. Kumaraswamy, S. Kumari, R.C. Choudhary, A. Pal, R. Raliya, P. Biswas, V. Saharan, Engineered chitosan based nanomaterials: Bioactivities, mechanisms and perspectives in plant protection and growth, *Int J Biol Macromol.* 113 (2018) 494–506. https://doi.org/10.1016/j.ijbiomac.2018.02.130

[42] J.L. de Oliveira, E.V.R. Campos, A.E.S. Pereira, T. Pasquoto, R. Lima, R. Grillo, D.J. de Andrade, F.A. dos Santos, L.F. Fraceto, Zein nanoparticles as eco-friendly carrier systems for botanical repellents aiming sustainable agriculture, *J Agric Food Chem.* 66 (2018) 1330–1340. https://doi.org/10.1021/acs.jafc.7b05552

[43] J.L. de Oliveira, E.V.R. Campos, T. Germano-Costa, R. Lima, J.F. Della Vechia, S.T. Soares, D.J. de Andrade, K.C. Gonçalves, J. do Nascimento, R.A. Polanczyk, L.F. Fraceto, Association of zein nanoparticles with botanical compounds for effective pest control systems, *Pest Manag Sci.* 75 (2019) 1855–1865. https://doi.org/10.1002/ps.5338

[44] Md.A. Wahab, N. Islam, M.E. Hoque, D.J. Young, Recent advances in silver nanoparticle containing biopolymer nanocomposites for infectious disease control—a mini review, *Curr Anal Chem.* 14 (2018) 198–202. https://doi.org/10.2174/1573411013666171009163829

[45] A.K. Srivastava, A. Dev, S. Karmakar, Nanosensors and nanobiosensors in food and agriculture, *Environ Chem Lett.* 16 (2018) 161–182. https://doi.org/10.1007/s10311-017-0674-7

[46] A. Ditta, Role of nanoclay polymers in agriculture: Applications and perspectives, in: *Nanohybrids in Environmental & Biomedical Applications*, CRC Press, Boca Raton, FL: CRC Press, Taylor & Francis Group, 2019: pp. 323–334. https://doi.org/10.1201/9781351256841-14

[47] J. Shi, J. Ye, H. Fang, S. Zhang, C. Xu, Effects of Copper Oxide Nanoparticles on paddy soil properties and components, *Nanomaterials*. 8 (2018) 839. https://doi.org/10.3390/nano8100839
[48] M.I. Rashid, T. Shahzad, M. Shahid, I.M.I. Ismail, G.M. Shah, T. Almeelbi, Zinc oxide nanoparticles affect carbon and nitrogen mineralization of Phoenix dactylifera leaf litter in a sandy soil, *J Hazard Mater*. 324 (2017) 298–305. https://doi.org/10.1016/j.jhazmat.2016.10.063
[49] M.I. Rashid, T. Shahzad, M. Shahid, M. Imran, J. Dhavamani, I.M.I. Ismail, J.M. Basahi, T. Almeelbi, Toxicity of iron oxide nanoparticles to grass litter decomposition in a sandy soil, *Sci Rep*. 7 (2017) 41965. https://doi.org/10.1038/srep41965
[50] V. Rajput, T. Minkina, S. Sushkova, A. Behal, A. Maksimov, E. Blicharska, K. Ghazaryan, H. Movsesyan, N. Barsova, ZnO and CuO nanoparticles: A threat to soil organisms, plants, and human health, *Environ Geochem Health*. 42 (2020) 147–158. https://doi.org/10.1007/s10653-019-00317-3.
[51] F. Wu, Y. You, D. Werner, S. Jiao, J. Hu, X. Zhang, Y. Wan, J. Liu, B. Wang, X. Wang, Carbon nanomaterials affect carbon cycle-related functions of the soil microbial community and the coupling of nutrient cycles, *J Hazard Mater*. 390 (2020) 122144. https://doi.org/10.1016/j.jhazmat.2020.122144.
[52] S. Mishra, X. Yang, H.B. Singh, Evidence for positive response of soil bacterial community structure and functions to biosynthesized silver nanoparticles: An approach to conquer nanotoxicity?, *J Environ Manage*. 253 (2020) 109584. https://doi.org/10.1016/j.jenvman.2019.109584
[53] P. Vega-Vásquez, N.S. Mosier, J. Irudayaraj, Nanoscale drug delivery systems: From medicine to agriculture, *Front Bioeng Biotechnol*. 8 (2020). https://doi.org/10.3389/fbioe.2020.00079
[54] P. Pramanik, P. Krishnan, A. Maity, N. Mridha, A. Mukherjee, V. Rai, Application of nanotechnology in agriculture, 2020: pp. 317–348. https://doi.org/10.1007/978-3-030-26668-4_9
[55] H. Chhipa, Applications of nanotechnology in agriculture, 2019: pp. 115–142. https://doi.org/10.1016/bs.mim.2019.01.002

10 Nanofertilizers for Sustainable Agriculture

Garima Chauhan, Juliana Espindola, Sehijpreet Kaur, Akashdeep Singh, Nitesh Kumar, Bharat Bhushan Rana, and Tarun Sharma
Corresponding author: Akashdeep Singh

10.1 INTRODUCTION

The world's population has been increasing since the twentieth century and reached 8 billion people in November 2022, a milestone in human development (United Nations, 2023). On the contrary, there is gradual increase in the production of foodgrains because of dwindling resources such as arable land and irrigation water, and overexploitation of the available resources such as fertilizers and pesticides, resulting in poor response to management practices. Optimizing fertilizer application is of paramount importance for enhancing crop productivity, sustaining yield, and food safekeeping. However, blanket application of fertilizers to enhance the productivity as well as the profitability of crops has brought higher consumption of nutrients, which not only results in major monetary and resource loss but is also liable to affect the environment (Pampolino et al., 2012; Stewart et al., 2005). The continuous use of conventional fertilizers might have helped in achieving food security, but they have also barely exceeded their nutrient use efficiency (NUE), that is, 30–35%, 18–20%, and 35–40% for N, P, and K, respectively, over the years (Subramanian et al., 2015; Thimiri Govinda Raj and Khan, 2016).

Nutrient losses not only reduce NUE but also cause environmental problems like nitrate-nitrogen (NO_3-N) and phosphorus entering rivers, lakes, and water bodies through leaching and contributing to eutrophication. NO_3-N, prone to leaching, causes groundwater to deteriorate, leading to major human disorders (Rahman et al., 2021; Yu et al., 2020). Stakeholders are primarily dependent on conventional inorganic fertilizers, and they are seeking new ways to overcome this problem. Besides the questionable use efficiencies of nutrients, the agriculture sector, which has been the backbone of society for centuries, is facing a plethora of problems, such as erratic climate conditions, declining productivity, an expensive and reduced labor force, and urbanization. Therefore, adopting strategies to replace inefficient nutrient contaminations of the environment is both necessary and represents a step towards strengthening agricultural systems. Alternatively, the use of engineered smart nanotools for the growth of agricultural systems can be an excellent strategy to revolutionize agriculture, minimizing the antagonistic effects of modern agriculture on the environment while enhancing both quality and quantity of yields (Liu and Lal, 2015; Prasad et al., 2017; Sekhon, 2014).

Nanotechnology, a new frontier in the agriculture sector, helps to overcome some of the problems with the use of nanofertilizers (NFs), and has attracted the attention of the scientific community because of its ability to increase production sustainably (Bulovic et al., 2007; Shang et al., 2019). It is an interdisciplinary field that combines principles from physics, chemistry, materials science, biology, and engineering. Its applications span a wide range of industries, including electronics, energy, medicine, and agriculture, and they hold great potential for solving some of the world's most pressing problems. The use of this technology is playing a major role in revolutionizing agriculture.

 DOI: 10.1201/9781003389408-10

Its applications are gradually converting theoretical possibilities into practical realities. The development of novel tools that can function at the nanometric level is facilitating research in molecular and cellular biology. The potential of nanotechnology in boosting agricultural productivity cannot be overstated. It can enhance plant and animal genetics, as well as provide cellular-level delivery of genes and chemical molecules to specific sites in both plants and animals (Maysinger, 2007; Preetha and Balakrishnan, 2017; Scott, 2007). This evolution in technology is making it possible to achieve higher crop yields, improved plant and animal health, and increased food security. Nanotechnology in agriculture is a prime example of how science and technology can work together to achieve a more sustainable and productive future. NFs carries nutrients of nanodimensions ranging from 1 to 100 nm (Suppan, 2017) but usually in the range of 30–40 nm. The higher surface area enables them to hold a generous amount of nutrient ions and release them slowly in line with the crop's requirements (Subramanian et al., 2015; Chhipa and Joshi, 2016).

Besides nourishment, if NFs are used correctly, they enhance NUE and reduce losses caused by volatilization, leaching, and other ecological risks (Kabiri et al., 2017; Manjunatha et al., 2016; Solanki et al., 2015). The sluggish release of nutrients by NFs over 40–50 days depends on polymer type, soil chemistry (pH), soil moisture, and temperature, whereas conventional fertilizers release in 4–10 days (Chen and Wei, 2018; Ghidan and Antary, 2019; Mittal et al., 2020; Singh et al., 2021). Therefore, NFs contribute to reducing environmental pollution, eutrophication, and groundwater pollution (Morales-Díaz et al., 2017). NFs help in the uplifting of physiological traits and crop yields due to their reduced dimensions of nutrient ions and, hence, greater absorptivity (Astaneh et al., 2021). In the context of more sustainable agriculture, NFs can increase fertilizer NUE, improve crop yield and quality, and minimize the adverse effects of synthetic fertilizers (Liu and Lal, 2015; Seleiman et al., 2020a). Foliar advocating of nanofertilizers has been found to help in increasing crop yield by 37.7% and 18–23% in pearl millet (Tarafdar et al., 2014) and cotton, respectively (Mahil and Baburai Nagesh, 2019), over conventional fertilizers. Hence, NFs are advocated for being ecologically sustainable and economically viable, having an extended role in crop production, and being environmentally safe. This chapter is an effort to summarize the available research reports on the role of nanotechnology produced fertilizers in meeting the existing challenges of sustainable crop production.

10.2 SYNTHESIS OF NANOFERTILIZERS

Due to several physical and chemical properties of nanosized particles, nanotechnology has gained attention and is being incorporated into several spheres. For a sustainable future, the agriculture sector has also taken the initiative and adopted nanotechnology. The performance of nanomaterials depends on the properties acquired while they are being synthesized. In general, there are two approaches for the synthesis of nanomaterials/nanofertilizers: top-down and bottom-up.

10.2.1 The Top-down Approach

The top-down approach is a simple approach wherein bulk material is broken down into simpler particles until a material of the desired property (size) is obtained (Figure 10.1). These methods are an extension of methods that have been used to create particles with a diameter of less than a micron. In order to obtain the nanodimension, this method requires substrates like zeolites or other materials that are ball-ground for several hours.

10.2.2 The Bottom-up Approach

The bottom-up approach is an alternative to the top-down approach that is more economical because it involves less waste of materials. The desired nanomaterial is built atom by atom (Figure 10.2).

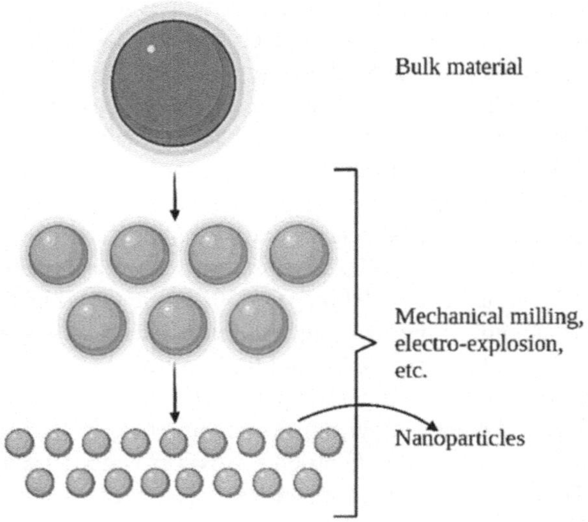

FIGURE 10.1 Top-down approach.

Certain chemical reactions begin with molecules in the solution and proceed through molecule association to create NPs, and hence the particle size can be regulated. Emulsion, co-precipitation, micelle formation, and reverse micelle formation are examples of regulated synthesis procedures for NP production that place a priority on minimizing coagulation or aggregation and producing homogeneous NPs. They must be characterized on the basis of their physical, chemical, and mechanical characteristics after being synthesized to determine their functionality in terms of solubility, dispersibility, and stability. It is critical to understand which qualities will be beneficial based on the type of nutrition used for each crop in order to achieve an adequate physiological interaction of NP with the plant. Along with physicochemical traits like shape, size, surface composition and charge, boiling point, melting point, pH variation, moisture, solubility, purity, soil type, stability, thermal and isoelectric properties, and heat and diffusion-controlled release, these traits also include chemical properties, surface functionality, thermal stability, and composition (Ribeiro and Carmo, 2019; Pradhan and Mailapalli, 2017; Ramzan et al., 2020).

10.3 NANOFERTILIZERS AS ADVANCED NUTRIENT DELIVERY SYSTEMS

Nanofertilizers represent a recent and pioneering strategy in plant nutrition that employs nanotechnology to enhance the delivery of vital nutrients to crops. They are categorized into three groups, namely macro nanofertilizers, micro nanofertilizers, and nanoparticulate fertilizers, which combine both macro and micro nutrients (Chhipa and Joshi, 2016). These fertilizers can be administered in powder or liquid form, with a diameter less than 100 nanometers (Jampílek and Kráľová, 2015). By supplying crops with easily accessible nutrients, nanofertilizers enhance nutrient absorption, resulting in improved plant growth and increased agricultural output (El-Saadony et al., 2021). This innovative approach to plant nutrition holds immense potential in addressing the escalating demand for sustainable food, feed, and fiber production.

Nanofertilizers are often referred to as deliverable agents or smart systems of nutrients due to their ability to efficiently transport nutrients to crops (Abdalla et al., 2022). These fertilizers have been

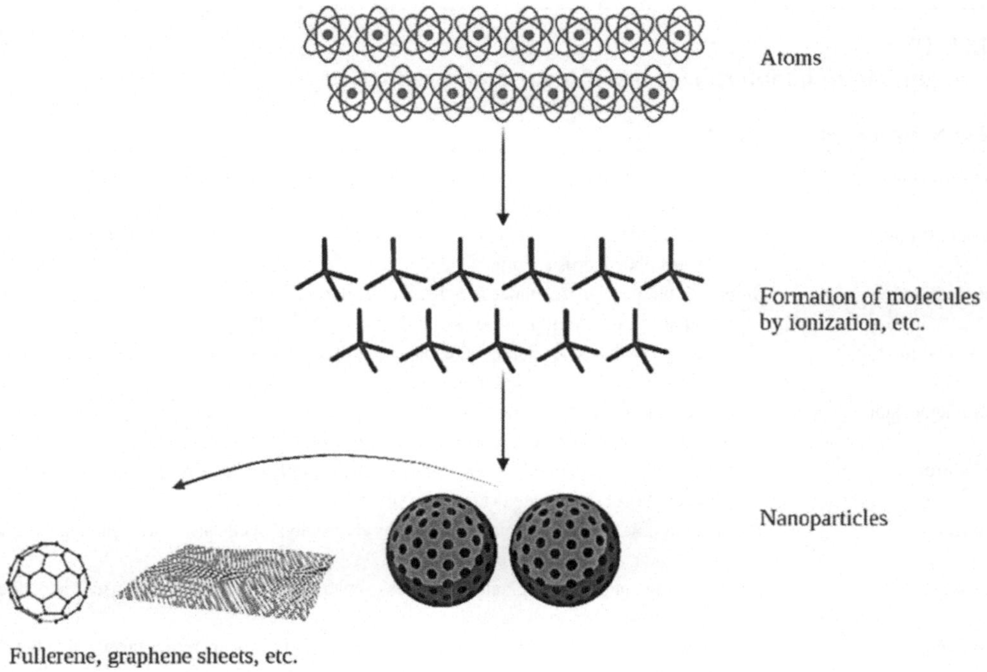

FIGURE 10.2 Bottom-up approach.

shown to enhance nutrient efficiency, improve plant nutrition, and boost soil fertility (Table 10.1) (Toksha et al., 2021). However, the effectiveness of nanofertilizers is largely dependent on the specific agroecosystem conditions, including the soil's physicochemical properties, moisture levels, other ecological factors (Jakhar et al., 2022), and methods of application.

Several methods for applying nanofertilizers in agriculture are available nowadays, each with its advantages and disadvantages. These methods include foliar application, soil application, seed coating, and fertigation. The choice of method depends on the specific requirements of the crop and the characteristics of the nanofertilizer being used. Overall, the selection of the application method plays a crucial role in determining the efficacy of the nanofertilizer and the outcomes attained by the producer.

10.3.1 Foliar Nanofertilizer Application

Foliar application is a method of plant nutrition where liquid fertilizers containing macro and micronutrients are sprayed directly onto the leaves of the plant. This technique should be used to complement and supplement soil fertilization as needed to target deficient nutrients and correct the imbalance quicker.

The utilization of nanomaterials as an ally in foliar fertilization provides numerous benefits, such as increased penetration capacity, larger surface area, and greater use efficiency. These properties reduce the potential for residual contamination in the environment, since the nanoparticles can have a size below 100 nm, which makes them highly effective as fertilizers for efficient nutrient management, promoting an ecofriendly approach and reducing environmental pollution. The development of these nanofertilizers with the help of nanotechnology might significantly impact the entire value chain of agricultural production (Al-Juthery and Saadoun, 2018; Morales-Díaz et al., 2017).

The utilization of foliar application for nanofertilizers has demonstrated positive effects on crop growth and yield. Studies have indicated favorable outcomes when employing foliar methods to

TABLE 10.1
Macro and micro nanofertilizer benefits for crop production

Type of Nanofertilizer	Benefits
Nano-nitrogen	Stimulates development and expansion of plants, enhances crop yield, minimizes nitrogen leaching, and reduces fertilizer application rates
Nano-phosphorus	Enhances plant root development, promotes seed germination, boosts crop yield, and mitigates phosphorus runoff
Nano-potassium	Improves plant stress resistance and disease resilience, facilitates nutrient uptake, enhances crop quality, and minimizes potassium loss through leaching
Nano-calcium	Promotes growth and development, enhances fruit quality, increases crop yield, and combats calcium deficiency disorders in plants
Nano-magnesium	Stimulates growth and development, enhances photosynthesis, increases crop yield, and alleviates magnesium deficiency disorders in plants
Nano-sulfur	Facilitates growth and development, aids in protein synthesis, enhances crop yield, and mitigates sulfur deficiency disorders in plants
Nano-iron	Corrects iron deficiencies in plants, promotes chlorophyll synthesis, fosters plant growth and development, and boosts crop yield
Nano-zinc	Prevents zinc deficiency in plants, promotes growth and development, increases yield, and enhances resistance to diseases and pests
Nano-manganese	Stimulates growth and development, improves photosynthesis, increases crop yield, and combats manganese deficiency disorders in plants
Nano-copper	Prevents copper deficiency in plants, promotes growth and development, increases yield, and enhances resistance to diseases and pests
Nano-boron	Enhances growth and development, facilitates flower and fruit formation, increases crop yield, and mitigates boron deficiency disorders in plants
Nano-molybdenum	Stimulates nitrogen fixation in legume plants, promotes growth and development, increases crop yield, and alleviates molybdenum deficiency disorders in plants

administer nanofertilizers. For example, Davarpanah et al. (2016) observed an increase in pomegranate fruit yield by utilizing a single foliar application of B or Zn nanofertilizers in relatively low doses (34 mg B per tree or 636 mg Zn per tree), resulting in a higher number of fruits per tree. Similarly, Abd-El-Aziz et al. (2021) reported significant enhancements in wheat and soybean yield through the use of nano-Zn, with yield increases of 37% and 33% respectively compared to soil-applied nano-Zn. These findings underscore the significance of identifying the most effective method of applying nanofertilizers for specific crops.

Furthermore, foliar nanofertilizers have demonstrated the ability to alleviate the negative impacts of salinity stress in cotton, as evidenced by Hussein and Abou-Baker (2018). Application of 100–200 ppm of nano-Zn via foliar spray led to increased dry weight in the roots, stems, leaves, and bolls of cotton plants, with a notable 34% increase in dry weight compared to the control (no nano-Zn) when considering the entire plant.

These studies highlight the potential of foliar nanofertilization in various crops to enhance yields and mitigate stress, providing a valuable tool for improving agricultural production. However, further research is necessary to fully comprehend the advantages and limitations of nanofertilization in different crop systems.

10.3.2 Soil Nanofertilizer Application

Nanofertilizers can be applied to the soil in various ways. One common method is through mixing the nanofertilizer with the soil before planting or incorporating it into the soil during cultivation.

Another method is to apply it as a top dressing, where the nanofertilizer is spread on top of the soil and allowed to mix with it naturally or through irrigation. The choice of application method depends on various factors such as crop type, soil structure, climate, and local regulations.

The application of conventional fertilizers in soil represents an effective approach for supplying crops with vital nutrients. However, the uptake efficiency of these nutrients by plants can be influenced by various factors, including soil type, soil pH, and the method of fertilizer application. According to Preetha and Balakrishnan (2017), conventional fertilizer-use efficiencies hardly exceed 30–35% (N), 18–20% (P), and 35–40% (K), and these data have not changed in recent years because research has not been successful in improving this uptake.

Nanofertilizers offer a promising solution to enhance soil fertilization, as their small size provides a larger surface area for nutrient absorption. This increased surface area allows for a greater concentration of nutrients, which are released slowly and consistently to the plant, providing a more stable and efficient supply of essential nutrients that can easily disperse into the soil structure. The customization of nanofertilizers to meet specific crop needs can also minimize any negative impacts on the environment.

For customization, there are materials used to make these nanofertilizers available to the plant during its cycle. Some examples of nanofertilizers are zeolite, sulfur nanocoatings, and formulations of biofertilizer inocula. Zeolite releases nutrients slowly to the crops, preventing loss through various processes in the soil. Sulfur nanocoatings provide sulfur elements in addition to primary nutrients and have a reduced fertilizer dissolution rate. Biofertilizer inocula are formulated into polymeric nanoparticles for coating biofertilizer preparations to increase resistance to desiccation. Additionally, other compounds like nanotubes of carbon, clay montmorillonite urea, nanoparticles of SiO^2, and nanoporous or mesoporous of silica can also be used to enhance fertilizer efficiency and slow release (Ditta and Arshad, 2016; Mahanta et al., 2019; Wilson et al., 2008).

The specific compounds used for nanofertilizer in soil application to improve crop yields and growth greatly depend on a variety of factors including crop needs, soil structure, and environmental conditions. In order to achieve optimal results, it is important to choose the right nanofertilizer for each situation. For example, in soil with high levels of clay, a nanofertilizer containing clay montmorillonite urea may be best suited to improve the uptake of nutrients. On the other hand, in areas with limited water resources, a nanofertilizer containing nanoparticles of SiO^2 may be more effective in helping crops retain moisture.

In a study conducted by Manikandan and Subramanian (2016), the application of nanofertilizer (nano-zeourea) in the soil for maize grain production resulted in an approximately 8% increase in grain yield in inceptisol and a significant 39% increase in alfisols compared to the control group (urea). Similarly, Zakzok et al. (2018) investigated the impact of 50% nano-NPK and 50% bio-organic fertilizer on cotton yield and fiber properties. They found that this combination led to a noteworthy ~10% increase in cotton seed yield and an approximately ~14% increase in cotton fiber yield compared to the conventional NPK recommendation.

10.3.3 Seed Nanofertilizer Coating

Seeds are the key to the growth and development of an entire plant. They contain food reserves stored in the form of starch which nourishes the embryo leading to germination. The application of nanofertilizers in seed coating holds immense potential for promoting crop growth and development. The small size of nanofertilizer particles allows for uniform distribution on the seed surface, leading to more efficient delivery of nutrients to the plant. This enhances the germination process and provides early nourishment to the developing plant, resulting in a stronger and healthier root system. The enhancement in the germination process might be due to the ability of the nanofertilizers to penetrate the thick seed coat, which makes them capable of enhancing water absorption by regulating the aquaporin gene expression and reducing the negative impact of salinity, drought, and other conditions. In addition, the slow-release characteristic of nanofertilizers can also reduce the risk of leaching and volatilization of nutrients, ensuring a consistent supply of nutrients throughout the growth period. By combining the benefits of both seed coating and slow-release fertilization, nanofertilizers applied in seed coating offer a highly efficient and effective method of soil fertilization.

Seed coating with nanoparticles can be done through different methods depending on the types of nanoparticles and their intended application. The process may involve mixing the nanoparticles with a liquid, such as a suspension, to create a slurry. Alternatively, the nanoparticles may be incorporated into a dry powder form which is then mixed with the seeds and applied using equipment made for seed treatment in general.

There are several benefits to having nanoparticles attached to the seed instead of just out in the soil. Firstly, the seed coatings allow a more precise and consistent application of the fertilizer, ensuring that each seed receives an optimal amount of nutrients. This leads to more uniform plant growth and development. Secondly, seed coatings reduce the amount of fertilizer that is lost to the environment through volatilization, leaching, or denitrification. This results in more efficient use of fertilizer resources, as well as reduced environmental impacts. Additionally, seed coatings offer improved stability and longer shelf life for the fertilizer, as they protect the nutrients from degradation and weathering. The utilization of seed coating not only facilitates the delivery of nutrients but also provides an effective method of administering biostimulants and plant growth regulators. This comprehensive approach further enhances the growth and development of plants.

Research investigating the effectiveness of seed nanofertilizers has yielded encouraging findings. For instance, Badran and Savin (2018) observed notable improvements in stem length, stem diameter, elongation of primary and secondary roots, and an increase in the number of secondary roots when utilizing nanofertilizers (specifically, nano-urea modified with hydroxyapatite nanoparticles) under saline conditions. Similarly, in a separate study, seed treatment with nanofertilizer demonstrated enhanced crop growth and grain yield in little millet, showcasing its favorable economic benefits (Rajput et al., 2022).

The utilization of a combination of different application methods for nanofertilizers has shown the potential to enhance nutrient uptake, as demonstrated by Pruthvi and Chandrashekara (2021) in their study of cotton plants and a combination of nanofertilizer applications. They found that seed treatment with nano-Zn combined with foliar application of nano-Zn resulted in improved crop growth and increased nutrient uptake.

10.3.4 Fertigation with Nanofertilizer

Fertigation, the application of fertilizers through irrigation systems, is another option for nanofertilization. This method allows for precise control over the delivery of nutrients to the plant, reducing the risk of overfertilization and waste. Fertigation can be especially useful in arid regions where water is limited and soil nutrients are depleted quickly. It can also be used in greenhouse and hydroponic systems where the soil structure may be too poor to support plant growth. When using fertigation for nanofertilization, it is important to monitor soil nutrient levels regularly and adjust the fertilizer application accordingly.

The extent of penetration of nanofertilizers is influenced by various factors, including soil properties, water flow rate, and the uniformity of the irrigation system's distribution. Generally, nanofertilizers exhibit superior soil penetration compared to traditional fertilizers due to their smaller particle size, enabling them to pass through soil pores more effectively. This enhanced penetration facilitates improved nutrient uptake by crop roots and a more even distribution of nutrients within the soil (Morales-Díaz et al., 2017; Singh et al., 2017). However, the efficacy of nanofertilizers in fertigation is also influenced by the chemical and physical characteristics of the soil, the specific nutrient requirements of the crop species, and the prevailing environmental conditions.

TABLE 10.2
Methods of application used in each nanofertilizer category

Category of nanofertilizer	Common methods of application
Macro nanofertilizer	Soil application, seed treatment, fertigation
Micro nanofertilizer	Foliar application, soil application, seed treatment
Nanoparticulate fertilizer	Soil application, fertigation

Research studies have shown promising outcomes regarding the utilization of nanofertilizers in irrigation systems. For instance, Hayyawi and Qusay (2019) reported a 20% increase in vegetative dry matter yield (mg ha-1) for potatoes when utilizing nano-NPK in a dripping irrigation setup compared to the control group. Similarly, Hayyawi and Estabraq (2020) observed a 50% increase in vegetative dry matter yield through a combination of nano-N and nano-Mo+B foliar applications. These studies provide evidence for the efficacy of nanofertilizers in irrigation systems, showcasing their adaptability to various fertilization methods. Table 10.2 presents the most commonly used application methods for each category of nanofertilizers.

Numerous studies have provided evidence of the transformative potential of nanofertilizers in revolutionizing crop management practices. However, certain challenges still need to be addressed for widespread adoption, including the relatively high production cost compared to conventional fertilizers, which may hinder large-scale implementation by farmers. The effectiveness of nanofertilizers is also contingent upon specific environmental conditions such as soil properties, moisture levels, and other ecological factors. Furthermore, while several studies have demonstrated the potential benefits of nanofertilizers, further research is required to fully comprehend their environmental impact and ensure the long-term sustainability of the food system.

Despite these limitations, it is important to note that technology is constantly evolving and the cost of producing nanofertilizers may decrease with time, making it more economically accessible to farmers. As more research is conducted, a greater understanding of the benefits and limitations of nanofertilizers will emerge, allowing more informed decisions to be made regarding their use. Ultimately, with continued investment in research and development, the goal is to make this innovative technology more accessible to farmers and to improve crop production in a sustainable manner.

10.4 ENHANCING CROP PRODUCTIVITY USING NANOFERTILIZERS

In modern agriculture, agricultural productivity is a major concern. The demand for food has increased due to the growing global population, and traditional farming practices are not sufficient to keep pace with population growth. There has been an increasing need to create innovative technology for the agricultural industry in order to boost crop productivity. Utilizing fertilizers effectively can increase agricultural output, but conventional fertilizers typically lose 40 to 70% of their nitrogen (N), 80 to 90% of their phosphorus (P), and 50 to 90% of their potassium (K) content, making them unavailable to crops (Avila-Quezada et al., 2022). The utilization of nanotechnology is one of the new methods that has shown promising results in raising crop productivity.

The use of nanofertilizers is one of the potential applications of nanotechnology in agriculture. Nanofertilizers contain vital nutrients for plant growth, and they are designed to deliver specific nutrients in a controlled manner. Nanofertilizers have a lot more surface area than conventional fertilizers, and their particle size is smaller than the pore size of a plant's leaf, allowing them to easily penetrate plant tissues and provide nutrients directly to plant cells (Qureshi et al., 2018). They are also designed in such a way that they release the nutrients gradually over an extended period rather than in one go, which makes them more efficient. By controlling the availability of fertilizers in the rhizosphere, nanofertilizers may strengthen plant defense systems, increase nutrient uptake, and extend the capacity of plants to withstand stress (Verma et al., 2022). Unique characteristics of nanofertilizers include superabsorbency, higher output, high photosynthetic activity, and substantial growth in the leaf surface area (Iran Nanotechnology Initiative Council, 2009). Nanomaterials are used to increase elemental efficiency, which improves the photosynthetic rate, chlorophyll formation, dry matter production, and eventually overall plant growth (Singh and Khangarot, 2021; Suriyaprabha et al., 2012). It decreases soil toxicity, at least in terms of the negative consequences of high fertilizer consumption, and decreases the frequency of fertilizer application (Naderi and Shahraki, 2013).

Nanofertilizers can be applied to plants through various methods, including foliar spray, seed treatment, soil application, and hydroponic systems. Foliar spray is useful when crops need a quick nutrient boost, and by this method, the nutrients can be easily absorbed by the plant's leaves and distributed to the rest of the plant. Black gram plants grew taller and had more branches after receiving foliar spraying of nanoformulations of an NPK and micronutrient mixture (Marimuthu and Surendran, 2015). Nano-NPK also exhibited increased growth of wheat leaves, as nutrient availability is enhanced by NPK nanoformulation penetration of leaf stomata through gas absorption (Al-Juthery et al., 2018). Another study examined the results of spraying tomato plants with copper oxide nanoparticles. According to the findings (López-Vargas et al., 2018), the production of firmer fruits increased as a result of the application of Cu nanofertilizers.

The utilization of nanofertilizers in seed treatment can serve as an effective method to enhance plant growth and nutrient absorption. Nanopriming of seeds brings about changes in metabolic and signaling pathways, influencing various stages of plant life, including germination and seedling establishment (do Espirito Santo Pereira et al., 2021). A study by Itroutwar et al. (2020) investigated the effects of biogenic zinc oxide nanoparticles on maize seed germination and seedling vigor. The researchers observed that the application of nanofertilizer significantly boosted the rate of seed germination, seedling growth, and biomass production compared to the control group. Similarly, the treatment of sorghum seeds with nano-iron(III) resulted in improved germination, enhanced chlorophyll content, and better growth, even under saline conditions (Maswada et al., 2018). Furthermore, wheat seeds treated with multiwalled carbon nanotubes exhibited increased root length, shoot length, germination, number of root hairs, number of stomata, plant weight, and average length of vascular systems (Joshi et al., 2018).

The mobility and stability of nanofertilizers appear to control their action in the soil. Depending on their makeup and interactions with the organic and inorganic soil components, nanofertilizers can be altered as they penetrate the soil. Nanofertilizers aggregate in the soil after application, which decreases their mobility in porous media. The nature of the soil, the distribution of pore sizes, and the characteristics of the nanofertilizers themselves all have an impact on how they behave in soil. Avila-Quezada et al. (2022) reported the successful synthesis of a chitosan-silicon nanofertilizer (CS-Si NF) with slow-release properties. The application of CS-Si NF showed significant improvements in the growth and yield of maize crops and demonstrated potential as a sustainable nanofertilizer for agriculture (Kumaraswamy et al., 2021). Hydroponically grown wheat was used to assess the impact of ZnO nanoparticles (NPs). ZnONPs were found to positively affect root number, seed germination, plant biomass, and overall shoot, root, and leaf growth (Awasthi et al., 2017).

The use of nanofertilizers has been shown in several studies to considerably boost crop productivity. Raliya and Tarafdar (2013) found that adding zinc oxide nanoparticles (ZnONPs) to the soil greatly boosted the activity of phosphorus-mobilizing enzymes, which in turn increased the gum content and crop production in cluster beans. A study of mulberry plants has shown that iron oxide nanoparticles (FeONPs) significantly improved sprouting percentage, number of leaves, aboveground and belowground biomass, chlorophyll content, and sugar content (Haydar et al.,

2021). Rock phosphate nanoparticles can be applied directly to crops; they may avoid soil fixation and enhance the amount of phosphorus that is available to plants (Devnita et al., 2018).

Despite the promising results of nanofertilizers in agriculture, there are still some concerns regarding their use. The potential for nanoparticles to accumulate in soil and water and cause environmental pollution is one of the major concerns. Nanoparticles can infiltrate the environment in several ways, such as wastewater discharge, soil erosion, and agricultural runoff. The possibility of nanofertilizers harming organisms that are not their actual target is another challenge. The interactions between nanoparticles and soil microbes might be harmful to the soil ecosystem. Moreover, nanoparticles can be ingested by plants and animals, which could be hazardous. The long-term impacts of nanofertilizers on human health are also a source of concern. In order to analyze the potential health concerns associated with the use of nanofertilizers, more research is required to better understand the toxicity of nanoparticles.

10.5 RESOURCE-USE EFFICIENCY UNDER NANOFERTILIZER-BASED PRACTICES

For a summary of the effects of different nanofertilizer formulations on cereals, see Table 10.3.

10.5.1 Rice

The utilization of Nano Complete fertilizer, which is a chelated cereal-specific fertilizer, has been shown to enhance the physiological parameters of rice plants. This is attributed to its ability to collectively improve the absorption of nutrients and water. When combined with Organic-DAP + micronutrient fertilizer, Nano Complete fertilizer exhibited the highest plant height and chlorophyll content in leaves (Al-Khuzai and Al-Juthery, 2020). Rose et al. (2015) suggested that the maximum plant height and chlorophyll content were achieved when Nano Complete fertilizer was used in combination with conventional fertilizers, even when applied at lower rates. The application of nanofertilizers (NFs) leads to increased crop growth by providing nutrients and aiding in the transport and absorption of available nutrients. NFs have demonstrated a remarkable advantage over conventional fertilizers in terms of nutrient absorption by plant cells, resulting in optimal growth. Additionally, NFs supply nutrients to crops in a manner tailored to meet specific plant needs, thereby improving NUE and reducing excessive input usage, which in turn benefits the economy and environment of rice cultivation (Al-Khuzai and Al-Juthery, 2020). Furthermore, the application of NFs has been found to promote rice growth and development, potentially increasing crop production and enhancing plant nutrition (Al-Khuzai and Al-Juthery, 2020; Rose et al., 2015). In a study conducted by Mahakham et al. (2017), the use of ZnO-based nanoparticles boosted the zinc content of rice grains by up to 50% compared to using traditional zinc fertilizers. In a similar approach, nutrient nanoencapsulation can enhance nutrient bioavailability and uptake, resulting in increased nutrient concentrations in rice grains. The use of NFs contributes to increased grain yield while reducing fertilizer costs and environmental pollution (Kheyri et al., 2019). Under better NUE, plant growth, and stress tolerance, both nano and non-nanofertilizers have been demonstrated to increase rice productivity. The content, rate of application, and environmental factors that these fertilizers are subject to can all affect their efficacy. In comparison to conventional fertilizers, research has shown that NFs may enhance rice yields by 10 to 25% (Lu et al., 2015). Liu et al. (2016) revealed that applying nanosized SiO_2 particles increased rice output by 9.6% compared to using traditional silicon fertilizers. In a field experiment, liquid nano-urea was found to enhance growth, yield attributes, and yield of rice when applied along with conventional urea. Nano-urea was applied as foliar spray at the tillering stage and panicle stage along with conventional urea as basal application at the rate of 50% of recommended dose of nitrogen (Sahu et al. 2022). In the case of phosphorus, another field study revealed that substituting 25% of conventional phosphorus source by nanophosphatic fertilizers

TABLE 10.3
Effects of different nanofertilizer formulations on cereals

Crop	Nanofertilizer	Effect	Reference
Rice	NF (N>1.2%; P_2O_5>0.001%; K_2O>0.0001%)	NFs application with dilution of 1 part NF to 1,000 parts of water increased 36.5% grain yield over control with better nutrient absorption by plant cells.	Rose et al., 2015
Rice	Nano-Zn	Nano-Zn sprayed @300g ha^{-1} increased grain yield by 14.5% over control with improvement in protein content in rice tissue.	Kheyri et al., 2019
Rice	Nano Si	Spray of Nano Si @300g ha^{-1} enhanced the grain yield by 10.5% over control and also adds structural color to plants, and improved the plant growth and overall yield.	Kheyri et al., 2019
Rice	Nano-chelated silicon and Nano Complete (nano-chelated cereal-specific fertilizer)	Application of Nano Si and Nano Complete @ 2 ml L^{-1} increased the rice yield by 38.1 and 34.8% over control, respectively, by increasing the chlorophyll content in leaves.	Al-Khuzai and Al-Juthery, 2020
Wheat	Super micro plus SMP nanofertilizer (N 5%, P 4%, K 2%, Mg 1%, Ca 1.5%, Fe 4%, Cu 1%, Zn 5%, Mn 2%, Mo 0.04% and B 0.06%)	Spray of SMP nanofertilizer enhanced the wheat yield by 47.7% over control with easy absorbance of NFs by the epidermis of leaves translocated to stems which enhanced the uptake of active molecules.	Aljutheri et al., 2018
Wheat	Nano-chelated super fertilizer (NCSF) (N 6%, P 3%, K 17%, Mg3%, Ca 1%, S 6%, Fe 4%, Zn 4%, Cu 0.5%, Mn 2%, B 0.5%, Mo 0.1%)	Spraying NCSF @1kg ha^{-1} yielded 30.7% more wheat grains compared to control treatment.	Al-Juthery et al., 2019
Wheat	Nano-NPK + Nano-K	Nano-NPK (liquid) sprays at 20, 30, & 45 DAS @ 3 ml/L of water + 2 Nano-K (L) sprays at grain development stage at 110 and 125 DAS @ 4 ml/L of water enhanced the grain yield significantly.	Mehta and Bharat, 2019
Wheat	Chitosan poly-methacrylic acid nanoparticles (NPK)	Nano-100 (NPK @500:60:400 ppm loaded on CS-PMAA) increased grain yield of wheat by 41.1% over control. Application of NFs resulted in a 23.5% reduction in the life cycle of wheat compared to normal-fertilized plants.	Abdel-Aziz et al., 2016
Wheat	Nano-Mn-oxide (30 nm)	Nano-MnO_2 enhanced wheat yield by 21.3% over control and also higher shoot and grain Mn contents was recorded.	Dimkpa et al., 2018
Barley	Nano-Fe/SiO_2	Nano-Fe/SiO_2 @ 25 mg kg^{-1} significantly enhanced the growth and yield of barley.	Najafi Disfani et al., 2017
Maize	Nano-Zn and Nano-B	Foliar spray of NFs @ of 2000 ppm at nine-leaf stage, stem elongation, heading had positive impact on the growth and development of maize.	Nouraein, 2019
Maize	Nano-Fe/SiO_2	Nano-Fe/SiO_2 spay @ 25 mg kg^{-1} increased the grain yield by 79.5% over control and also enhanced root dry weight in maize.	Najafi Disfani et al., 2017
Maize	Nano-zeourea	Spray of nano-zeourea yielded 8.6% more maize yield as compared to control by enhancing nutrient uptake.	Manikandan and Subramanian, 2016;
Maize	Nano-silica (SiO_2-NPs)	SiO_2-NPs spayed @10 g kg^{-1} had 29.5% more grain yield over control.	El-Naggar et al., 2020

Maize	Nano-K fertilizer	Application of 1.33 g MOP in 1.11 liter of nano polymer/15 kg soil increased 54.4% grain yield over control treatment though controlled release of K and supply of K for extended time as per the requirement of crop.	Babubhai et al., 2019
Maize	Complete NF	Application of complete NF three times at a concentration of 2,000 ppm during the 9-leaf stage, stem elongation, heading enhanced the maize yield by 22.0% over control.	Janmohammadi et al., 2016
Maize	Nano-NPK (slow-release fertilizer)	Slow-release fertilizers (SRF) @NPK:: 223:285:83kg ha^{-1} are more ecofriendly, easy to synchronize, and have greater benefits.	Rop et al., 2019
Sorghum	Nano-N fertilizer	Spraying 1.38 g urea in 1.15 liter nanopolymer enhanced the grain yield by 62.5% over control treatment.	Rani et al., 2019
Sorghum	Nano-ZnO (30 nm)	500ppm 36.4 reduced doses of fertilizers, consumption of fertilizers, and environmental hazards, and increased NUE.	Poornima and Koti, 2019
Sorghum	Nano-NPK (8% N, 5% P, 3% K, 10% micronutrients, 5% amino acids and 5% seaweed extract)	Application of Nano-NPK @5 L ha^{-1} led to a significant increase of 17.5% in grain yield compared to the control. Replacing 50% of the recommended NPK with Nano-NPK not only met the crop's nutrient requirements but also contributed to higher yields by suppressing weed growth.	Rehab et al., 2020
Pearl millet	Nano-Zn	Nano-Zn @ 16 L ha^{-1} increased plant dry matter (12.5%), grain yield (37.7%), and plant zinc concentration (10.4%) over the control.	Tarafdar et al., 2014
Pearl millet	$NP_{2.5}$: 2.5 times reduction of RDP through nanofertilizer	Enhancement of 81.9% grain yield compared to control treatment.	Dhansil et al., 2018
Pearl millet	ZnO	In the greenhouse, the application of Nano-ZnO @ 200 ppm resulted in notable enhancements in plant height and increased fresh and dry weight of seedlings.	Nandhini et al., 2019

resulted in enhanced yield levels as compared to 100% of the recommended dose of phosphorus (keeping 100% of nitrogen and potassium for both) (Poudel et al. 2023). This may be attributed to the larger surface area of nanoparticles, which accelerates interactions, resulting in increased production. This ultimately leads to a higher number of flowers, panicles, and grains, and higher overall yield (Al-Khuzai and Al-Juthery, 2020).

10.5.2 Wheat

The combined foliar application of nanobased fertilizers containing nitrogen, phosphorous, and potassium has been shown to effectively enhance the growth and yield parameters of wheat, even at lower concentrations (Abdel-Aziz et al., 2016). The easy absorption of nanofertilizers by the epidermal cells of leaves and their subsequent translocation to stems facilitate the uptake of active molecules, leading to improved growth and productivity of wheat (Abdel-Aziz et al., 2018). Numerous studies have demonstrated the positive impact of different nanofertilizers on various growth indicators of wheat, such as plant height, leaf area, root length, fresh weight, dry weight, and chlorophyll content (Abdel-Aziz et al., 2016; Aljutheri et al., 2018; Al-Juthery et al., 2019; Mehta and Bharat, 2019). Al-Juthery et al. (2018) studied the impact of nanofertilizer formulations and revealed that nanofertilizers for nitrogen, phosphorus, and potassium performed better than conventional or traditional fertilizer sources in terms of enhancing growth and yield of wheat. Nanofertilized wheat plants also exhibit a shorter life cycle compared to traditionally fertilized wheat plants, implying a reduction in the crop's duration (Abdel-Aziz et al., 2016). The smaller particle size of nanofertilizers results in a larger surface area, enabling enhanced absorptivity into plant tissues through applied surfaces. This leads to improved uptake and ultimately enhances NUE (Dimkpa et al., 2018; Qureshi et al., 2018). Nanofertilization of wheat crops has been found to promote the development of more effective tillers, longer ears, a greater number of grains per ear, and higher 1000-grain weight, all contributing to increased grain yield (Abdel-Aziz et al., 2016; Aljutheri et al., 2018; Al-Juthery et al., 2019; Mehta and Bharat, 2019). Foliar exposure to nanofertilizers also results in significant differences in yield attributes and overall yield of wheat, along with increased soil nitrate-N and shoot phosphorus (Dimkpa et al., 2018). Nano-NPK fertilizers enhance water activity when applied, allowing for the simultaneous absorption of nitrogen, phosphorus, and potassium along with the absorbed water. This leads to increased production (Liu and Liao, 2008). Application of nano-zinc has been found to enhance the protein content of wheat, thus improving the quality of the harvest (Ahmadian et al., 2021). Nano-zinc especially improves lipid, chlorophyll, and amino acid content of the crop. Nano-silica plays an important role in enhancing wheat crop yield levels, especially in times of drought. Silica improves the growth of wheat crop with maintenance of plant growth under times of drought. Such a positive impact of nano-silica was reflected as with significant improvement in water use efficiency of wheat (Ahmadian et al., 2021). The efficiency and durability of nanofertilizers contribute to their ability to supply nutrients to crops precisely when they are needed the most, resulting in improved plant performance.

10.5.3 Maize

The effectiveness of nanofertilizers on maize plants is attributed to the physiological and structural variability present among the plants (Najafi Disfani et al., 2017). Fe/SiO_2 nanoparticles have been found to decrease the germination time of maize, potentially due to an enhanced seed absorption capacity (Najafi Disfani et al., 2017). Adhikari et al. (2014) evaluated the efficacy of nano-rock phosphate (RP) for field application in maize; the findings revealed that the crop utilization of P from nano-RP was comparable to that of superphosphate (SSP), and it was a cheaper source of nutrient phosphorus than SSP. When maize plants are fertilized with nanofertilizer formulations, there is an observed increase in plant height, dry matter production, and the number of leaves per plant (Babubhai et al.,

2019; El-Naggar et al., 2020; Manikandan and Subramanian, 2016). The application of nano-zeourea as a fertilizer has shown consistently higher growth, yield, quality, and nutrient uptake compared to traditional urea fertilizers (El-Naggar et al., 2020; Manikandan and Subramanian, 2016). In addition to nutrient delivery, nanofertilizers have been found to enhance nutrient availability and NUE in plants (Babubhai et al., 2019). The use of nano-chelated micronutrients has demonstrated a more prominent improvement in grain yield compared to other fertilizer types, making them promising tools for achieving precise and desirable results (Janmohammadi et al., 2016).

10.5.4 Sorghum

Rani et al. (2019) reported that nano-nitrogenous fertilizers effectively enhance the growth parameters of sorghum, such as plant height, number of leaves, and dry matter accumulation. This improvement can be attributed to the controlled release of nano-nitrogen, gradually providing nutrients to the crop for an extended period as per its requirements. The application of nanofertilizers has shown positive effects on the yield, yield attributes, and quality characteristics of sorghum, while also increasing nutrient content and uptake (Rani et al., 2019). The enhanced nitrogen availability resulting from the use of nitrogenous nanofertilizers is likely responsible for these positive outcomes. The foliar application of nanofertilizers requires smaller amounts of fertilizer, which can potentially reduce the overall application doses, minimize fertilizer wastage, mitigate environmental risks, and improve NUE (Poornima and Koti, 2019).

10.5.5 Pearl Millet

In a study by Tarafdar et al. (2014), significant improvements were observed in shoot length, root length, and root area when zinc nanofertilizer was applied via foliar application at a concentration of 10 mg L^{-1} during critical growth stages of the crop. This application also led to an increase in photosynthetic pigments (chlorophyll) and total soluble leaf protein, likely due to the essential role of zinc ions as a micronutrient element in plants. The use of zinc nanoparticles resulted in enhanced plant dry matter (12.5%), grain yield (37.7%), and plant zinc concentration (10.4%) compared to the control (Table 10.4). The improved grain yield at crop maturity was attributed to the application of zinc nanofertilizer, which facilitated better nutrient mobilization and nutrient availability for plant uptake (Tarafdar et al., 2014). Furthermore, the growth and development characteristics of pearl millet were improved through seed priming and foliar application of nano-ZnO in a greenhouse study, with foliar spray showing better results than seed priming (Nandhini et al., 2019). Therefore, it is recommended that foliar application should be preferred over priming to achieve better results with nanofertilizers. Nanoparticles are primarily adsorbed onto the cell surface and are rapidly and efficiently translocated to the sink, as indicated by studies by Lin and Xing (2008).

TABLE 10.4
Effect of zinc NF on percent increase of grain yield, dry biomass, and zinc concentration of pearl millet

Treatment	% increase of grain yield	% increase of dry biomass	% increase of zinc concentration
Control	—	—	—
Ordinary Zn	14.3	0.4	12.1
Nano-Zn	37.7	12.5	10.4

Source: Tarafdar et al., 2014

10.5.6 Nutrient use Efficiency (NUE)

The controlled release of nutrients based on plant requirements and the improved NUE are notable advantages of coated or encapsulated nanofertilizers (Qureshi et al., 2018). Their large surface area allows for maximal reactivity, leading to increased nutrient availability and improved NUE in plants (Liscano et al., 2000; Siddiqi and Husen, 2017). Unlike conventional fertilizers, which tend to adsorb and fix strongly to soil particles due to their high solubility, nanofertilizers dissolve in water, facilitating the dispersion of nutrients in the soil and increasing their availability to plants. Encapsulation of fertilizers in nanoparticles promotes plant uptake and availability while reducing bulk demands (Chhipa, 2017).

Nanostructured fertilizers may utilize targeted and controlled release mechanisms in response to environmental cues and biological needs to improve NUE. These strategies have the potential to significantly enhance NUE, allowing plants to better withstand environmental stressors. Nanofertilizers have been found to stimulate various biological processes, including seed germination, seedling development, nitrogen metabolism, photosynthetic activity, protein synthesis, and antioxidant systems, thereby increasing agricultural production (Hossain, 2018; Siddiqi and Husen, 2017). In terms of phosphorus use efficiency, nanoencapsulated phosphorus has shown 14.5% higher NUE compared to conventional phosphorus fertilizers (Hagab et al., 2018). Additionally, the application of nano-KH_2PO_4 (potassium dihydrogen phosphate) for phosphorus has led to greater physiological efficiency in shoots and roots, resulting in higher phosphorus accumulations in these plant organs (Miranda-Villagómez et al., 2019). According to Al-Uthery and Al-Shami (2019), single fertigation of nano-nitrogen, nano-phosphorus, and nano-potassium recorded NUE values of 85.30%, 94.34%, and 94.06%, respectively, while traditional fertilizers generally exhibit lower NUE.

10.6 ROLE OF NANOFERTILIZERS IN STRESS MANAGEMENT

Nanofertilizers, in comparison to their counterparts, are more efficient in delivering nutrients in order to improve crop growth and productivity. Their efficiency in nutrient delivery is primarily due to their minuscule size offering greater surface area and higher adhesion. These characteristics ensure that the nutrients are released in a controlled manner and when they are most required. The losses due to excessive release are thus minimized. Besides minimizing the losses, the nanoscale size of the nanofertilizers is capable of moderating the morphological and physiological functions of crop plants (Zhang et al., 2021; Shang et al., 2019; Bhatt et al., 2020). These functions include the photosynthetic rate and its efficiency, nutrient uptake, and enhanced defense against biotic and abiotic stresses.

Agriculture in any part of the world is under the direct influence of the environment, which in the present circumstances is undergoing rapid changes resulting in a decline in crop productivity (Ashraf et al., 2018; Benevenuto et al., 2017; Suzuki et al., 2014). The setbacks to the agriculture community are due to the various abiotic stresses compounded with biotic stresses. The abiotic stresses include temperature variations, erratic rainfall, drought, and salinity (dos Santos et al., 2022). The response of plants to an increasingly stressed environment is characterized by alterations in enzymatic activities (the reduction in activity can also be attributed to the destruction of the functional quaternary structures of the enzymes due to a rise in the ambient temperature), alterations in gene expression, increased production of reactive oxygen/oxidative species (ROSs), and damage to deoxyribonucleic acids (DNA) and cellular proteins (Menezes-Silva et al., 2017; Bryant et al., 2021; Choudhury et al., 2017).

As discussed earlier, the size of the nanoparticles is the main feature of nanofertilizers that has the ability to mitigate stresses by increasing nutrient uptake, hence the reduction in the development of ROSs. As the species population declines, the damage to the plasma membrane can be minimized. This helps crop plants to redirect to the sink the photosynthates which otherwise would have been utilized to protect the plant physiology from failing.

10.7 POTENTIAL OF NANOFERTILIZERS IN CROP PRODUCTION UNDER CLIMATE CHANGE

The potential risk to food security is evident from the rapid rise in the global population, which reached 8 billion people in 2022. It is predicted that by the end of 2080, the global population will peak and reach ~10 billion people (United Nations, 2023; Fahad et al., 2017). The only possible solutions are strengthening agricultural production systems, and improvising or diversifying existing cropping systems. Climate change is another factor threatening the global food supply. Agriculture, unlike other sectors, is directly and indirectly influenced by changes in climate. The effect includes abiotic stresses accounting for losses of nearly 50–70%, and biotic stresses causing losses of up to 40–60% (Tiwari et al., 2020). Under these circumstances, management practices such as fertilizer management, weed management, and water management assume a pivotal role in maintaining food production. Nanofertilizers are particularly important for plant nutrition under stressed environmental conditions because of their attributes such as controlled and slow release of nutrients and high NUE.

The ability of nanofertilizers to mitigate stresses (biotic and abiotic) from plants is the result of several mechanisms which are influenced primarily by the nano size of the particles holding the nutrients. Nanofertilizers have the ability to improve morphological, biochemical, and physiological indices such as photosynthetic efficiency, photosynthetic rate, nutrient uptake and efficiency, and hormonal control (Verma et al., 2022). Nanofertilizers mitigate stresses by, for example, reducing the generation of ROS, increasing the activity of enzymes such as superoxide dismutase (SOD), promoting nutrient uptake, and decreasing the degradation of plasma membrane and chlorophyll (Sarraf et al., 2022). Thus, nanofertilizers are able to build resistance in plants against the rapidly changing climate. The role of nanofertilizers in reducing greenhouse gas emissions may be considered because it is able to forgo/reduce the loss of nitrogen in the form of nitrous oxide. N loss via denitrification in the form of N_2O, a potent greenhouse gas, is well known to the scientific community. It is the result of excessive use of synthetic fertilizers, which accounts for 12% of the loss by denitrification, present in the soil which crop plants fail to transport (IPCC, 2015). The controlled release due to the tiny size of nanoparticles helps to minimize nitrogen losses and, in the process, reduces greenhouse gas emissions.

10.8 FUTURE PROSPECTS

The application of nanofertilizers capable of minimizing nutrient losses as well as enhancing nutrient use efficiency in order to increase crop production and reduce the environmental risks is truly a boost for the agriculture sector. The controlled and slow release of the nutrients from nanofertilizers helps meet the nutrient demands of the crop, thereby helping to improve the growth of crop plants and nutrition. However, many questions about toxicology and the safety of these systems for human and ecosystem health remain unanswered. According to Du et al. (2019), ZnONPs were significantly more effective than $ZnSO_4$ in the germination and growth of wheat. They also discovered that, at higher doses, $ZnSO_4$ was more noxious than ZnONPs. Solid research is still lacking, and so this concern gains prime importance in the field of nanotechnology. The effectiveness of using nanomaterials as fertilizers in plant growth is determined by the plant species as well as other variables, such as the size, concentration, composition, and chemical properties of nanoparticles. A reduction in the size of nanomaterials facilitates the enhancement of the surface mass ratio of particles; as a consequence, a large number of nutrient ions are adsorbed and desorbed gradually and steadily for a protracted period. When assessing the risk of nanoparticles, it will be necessary to consider the size of each material at which the nano scale is identified. Any nanomaterials, including biological nanoparticles, have the potential to cause harm. Subsequently, if the risk assessment data are to be analogous, proper nanofertilizer characterization is necessary before risk assessment. Finally, because surface reactivity increases dramatically at the nano scale, most nanoparticles have increased solubility, making it difficult to distinguish between toxic effects caused by nanoparticles and those caused by dissolved ions. This last issue is especially acute for nanoparticles made of inorganic materials. Being an interesting and promising field, nanotechnology will revitalize agriculture with governments and researchers devoting efforts to developing effective nanofertilizers by mitigating the abovementioned and other related concerns.

10.9 CONCLUSIONS

Uncontrolled release of nutrients into the soil and the atmosphere compromises crop productivity and contributes to global warming. Advances in nanotechnology, in favor of agriculture, can help curb nutrient losses and improve crop performances. One of the United Nations sustainable development goals (SDGs) is to increase crop production, which can be fulfilled by optimizing the use of nanofertilizers in the agriculture sector. The slow or controlled release and carrier delivery properties of nanofertilizers enhance crop yields and soil health, and reduce nutrient loss compared with traditional fertilizers. Work on nanofertilizers is very scarce across the world, but the relevant literature confirms that nanofertilizers can sustain farm productivity.

REFERENCES

Abdalla, Z.F., El-Sawy, S., El-Bassiony, A.E.-M., Jun, H., Shedeed, S., Okasha, A.M., Bayoumi, Y., El-Ramady, H., and Prokisch, J. 2022. Smart fertilizers vs. nano-fertilizers: A pictorial overview. *Environment, Biodiversity and Soil Security* 6: 191–204. https://doi.org/10.21608/jenvbs.2022.153990.1184

Abd-El-Aziz, G.H., Abd El-Rahman, L.A., Ahmed, S.S., and Mahrous, S.E.M. 2021. Efficacy of ZnO nanoparticles as a remedial zinc fertilizer for soya bean and wheat corps. *Journal of Soil Sciences and Agricultural Engineering* 12: 573–582. https://doi.org/10.21608/jssae.2021.197970

Abdel-Aziz, H., Hasaneen, M.N., and Omar, A. 2018. Effect of foliar application of nano chitosan NPK fertilizer on the chemical composition of wheat grains. *Egyptian Journal of Botany*. https://doi.org/10.21608/ejbo.2018.1907.1137

Abdel-Aziz, H.M.M., Hasaneen, M.N.A., and Ome, A.M. 2016. Nano chitosan-NPK fertilizer enhances the growth and productivity of wheat plants grown in sandy soil. *Spanish Journal of Agricultural Research* 14: 9.

Adhikari, T., Kundu, S., Meena, V., and Rao, A.S. 2014. Utilization of nano rock phosphate by maize (*Zea mays* L.) crop in a vertisol of Central India. *Journal of Agriculture, Science and Technology* 4: 384–394.

Ahmadian, K., Jalilian, J., and Pirzad, A. 2021. Nano-fertilizers improved drought tolerance in wheat under deficit irrigation. *Agricultural Water Management* 244: 106544.

Aljutheri, H., Ali, N. (Nooreldeen), Al-Taey, D., and Ali, E. 2018. *The impact of foliar application of nanofertilizer, seaweed and hypertonic on yield of potato*. https://doi.org/10.13140/RG.2.2.16557.33763

Al-Juthery, H.W., Habeeb, K.H., Altaee, F.J.K., AL-Taey, D.K., and Al-Tawaha, A.R.M. 2018. Effect of foliar application of different sources of nano-fertilizers on growth and yield of wheat. *Bioscience Research* (4): 3976–3985.

Al-Juthery, H.W.A., and Saadoun, S.F. 2018. Impact of foliar application of some micronutrients nanofertilizer on growth and yield of Jerusalem Artichoke. *Iraqi Journal of Agricultural Sciences* 49. https://doi.org/10.36103/ijas.v49i4.65

Al-Juthery, H.W.A., Hardan, H.M., Al-Swedi, F.G.A., Obaid, M.H., and Al-Shami, Q.M.N. 2019. Effect of foliar nutrition of nano-fertilizers and amino acids on growth and yield of wheat. *IOP Conference Series Earth Environmental Science* 388, 012046. https://doi.org/10.1088/1755-1315/388/1/012046

Al-Khuzai, A.H.G., and Al-Juthery, H.W.A. 2020. Effect of DAP fertilizer source and nano fertilizers (Silicon and Complete) spray on some growth and yield indicators of rice (Oryza sativa L. cv. Anber 33). *IOP Conference Series: Earth and Environmental Science* 553, 012008. https://doi.org/10.1088/1755-1315/553/1/012008

Al-Uthery, H.W., and Al-Shami, Q.M. 2019. Impact of fertigation of nano NPK fertilizers, nutrient use efficiency and distribution in soil of potato (Solanum tuberosum L.). *Plant Archives* 19: 1087–1096.

Ashraf M.A., Akbar A., Askari S.H., Iqbal M., Rasheed R., and Hussain I. 2018. Recent advances in abiotic stress tolerance of plants through chemical priming: An overview. In: Rakshit, A., and Singh, H. (eds), *Advances in Seed Priming*. Springer, Singapore, pp. 51–79.

Astaneh, N., Bazrafshan, F., Zare, M., Amiri, B., Bahrani, A., Astaneh, N., Bazrafshan, F., Zare, M., Amiri, B., and Bahrani, A. 2021. Nano-fertilizer prevents environmental pollution and improves physiological traits of wheat grown under drought stress conditions. *Scientia Agropecauria* 12: 41–47. https://doi.org/10.17268/sci.agropecu.2021.005

Avila-Quezada, G.D., Ingle, A.P., Golińska, P., and Rai, M. 2022. Strategic applications of nano-fertilizers for sustainable agriculture: Benefits and bottlenecks. *Nanotechnology Reviews* 11(1): 2123–2140.

Awasthi, A., Bansal, S., Jangir, L.K., Awasthi, G., Awasthi, K.K., and Awasthi, K. 2017. Effect of ZnO nanoparticles on germination of *Triticum aestivum* seeds. *Macromolecular Symposia* 376 (1): 1700043.

Babubhai, B.D., Solanki, M.S., Rani, B., and Femi, V. 2019. Effect of different levels of chemical and nano potassic fertilizer on yield and yield attribute of maize crop (Zea mays L.) cv. Amber. *Journal of Pharmacognosy and Phytochemistry* 8: 58–61.

Badran, A., and Savin, I. 2018. Effect of nano-fertilizer on seed germination and first stages of bitter almond seedlings' growth under saline conditions. *BioNanoScience* 8: 742–751. https://doi.org/10.1007/s12668-018-0531-6

Benevenuto R.F., Agapito-Tenfen S.Z., Vilperte V., Wikmark O.G., Van Rensburg P.J., and Nodari R.O. 2017. Molecular responses of genetically modified maize to abiotic stresses as determined through proteomic and metabolomic analyses. *PLoS One* 12: e0173069. DOI: 10.1371/journal.pone.0173069

Bhatt, D., Bhatt, M.D., Nath, M., Dudhat, R., Sharma, M., and Bisht, D.S. 2020. Application of nanoparticles in overcoming different environmental stresses. In: Roychoudhury A. and Tripathi D.K. (eds), *Protective chemical agents in the amelioration of plant abiotic stress: Biochemical and molecular perspectives.* Wiley-Blackwell, Hoboken, 635–654.

Bryant, C., Fuenzalida, T.I., Brothers, N., Mencuccini, M., Sack, L., Binks, O., and Ball, M.C. 2021. Shifting access to pools of shoot water sustains gas exchange and increases stem hydraulic safety during seasonal atmospheric drought. *Plant, Cell & Environment* 44: 2898–2911.

Bulovic, V., Mandell, A., and Perlman, A. 2007. *Molecular memory device.* US7157750B2.

Chen, J., and Wei, X. 2018. Controlled-release fertilizers as a means to reduce nitrogen leaching and runoff in container-grown plant production, nitrogen in agriculture – Updates. *IntechOpen.* https://doi.org/10.5772/intechopen.73055

Chhipa, H. 2017. Nanofertilizers and nanopesticides for agriculture. *Environmental Chemistry Letters* 15: 15–22. https://doi.org/10.1007/s10311-016-0600-4

Chhipa, H., and Joshi, P. 2016. Nanofertilisers, nanopesticides and nanosensors in agriculture, in: Ranjan, S., Dasgupta, N., and Lichtfouse, E. (Eds.), *Nanoscience in food and agriculture 1, Sustainable agriculture reviews*. Springer International Publishing, Cham, pp. 247–282. https://doi.org/10.1007/978-3-319-39303-2_9

Choudhury, F.K., Rivero, R.M., Blumwald, E., and Mittler, R. 2017. Reactive oxygen species, abiotic stress and stress combination. *Plant Journal* 90: 856–867.

Council, I.N.I. 2009. *First nano-organic Iron chelated fertilizer invented in Iran*. The Free Library, www.thefreelibrary.com/First Nano-Organic Iron Chelated Fertilizer Invented in Iran.-a0211779457

Davarpanah, S., Tehranifar, A., Davarynejad, G., Abadía, J., and Khorasani, R. 2016. Effects of foliar applications of zinc and boron nano-fertilizers on Pomegranate (*Punica granatum* cv. Ardestani) fruit yield and quality. *Scientia Horticulturae* 210: 57–64. https://doi.org/10.1016/j.scienta.2016.07.003

Devnita, R., Joy, B., Arifin, M., Hudaya, R., and Oktaviani, N. 2018. Application of nanoparticle of rock phosphate and biofertilizer in increasing some soil chemical characteristics of variable charge soil. In *AIP conference proceedings* (Vol. 1927, No. 1, p. 030027). AIP Publishing LLC.

Dhansil, A., Zalawadia, N.M., Prajapat, B.S., and Yadav, K. 2018. Effect of nano phosphatic fertilizer on nutrient content and uptake by pearl millet (Pennisetum glaucum L.) crop. *International Journal of Current Microbiology and Applied Sciences* 7: 2327–2337. https://doi.org/10.20546/ijcmas.2018.712.264

Dimkpa, C.O., Singh, U., Adisa, I.O., Bindraban, P.S., Elmer, W.H., Gardea-Torresdey, J.L., and White, J.C. 2018. Effects of manganese nanoparticle exposure on nutrient acquisition in wheat (Triticum aestivum L.). *Agronomy* 8: 158. https://doi.org/10.3390/agronomy8090158

Ditta, A., and Arshad, M. 2016. Applications and perspectives of using nanomaterials for sustainable plant nutrition. *Nanotechnology Reviews* 5: 209–229. https://doi.org/10.1515/ntrev-2015-0060

do Espirito Santo Pereira, A., Caixeta Oliveira, H., Fernandes Fraceto, L., and Santaella, C. 2021. Nanotechnology potential in seed priming for sustainable agriculture. *Nanomaterials* 11(2): 267.

dos Santos T.B., Ribas, A.F., de Souza S.G.H., Budzinski, I.G.F., and Domingues, D.S. 2022. Physiological responses to drought, salinity, and heat stress in plants: A review. *Stresses* 2(1): 113–135. https://doi.org/10.3390/stresses2010009

Du, W., Yang, J., Peng, Q., Liang, X., and Hui, M. 2019. Comparison study of zinc nanoparticles and zinc sulphate on wheat growth: From toxicity and zinc biofortification. *Chemosphere* 227. https://doi.org/10.1016/j.chemosphere.2019.03.168

El-Naggar, M.E., Abdelsalam, N.R., Fouda, M.M.G., Mackled, M.I., Al-Jaddadi, M.A.M., Ali, H.M., Siddiqui, M.H., and Kandil, E.E. 2020. Soil application of nano silica on maize yield and its insecticidal activity against some stored insects after the post-harvest. *Nanomaterials* 10: 739. https://doi.org/10.3390/nano10040739

El-Saadony, M.T., ALmoshadak, A.S., Shafi, M.E., Albaqami, N.M., Saad, A.M., El-Tahan, A.M., Desoky, E.-S.M., Elnahal, A.S.M., Almakas, A., Abd El-Mageed, T.A., Taha, A.E., Elrys, A.S., and Helmy, A.M. 2021. Vital roles of sustainable nano-fertilizers in improving plant quality and quantity-An updated review. *Saudi Journal of Biological Sciences* 28: 7349–7359. https://doi.org/10.1016/j.sjbs.2021.08.032

Fahad, S., Bajwa, A.A., Nazir, U., Anjum, S.A., Farooq, A., Zohaib, A., Sadia, S., Nasim, W., Adkins, S., and Saud, S. 2017. Crop production under drought and heat stress: Plant responses and management options. *Frontiers in Plant Science* 8: 1147.

Ghidan, A.Y., and Antary, T.M.A. 2019. Applications of nanotechnology in agriculture, applications of nanobiotechnology. In: Stoytcheva, M., and Zlatev, R. (eds), *Applications of biotechnology*, London: IntechOpen. https://doi.org/10.5772/intechopen.88390

Hagab, R.H., Kotp, Y.H., and Eissa, D., 2018. Using nanotechnology for enhancing phosphorus fertilizer use efficiency of peanut bean grown in sandy soils. *Journal of Advanced Pharmacy Education & Research* 8: 59–67.

Haydar, M.S., Ghosh, S., and Mandal, P. 2021. Application of iron oxide nanoparticles as micronutrient fertilizer in mulberry propagation. *Journal of Plant Growth Regulation* 41: 1726–1746.

Hayyawi, W.A.A., and Estabraq, H.O.A.-M. 2020. Effect of urea and nano-nitrogen fertigation and foliar application of nano-boron and molybdenum on some growth and yield parameters of potato. *Al-Qadisiyah Journal for Agriculture Sciences* 10: 253–263. www.iasj.net/iasj/download/2cbd1c3985447615

Hayyawi, W.A.A., and juthery, Qusay, M.N.A.-S. 2019. The effect of fertigation with nano NPK fertilizers on some parameters of growth and yield of potato (*Solanum tuberosum* L.). Al-*Qadisiyah Journal for Agriculture Sciences* 9: 225–232. https://doi.org/10.33794/qjas.Vol9.Iss2.93

Hossain, Md.F. 2018. Evaluation of nitrogen, phosphorus and potassium nano-fertilizers on yield, yield components and fiber properties of Egyptian cotton (*Gossypium Bar-badense L.*). *Journal of Plant Sciences and Crop Protection* 1: 208.

Hussein, M.M., and Abou-Baker, N.H. 2018. The contribution of nano-zinc to alleviate salinity stress on cotton plants. *Royal Society Open Science* 5: 171809. https://doi.org/10.1098/rsos.171809

IPCC. 2015. Climate change 2014: Mitigation of climate change. In: Edenhofer, O., Pichs-Madruga, R., Sokona, Y., Minx, J.C., Farahani, E., and Kadneret, S. Geneva (Eds.), *Summary for policymakers technical summary; Part of the Working Group III contribution to the fifth assessment report of the intergovernmental panel on climate change*, Switzerland: Intergovernmental Panel on Climate Change.

Itroutwar, P.D., Kasivelu, G., Raguraman, V., Malaichamy, K., and Sevathapandian, S.K. 2020. Effects of biogenic zinc oxide nanoparticles on seed germination and seedling vigor of maize (*Zea mays*). *Biocatalysis and Agricultural Biotechnology* 29: 101778.

Jakhar, A.M., Aziz, I., Kaleri, A.R., Hasnain, M., Haider, G., Ma, J., and Abideen, Z. 2022. Nano-fertilizers: A sustainable technology for improving crop nutrition and food security. *NanoImpact* 27: 100411. https://doi.org/10.1016/j.impact.2022.100411

Jampílek, J., and Kráľová, K. 2015. Application of nanotechnology in agriculture and food industry, its prospects and risks. *Ecological Chemistry and Engineering S* 22: 321–361. https://doi.org/10.1515/eces-2015-0018

Janmohammadi, M., Navid, A., Ebadi, A., and Sabaghnia, N., 2016. Impact of nano-chelated micronutrients and biological fertilizers on growth performance and grain yield of maize under deficit irrigation condition. *Biologija* 62. https://doi.org/10.6001/biologija.v62i2.3339

Joshi, A., Kaur, S., Dharamvir, K., Nayyar, H., and Verma, G. 2018. Multi-walled carbon nanotubes applied through seed-priming influence early germination, root hair, growth and yield of Bread Wheat (*Triticum aestivum* L.). *Journal of the Science of Food and Agriculture* 98(8): 3148–3160.

Kabiri, S., Degryse, F., Tran, D.N.H., da Silva, R.C., McLaughlin, and M.J., Losic, D. 2017. Graphene oxide: A new carrier for slow release of plant micronutrients. *ACS Applied Materials & Interfaces* 9: 43325–43335. https://doi.org/10.1021/acsami.7b07890

Kheyri, N., Norouzi, H.A., Mobasser, H.R., and Torabi, B. 2019. Effects of silicon and zinc nanoparticles on growth, yield, and biochemical characteristics of rice. *Agronomy Journal* 111: 3084–3090. https://doi.org/10.2134/agronj2019.04.0304

Kumaraswamy, R.V., Saharan, V., Kumari, S., Choudhary, R.C., Pal, A., Sharma, S.S., and Biswas, P. 2021. Chitosan-silicon nanofertilizer to enhance plant growth and yield in maize (*Zea mays* L.). *Plant Physiology and Biochemistry* 159: 53–66.

Lin, D., and Xing, B. 2008. Root uptake and phytotoxicity of ZnO nanoparticles. *Environmental Science and Technology* 42: 5580–5585. https://doi.org/10.1021/es800422x

Liscano, J.F., Wilson, C.E., Norman, R.J., and Slaton, N.A. 2000. Zinc availability to rice from seven granular fertilizers. AAES Research Bulletin 963, Fayetteville, AR: Arkansas Agricultural Experiment Station.

Liu, A.X., and Liao, Z.W. 2008. Effects of nano-materials on water clusters. *Journal of Anhui Agricultural Sciences* 36: 15780–15781.

Liu, R., and Lal, R. 2015. Potentials of engineered nanoparticles as fertilizers for increasing agronomic productions. *Science of the Total Environment* 514: 131–139. https://doi.org/10.1016/j.scitotenv.2015.01.104

Liu., R, Lal, R., and Wiebe, K.D. 2016. Effects of nano-SiO_2 on the growth and photosynthesis of paddy rice. *Journal of Plant Nutrition* 39(10): 1377–1385.

López-Vargas, E.R., Ortega-Ortíz, H., Cadenas-Pliego, G., de Alba Romenus, K., Cabrera de la Fuente, M., Benavides-Mendoza, A., and Juárez-Maldonado, A. 2018. Foliar application of copper nanoparticles increases the fruit quality and the content of bioactive compounds in tomatoes. *Applied Sciences* 8(7): 1020.

Lu, J., Li, X., Yuan, S., Sun, Y., Sheng, G., and Yu, H. 2015. Enhanced rice production but greatly reduced carbon emission following biochar amendment in a metal-polluted rice paddy. *Environmental Science and Technology* 49(16): 9754–9762.

Mahakham, W., Sarmah, A.K., Maensiri, S., and Theerakulpisut, P. 2017. Nanopriming technology for enhancing germination and starch metabolism of aged rice seeds using phytosynthesized silver nanoparticles. *Scientific Reports* 7(1): 8263.

Mahanta, N., Dambale, A., and Rajkhowa, M. 2019. Nutrient use efficiency through nano fertilizers. *International Journal of Chemical Studies* 7(3): 2839–2842. www.researchgate.net/publication/344255688_Nutrient_use_efficiency_through_Nano_fertilizers#fullTextFileContent

Mahil, E.I., and Baburai Nagesh, A.K. 2019. Foliar application of nanofertilizers in agricultural crops: A review. *Journal of Farm Science* 32(3): 239–249.

Manikandan, A., and Subramanian, K.S. 2016. Evaluation of zeolite based nitrogen nano-fertilizers on maize growth, yield and quality on inceptisols and alfisols. *International Journal of Plant & Soil Science* 9(4): 1–9. https://doi.org/10.9734/IJPSS/2016/22103

Manjunatha, S.B., Biradar, D.P., and Aladakatti, Y.R. 2016. Nanotechnology and its applications in agriculture: A review. *Journal of Farm Sciences* 29: 1–13.

Marimuthu, S., and Surendran, U. 2015. Effect of nutrients and plant growth regulators on growth and yield of black gram in sandy loam soils of Cauvery new delta zone, India. *Cogent Food & Agriculture* 1(1): 1010415.

Maswada, H.F., Djanaguiraman, M., and Prasad, P.V.V. 2018. Seed treatment with Nano-iron (III) oxide enhances germination, seeding growth and salinity tolerance of Sorghum. *Journal of Agronomy and Crop Science* 204(6): 577–587.

Maysinger, D. 2007. Nanoparticles and cells: good companions and doomed partnerships. *Organic and Biomolecular Chemistry Journal* 5: 2335–2342. https://doi.org/10.1039/B704275B

Mehta, S., and Bharat, R. 2019. Effect of integrated use of nano and non-nano fertilizers on yield and yield attributes of wheat (Triticum aestivum L.). *International Journal of Current Microbiology and Applied Sciences* 8: 598–606. https://doi.org/10.20546/ijcmas.2019.812.078

Menezes-Silva, P.E., Sanglard, L.M., Ávila, R.T., Morais, L.E., Martins, S.C., Nobres, P., Patreze, C.M., Ferreira, M.A., Araújo, W.L., and Fernie, A.R. 2017. Photosynthetic and metabolic acclimation to repeated drought events play key roles in drought tolerance in coffee. *Journal of Experimental Botany* 68: 4309–4322.

Miranda-Villagómez, E., Trejo-Téllez, L.I., Gómez-Merino, F.C., Sandoval-Villa, M., Sánchez-García, P., and Aguilar-Méndez, M.Á. 2019. Nanophosphorus fertilizer stimulates growth and photosynthetic activity and improves P status in Rice. *Journal of Nanomaterials* 2019: e5368027. https://doi.org/10.1155/2019/5368027

Mittal, D., Kaur, G., Singh, P., Yadav, K., and Ali, S.A. 2020. Nanoparticle-based sustainable agriculture and food science: Recent advances and future outlook. *Frontiers in Nanotechnology* 2: 10. https://doi.org/10.3389/fnano.2020.579954

Morales-Díaz, A.B., Ortega-Ortíz, H., Juárez-Maldonado, A., Cadenas-Pliego, G., González-Morales, S., and Benavides-Mendoza, A. 2017. Application of nanoelements in plant nutrition and its impact in ecosystems. *Advances in Natural Sciences: Nanoscience and Nanotechnology* 8: 013001. https://doi.org/10.1088/2043-6254/8/1/013001

Naderi, M.R., and Danesh-Shahraki, A. 2013. Nanofertilizers and their roles in sustainable agriculture. *International Journal of Agriculture and Crop Sciences* 5(19): 2229–2232.

Najafi Disfani, M., Mikhak, A., Kassaee, M.Z., and Maghari, A. 2017. Effects of nano Fe/SiO2 fertilizers on germination and growth of barley and maize. *Archives of Agronomy and Soil Science* 63: 817–826. https://doi.org/10.1080/03650340.2016.1239016

Nandhini, M., Rajini, S.B., Udayashankar, A.C., Niranjana, S.R., Lund, O.S., Shetty, H.S., and Prakash, H.S. 2019. Biofabricated zinc oxide nanoparticles as an eco-friendly alternative for growth promotion and management of downy mildew of pearl millet. *Crop Protection* 121: 103–112. https://doi.org/10.1016/j.cropro.2019.03.015

Nouraein, M. 2019. Effect of nanofertilizers and biofertilizers on yield of maize: Biplot analysis. *Botanica* 25: 121–130. https://doi.org/10.2478/botlit-2019-0014

Pampolino, M.F., Witt, C., Pasuquin, J.M., Johnston, A., and Fisher, M.J. 2012. Development approach and evaluation of the nutrient expert software for nutrient management in cereal crops. *Computers and Electronics in Agriculture* 88: 103–110. https://doi.org/10.1016/j.compag.2012.07.007

Poornima, R., and Koti, R.V. 2019. Effect of nano zinc oxide on growth, yield and grain zinc content of sorghum (Sorghum bicolor). *Journal of Pharmacognosy and Phytochemistry* 8: 727–731.

Poudel, A., Singh, S.K., Jiménez-Ballesta, R., Jatav, S.S., Patra, A., and Pandey, A. 2023. Effect of nanophosphorus formulation on growth, yield and nutritional quality of wheat under semi-arid climate. *Agronomy* 13(3): 768.

Pradhan, S., and Mailapalli, D.R. 2017. Interaction of engineered nanoparticles with the agri-environment. *Journal of Agricultural and Food Chemistry* 65: 8279–94.

Prasad, R., Bhattacharyya, A., and Nguyen, Q.D. 2017. Nanotechnology in sustainable agriculture: Recent developments, challenges, and perspectives. *Frontiers in Microbiology* 8: 1014. https://doi.org/10.3389/fmicb.2017.01014

Preetha, P.S., and Balakrishnan, N. 2017. A review of nano fertilizers and their use and functions in soil. *International Journal of Current Microbiology and Applied Sciences* 6: 3117–3133. https://doi.org/10.20546/ijcmas.2017.612.364

Pruthvi, R.N., and Chandrashekara, C.P. 2021. Soil and plant nutrient status of cotton as influenced by different methods of application of nano ZnO. *Journal of Pharmacognosy and Phytochemistry* 10: 1386–1389. https://doi.org/10.22271/phyto.2021.v10.i2r.14011

Qureshi, A., Singh, D.K., and Dwivedi, S. 2018. Nano-fertilizers: A novel way for enhancing nutrient use efficiency and crop productivity. *International Journal of Current Microbiology and Applied Sciences* 7(2): 3325–3335.

Rahman, A., Mondal, N.C., and Tiwari, K.K. 2021. Anthropogenic nitrate in groundwater and its health risks in the view of background concentration in a semi arid area of Rajasthan, India. *Scientific Reports* 11: 9279. https://doi.org/10.1038/s41598-021-88600-1

Rajput, J.S., Thakur, A., Nag, N., Chandrakar, T., and Singh, D. 2022. Effect of nano fertilizer in relation to growth, yield and economics of little millet (*Panicum sumatrense* roth) under rainfed conditions. *Pharma Innovation Journal* 11: 153–156.

Raliya, R., and Tarafdar, J.C. 2013. ZnO nanoparticle biosynthesis and its effect on phosphorous-mobilizing enzyme secretion and gum contents in clusterbean (*Cyamopsis tetragonoloba* L.). *Agricultural Research* 2: 48–57.

Ramzan, M., Arshad, M.I., Mahmood, K., Amin, N., Khan, M.I., and Iqbal, F. 2020. Investigation of structural and optical properties of Pr3+ -substituted M-type Ba–Ni nano-ferrites. *Journal of Superconductivity and Novel Magnetism* 34:1759–64.

Rani, B., Zalawadia, D.N., Buha, D., and Rushang, K. 2019. Effect of different levels of chemical and nano nitrogenous fertilizers on content and uptake of N, P, K by sorghum crop cv. Gundari. *Journal of Pharmacognosy and Phytochemistry* 8: 454–458.

Rehab, I.F., Kordy, A.M., and Salim, B. 2020. Assessment of Sorghum (Sorghum bicolor L.) productivity under different weed control methods, mineral and nano fertilization. *Egyptian Academic Journal of Biological Sciences, H. Botany* 11: 1–11.

Ribeiro, C., and Carmo, M. 2019. Why nonconventional materials are answers for sustainable agriculture. *MRS Energy & Sustainability.* pp. 6.

Rop, K., Karuku, G.N., Mbui, D., Njomo, N., and Michira, I. 2019. Evaluating the effects of formulated nano-NPK slow release fertilizer composite on the performance and yield of maize, kale and capsicum. *Annals of Agricultural Sciences* 64: 9–19. https://doi.org/10.1016/j.aoas.2019.05.010

Rose, H., Benzon, L., Rubenecia, M.R., Rubenecia, U., Ultra, V., and Lee, S.C. 2015. Nano-fertilizer affects the growth, development, and chemical properties of rice. *International Journal of Agronomy and Agricultural Research* 7: 105–117.

Sahu, K.B., Sharma, G., Pandey, D., Keshry, P.K., and Chaure, N.K. 2022. Effect of nitrogen management through nano-fertilizer in rice (Oryza sativa L.). *International Journal of Chemical Research and Development* 4(1): 25–27.

Sarraf, M., Vishwakarma, K., Kumar, V., Arif, N., Das, S., Johnson, R., Janeeshma, E., Puthur, J.T., Aliniaeifard, S., and Chauhan, D.K. 2022. Metal/metalloid-based nanomaterials for plant abiotic stress tolerance: An overview of the mechanisms. *Plants* 11: 316.

Scott, N.R. 2007. Nanoscience in veterinary medicine. *Veterinary Research Communications* 31: 139–144. https://doi.org/10.1007/s11259-007-0083-7

Sekhon, B.S. 2014. Nanotechnology in agri-food production: An overview. *Nanotechnology Science and Applications* 7: 31–53. https://doi.org/10.2147/NSA.S39406

Seleiman, M.F., Alotaibi, M.A., Alhammad, B.A., Alharbi, B.M., Refay, Y., and Badawy, S.A. 2020. Effects of ZnO nanoparticles and biochar of rice straw and cow manure on characteristics of contaminated soil and sunflower productivity, oil quality, and heavy metals uptake. *Agronomy* 10: 790. https://doi.org/10.3390/agronomy10060790

Shang, Y., Hasan, Md.K., Ahammed, G.J., Li, M., Yin, H., and Zhou, J. 2019. Applications of nanotechnology in plant growth and crop protection: A review. *Molecules* 24: 2558. https://doi.org/10.3390/molecules24142558

Siddiqi, K.S., and Husen, A. 2017. Plant response to engineered metal oxide nanoparticles. *Nanoscale Research Letters* 12: 92. https://doi.org/10.1186/s11671-017-1861-y

Singh, H., Sharma, A., Bhardwaj, S.K., Arya, S.K., Bhardwaj, N., and Khatri, M. 2021. Recent advances in the applications of nano-agrochemicals for sustainable agricultural development. *Environmental Science: Processes & Impacts* 23: 213–239. https://doi.org/10.1039/D0EM00404A

Singh, M.D., Chirag, G., Prakash, P.O., Mohan, and M.H. 2017. Nano fertilizers is a new way to increase nutrients use efficiency in crop production. *International Journal of Agriculture Sciences* 9(7): 3831–3833.

Singh, P., and Khangarot, A.K. 2021. Nano-fertilizers a technology to increase crop production. *Agriculture & Food: e-newsletter* 3(11): 379–381.

Solanki, P., Bhargava, A., Chhipa, H., Jain, N., and Panwar, J. 2015. Nano-fertilizers and their smart delivery system, in: Rai, M., Ribeiro, C., Mattoso, L., Duran, and N. (Eds.), *Nanotechnologies in food and agriculture*. Springer International Publishing, Cham, pp. 81–101. https://doi.org/10.1007/978-3-319-14024-7_4

Stewart, W.M., Dibb, D.W., Johnston, A.E., and Smyth, T.J. 2005. The contribution of commercial fertilizer nutrients to food production. *Agronomy Journal* 97:a 1–6. https://doi.org/10.2134/agronj2005.0001

Subramanian, K.S., Manikandan, A., Thirunavukkarasu, M., and Rahale, C.S. 2015. Nano-fertilizers for balanced crop nutrition. *Nanotechnologies in Food and Agriculture* 69–80. https://doi.org/10.1007/978-3-319-14024-7_3

Suppan, S. 2017. *Applying nanotechnology to fertilizer: Rationales, research, risks and regulatory challenges.* IATP Report October 2017, Minneapolis, MN: Institute for Agriculture and Trade Policy.

Suriyaprabha, R., Karunakaran, G., Yuvakkumar, R., Prabu, P., Rajendran, V., and Kannan, N. 2012. Growth and physiological responses of Maize (*Zea mays* L.) to porous silica nanoparticles in soil. *Journal of Nanoparticle Research* 14: 1–14.

Suzuki, N., Rivero, R.M., Shulaev, V., Blumwald, E., and Mittler, R. 2014. Abiotic and biotic stress combinations. *New Phytologist* 203: 32–43. DOI: 10.1111/nph.12797

Tarafdar, J.C., Raliya, R., Mahawar, H., and Rathore, I. 2014. Development of zinc nanofertilizer to enhance crop production in Pearl Millet (Pennisetum americanum). *Agricultural Research* 3: 257–262. https://doi.org/10.1007/s40003-014-0113-y

Thimiri Govinda Raj, D.B., and Khan, N.A. 2016. Designer nanoparticle: nanobiotechnology tool for cell biology. *Nano Convergence* 3: 22. https://doi.org/10.1186/s40580-016-0082-x

Tiwari, R.K., Lal, M.K., Naga, K.C., Kumar, R., Chourasia, K.N., Subhash, S., Kumar, D., and Sharma, S. 2020. Emerging roles of melatonin in mitigating abiotic and biotic stresses of horticultural crops. *Scientia Horticulturae* 272: 109592.

Toksha, B., Sonawale, V.A.M., Vanarase, A., Bornare, D., Tonde, S., Hazra, C., Kundu, D., Satdive, A., Tayde, S., and Chatterjee, A. 2021. Nanofertilizers: A review on synthesis and impact of their use on crop yield and environment. *Environmental Technology & Innovation* 24: 101986. https://doi.org/10.1016/j.eti.2021.101986

United Nations. 2023. Global issues: Population. www.un.org/en/global-issues/

Verma, K.K., Song, X.P., Joshi, A., Tian, D.D., Rajput, V.D., Singh, M., and Li, Y.R. 2022. Recent trends in nano-fertilizers for sustainable agriculture under climate change for global food security. *Nanomaterials* 12(1): 173.

Wilson, M.A., Tran, N.H., Milev, A.S., Kannangara, G.S.K., Volk, H., and Lu, G.Q.M. 2008. Nanomaterials in soils. *Geoderma* 146: 291–302. https://doi.org/10.1016/j.geoderma.2008.06.004

Yu, G., Wang, J., Liu, L., Li, Y., Zhang, Y., and Wang, S. 2020. The analysis of groundwater nitrate pollution and health risk assessment in rural areas of Yantai, China. *BMC Public Health* 20: 437. https://doi.org/10.1186/s12889-020-08583-y

Zakzok, A.K., Abdrabou, R.T., Arafa, A.S., and Abd-Elsamad, G.A.A. 2018. Response of cotton yield and lint properties to mineral n.p.k. nano-fertilization. *Arab Universities Journal of Agricultural Sciences* 26: 1029–1039. https://doi.org/10.21608/ajs.2018.28290

Zhang, H., Zhu, J., Gong, Z., and Zhu, J. 2021. Abiotic stress responses in plants. *Nature Reviews Genetics* 23: 104–119.

11 Recent Advances in Nanotechnology-Based Crop Management

Strategies for Sustainable Agriculture

Vishwa Jyoti Baruah, Saima Masood Wani, Aabid Hussain, Manny Saluja, and Bhaswati Sarmah
Corresponding author: Bhaswati Sarmah

11.1 INTRODUCTION

Nanotechnology has gained recognition by the European Commission, which considers it a highly promising and innovative tool in the current context (Parisi et al., 2015). Over recent years, nanotechnology has played a vital role in diverse sectors, including biotechnology, electronics, physical sciences, chemical sciences, life sciences, and pharmacy (Prasad et al., 2017; Rajput et al., 2018). Amidst the challenges posed by climate change, polluted environments, global food insecurity, and declining agricultural productivity observed in recent times (Arora, 2019), there is an increasing need for advances in the agricultural sector. This is largely due to the diminishing agricultural land area with a growing population and escalating environmental stresses (Usman et al., 2020). Moreover, the ever-expanding global population demands food security that ensures access to high-quality food capable of meeting people's nutritional requirements (Capone et al., 2014). To improve agricultural productivity, there has been an enhanced use of agrochemicals, including fertilizers and pesticides; however, indiscriminate use of these chemicals has adversely affected our environment and disturbed our ecological balance, with an impact on consumers and every productive sector directly or indirectly connected to the agriculture value chain (Abhilash & Singh, 2009). In addition, the growing number of new pathogens resistant to various pesticides in use has highlighted the demand for improved crop management techniques (Thrall et al., 2011; Tilman et al., 2002). While plant breeding, synthetic chemicals, and genetic engineering have contributed to enhancing agricultural productivity to some degree, they possess inherent limitations. To meet the increasing global demand for food and to attain economic, environmental, and health advantages, there is a critical requirement for ongoing innovation in agriculture (Fincheira et al., 2020). This is precisely where nanotechnology and its applications can emerge as a pivotal alternative to achieve the objective of sustainable agriculture. Hence, the application of nanotechnology in agriculture has been strongly advocated to address these limitations and further enhance agricultural productivity (Rastogi et al., 2019). Nanotechnology offers unique properties that can be exploited to improve plant growth, nutrient uptake, pest and disease control, and stress tolerance (Alabresm et al., 2018; Tripathi et al., 2022).

Nanoparticles are being increasingly explored for their potential applications in agriculture, owing to their ability to cross cellular barriers and influence plant growth and productivity (Singh et al.,

DOI: 10.1201/9781003389408-11

2021). Nanoparticles, which are materials with dimensions ranging from 1 to 100 nm, play a significant role in nanotechnology and possess distinctive physico-chemical properties like crystallinity, catalytic activities, porosity, and aggregating properties (Khan et al., 2019; Ma et al., 2010). These unique attributes enable nanoparticles to penetrate cellular barriers and interact with living organisms, making them suitable for diverse applications. The broader utilization of nanoparticles in crop improvement and abiotic stress management has long been anticipated. Nanotechnology-based approaches, such as employing nanofertilizers, nanopesticides, and nanosensors, hold the potential to enhance crop productivity, nutrient uptake, and stress tolerance (Kaphle et al., 2018; Martínez-Ballesta et al., 2018). Moreover, incorporating nanotechnology into agriculture can help reduce the environmental impact and promote sustainable agricultural practices (Usman et al., 2020; Zulfiqar et al., 2019).

To optimize the application of nanoparticles in agriculture, they can be administered through various methods, including seed incubation, root treatment, pollen treatment, isolated cell treatment, protoplast treatment, foliar spraying, irrigation, direct injection, hydroponic treatment, and delivery by biolistics (Sheikh Mohamed & Sakthi Kumar, 2016; Singh et al., 2021). Once nanoparticles come into contact with plant cells, they are taken up through plasmodesmatal or endocytosed pathways and transported via apoplastic and/or symplastic routes (Shukla et al., 2016; Singh et al., 2021). The advantageous effects of nanoparticles on plant growth and productivity include increased photosynthetic rate, biomass, chlorophyll content, sugar levels, accumulation of osmolytes and antioxidants, and enhanced nitrogen metabolism (Ghani et al., 2022). However, it is crucial to acknowledge that the use of nanoparticles in agriculture is still at an early stage, and further research is required to evaluate their safety and potential risks to the environment and human health. Therefore, it is of the utmost importance to regulate and monitor the use of nanoparticles in agriculture to minimize any potential negative impacts (Mnyusiwalla et al., 2003; Nair et al., 2010). Nevertheless, NPs have several applications in agriculture which will be discussed in this chapter. Some of the crucial applications of NPs in agriculture are as follows:

(1) Controlled-release nanofertilizers: These are designed to gradually release nutrients to plants over time, leading to improved crop growth and yields. They can also reduce nutrient loss due to leaching or volatilization, and minimize environmental impacts.
(2) Targeted gene transfer: Nanoparticles can be used to deliver genetic material to plant cells, allowing for more precise and efficient genetic modification. This technology has the potential to help develop crops with improved traits such as disease resistance, drought tolerance, and higher yields.
(3) Nanopesticides: These are designed to be more effective and less harmful to the environment than traditional pesticides. Nanoparticles can be used to deliver pesticides directly to the target site, reducing the amount needed and minimizing off-target effects.
(4) Nanosensors and precision farming: By employing nanosensors, it becomes possible to monitor soil conditions, nutrient levels, and other factors influencing plant growth. This valuable information can then be utilized to optimize the usage of fertilizers and water, resulting in increased crop yields and a decreased environmental footprint.
(5) Plant stress tolerance and soil enhancement: Nanomaterials can be used to improve soil quality and plant health by enhancing nutrient uptake, promoting root growth, and increasing tolerance to environmental stressors such as drought, heat, and salinity.
(6) Upregulation of the expression of stress-related genes: Nanomaterials can be used to help plants to cope with abiotic and biotic stress.

11.2 AGRICULTURE-BASED NANOTECHNOLOGY STRATEGIES

To achieve a sustainable agriculture, nanotechnology is a promising tool that can significantly advance and improve agriculture practices. This may be attributed to the smart controlled release of agrochemicals and fertilizers, increased bioavailability, and targeted delivery of active substances, all of which may contribute to the development of precision and sustainable agriculture.

11.2.1 Nanofertilizers

To achieve the optimum growth and increased yield, plants require nutrients in appropriate amounts, but existing farming techniques cannot meet the increasing demand for food without relying on fertilizers for sustainable productivity (Havlin, 2020). However, frequent application of fertilizers by farmers through the soil in different ways, such as surface spreading, mixing with irrigation water, or subsurface application, causes a sizable amount of the fertilizer to be lost to the air or surface water bodies, thereby damaging ecosystems. Supporting intensive agriculture while limiting its negative environmental effects is a critical challenge that modern agriculture is facing (Lee et al., 2006). Therefore, nanofertilizers could represent a significant advance in environmental protection because of their ability to be used in smaller amounts than conventional fertilizers.

Nanofertilizers are referred to as materials having particle dimensions in nano range (<100 nm), with macro and micronutrients that are supplied to crops in a regulated manner (DeRosa et al., 2010). In comparison to traditional bulky chemical fertilizers, nanofertilizers because of their ultra-small size have a larger surface area to volume that increases their ability for absorption and retention (Nongbet et al., 2022). There are three types of nanofertilizers distinguished based on their formulation: nanoscale fertilizer, conventional fertilizer in the form of nanoparticles scaled down through a top-down approach; nanoscale additive fertilizer, also a conventional fertilizer that contains nanomaterials as a supplement; and nanoscale coating fertilizer, which is composed of nutrients that have been intercalated into nanoscale pores of a host material or encapsulated by nanofilms (Mastronardi et al., 2015; Mejias et al., 2021).

Nanofertilizers have an excellent capacity to improve nutrient use efficiency, and therefore play a role in the enhancement of nutritive characteristics. Unlike in conventional fertilizers, the nutrients applied are bonded to nanodimensional adsorbents, which cause the release of nutrients in a gradual manner, thereby reducing nutrient leaching into ground water with improved nutrient use efficiency (Adisa et al., 2019; Mejias et al., 2021). It has been shown that the use of nanofertilizers improves crops' ability to absorb nutrients, producing outputs of higher quality (Abdel-Aziz et al., 2016). Besides having role in increasing productivity, nanofertilizers can also be utilized to increase tolerance towards abiotic stress such as drought stress, salinity stress, and cold stress (Badran & Savin, 2018; Mejias et al., 2021; Qados & Moftah, 2015; Qi et al., 2013; Seghatoleslami et al., 2015;). Furthermore, nanofertilizers improve the health of soil in several ways, such as reducing the amount of chemicals present in the soil, and enhancing microbial activity and native nutrient mobilization in the rhizosphere (Raliya et al., 2018). Also, nanofertilizers, when combined with microorganisms called nanobiofertilizers, offer numerous additional advantages (Sharma et al., 2022).

Therefore, nanofertilizers have shown encouraging results in the agriculture sector through a considerable impact in, for example, increasing the yield and tolerance of crops to abiotic stressors and boosting soil health, which has sparked a lot of interest in enhancing agricultural output under the global climate change scenario as well. However, despite the positive outcomes, there is still a lot of ambiguity in their application at a commercial level, the main concern being their accumulation in the environment and their effects on human health either due to the consumption of crops or through the environment. Therefore, an in-depth evaluation of how nanofertilizers affect the ecosystem must be a research priority so that we can work on solutions that manage their accumulation and potential dangers for human health while adopting this technology.

11.2.2 Nanopesticides

Crop yield is negatively impacted by biotic stressors, and so the protection of plants from harmful pests, diseases, and weeds necessitates the use of pesticides in modern agriculture. The climate change scenario has led to an increase in the number and diversity of phytopathogens. Furthermore, indiscriminate use of pesticides has adversely affected the agroecosystem. Due to the off-target use of pesticides, beneficial insects and microflora in the soil have been destroyed, influencing the soil health and leading to the development of pesticide resistance over time. Thus, alternative techniques and agrochemicals are needed to overcome these consequences. One potential approach to achieving these goals is nanotechnology, which can be used to obtain pesticide formulations in the nano range, referred to as nanopesticides. To overcome the potential risks due to the use of pesticides, scientists have suggested the use of nanopesticides as a substitute for conventional pesticides. Nanopesticides contain engineered nanoparticles as active ingredients and have biocidal capabilities, either as a whole or as part of the designed structure. They have certain advantages over conventional pesticides owing to their size and higher surface area to volume ratio. Some of their advantages are that they potentially enhance water-solubility and bioavailability of the active ingredient, increase stability by shielding agrochemicals from environmental degradation, and decrease the environmental load.

For nanopesticide preparation, a variety of materials may be used, such as metal oxides, proteins, lipids, and dendrimers. Nanopesticides can be either physically included into the formulation or chemically linked, and accordingly they can be prepared as nanoemulsions, nanocapsules, nanosuspensions, solid lipid nanoparticles, nanospheres, mesoporous nanoparticles, and nanoclays. Among these, nanoemulsions and suspensions increase the water-solubility of active ingredient, so that otherwise toxic organic solvents can be eliminated. Similarly, the encapsulation of pesticides in the form of nanocapsules and nanospheres prevents the sensitive chemicals from degradation, improves stability, and increases longevity of pesticides (Chaud et al., 2021). Furthermore, the effective nanoparticles for the controlled release and delivery methods for the hydrophobic and hydrophilic insecticides are mesoporous nanoparticles such as nanoclay, activated carbon, and porous hollow silica, which have excellent biocompatibility, high drug-loading capacity, and minimal toxicity (Li et al., 2006).

Despite the advantages offered by loading active ingredients of pesticides into nanoparticles, a new form of environmental trash is being generated that gets concentrated over time. The biomimetic properties of nanoparticles and their significant ability for bioaccumulation and diffusion in food, aquatic ecosystems, and soil have been noted to have toxicological impacts on all living things, notably mammals. The adverse effects due to prolonged exposure to nanoparticles affect various physiological systems such as the cardiovascular, respiratory, and nervous systems. In fact, for the safe application of nanoparticles in agriculture, a thorough analysis of the factors that affect the activity and toxicity of nanopesticides is essential.

11.2.3 Nanobiosensors

Ever-increasing demand for food for the growing population, along with use of agrochemicals, has led to an escalation of these toxic chemicals into our environment as seen in terms of water pollution and contaminated soils. Therefore, effective detection of such toxic materials is an important aspect to devise strategies for their efficient removal to restore soil health and fertility. By utilizing unique qualities of nanoparticles, nanotechnology holds a lot of potential in this regard in the form of nanobiosensors, which provide a low-cost detection system with higher selectivity and sensitivity. A biosensor is an integrated system that works on the coupled interaction of receptor molecules with high-affinity detectable biomolecules or analytes (much like what happens in a biological system), wherein physico-chemical transducers are combined with the recognition elements to generate quantitative data (Bhalla et al., 2016). Based on the type of transducers used, the biosensors can be electrochemical, optical, calorimetric, or piezoelectric; while based on biological interaction, the biosensors may be classified as immunosensors, enzymatic biosensors, or aptasensors (Mondal et al., 2022). Nanobiosensors, generated by combining nanomaterials with biosensors, are sensitive, noninvasive devices created specifically for the monitoring of a wide range of environmental samples (Kalyani et al., 2021). Nanoparticles help in signal amplification and sensor sensitivity by decreasing the detection limit in order to produce a reliable sensing analysis. Various nanoparticles are used for making biosensors in the agriculture industry due to their versatile properties, such as low toxicity and good biocompatibility of carbon nanoparticles (Shojaei et al., 2019), distinct sensing properties of noble metals (gold, platinum, etc.) (Jiang et al., 2018), and highly biocompatible graphene due to its suitability for forming metal composites (Verma et al., 2022).

Nanobiosensors have a lot of applications in the entire agrifood supply chain, from the detection of soil condition to the management of severe pathogens. Soil structure is a very dynamic in nature, and real-time detection of various parameters such as pH, moisture, and nutrients is important for precision agriculture. For detection of soil nitrite, a lipid-based nanobiosensor has been developed that can detect 2.1 μg/L (Siontorou & Georgopoulos, 2016). Similarly, sensors have been developed that can rapidly sense soil moisture in situ (Surya et al., 2020). In addition to this, nanobiosensors have been developed to detect plant pathogens such as *Pseudomonas syringae* (Vaseghi et al., 2013), *Ralstonia solanacearum* (Khaledian et al., 2017), and *Pseudomonas aeruginosa* (Soundy & Day, 2017). In agroecosystems, nanosensors have also been used to detect residual pesticides (Songa & Okonkwo, 2016; Talan et al., 2018) and heavy metals such as Hg and Cd (Ahmed et al., 2020; Jin et al., 2018; Xiao et al., 2018). Recently, wearable sensors have become a new frontier in the detection of dangerous chemicals in plants that can be used directly on the surfaces of plant leaves and stems. Such smart nanobiosensors have been developed that can detect pesticides, biotic and abiotic stressors, and real-time water transport in plants. They have generated a lot of interest because of their low cost, versatility, light weight, and biocompatibility, and they may aid in management of plant pathogens, maintenance of soil health, and protection against abiotic stress (Nassar et al., 2018; Oren et al., 2017; Zhao et al., 2020).

11.2.4 Nanobioremediation

The increased use of pesticides in agriculture started soon after the green revolution, and every year several million tons of these hazardous chemicals are pumped into the environment, leading to their accumulation and posing threats to human and animal life, as well as to our ecosystem. A natural way for their removal exists in the form of bioremediation wherein living organisms are utilized to overcome the menace by converting harmful chemicals to harmless ones. However, the process is less efficient when the load is high or the area to be decontaminated is large. A solution to this is the integration of nanotechnology with bioremediation (use of biological organisms) such that the bioremediation is efficiently enhanced by means of nanoparticles, a process also known as nanobioremediation (Bhatt et al., 2022). The extremely small size of the nanoparticles allows for deeper interactions, and because they have a huge surface area per unit of mass, more nanoparticles can interact with the environment, which increases the effectiveness and speed of bioremediation.

Microbial nanoremediation and nanophytoremediation are two types of nanobioremediation (Singh & Saxena, 2022). While nanophytoremediation refers to the application of plants for decontamination of soils and so on, in microbial nanoremediation, microbes are used. Therefore, various factors must be taken into consideration for selecting the plants and microbes, such as the interaction of nanoparticles with the system and the contaminant throughout the process. In contrast to nanophytoremediation, microbe-mediated remediation is better in that different microorganisms such as algae and fungi have shorter generation times and can survive over a range of temperatures.

Microbial nanoremediation is the process through which soil microorganisms and nanoparticles are combined to increase the efficacy of biodegradation. It occurs in two steps: a abiotic process that involves the entry of nanoparticles into the system and the pollutants being subjected to processes such as absorption, adsorption, dissolution, and photocatalysis; and a biotic process wherein particles are eliminated from the system through biostimulation and biotransformation (Abebe et al., 2018; Desiante et al., 2021; Usman et al., 2020). It has been seen that microbial nanoremediation can play an important role in removal of heavy metals such as cadmium, lead, and nickel from the soil (Alabresm et al., 2018; Baragaño et al., 2020; Cao et al., 2020; Laslo et al., 2022) and in remediation of organic pollutants (Ansari et al., 2009; Bokare et al., 2012; Patil et al., 2016). Through this process, selective microbes consume heavy metal pollutants generating bioorganic nanoparticles, followed by their removal from the environment, thus producing value from the waste. The organic remediation not only helps in degrading persistent pollutants in the environment but also enhances soil health, as has been reported for silicon nanoparticles which improve the biomass and soil microflora by increasing the growth of rhizospheric microorganisms (Rajput et al., 2022a). Although various trials involving the incorporation of nanoparticles with the bioremediation process have demonstrated encouraging possibilities, further investigation and testing are still needed in this field. The effectiveness of bioremediation has been improved by many nanoparticles investigated, but there are still questions about how safe these particles are for the environment and the food chain. As a result, in-depth study is needed in the field of nanoparticle safety analysis.

11.2.5 Green Nanotechnology

While the advantages of nanotechnology in agriculture are many, the use of nanostructured materials has raised serious concerns pertaining to human health and environmental safety, mainly due to bioaccumulation in the food chain, persistence in the environment, and increased stability (Figueiredo et al., 2019; Santos et al., 2020). These concerns have led to support for a greener approach to circumvent the toxic effects and produce ecofriendly nanoparticles, in addition to being cost-effective (Gottardo et al., 2021; Kaczerewska et al., 2020). The synthesis of nanoparticles may involve physical, chemical, or biological routes. The physical methods require a lot of energy and space for big machines, such as furnaces, and are therefore unsuitable for nanoparticle preparation. Chemical synthesis makes use of organic solvents and reducing agents, both of which are toxic to the environment and human health. Thus, for these reasons, the biological method of synthesizing the nanoparticles is preferred. Green nanotechnology generally refers to the use of plants, microbes, or other naturally occurring substances for producing nanoparticles, thus alleviating the hazardous effects on the environment or human health. Green chemistry is based on a similar approach of using a water or a solvent-free system for catalysis in many chemical reactions that can either minimize or eliminate the production of hazardous substances or intermediates during the reaction (Das & Maji, 2021; Li et al., 2012). Green synthesis utilizing plants or their parts for reducing metal ions to obtain nanoparticles is more efficient, more economical, and simpler. Green synthesis usually follows a bottom-up approach wherein the reducing power is provided by plant extracts instead of chemical reagents. The nanoparticles generated this way are more stable due to the presence of biomolecule coating (proteins, lipids, etc.), and cost-effective because of the low energy consumption. In a comparative study of silver nanoparticle synthesis, it was seen that green-synthesized silver nanoparticles had 20 times more antibacterial activity than chemically synthesized ones (Sintubin et al., 2011). In short, by producing nanomaterials and nanoproducts without endangering the environment or human health, green nanotechnology makes a substantial contribution to environmental sustainability.

11.3 NANOTECHNOLOGY FOR RESTORATION OF POLLUTED SOIL

Growing human and technological activities are enriching the soil with extra nutrients, possibly harmful metals, and agrochemicals (Midhat et al., 2019). Crop production is based on soil because it helps plants absorb nutrients. Agriculture defines and supports human existence. It has long been shown in the literature that there is a conflict between the conservation of agricultural land and its sustainable management. Sustainable development, quality of life, and livelihoods are all threatened by land contamination (Pretty & Bharucha, 2014; Struik & Kuyper, 2017). Given the demands of an expanding population and the reduction of arable land due to technological activity, soil conservation is thus of the highest importance for the current era (Martin et al., 2016). The development of nanotechnology creates a global opportunity for the efficient remediation or restoration of damaged soil (Martin et al., 2016; Pretty & Bharucha, 2014). According to some research, nanotechnology offers a lot of potential for making the environment cleaner, especially by reducing the toxicity of certain metals and metalloids (Kumari et al., 2022; Raffa et al., 2021). Moreover, the use of nanotechnology for soil remediation has been identified as an effective method for removing environmental contaminants. One of the key areas where nanotechnological methods have been applied extensively in this context is soil remediation. As a result of their huge specific surface area, nanoparticles are being employed to remove pollutants in several methods, including adsorption, redox reactions, precipitation, and coprecipitation (Raffa et al., 2021). Because of the tiny size (1–100 nm) of nanoparticles and their large surface area and high reactivity, using them in bioremediation methods may have several advantages (Alazaiza et al., 2021).

Increased human activity is primarily to blame for the presence of heavy metals (HMs) in the environment; however, they are also added to the environment due to disrupted biogeochemical cycles. There are no biological tasks for elements like arsenic, cadmium, chromium, mercury, and lead to carry out in the biological system. Even at low concentrations, heavy metals have significant harmful effects on biota, making them a prominent component of inorganic contaminants (Alissa & Ferns, 2011; Jaishankar et al., 2014). The bioavailability and absorption of HMs also determines their toxicity (Rasmussen et al., 2000). As the production of reactive oxygen species (ROS) leads to the degradation of macromolecules such as nucleic acids and proteins, heavy metals predominantly impact plants and lower soil creatures. The impact of heavy metals varies depending on the kind of crop, the chemical composition of soil, and the type of heavy metal (Chaplygin et al., 2021). ROS production is the general mechanism by which HMs cause toxicity in agricultural plants, affecting the cell membrane and organelles and thereby affecting the structural integrity of plants (Chaplygin et al., 2021; Huang et al., 2019). Moreover, it has been noted to have an impact on respiration and photosynthesis, decrease enzyme activities, increase oxidative stress, decrease biomass, lower crop yields, and alter the activity and genetic makeup of beneficial soil microflora (Arif et al., 2016; Enez et al., 2018). Site stabilization, which immobilizes HMs at a given site in soil, leading to a reduction in their mobility and availability, and prevents them from leaking across sites, is one of the primary approaches for the removal of HMs (Suman et al., 2018). The elimination of HMs has attracted a lot of interest in the use of other NPs.

11.3.1 Hyperaccumulator Plants and Nanoparticles for Remediation of Heavy Metal Contaminated Soils

Some plant species, like *Amaranthus spinosus*, *Pedioplanis burchelli*, and *Alternanthera pungens*, have developed mechanisms to avoid heavy metal absorption in their environment and have persisted at an ideal level of heavy metals through rhizofiltration (Rizwan et al., 2019). Above the recommended dosage, heavy metals negatively affect both plant and human health (Sun et al., 2016). Hyperaccumulating plant species, on the other hand, transport and collect metals aerial organs in larger concentrations than nonhyperaccumulating species, with no toxic effect on the organisms (Jabeen et al., 2009; Rascio & Navari-Izzo, 2011). Due to the potential for phytoremediation in the form of phytostabilization or phytoextraction, plant hyperaccumulators for HMs are active. The process of phytostabilization and phytoextraction gives hyperaccumulator plants the ability to have a bioconcentration factor (BCF) greater than 1 for HMs. When the BCF and translocation factor (TF) are greater than 1, phytostabilization is shown by the plant (Usman et al., 2019). Similarly, the processes of phytostabilization and phytoextraction were studied in *Sacrum munja*, *Parthenium hysterophorus*, and *Ipomoea carnea*. The authors found that BCF and TF greater than 1 for chromium, nickel, cadmium, and lead implied phytostabilization mechanism, whereas BCF of more than 1 and TF of less than 1 for Zn and Mn revealed a phytoextraction process for HMs (Kisku et al., 2018). Rhizodegradation is the process of introduction of pollutants into the soil's rhizosphere via microbial activity, where bacteria use them as food and energy. With this technique, bacteria can transform harmful contaminants into safe, nontoxic byproducts (Ojuederie & Babalola, 2017). In order to provide the bacteria with more nutrients, plant roots produce naturally occurring carbon containing compounds including sugars, alcohol, and acids, enhancing rhizodegradation activities (Rai et al., 2021).

Various carbon or noncarbon nanoparticles are used frequently to treat HM-contaminated soil. CNTs (carbon nanotubes), nZVI (nano zero-valent iron), TiO_2NPs, and AgNPs have been thoroughly studied for their efficiency to treat soil (Mubarak et al., 2016). The pH, temperature, contact time, and CNT dosage of the surrounding environment all affect how effectively HM is removed by CNTs (Rai et al., 2021). FeSNPs eliminated 99.65% of Cr^{6+} at 6.0 pH, but their elimination rate declined at high pH (>10.0) (Wu et al., 2021). The CNTs (0.05 g) removed 99.9% of Zn^{2+} at 10.0 pH and significant amount of Cd^{2+} at 3.0 pH. Temperature variations support the equilibrium of NPs' adsorption capacity. Although chitosan-alginate NPs remove Hg^{2+} at 30 °C, Pb^{2+} is adsorbable by Fe_3O_4NPs between 298 and 328 K (Nassar, 2010). Similarly, TiO_2 nanoparticles are known to have strong reactivity and photocatalytic activity, which allows contaminants to bind to their surface area and reduces the toxicity of HMs. At 7.0 pH, it efficiently absorbed 88.01% of Cu^{2+} and 70.67% of Cd^{2+} (Mahmoud et al., 2016).

11.3.2 Restoring Metalloid-contaminated Soils

Metalloids are six elements of the periodic table: antimony (Sb), arsenic (As), boron (B), germanium (Ge), tellurium (Te), and silicon (Si) (Sekhon, 2013). Due to the environmental toxicity of these substances, metalloid soil contamination is a problem that affects not only plants and animals but also people and their health, being unsafe even at very low concentrations (Raffa et al., 2021). Among NPs, iron oxides NPs are investigated to remediate contaminants like HMs from soils because of their ability to absorb these pollutants (Kumpiene et al., 2008). In this regard, a study demonstrated the efficiency of iron-based nanoparticles, namely nZVI, FeS, and Fe_3O_4 particles, for immobilizing arsenic in polluted soils (Zhang et al., 2010). Fe_3O_4NPs were among them, and it was claimed that they were superior to other NPs in terms of immobilizing As. Other NPs besides iron oxides are also recommended for regenerating soils contaminated with metals and metalloids (Rajput et al., 2022b). Some well-known iron nanoparticles, such as nZVI, are used to treat water that has been contaminated by metals or metalloids in addition to magnetite. In a study, the photolytic system was employed to convert chromium from its poisonous form, Cr^{6+}, to its nontoxic form, Cr^{3+} in aqueous. The Cr^{6+} reduction efficiency in the same research was up to 90%. TiO_2, MgO, and ZnO were discovered to be effective adsorbents for removing Cr ions from soil treated with leather manufacturing waste (Taghipour & Jalali, 2015).

11.3.3 Various Nanomaterials for Adsorption of Metals and Metalloids

Iron oxides, such as goethite (α-FeOOH; prevalent in temperate climatic areas), maghemite (γ-Fe_2O_3), magnetite (Fe_3O_4), and hematite (α-Fe_2O_3; occurring mainly in warm-dry climate zones), constitute soil components in nature and occur in diverse forms. Owing to their tendency to adsorb potentially toxic elements including metals and metalloids, they are highly significant in the soil system (Kumpiene et al., 2008; Zhang et al., 2010). Therefore, they represent a highly interesting area to develop synthetic iron-based NMs that can be utilized for detoxifying water and soil systems contaminated with metals and metalloids, or their stabilization therein (Mohapatra & Anand, 2010; Waychunas et al., 2005). They can also be applied in the form of either their nanooxides or precursors such as (nano zero-valent iron, nZVI) (Komárek et al., 2015). One potent feature of using iron-based NMs is that the in situ immobilization of metal contaminants minimizes their leaching into the soil by preventing their penetration into deeper layers of soil. One of the synthetic iron nanoparticles, synthetic nanogoethite (nFeOOH), has a high adsorption capacity for Cu and as such has shown promising results for removing Cu from aqueous solutions (Chen & Li, 2010).

Maghemite is formed through various routes, either due to weathering or oxidation of magnetite, in soils, and is magnetic in nature. One of the promising materials for removing inorganic contaminants is synthetic maghemite (nFe_2O_3), which is readily available and inexpensive (Komárek et al., 2015). In addition, nanomaghemite is highly efficient in removing the most toxic form of As (arsenite, As (III) (Akhbarizadeh et al., 2014; Auffan et al., 2008; Lin et al., 2012). It has also been shown to act as a scavenger of Cd(II), Pb(II), Cu(II), Cr(VI), and Zn(II) from solutions and hence can be utilized as a sorbent for soil and water remediation (Komárek et al., 2015). Interestingly, nFe_2O_3 immobilizes Zn from soil pore water (available to plants), resulting in reduction of toxic effects to roots and aerial parts, and therefore aiding in promoting plant growth. The manner in which adsorption occurs varies; that of Pb(II) occurs mostly via the inner-sphere complex formation, whereas that of Cd(II) occurs mainly through a mixture of inner- and outer-sphere complexes (Kisku et al., 2018).

Magnetite, one of the main ores of iron, contains both Fe^{2+} and Fe^{3+} contributing to its mixed-valency and ferrimagnetic properties. Its formation occurs in the soil via weathering of ferrihydrite. The nanoform of magnetite, nanomagnetite (nFe_3O_4), is more efficient in immobilizing As in soils than iron sulfide or nZVI, besides having a higher stabilization efficiency (Zhang et al., 2010). Furthermore, nFe_3O_4 has also been reported as efficient in Pb removal from aqueous solutions showing greater adsorption capacity of 36 mg Pb g-1 than goethite (Nassar, 2010). A lot of work pertaining to the synthesis and applications of iron NMs (nFeOx) has been done, highlighting their novel attributes and applicability, specifically due to their nanosize and magnetic properties (Afkhami et al., 2010; Mohapatra & Anand, 2010). Normally, the preparation of iron NMs involves a chemical route via precipitation (Nazari et al., 2014), sol-gel, hydrothermal, dry vapor deposition, surfactant mediation, microemulsion, electro-deposition, and sonochemical methods. Various composites of iron oxide, such as clay-iron oxide composites and magnetic zeolites, can also be prepared and may find applications in removing metal contaminants from water (Tosco et al., 2014).

11.3.4 Nano Zero-valent Iron (nZVI)

The particles of nZVI consist of a core–shell structure consisting of 98% Fe inside, surrounded by a shell made of iron oxides and hydroxides (FeO, Fe_2O_3, FeOOH). The chemical composition confers upon them the typical characteristics of sorption and reduction. Additionally, the surface of nZVI provides them with stability, thereby preventing their rapid oxidation in the environment. Also, the increased surface area makes them efficient and highly reactive; however, the particles less than 20 nm are so fast-reactive that they lose their activity even before reaching the contaminant. Therefore, a proper balance is to be maintained between the reactivity and half-life of nZVI for future in situ applications (Crane & Scott, 2012; Tosco et al., 2014; Zhang & Somasundaran, 2006).

11.3.5 Manganese-based Materials

Both manganese and Fe oxides occur naturally in soil as breakdown products due to weathering. They exist mainly on the surface of soil particles and as pores or in the form of concretions and nodules with near-amorphous structure (Post, 1999). Although Mn oxides are less abundant than

Fe oxides, they are more efficient in immobilizing some metals (Dong et al., 2000; O'Reilly & Hochella Jr, 2003). This may be attributed to their large specific surface and low pH value responsible for conferring negative surface charge in normal soil conditions (Essington, 2015). One of the characteristic features of their structure, formed mostly by layers or tunnels, is that it can accommodate water molecules or various cations in these regions (Post, 1999). Owing to their strong oxidative potential, Mn oxides can participate in redox and cation exchange reactions. For this reason, Mn oxides are not suitable in cases where Cr exists as a contaminant resulting in conversion of Cr(III) to Cr(VI) that is mobile and more toxic (Fandeur et al., 2009; Feng et al., 2006). However, the oxidizing behavior of Mn oxides can be harnessed for removing As contaminants where it is beneficial by oxidizing the more mobile and toxic As(III) to As(V) (Chen et al., 2006; Driehaus et al., 1995; Watanabe et al., 2013). Manganese oxide NPs can be synthesized by both the classical chemical routes and biotechnological means, via microbial action like that of bacteria or fungi. In fact, biogenic oxidation of Mn(II) represents the prevailing route for the formation of Mn oxides in soil. Application of Mn nanooxides for soil remediation appears relatively safe, as nanoscale biogenic Mn oxides are natural and ubiquitous soil components. Although various studies have reported the adsorption performance of Mn-based NPs, their direct application for remediation of contaminated soil, together with assisted phytoremediation, is still rather rare.

Pollutants are rapidly released into the environment, posing a serious risk to human as well as environmental health. As a result, limiting their detrimental effects will depend on developing efficient methods for removing them from different environments. Novel ways are consequently needed to eradicate pollutants to the greatest degree feasible, because many existing processes are incapable of effectively eliminating various categories of contaminants. According to the findings of studies in this area, nanotechnology is a highly promising source of solutions. Therefore, it will be helpful for guiding future soil cleaning initiatives, which are occasionally required and cannot wait for experimental findings to be acquired. In addition, most of the existing studies on nanobioremediation are restricted to lab tests. Before extending the use of nanobioremediation techniques, it is important to carefully analyze the fate of NPs in the environment.

11.4 ROLE OF NANOPARTICLES IN PLANT MECHANISMS

Beneficial nanoparticles have surfaced as a propitious alternative to traditional fertilizers, primarily due to their unique properties and positive impact on crop plants (Zulfiqar et al., 2019). They can serve as a nutrient source or act as nanocarriers for fertilizers, offering several advantages, such as improved performance, controlled release, and reduced environmental pollution (Jakhar et al., 2022; Kopittke et al., 2019). Nanoparticles' small size and large surface area enable easy solubility and absorption, making them effective in enhancing nutrient uptake, biomass production, and biosynthesis of secondary metabolites (Nile et al., 2022). Additionally, the introduction of beneficial nanoparticles can modify biochemical parameters, like chlorophyll and protein content, and enhance plant tolerance against environmental stresses (Singh et al., 2021). Some extensively studied beneficial NPs include carbon nanotubes, metal-based NPs, crystalline powders of nanoscale, and metal oxide NPs. However, it is crucial to emphasize that further research is necessary to comprehensively understand the effects of these NPs on plants and the environment (Arora, 2019; Gopinath et al., 2014; Gruyer et al., 2014). The utilization of beneficial NPs in agriculture holds the potential to boost yields while minimizing environmental impact, making it a promising area of research for sustainable agriculture.

11.4.1 Impact on Morphological Attributes

Numerous studies have demonstrated the effects of various nanoparticles on the growth, yield, and morphology of diverse crop plants. The utilization of carbon nanotubes (CNTs), multiwalled carbon nanotubes (MWCNTs), fullerols, silver (Ag) NPs, titanium (Ti) NPs, gold (Au) NPs, copper (Cu)

NPs, iron oxide NPs, zinc oxide (ZnO) NPs, silicon dioxide (SiO_2) NPs, and cerium dioxide (CeO_2) NPs has been shown to improve crop growth, yield, and quality in a range of crops, including tomato, bitter melon, potato, common bean, maize, borage, basil, wheat, *Spinacea*, *Triticum aestivum*, lettuce, soybean, mung bean, onion, and maize (Singh et al., 2021). While nanoparticles hold significant potential for crop improvement, it is crucial to recognize that their use in agriculture is still a relatively new field, and further research is needed to fully comprehend the potential benefits and risks associated with their application.

11.4.2 Impact on Nutrient Uptake

It is important to recognize that the impact of nanoparticles on plant nutrition can vary depending on the specific plant species and the physical and chemical properties of the nanoparticles. While the use of nanoparticles in agriculture can enhance nutrient uptake in plants, it offers several benefits, such as a controlled release of nutrients that minimizes leaching, easy absorption by the porous soil, and enrichment of soil nutrients (Dehkourdi & Mosavi, 2013; Shukla et al., 2016). Moreover, nanoparticles can be absorbed and accumulated by plants, effectively addressing nutrient deficiencies and ultimately leading to increased crop production (Tiwari et al., 2014).

11.4.3 Impact on Seed Germination and Seed Vigor

The application of beneficial nanoparticles has a positive impact on various physiological attributes in plants, including seed germination, germination rate, seed vigor index, accumulation of osmolytes, adjustments in photosynthetic pigments, and transpiration and stomatal conductance (Omasa et al., 2007). These attributes play a vital role in enabling plants to cope with abiotic stresses, stimulate water channels, and improve water uptake and seed germination in different crops like soybean, maize, barley, wheat, peanut, and garlic. Notably, carbon nanotubes and AgNPs have demonstrated the ability to improve seed germination in various plant species (Lahiani et al., 2015; Yin et al., 2012). Similarly, nano-TiO_2 has been reported to enhance seed germination and vigor in *Spinacia oleracea* seeds (Zheng et al., 2005), while nanocrystalline powders of Fe, Co, and Cu have shown increased germination frequencies in soybean seeds (Ngo et al., 2014). Furthermore, SiO_2NPs have been found to enhance seed germination in tomatoes, while AuNPs and alumina NPs have shown similar effects when used at a concentration of 50 ppm (Barazzouk et al., 2005; Saeid Hojjat, 2019).

11.4.4 Impact on Photosynthetic Parameters

It is truly fascinating to discover how nanoparticles can enhance the activity of carbonic anhydrase and promote the synthesis of photosynthetic pigments, ultimately leading to an increase in the rate of photosynthesis and chlorophyll content (Singh et al., 2021). The ability of nanoparticles to protect chloroplasts and enhance the number of light-harvesting complexes is equally impressive. Remarkably, SiO_2NPs and ZnONPs have been found to improve photosynthesis and chlorophyll content by increasing the activity of carbonic anhydrase, enhancing gas exchange, and improving chlorophyll fluorescence (Ali et al., 2021). Furthermore, the application of nanoparticles can significantly assist plants in combating abiotic stresses, such as salinity stress, by enhancing their stress tolerance and improving their ability to scavenge ROS (Qi et al., 2013). The overall impact of nanoparticles on photosynthesis and chlorophyll content is substantial, and it has the potential to enhance crop yield and foster plant growth across various environmental conditions (Siddiqui et al., 2012).

11.4.5 Impact on Nitrogen Fixation

It is crucial to acknowledge that findings on the effects of nanoparticles on N_2 fixation are mixed, with some studies reporting positive outcomes, while others have indicated negative effects or no significant impact. For instance, certain studies have demonstrated that nanoparticles exposure can hinder N_2 fixation by disrupting the function of nitrogenase enzymes (Lin et al., 2009). Consequently, further research is necessary to comprehensively understand the effects of nanoparticles on N_2 fixation and to optimize their application in agriculture.

11.4.6 Impact on Antioxidant and Cellular Reductants

The application of nanoparticles has been found to result in the upregulation of several antioxidative enzymes, such as peroxidase enzyme (POD), superoxide dismutase (SOD) enzyme, and catalase (CAT) enzyme (Nair & Chung, 2015; Ngo et al., 2014). This increased enzyme activity has been observed to enhance tolerance to environmental stresses in various plant species. For instance, wheat seedlings under drought stress, maize plants under heavy metal stress, and tomato plants under salt stress have exhibited increased stress tolerance due to the upregulation of antioxidative enzymes (Kaveh et al., 2013). Furthermore, the treatment of plants with nanoparticles has significantly increased the content of proline, ascorbate, and glutathione, which in turn contributes to enhanced tolerance to drought stress (Lee et al., 2012). Overall, the application of nanoparticles in plants has demonstrated promising results in augmenting stress tolerance through the upregulation of antioxidant enzymes and other biochemical pathways.

11.4.7 Impact of Nanoparticles on Molecular Attributes

Collectively, the research studies indicate that exposure to different types of nanoparticles can induce alterations in gene expression and protein activity in plants, particularly those involved in stress-response and defense mechanisms (Lee et al., 2012; Nair & Chung, 2015; Siddiqui et al., 2012). Certain nanoparticles were found to upregulate genes related to antioxidative enzymes, photosynthesis, and stress tolerance, while others downregulated genes associated with energy pathways and cell organization (Dugas & Bartel, 2008). Additionally, exposure to specific nanoparticles was shown to enhance the activities of particular enzymes and increase the levels of certain metabolites in plants (Banerjee et al., 2010; Banerjee & Maiti, 2010; Kaveh et al., 2013). These findings suggest that nanoparticles hold potential as a tool to enhance stress resistance and productivity of crops in agriculture.

11.5 ROLE OF NANOPARTICLES IN ABIOTIC STRESS MANAGEMENT

The expansion in our knowledge regarding the important role of nanoparticles in cellular mechanisms has encouraged scientists to use them for stress management. Several studies have shown that the foliar application of nanoparticles can modulate the physiological and biochemical response of plants under abiotic stresses such as drought and salt stress (Dimkpa et al., 2018; Heikal et al., 2022; Upadhyaya et al., 2018; Ye et al., 2019).

Abiotic stresses such as drought and salt stress are known to increase ROS production, leading to phytotoxicity and other physiological impacts that reduce plant growth. Nanoparticles can influence function of antioxidant enzymes such as catalase, superoxide dismutase, and peroxidase, and can be used to alleviate negative impacts of drought stress and help reduce salt toxicity. TiO_2 nanoparticles have been shown to increase the activity of superoxide dismutase in a dose-dependent manner in onion seedlings. Multiple researchers have reported that AgNPs can increase antioxidant response to improve salt and drought stress tolerance in plants (Saeid Hojjat, 2019; Soares et al., 2018). Similarly, the impact of nanoparticles such as silica (SiO_2), zinc, and iron in reducing salt toxicity has been studied in various crops such as maize, tomato, *Brassica napus*, basil, and peppermint (Gao et al., 2006; Rossi et al., 2016; Savvas et al., 2009).

Additionally, it has been shown that AgNPs can bind with calcium ion receptors (Goyer, 1995; Mirzajani et al., 2014) of calcium-binding proteins which are known to initiate cell signaling to activate stress-responsive genes under abiotic stresses. Nanoparticles have also been reported to improve stress tolerance by modifying plant morphology. CuO was reported to increase lignification in cell walls of mustard and *Arabidopsis*, which altered water flow and regulated root growth under drought (Yang et al., 2018). It has also been reported that carbon nanotubes can change properties of root plasma membrane, which led to increased growth and photosynthetic ability in broccoli under salt stress (Martínez-Ballesta et al., 2016).

As the adverse effects on the environment are increasing, extreme climate conditions such as cold or high heat are becoming more frequent and severe. The potential of nanoparticles to mitigate adverse effects of cold and heat stress have also been studied. The application of ZnONPs has been found to increase chlorophyll activity and yield under high temperature in mung bean (Kareem et al., 2022). ZnONPs have also been reported to improve cold tolerance in rice by improving photosynthetic ability and reducing oxidative stress (Song et al., 2021). Similarly, foliar application of SiO_2, Se, and GNR nanoparticles has been reported to improve cold tolerance in sugarcane (Soares et al., 2018).

Nanoparticles also have impact at the molecular level under abiotic stresses. For example, CeO_2 and carbon nanotubes have been reported to increase the expression of heat shock proteins which help to stabilize and protect other cellular proteins under heat stress (Shafqat et al., 2021; Wang et al., 2022).

The potential of nanoparticles for stress management has been studied more for drought and salt stress than for temperature and cold stress management. Overall, nanoparticles can provide effective strategies to alleviate the negative impact of abiotic stresses, but the impact is dose-dependent. Higher levels of nanoparticles can result in toxicity, and therefore concentrations of nanoparticles need careful consideration before using nanoparticles for stress management (Rajput et al., 2021; Wang et al., 2016).

11.6 CONCLUSIONS

Nanotechnology-based approaches, such as utilizing nanofertilizers, nanopesticides, and nanosensors, show great potential in increasing crop yields and enhancing food security. Moreover, applying nanotechnology in agriculture can contribute to reducing the environmental impact by minimizing the use of synthetic chemicals and nutrient loss to the environment. By addressing challenges in the agricultural sector, including sustainable food production, environmental protection, and human health, nanotechnology holds a significant role. However, it is essential to acknowledge that the use of nanotechnology in agriculture also comes with potential risks. Excessive use of certain nanoparticles can lead to their accumulation in the environment and potential harm to organisms. Therefore, continuous research and innovation in this field are vital to ensure the safe and effective application of nanotechnology in agriculture. Careful evaluation of the benefits and risks associated with using nanoparticles in agricultural practices is essential, and it is crucial to employ them responsibly and sustainably. As such, more research is needed to fully comprehend the mechanisms underlying these changes and to thoroughly assess the potential benefits and risks related to the use of nanoparticles in agriculture.

REFERENCES

Abdel-Aziz, H. M. M., Hasaneen, M. N. A., & Omer, A. M. (2016). Nano chitosan-NPK fertilizer enhances the growth and productivity of wheat plants grown in sandy soil. *Spanish Journal of Agricultural Research*, *14*(1). https://doi.org/10.5424/sjar/2016141-8205

Abebe, B., Murthy, H. C. A., & Amare, E. (2018). Summary on adsorption and photocatalysis for pollutant remediation: Mini review. *Journal of Encapsulation and Adsorption Sciences*, *08*(04), 225–255. https://doi.org/10.4236/jeas.2018.84012

Abhilash, P. C., & Singh, N. (2009). Pesticide use and application: An Indian scenario. *Journal of Hazardous Materials*, *165*(1–3), 1–12. https://doi.org/10.1016/j.jhazmat.2008.10.061

Adisa, I. O., Pullagurala, V. L. R., Peralta-Videa, J. R., Dimkpa, C. O., Elmer, W. H., Gardea-Torresdey, J. L., & White, J. C. (2019). Recent advances in nano-enabled fertilizers and pesticides: A critical review of mechanisms of action. *Environmental Science: Nano*, *6*(7), 2002–2030. https://doi.org/10.1039/c9en00265k

Afkhami, A., Saber-Tehrani, M., & Bagheri, H. (2010). Modified maghemite nanoparticles as an efficient adsorbent for removing some cationic dyes from aqueous solution. *Desalination*, *263*(1–3), 240–248. https://doi.org/10.1016/j.desal.2010.06.065

Ahmed, F., Dwivedi, S., Shaalan, N. M., Kumar, S., Arshi, N., Alshoaibi, A., & Husain, F. M. (2020). Development of selenium nanoparticle based agriculture sensor for heavy metal toxicity detection. *Agriculture (Switzerland)*, *10*(12), 1–11. https://doi.org/10.3390/agriculture10120610

Akhbarizadeh, R., Shayestefar, M. R., & Darezereshki, E. (2014). Konkurrierende Entfernung von Metallen aus Abwasser mit Hilfe von Maghemit- Nanopartikeln: Ein Vergleich zwischen künstlichem Abwasser und Sauerwasser. *Mine Water and the Environment*, *33*(1), 89–96. https://doi.org/10.1007/s10230-013-0255-3

Alabresm, A., Chen, Y. P., Decho, A. W., & Lead, J. (2018). A novel method for the synergistic remediation of oil-water mixtures using nanoparticles and oil-degrading bacteria. *Science of the Total Environment*, *630*, 1292–1297. https://doi.org/10.1016/j.scitotenv.2018.02.277

Alazaiza, M. Y. D., Albahnasawi, A., Ali, G. A. M., Bashir, M. J. K., Copty, N. K., Amr, S. S. A., Abushammala, M. F. M., & Al Maskari, T. (2021). Recent advances of nanoremediation technologies for soil and groundwater remediation: A review. *Water (Switzerland)*, *13*(16). https://doi.org/10.3390/w13162186

Ali, S., Mehmood, A., & Khan, N. (2021). Uptake, translocation, and consequences of nanomaterials on plant growth and stress adaptation. *Journal of Nanomaterials*, *2021*. https://doi.org/10.1155/2021/6677616

Alissa, E. M., & Ferns, G. A. (2011). Heavy metal poisoning and cardiovascular disease. *Journal of Toxicology*, *2011*. https://doi.org/10.1155/2011/870125

Ansari, F., Grigoriev, P., Libor, S., Tothill, I. E., & Ramsden, J. J. (2009). DBT degradation enhancement by decorating Rhodococcus erythropolis IGST8 with magnetic Fe3O4 nanoparticles. *Biotechnology and Bioengineering*, *102*(5), 1505–1512. https://doi.org/10.1002/bit.22161

Arif, N., Yadav, V., Singh, S., Singh, S., Ahmad, P., Mishra, R. K., Sharma, S., Tripathi, D. K., Dubey, N. K., & Chauhan, D. K. (2016). Influence of high and low levels of plant-beneficial heavy metal ions on plant growth and development. *Frontiers in Environmental Science*, *4*(NOV). https://doi.org/10.3389/fenvs.2016.00069

Arora, N. K. (2019). Impact of climate change on agriculture production and its sustainable solutions. *Environmental Sustainability*, *2*(2), 95–96. https://doi.org/10.1007/s42398-019-00078-w

Auffan, M., Rose, J., Proux, O., Borschneck, D., Masion, A., Chaurand, P., Hazemann, J. L., Chaneac, C., Jolivet, J. P., Wiesner, M. R., Van Geen, A., & Bottero, J. Y. (2008). Enhanced adsorption of arsenic onto maghemites nanoparticles: As(III) as a probe of the surface structure and heterogeneity. *Langmuir*, *24*(7), 3215–3222. https://doi.org/10.1021/la702998x

Badran, A., & Savin, I. (2018). Effect of nano-fertilizer on seed germination and first stages of bitter almond seedlings' growth under saline conditions. *BioNanoScience*, *8*(3), 742–751. https://doi.org/10.1007/s12668-018-0531-6

Banerjee, J., Das, N., Dey, P., & Maiti, M. K. (2010). Transgenically expressed rice germin-like protein1 in tobacco causes hyper-accumulation of H 2O 2 and reinforcement of the cell wall components. *Biochemical and Biophysical Research Communications*, *402*(4), 637–643. https://doi.org/10.1016/j.bbrc.2010.10.073

Banerjee, J., & Maiti, M. K. (2010). Functional role of rice germin-like protein1 in regulation of plant height and disease resistance. *Biochemical and Biophysical Research Communications*, *394*(1), 178–183. https://doi.org/10.1016/j.bbrc.2010.02.142

Baragaño, D., Forján, R., Welte, L., & Gallego, J. L. R. (2020). Nanoremediation of As and metals polluted soils by means of graphene oxide nanoparticles. *Scientific Reports*, *10*(1). https://doi.org/10.1038/s41598-020-58852-4

Barazzouk, S., Kamat, P. V., & Hotchandani, S. (2005). Photoinduced electron transfer between chlorophyll a and gold nanoparticles. *Journal of Physical Chemistry B*, *109*(2), 716–723. https://doi.org/10.1021/jp046474s

Bhalla, N., Jolly, P., Formisano, N., & Estrela, P. (2016). Introduction to biosensors. *Essays in Biochemistry*, *60*(1), 1–8. https://doi.org/10.1042/EBC20150001

Bhatt, P., Pandey, S. C., Joshi, S., Chaudhary, P., Pathak, V. M., Huang, Y., Wu, X., Zhou, Z., & Chen, S. (2022). Nanobioremediation: A sustainable approach for the removal of toxic pollutants from the environment. *Journal of Hazardous Materials*, *427*. https://doi.org/10.1016/j.jhazmat.2021.128033

Bokare, V., Murugesan, K., Kim, J. H., Kim, E. J., & Chang, Y. S. (2012). Integrated hybrid treatment for the remediation of 2,3,7,8-tetrachlorodibenzo-p-dioxin. *Science of the Total Environment*, *435–436*, 563–566. https://doi.org/10.1016/j.scitotenv.2012.07.079

Cao, F., Yu, D., Ma, W., Xu, X., Cai, B., Yang, Y. M., Liu, S., He, L., Ke, Y., Lan, S., Choy, K. L., & Zeng, H. (2020). Shining emitter in a stable host: Design of Halide perovskite scintillators for X-ray imaging from commercial concept. *ACS Nano*, *14*(5), 5183–5193. https://doi.org/10.1021/acsnano.9b06114

Capone, R., El Bilali, H., Debs, P., Cardone, G., & Driouech, N. (2014). Mediterranean food consumption patterns sustainability: Setting up a common ground for future research and action. *American Journal of Nutrition and Food Science*, *1*(2), 37. https://doi.org/10.12966/ajnfs.04.04.2014

Chaplygin, V. A., Minkina, T. M., Mandzhieva, S. S., Nazarenko, O. G., Zimulina, I. V., Bauer, T. V., Litvinov, Y. A., & Rajput, V. (2021). Heavy metals in agricultural crops of Rostov region through the example of soft wheat (Triticum aestivum). *IOP Conference Series: Earth and Environmental Science*, *624*(1). https://doi.org/10.1088/1755-1315/624/1/012204

Chaud, M., Souto, E. B., Zielinska, A., Severino, P., Batain, F., Oliveira-Junior, J., & Alves, T. (2021). Nanopesticides in agriculture: Benefits and challenge in agricultural productivity, toxicological risks to human health and environment. *Toxics*, *9*(6). https://doi.org/10.3390/toxics9060131

Chen, Y. H., & Li, F. A. (2010). Kinetic study on removal of copper(II) using goethite and hematite nano-photocatalysts. *Journal of Colloid and Interface Science*, *347*(2), 277–281. https://doi.org/10.1016/j.jcis.2010.03.050

Chen, Z., Kim, K. W., Zhu, Y. G., McLaren, R., Liu, F., & He, J. Z. (2006). Adsorption (AsIII,V) and oxidation (AsIII) of arsenic by pedogenic Fe-Mn nodules. *Geoderma*, *136*(3–4), 566–572. https://doi.org/10.1016/j.geoderma.2006.04.012

Crane, R. A., & Scott, T. B. (2012). Nanoscale zero-valent iron: Future prospects for an emerging water treatment technology. *Journal of Hazardous Materials*, *211–212*, 112–125. https://doi.org/10.1016/j.jhazmat.2011.11.073

Das, A., & Maji, B. (2021). The emergence of palladium-catalyzed C(sp3)–H functionalization of free carboxylic acids. *Chemistry – An Asian Journal*, *16*(5), 397–408. https://doi.org/10.1002/asia.202001440

Dehkourdi, E. H., & Mosavi, M. (2013). Effect of anatase nanoparticles (TiO2) on parsley seed germination (petroselinum crispum) in vitro. *Biological Trace Element Research*, *155*(2), 283–286. https://doi.org/10.1007/s12011-013-9788-3

DeRosa, M. C., Monreal, C., Schnitzer, M., Walsh, R., & Sultan, Y. (2010). Nanotechnology in fertilizers. *Nature Nanotechnology*, *5*(2), 91. https://doi.org/10.1038/nnano.2010.2

Desiante, W. L., Minas, N. S., & Fenner, K. (2021). Micropollutant biotransformation and bioaccumulation in natural stream biofilms. *Water Research*, *193*. https://doi.org/10.1016/j.watres.2021.116846

Dimkpa, C. O., Singh, U., Adisa, I. O., Bindraban, P. S., Elmer, W. H., Gardea-Torresdey, J. L., & White, J. C. (2018). Effects of manganese nanoparticle exposure on nutrient acquisition in wheat (Triticum aestivum L.). *Agronomy*, *8*(9). https://doi.org/10.3390/agronomy8090158

Dong, D., Nelson, Y. M., Lion, L. W., Shuler, M. L., & Ghiorse, W. C. (2000). Adsorption of Pb and Cd onto metal oxides and organic material in natural surface coatings as determined by selective extractions: New evidence for the importance of Mn and Fe oxides. *Water Research*, *34*(2), 427–436. https://doi.org/10.1016/S0043-1354(99)00185-2

Dos Santos, J. V. N., Martins, R., Fontes, M. K., de Campos, B. G., E Silva, M. B. M. D. P., Maia, F., Abessa, D. M. de S., & Perina, F. C. (2020). Can encapsulation of the biocide DCOIT affect the anti-fouling efficacy and toxicity on tropical bivalves? *Applied Sciences (Switzerland)*, *10*(23), 1–12. https://doi.org/10.3390/app10238579

Driehaus, W., Seith, R., & Jekel, M. (1995). Oxidation of arsenate(III) with manganese oxides in water treatment. *Water Research*, *29*(1), 297–305. https://doi.org/10.1016/0043-1354(94)E0089-O

Dugas, D. V., & Bartel, B. (2008). Sucrose induction of Arabidopsis miR398 represses two Cu/Zn superoxide dismutases. *Plant Molecular Biology*, *67*(4), 403–417. https://doi.org/10.1007/s11103-008-9329-1

Enez, A., Hudek, L., & Bräu, L. (2018). Reduction in trace element mediated oxidative stress towards cropped plants via beneficial microbes in irrigated cropping systems: A review. *Applied Sciences (Switzerland)*, *8*(10). https://doi.org/10.3390/app8101953

Essington, M. E. (2015). Soil and water chemistry: An integrative approach, 2nd ed., 1–636. https://doi.org/10.1201/b18385

Fandeur, D., Juillot, F., Morin, G., Livi, L., Cognigni, A., Webb, S. M., Ambrosi, J. P., Fritsch, E., Guyot, F., & Brown, G. E. (2009). XANES evidence for oxidation of Cr(III) to Cr(VI) by Mn-oxides in a lateritic regolith developed on serpentinized ultramafic rocks of New Caledonia. *Environmental Science and Technology*, *43*(19), 7384–7390. https://doi.org/10.1021/es900498r

Feng, X. H., Zhai, L. M., Tan, W. F., Zhao, W., Liu, F., & He, J. Z. (2006). The controlling effect of pH on oxidation of Cr(III) by manganese oxide minerals. *Journal of Colloid and Interface Science*, *298*(1), 258–266. https://doi.org/10.1016/j.jcis.2005.12.012

Figueiredo, J., Oliveira, T., Ferreira, V., Sushkova, A., Silva, S., Carneiro, D., Cardoso, D. N., Gonçalves, S. F., Maia, F., Rocha, C., Tedim, J., Loureiro, S., & Martins, R. (2019). Toxicity of innovative anti-fouling nano-based solutions to marine species. *Environmental Science: Nano*, *6*(5), 1418–1429. https://doi.org/10.1039/c9en00011a

Fincheira, P., Tortella, G., Duran, N., Seabra, A. B., & Rubilar, O. (2020). Current applications of nanotechnology to develop plant growth inducer agents as an innovation strategy. *Critical Reviews in Biotechnology*, *40*(1), 15–30. https://doi.org/10.1080/07388551.2019.1681931

Gao, X., Zou, C., Wang, L., & Zhang, F. (2006). Silicon decreases transpiration rate and conductance from stomata of maize plants. *Journal of Plant Nutrition*, *29*(9), 1637–1647. https://doi.org/10.1080/01904160600851494

Ghani, M. I., Saleem, S., Rather, S. A., Rehmani, M. S., Alamri, S., Rajput, V. D., Kalaji, H. M., Saleem, N., Sial, T. A., & Liu, M. (2022). Foliar application of zinc oxide nanoparticles: An effective strategy to mitigate drought stress in cucumber seedling by modulating antioxidant defense system and osmolytes accumulation. *Chemosphere*, *289*. https://doi.org/10.1016/j.chemosphere.2021.133202

Gopinath, K., Gowri, S., Karthika, V., & Arumugam, A. (2014). Green synthesis of gold nanoparticles from fruit extract of Terminalia arjuna, for the enhanced seed germination activity of Gloriosa superba. *Journal of Nanostructure in Chemistry*, *4*(3). https://doi.org/10.1007/s40097-014-0115-0

Gottardo, S., Mech, A., Drbohlavová, J., Małyska, A., Bøwadt, S., Riego Sintes, J., & Rauscher, H. (2021). Towards safe and sustainable innovation in nanotechnology: State-of-play for smart nanomaterials. *NanoImpact*, *21*. https://doi.org/10.1016/j.impact.2021.100297

Goyer, R. A. (1995). Nutrition and metal toxicity. *American Journal of Clinical Nutrition*, *61*(3 SUPPL.). https://doi.org/10.1093/ajcn/61.3.646S

Gruyer, N., Dorais, M., Bastien, C., Dassylva, N., & Triffault-Bouchet, G. (2014). Interaction between silver nanoparticles and plant growth. *Acta Horticulturae*, *1037*, 795–800. https://doi.org/10.17660/ActaHortic.2014.1037.105

Havlin, J. L. (2020). Soil: Fertility and nutrient management. *Landscape and Land Capacity*, 251–265. https://doi.org/10.1201/9780429445552-34

Heikal, Y. M., El-Esawi, M. A., El-Ballat, E. M., & Abdel-Aziz, H. M. M. (2022). Applications of nanoparticles for mitigating salinity and drought stress in plants: An overview on the physiological, biochemical and molecular genetic aspects. *New Zealand Journal of Crop and Horticultural Science*. https://doi.org/10.1080/01140671.2021.2016870

Huang, H., Ullah, F., Zhou, D. X., Yi, M., & Zhao, Y. (2019). Mechanisms of ROS regulation of plant development and stress responses. *Frontiers in Plant Science*, *10*. https://doi.org/10.3389/fpls.2019.00800

Jabeen, R., Ahmad, A., & Iqbal, M. (2009). Phytoremediation of heavy metals: Physiological and molecular mechanisms. *Botanical Review*, *75*(4), 339–364. https://doi.org/10.1007/s12229-009-9036-x, https://doi.org/10.1007/s11306-007-0056-4

Jaishankar, M., Tseten, T., Anbalagan, N., Mathew, B. B., & Beeregowda, K. N. (2014). Toxicity, mechanism and health effects of some heavy metals. *Interdisciplinary Toxicology*, *7*(2), 60–72. https://doi.org/10.2478/intox-2014-0009

Jakhar, A. M., Aziz, I., Kaleri, A. R., Hasnain, M., Haider, G., Ma, J., & Abideen, Z. (2022). Nano-fertilizers: A sustainable technology for improving crop nutrition and food security. *NanoImpact*, *27*. https://doi.org/10.1016/j.impact.2022.100411

Jiang, P., Wang, Y., Zhao, L., Ji, C., Chen, D., & Nie, L. (2018). Applications of gold nanoparticles in non-optical biosensors. *Nanomaterials*, *8*(12). https://doi.org/10.3390/nano8120977

Jin, B., Yang, Y., He, R., Park, Y. Il, Lee, A., Bai, D., Li, F., Lu, T. J., Xu, F., & Lin, M. (2018). Lateral flow aptamer assay integrated smartphone-based portable device for simultaneous detection of multiple targets using upconversion nanoparticles. *Sensors and Actuators, B: Chemical*, *276*, 48–56. https://doi.org/10.1016/j.snb.2018.08.074

Kaczerewska, O., Sousa, I., Martins, R., Figueiredo, J., Loureiro, S., & Tedim, J. (2020). Gemini surfactant as a template agent for the synthesis of more eco-friendly silica nanocapsules. *Applied Sciences (Switzerland)*, *10*(22), 1–13. https://doi.org/10.3390/app10228085

Kalyani, N., Goel, S., & Jaiswal, S. (2021). On-site sensing of pesticides using point-of-care biosensors: a review. *Environmental Chemistry Letters*, *19*(1), 345–354. https://doi.org/10.1007/s10311-020-01070-1

Kaphle, A., Navya, P. N., Umapathi, A., & Daima, H. K. (2018). Nanomaterials for agriculture, food and environment: Applications, toxicity and regulation. *Environmental Chemistry Letters*, *16*(1), 43–58. https://doi.org/10.1007/s10311-017-0662-y

Kareem, H. A., Saleem, M. F., Saleem, S., Rather, S. A., Wani, S. H., Siddiqui, M. H., Alamri, S., Kumar, R., Gaikwad, N. B., Guo, Z., Niu, J., & Wang, Q. (2022). Zinc oxide nanoparticles interplay with physiological and biochemical attributes in terminal heat stress alleviation in Mungbean (Vigna radiata L.). *Frontiers in Plant Science*, *13*. https://doi.org/10.3389/fpls.2022.842349

Kaveh, R., Li, Y. S., Ranjbar, S., Tehrani, R., Brueck, C. L., & Van Aken, B. (2013). Changes in Arabidopsis thaliana gene expression in response to silver nanoparticles and silver ions. *Environmental Science and Technology*, *47*(18), 10637–10644. https://doi.org/10.1021/es402209w

Khaledian, S., Nikkhah, M., Shams-bakhsh, M., & Hoseinzadeh, S. (2017). A sensitive biosensor based on gold nanoparticles to detect Ralstonia solanacearum in soil. *Journal of General Plant Pathology*, *83*(4), 231–239. https://doi.org/10.1007/s10327-017-0721-z

Khan, I., Saeed, K., & Khan, I. (2019). Nanoparticles: Properties, applications and toxicities. *Arabian Journal of Chemistry*, *12*(7), 908–931. https://doi.org/10.1016/j.arabjc.2017.05.011

Kisku, G. C., Kumar, V., Sahu, P., Kumar, P., & Kumar, N. (2018). Characterization of coal fly ash and use of plants growing in ash pond for phytoremediation of metals from contaminated agricultural land. *International Journal of Phytoremediation*, *20*(4), 330–337. https://doi.org/10.1080/15226514.2017.1381942

Komárek, M., Koretsky, C. M., Stephen, K. J., Alessi, D. S., & Chrastný, V. (2015). Competitive adsorption of Cd(II), Cr(VI), and Pb(II) onto nanomaghemite: A spectroscopic and modeling approach. *Environmental Science and Technology*, *49*(21), 12851–12859. https://doi.org/10.1021/acs.est.5b03063

Kopittke, P. M., Lombi, E., Wang, P., Schjoerring, J. K., & Husted, S. (2019). Nanomaterials as fertilizers for improving plant mineral nutrition and environmental outcomes. *Environmental Science: Nano*, *6*(12), 3513–3524. https://doi.org/10.1039/c9en00971j

Kumari, A., Kumari, P., Rajput, V. D., Sushkova, S. N., & Minkina, T. (2022). Metal(loid) nanosorbents in restoration of polluted soils: Geochemical, ecotoxicological, and remediation perspectives. *Environmental Geochemistry and Health*, *44*(1), 235–246. https://doi.org/10.1007/s10653-021-00996-x

Kumpiene, J., Lagerkvist, A., & Maurice, C. (2008). Stabilization of As, Cr, Cu, Pb and Zn in soil using amendments – A review. *Waste Management*, *28*(1), 215–225. https://doi.org/10.1016/j.wasman.2006.12.012

Lahiani, M. H., Chen, J., Irin, F., Puretzky, A. A., Green, M. J., & Khodakovskaya, M. V. (2015). Interaction of carbon nanohorns with plants: Uptake and biological effects. *Carbon*, *81*(1), 607–619. https://doi.org/10.1016/j.carbon.2014.09.095

Laslo, V., Pinzaru, S. C., Zaguła, G., Kluz, M., Vicas, S. I., & Cavalu, S. (2022). Synergic effect of selenium nanoparticles and lactic acid bacteria in reduction cadmium toxicity. *Journal of Molecular Structure*, *1247*. https://doi.org/10.1016/j.molstruc.2021.131325

Lee, D. R., Barrett, C. B., & McPeak, J. G. (2006). Policy, technology, and management strategies for achieving sustainable agricultural intensification. *Agricultural Economics*, *34*(2), 123–127. https://doi.org/10.1111/j.1574-0864.2006.00112.x

Lee, S. H., Pie, J. E., Kim, Y. R., Lee, H. R., Son, S. W., & Kim, M. K. (2012). Effects of zinc oxide nanoparticles on gene expression profile in human keratinocytes. *Molecular and Cellular Toxicology*, *8*(2), 113–118. https://doi.org/10.1007/s13273-012-0014-8

Li, B., Devaraj, K., Darcel, C., & Dixneuf, P. H. (2012). Catalytic C-H bond arylation of aryl imines and oxazolines in water with ruthenium(II)-acetate catalyst. *Tetrahedron*, *68*(26), 5179–5184. https://doi.org/10.1016/j.tet.2012.03.117

Li, Z. Z., Xu, S. A., Wen, L. X., Liu, F., Liu, A. Q., Wang, Q., Sun, H. Y., Yu, W., & Chen, J. F. (2006). Controlled release of avermectin from porous hollow silica nanoparticles: Influence of shell thickness on loading efficiency, UV-shielding property and release. *Journal of Controlled Release*, *111*(1–2), 81–88. https://doi.org/10.1016/j.jconrel.2005.10.020

Lin, S., Lu, D., & Liu, Z. (2012). Removal of arsenic contaminants with magnetic γ-Fe2O3 nanoparticles. *Chemical Engineering Journal*, *211–212*, 46–52. https://doi.org/10.1016/j.cej.2012.09.018

Lin, S., Reppert, J., Hu, Q., Hudson, J. S., Reid, M. L., Ratnikova, T. A., Rao, A. M., Luo, H., & Ke, P. C. (2009). Uptake, translocation, and transmission of carbon nanomaterials in rice plants. *Small*, *5*(10), 1128–1132. https://doi.org/10.1002/smll.200801556

Ma, X., Geiser-Lee, J., Deng, Y., & Kolmakov, A. (2010). Interactions between engineered nanoparticles (ENPs) and plants: Phytotoxicity, uptake and accumulation. *Science of the Total Environment*, *408*(16), 3053–3061. https://doi.org/10.1016/j.scitotenv.2010.03.031

Mahmoud, M. E., Abdou, A. E. H., Mohamed, S. M. S., & Osman, M. M. (2016). Engineered staphylococcus aureus via immobilization on magnetic Fe3O4-phthalate nanoparticles for biosorption of divalent ions from aqueous solutions. *Journal of Environmental Chemical Engineering*, *4*(4), 3810–3824. https://doi.org/10.1016/j.jece.2016.08.022

Martin, J. L., Maris, V., & Simberloff, D. S. (2016). The need to respect nature and its limits challenges society and conservation science. *Proceedings of the National Academy of Sciences of the United States of America*, *113*(22), 6105–6112. https://doi.org/10.1073/pnas.1525003113

Martínez-Ballesta, M. C., Zapata, L., Chalbi, N., & Carvajal, M. (2016). Multiwalled carbon nanotubes enter broccoli cells enhancing growth and water uptake of plants exposed to salinity. *Journal of Nanobiotechnology*, *14*(1). https://doi.org/10.1186/s12951-016-0199-4

Martínez-Ballesta, Mc., Gil-Izquierdo, Á., García-Viguera, C., & Domínguez-Perles, R. (2018). Nanoparticles and controlled delivery for bioactive compounds: Outlining challenges for new "smart-foods" for health. *Foods*, *7*(5). https://doi.org/10.3390/foods7050072

Mastronardi, E., Tsae, P., Zhang, X., Monreal, C., & DeRosa, M. C. (2015). Strategic role of nanotechnology in fertilizers: Potential and limitations. *Nanotechnologies in Food and Agriculture*, 25–67. https://doi.org/10.1007/978-3-319-14024-7_2

Mejias, J. H., Salazar, F., Pérez Amaro, L., Hube, S., Rodriguez, M., & Alfaro, M. (2021). Nanofertilizers: A cutting-edge approach to increase nitrogen use efficiency in grasslands. *Frontiers in Environmental Science*, *9*. https://doi.org/10.3389/fenvs.2021.635114

Midhat, L., Ouazzani, N., Hejjaj, A., Ouhammou, A., & Mandi, L. (2019). Accumulation of heavy metals in metallophytes from three mining sites (Southern Centre Morocco) and evaluation of their phytoremediation potential. *Ecotoxicology and Environmental Safety*, *169*, 150–160. https://doi.org/10.1016/j.ecoenv.2018.11.009

Mirzajani, F., Askari, H., Hamzelou, S., Schober, Y., Römpp, A., Ghassempour, A., & Spengler, B. (2014). Proteomics study of silver nanoparticles toxicity on Oryza sativa L. *Ecotoxicology and Environmental Safety*, *108*, 335–339. https://doi.org/10.1016/j.ecoenv.2014.07.013

Mnyusiwalla, A., Daar, A. S., & Singer, P. A. (2003). "Mind the gap": Science and ethics in nanotechnology. *Nanotechnology*, *14*(3). https://doi.org/10.1088/0957-4484/14/3/201

Mohapatra, M., & Anand, S. (2010). Synthesis and applications of nano-structured iron oxides/hydroxides — A review. *International Journal of Engineering, Science and Technology*, *2*(8). https://doi.org/10.4314/ijest.v2i8.63846

Mondal, R., Dam, P., Chakraborty, J., Paret, M. L., Katı, A., Altuntas, S., Sarkar, R., Ghorai, S., Gangopadhyay, D., Mandal, A. K., & Husen, A. (2022). Potential of nanobiosensor in sustainable agriculture: The state-of-art. *Heliyon*, *8*(12). https://doi.org/10.1016/j.heliyon.2022.e12207

Mubarak, N. M., Sahu, J. N., Abdullah, E. C., Jayakumar, N. S., & Ganesan, P. (2016). Microwave-assisted synthesis of multi-walled carbon nanotubes for enhanced removal of Zn(II) from wastewater. *Research on Chemical Intermediates*, *42*(4), 3257–3281. https://doi.org/10.1007/s11164-015-2209-9

Nair, P. M. G., & Chung, I. M. (2015). Study on the correlation between copper oxide nanoparticles induced growth suppression and enhanced lignification in Indian mustard (Brassica juncea L.). *Ecotoxicology and Environmental Safety*, *113*, 302–313. https://doi.org/10.1016/j.ecoenv.2014.12.013

Nair, R., Varghese, S. H., Nair, B. G., Maekawa, T., Yoshida, Y., & Kumar, D. S. (2010). Nanoparticulate material delivery to plants. *Plant Science*, *179*(3), 154–163. https://doi.org/10.1016/j.plantsci.2010.04.012

Nassar, J. M., Khan, S. M., Villalva, D. R., Nour, M. M., Almuslem, A. S., & Hussain, M. M. (2018). Compliant plant wearables for localized microclimate and plant growth monitoring. *NPJ Flexible Electronics*, *2*(1). https://doi.org/10.1038/s41528-018-0039-8

Nassar, N. N. (2010). Rapid removal and recovery of Pb(II) from wastewater by magnetic nanoadsorbents. *Journal of Hazardous Materials*, *184*(1–3), 538–546. https://doi.org/10.1016/j.jhazmat.2010.08.069

Nazari, M., Ghasemi, N., Maddah, H., & Motlagh, M. M. (2014). Synthesis and characterization of maghemite nanopowders by chemical precipitation method. *Journal of Nanostructure in Chemistry*, *4*(2). https://doi.org/10.1007/s40097-014-0099-9

Ngo, Q. B., Dao, T. H., Nguyen, H. C., Tran, X. T., Van Nguyen, T., Khuu, T. D., & Huynh, T. H. (2014). Effects of nanocrystalline powders (Fe, Co and Cu) on the germination, growth, crop yield and product quality of soybean (Vietnamese species DT-51). *Advances in Natural Sciences: Nanoscience and Nanotechnology*, *5*(1). https://doi.org/10.1088/2043-6262/5/1/015016

Nile, S. H., Thiruvengadam, M., Wang, Y., Samynathan, R., Shariati, M. A., Rebezov, M., Nile, A., Sun, M., Venkidasamy, B., Xiao, J., & Kai, G. (2022). Nano-priming as emerging seed priming technology for sustainable agriculture—recent developments and future perspectives. *Journal of Nanobiotechnology*, *20*(1). https://doi.org/10.1186/s12951-022-01423-8

Nongbet, A., Mishra, A. K., Mohanta, Y. K., Mahanta, S., Ray, M. K., Khan, M., Baek, K.-H., & Chakrabartty, I. (2022). Nanofertilizers: A smart and sustainable attribute to modern agriculture. *Plants*, *11*(19): 2587.

O'Reilly, S. E., & Hochella, M. F. (2003). Lead sorption efficiencies of natural and synthetic Mn and Fe-oxides. *Geochimica et Cosmochimica Acta*, *67*(23), 4471–4487. https://doi.org/10.1016/S0016-7037(03)00413-7

Ojuederie, O. B., & Babalola, O. O. (2017). Microbial and plant-assisted bioremediation of heavy metal polluted environments: A review. *International Journal of Environmental Research and Public Health*, *14*(12). https://doi.org/10.3390/ijerph14121504

Omasa, K., Hosoi, F., & Konishi, A. (2007). 3D lidar imaging for detecting and understanding plant responses and canopy structure. *Journal of Experimental Botany*, *58*(4), 881–898. https://doi.org/10.1093/jxb/erl142

Oren, S., Ceylan, H., Schnable, P. S., & Dong, L. (2017). High-resolution patterning and transferring of graphene-based nanomaterials onto tape toward roll-to-roll production of tape-based wearable sensors. *Advanced Materials Technologies*, *2*(12). https://doi.org/10.1002/admt.201700223

Parisi, C., Vigani, M., & Rodríguez-Cerezo, E. (2015). Agricultural nanotechnologies: What are the current possibilities? *Nano Today*, *10*(2), 124–127. https://doi.org/10.1016/j.nantod.2014.09.009

Patil, S. S., Shedbalkar, U. U., Truskewycz, A., Chopade, B. A., & Ball, A. S. (2016). Nanoparticles for environmental clean-up: A review of potential risks and emerging solutions. *Environmental Technology and Innovation*, *5*, 10–21. https://doi.org/10.1016/j.eti.2015.11.001

Post, J. E. (1999). Manganese oxide minerals: Crystal structures and economic and environmental significance. *Proceedings of the National Academy of Sciences of the United States of America*, *96*(7), 3447–3454. https://doi.org/10.1073/pnas.96.7.3447

Prasad, R., Bhattacharyya, A., & Nguyen, Q. D. (2017). Nanotechnology in sustainable agriculture: Recent developments, challenges, and perspectives. *Frontiers in Microbiology*, *8*(JUN). https://doi.org/10.3389/fmicb.2017.01014

Pretty, J., & Bharucha, Z. P. (2014). Sustainable intensification in agricultural systems. *Annals of Botany*, *114*(8), 1571–1596. https://doi.org/10.1093/aob/mcu205

Qados, A., & Moftah, A. (2015). Influence of silicon and nano-silicon on germination, growth and yield of Faba Bean (Vicia faba L.) under salt stress conditions. *American Journal of Experimental Agriculture*, *5*(6), 509–524. https://doi.org/10.9734/ajea/2015/14109

Qi, M., Liu, Y., & Li, T. (2013). Nano-TiO2 improve the photosynthesis of tomato leaves under mild heat stress. *Biological Trace Element Research*, *156*(1–3), 323–328. https://doi.org/10.1007/s12011-013-9833-2

Raffa, C. M., Chiampo, F., & Shanthakumar, S. (2021). Remediation of metal/metalloid-polluted soils: A short review. *Applied Sciences (Switzerland)*, *11*(9). https://doi.org/10.3390/app11094134

Rai, G. K., Bhat, B. A., Mushtaq, M., Tariq, L., Rai, P. K., Basu, U., Dar, A. A., Islam, S. T., Dar, T. U. H., & Bhat, J. A. (2021). Insights into decontamination of soils by phytoremediation: A detailed account on heavy metal toxicity and mitigation strategies. *Physiologia Plantarum*, *173*(1), 287–304. https://doi.org/10.1111/ppl.13433

Rajput, V. D., Minkina, T. M., Behal, A., Sushkova, S. N., Mandzhieva, S., Singh, R., Gorovtsov, A., Tsitsuashvili, V. S., Purvis, W. O., Ghazaryan, K. A., & Movsesyan, H. S. (2018). Effects of zinc-oxide nanoparticles on soil, plants, animals and soil organisms: A review. *Environmental Nanotechnology, Monitoring and Management*, *9*, 76–84. https://doi.org/10.1016/j.enmm.2017.12.006

Rajput, V. D., Minkina, T., Kumari, A., Harish, Singh, V. K., Verma, K. K., Mandzhieva, S., Sushkova, S., Srivastava, S., & Keswani, C. (2021). Coping with the challenges of abiotic stress in plants: New dimensions in the field application of nanoparticles. *Plants*, *10*(6). https://doi.org/10.3390/plants10061221

Rajput, V. D., Minkina, T., Kumari, A., Shende, S. S., Ranjan, A., Faizan, M., Barakvov, A., Gromovik, A., Gorbunova, N., Rajput, P., Singh, A., Khabirov, I., Nazarenko, O., Sushkova, S., & Kızılkaya, R. (2022a). A review on nanobioremediation approaches for restoration of contaminated soil. *Eurasian Journal of Soil Science*, *11*(1), 43–60. https://doi.org/10.18393/EJSS.990605

Rajput, V. D., Minkina, T., Upadhyay, S. K., Kumari, A., Ranjan, A., Mandzhieva, S., Sushkova, S., Singh, R. K., & Verma, K. K. (2022b). Nanotechnology in the restoration of polluted soil. *Nanomaterials*, *12*(5): 769.

Raliya, R., Saharan, V., Dimkpa, C., & Biswas, P. (2018). Nanofertilizer for precision and sustainable agriculture: Current state and future perspectives. *Journal of Agricultural and Food Chemistry*, *66*(26), 6487–6503. https://doi.org/10.1021/acs.jafc.7b02178

Rascio, N., & Navari-Izzo, F. (2011). Heavy metal hyperaccumulating plants: How and why do they do it? And what makes them so interesting? *Plant Science*, *180*(2), 169–181. https://doi.org/10.1016/j.plantsci.2010.08.016

Rasmussen, L. D., Sørensen, S. J., Turner, R. R., & Barkay, T. (2000). Application of a mer-lux biosensor for estimating bioavailable mercury in soil. *Soil Biology and Biochemistry*, *32*(5), 639–646. https://doi.org/10.1016/S0038-0717(99)00190-X

Rastogi, A., Tripathi, D. K., Yadav, S., Chauhan, D. K., Živčák, M., Ghorbanpour, M., El-Sheery, N. I., & Brestic, M. (2019). Application of silicon nanoparticles in agriculture. *3 Biotech*, *9*(3). https://doi.org/10.1007/s13205-019-1626-7

Rizwan, M., Ali, S., Zia ur Rehman, M., Adrees, M., Arshad, M., Qayyum, M. F., Ali, L., Hussain, A., Chatha, S. A. S., & Imran, M. (2019). Alleviation of cadmium accumulation in maize (Zea mays L.) by foliar spray of zinc oxide nanoparticles and biochar to contaminated soil. *Environmental Pollution*, *248*, 358–367. https://doi.org/10.1016/j.envpol.2019.02.031

Rossi, L., Zhang, W., Lombardini, L., & Ma, X. (2016). The impact of cerium oxide nanoparticles on the salt stress responses of Brassica napus L. *Environmental Pollution*, *219*, 28–36. https://doi.org/10.1016/j.envpol.2016.09.060

Saeid Hojjat, S. (2019). Effect of interaction between Ag nanoparticles and salinity on germination stages of Lathyrus Sativus l. *Open Access Journal of Environmental and Soil Sciences*, *2*(2). https://doi.org/10.32474/oajess.2019.02.000132

Savvas, D., Giotis, D., Chatzieustratiou, E., Bakea, M., & Patakioutas, G. (2009). Silicon supply in soilless cultivations of zucchini alleviates stress induced by salinity and powdery mildew infections. *Environmental and Experimental Botany*, *65*(1), 11–17. https://doi.org/10.1016/j.envexpbot.2008.07.004

Seghatoleslami, M., Feizi, H., Mousavi, G., & Berahmand, A. (2015). Effect of magnetic field and silver nanoparticles on yield and water use efficiency of Carum copticum under water stress conditions. *Polish Journal of Chemical Technology*, *17*(1), 110–114. https://doi.org/10.1515/pjct-2015-0016

Sekhon, B. S. (2013). Metalloid compounds as drugs. *Research in Pharmaceutical Sciences*, *8*(3), 145.

Shafqat, W., Jaskani, M. J., Maqbool, R., Chattha, W. S., Ali, Z., Naqvi, S. A., Haider, M. S., Khan, I. A., & Vincent, C. I. (2021). Heat shock protein and aquaporin expression enhance water conserving behavior of citrus under water deficits and high temperature conditions. *Environmental and Experimental Botany*, *181*. https://doi.org/10.1016/j.envexpbot.2020.104270

Sharma, P., Yadav, D., & Gaur, R. K. (2022). Bioinformatics in agriculture: Next generation sequencing era. *Bioinformatics in Agriculture: Next Generation Sequencing Era*, 1–677. https://doi.org/10.1016/C2020-0-03112-9

Sheikh Mohamed, M., & Sakthi Kumar, D. (2016). Methods of using nanoparticles. *Plant Nanotechnology: Principles and Practices*, 65–93. https://doi.org/10.1007/978-3-319-42154-4_4

Shojaei, T. R., Salleh, M. A. M., Tabatabaei, M., Mobli, H., Aghbashlo, M., Rashid, S. A., & Tan, T. (2019). Applications of nanotechnology and carbon nanoparticles in agriculture. *Synthesis, Technology and Applications of Carbon Nanomaterials*, 247–277. https://doi.org/10.1016/B978-0-12-815757-2.00011-5

Shukla, P. K., Misra, P., & Kole, C. (2016). Uptake, translocation, accumulation, transformation, and generational transmission of nanoparticles in plants. *Plant Nanotechnology: Principles and Practices*, 183–218. https://doi.org/10.1007/978-3-319-42154-4_8

Siddiqui, M. H., Mohammad, F., Khan, M. M. A., & Al-Whaibi, M. H. (2012). Cumulative effect of nitrogen and sulphur on Brassica juncea L. genotypes under NaCl stress. *Protoplasma*, *249*(1), 139–153. https://doi.org/10.1007/s00709-011-0273-6

Singh, A., Tiwari, S., Pandey, J., Lata, C., & Singh, I. K. (2021). Role of nanoparticles in crop improvement and abiotic stress management. *Journal of Biotechnology*, *337*, 57–70. https://doi.org/10.1016/j.jbiotec.2021.06.022

Singh, Y., & Saxena, M. K. (2022). Insights into the recent advances in nano-bioremediation of pesticides from the contaminated soil. *Frontiers in Microbiology*, *13*. https://doi.org/10.3389/fmicb.2022.982611

Sintubin, L., De Gusseme, B., Van Der Meeren, P., Pycke, B. F. G., Verstraete, W., & Boon, N. (2011). The antibacterial activity of biogenic silver and its mode of action. *Applied Microbiology and Biotechnology*, *91*(1), 153–162. https://doi.org/10.1007/s00253-011-3225-3

Siontorou, C. G., & Georgopoulos, K. N. (2016). A biosensor platform for soil management: The case of nitrites. *Journal of Cleaner Production*, *111*, 133–142. https://doi.org/10.1016/j.jclepro.2015.07.038

Soares, C., Branco-Neves, S., de Sousa, A., Azenha, M., Cunha, A., Pereira, R., & Fidalgo, F. (2018). SiO2 nanomaterial as a tool to improve Hordeum vulgare L. tolerance to nano-NiO stress. *Science of the Total Environment*, *622–623*, 517–525. https://doi.org/10.1016/j.scitotenv.2017.12.002

Song, Y., Jiang, M., Zhang, H., & Li, R. (2021). Zinc oxide nanoparticles alleviate chilling stress in rice (Oryza sativa l.) by regulating antioxidative system and chilling response transcription factors. *Molecules*, *26*(8). https://doi.org/10.3390/molecules26082196

Songa, E. A., & Okonkwo, J. O. (2016). Recent approaches to improving selectivity and sensitivity of enzyme-based biosensors for organophosphorus pesticides: A review. *Talanta*, *155*, 289–304. https://doi.org/10.1016/j.talanta.2016.04.046

Soundy, J., & Day, D. (2017). Selection of DNA aptamers specific for live Pseudomonas aeruginosa. *PLoS ONE*, *12*(9). https://doi.org/10.1371/journal.pone.0185385

Struik, P. C., & Kuyper, T. W. (2017). Sustainable intensification in agriculture: the richer shade of green. A review. *Agronomy for Sustainable Development*, *37*(5). https://doi.org/10.1007/s13593-017-0445-7

Suman, J., Uhlik, O., Viktorova, J., & Macek, T. (2018). Phytoextraction of heavy metals: A promising tool for clean-up of polluted environment? *Frontiers in Plant Science*, *871*. https://doi.org/10.3389/fpls.2018.01476

Sun, T. Y., Bornhöft, N. A., Hungerbühler, K., & Nowack, B. (2016). Dynamic probabilistic modeling of environmental emissions of engineered nanomaterials. *Environmental Science and Technology*, *50*(9), 4701–4711. https://doi.org/10.1021/acs.est.5b05828

Surya, S. G., Yuvaraja, S., Varrla, E., Baghini, M. S., Palaparthy, V. S., & Salama, K. N. (2020). An in-field integrated capacitive sensor for rapid detection and quantification of soil moisture. *Sensors and Actuators, B: Chemical*, *321*. https://doi.org/10.1016/j.snb.2020.128542

Taghipour, M., & Jalali, M. (2015). Effect of clay minerals and nanoparticles on chromium fractionation in soil contaminated with leather factory waste. *Journal of Hazardous Materials*, *297*, 127–133.

Talan, A., Mishra, A., Eremin, S. A., Narang, J., Kumar, A., & Gandhi, S. (2018). Ultrasensitive electrochemical immuno-sensing platform based on gold nanoparticles triggering chlorpyrifos detection in fruits and vegetables. *Biosensors and Bioelectronics*, *105*, 14–21. https://doi.org/10.1016/j.bios.2018.01.013

Thrall, P. H., Oakeshott, J. G., Fitt, G., Southerton, S., Burdon, J. J., Sheppard, A., Russell, R. J., Zalucki, M., Heino, M., & Ford Denison, R. (2011). Evolution in agriculture: The application of evolutionary approaches to the management of biotic interactions in agro-ecosystems. *Evolutionary Applications*, *4*(2), 200–215. https://doi.org/10.1111/j.1752-4571.2010.00179.x

Tilman, D., Cassman, K. G., Matson, P. A., Naylor, R., & Polasky, S. (2002). Agricultural sustainability and intensive production practices. *Nature*, *418*(6898), 671–677. https://doi.org/10.1038/nature01014

Tiwari, D. K., Dasgupta-Schubert, N., Villaseñor Cendejas, L. M., Villegas, J., Carreto Montoya, L., & Borjas García, S. E. (2014). Interfacing carbon nanotubes (CNT) with plants: Enhancement of growth, water and ionic nutrient uptake in maize (Zea mays) and implications for nanoagriculture. *Applied Nanoscience (Switzerland)*, *4*(5), 577–591. https://doi.org/10.1007/s13204-013-0236-7

Tosco, T., Petrangeli Papini, M., Cruz Viggi, C., & Sethi, R. (2014). Nanoscale zerovalent iron particles for groundwater remediation: A review. *Journal of Cleaner Production*, *77*, 10–21. https://doi.org/10.1016/j.jclepro.2013.12.026

Tripathi, D., Singh, M., & Pandey-Rai, S. (2022). Crosstalk of nanoparticles and phytohormones regulate plant growth and metabolism under abiotic and biotic stress. *Plant Stress*, *6*. https://doi.org/10.1016/j.stress.2022.100107

Upadhyaya, H., Dutta, B. K., & Panda, S. K. (2018). Impact of zinc on dehydration and rehydration responses in tea. *Biologia Plantarum*, *62*(2), 395–399. https://doi.org/10.1007/s10535-017-0758-z

Usman, K., Al-Ghouti, M. A., & Abu-Dieyeh, M. H. (2019). The assessment of cadmium, chromium, copper, and nickel tolerance and bioaccumulation by shrub plant Tetraena qataranse. *Scientific Reports*, *9*(1). https://doi.org/10.1038/s41598-019-42029-9

Usman, M., Farooq, M., Wakeel, A., Nawaz, A., Cheema, S. A., Rehman, H. ur, Ashraf, I., & Sanaullah, M. (2020). Nanotechnology in agriculture: Current status, challenges and future opportunities. *Science of the Total Environment*, *721*. https://doi.org/10.1016/j.scitotenv.2020.137778

Vaseghi, A., Safaie, N., Bakhshinejad, B., Mohsenifar, A., & Sadeghizadeh, M. (2013). Detection of Pseudomonas syringae pathovars by thiol-linked DNA-Gold nanoparticle probes. *Sensors and Actuators, B: Chemical*, *181*, 644–651. https://doi.org/10.1016/j.snb.2013.02.018

Verma, M. L., Sukriti, Dhanya, B. S., Saini, R., Das, A., & Varma, R. S. (2022). Synthesis and application of graphene-based sensors in biology: a review. *Environmental Chemistry Letters*, *20*(3), 2189–2212. https://doi.org/10.1007/s10311-022-01404-1

Wang, H., Liu, J., Xu, X., Huang, Q., Chen, S., Yang, P., Chen, S., & Song, Y. (2016). Fully-automated high-throughput NMR system for screening of haploid kernels of maize (corn) by measurement of oil content. *PLoS ONE*, *11*(7), e0159444. https://doi.org/10.1371/journal.pone.0159444

Wang, S., Fu, Y., Zheng, S., Xu, Y., & Sun, Y. (2022). Phytotoxicity and accumulation of copper-based nanoparticles in brassica under Cadmium stress. *Nanomaterials*, *12*(9). https://doi.org/10.3390/nano12091497

Watanabe, J., Tani, Y., Chang, J., Miyata, N., Naitou, H., & Seyama, H. (2013). As(III) oxidation kinetics of biogenic manganese oxides formed by Acremonium strictum strain KR21-2. *Chemical Geology*, *347*, 227–232. https://doi.org/10.1016/j.chemgeo.2013.03.012

Waychunas, G. A., Kim, C. S., & Banfield, J. F. (2005). Nanoparticulate iron oxide minerals in soils and sediments: Unique properties and contaminant scavenging mechanisms. *Journal of Nanoparticle Research*, *7*(4–5), 409–433. https://doi.org/10.1007/s11051-005-6931-x

Wu, X., Hu, J., Wu, F., Zhang, X., Wang, B., Yang, Y., Shen, G., Liu, J., Tao, S., & Wang, X. (2021). Application of TiO2 nanoparticles to reduce bioaccumulation of arsenic in rice seedlings (Oryza sativa L.): A mechanistic study. *Journal of Hazardous Materials*, *405*. https://doi.org/10.1016/j.jhazmat.2020.124047

Xiao, M., Fu, Q., Shen, H., Chen, Y., Xiao, W., Yan, D., Tang, X., Zhong, Z., & Tang, Y. (2018). A turn-on competitive immunochromatographic strips integrated with quantum dots and gold nano-stars for cadmium ion detection. *Talanta*, *178*, 644–649. https://doi.org/10.1016/j.talanta.2017.10.002

Yang, K. Y., Doxey, S., McLean, J. E., Britt, D., Watson, A., Al Qassy, D., Jacobson, A., & Anderson, A. J. (2018). Remodeling of root morphology by CuO and ZnO nanoparticles: Effects on drought tolerance for plants colonized by a beneficial pseudomonad. *Botany*, *96*(3), 175–186. https://doi.org/10.1139/cjb-2017-0124

Ye, Y., Medina-Velo, I. A., Cota-Ruiz, K., Moreno-Olivas, F., & Gardea-Torresdey, J. L. (2019). Can abiotic stresses in plants be alleviated by manganese nanoparticles or compounds? *Ecotoxicology and Environmental Safety*, *184*. https://doi.org/10.1016/j.ecoenv.2019.109671

Yin, L., Colman, B. P., McGill, B. M., Wright, J. P., & Bernhardt, E. S. (2012). Effects of silver nanoparticle exposure on germination and early growth of eleven wetland plants. *PLoS One*, *7*(10). https://doi.org/10.1371/journal.pone.0047674

Zhang, M. Y., Wang, Y., Zhao, D. Y., & Pan, G. (2010). Immobilization of arsenic in soils by stabilized nanoscale zero-valent iron, iron sulfide (FeS), and magnetite (Fe3O4) particles. *Chinese Science Bulletin*, *55*(4), 365–372. https://doi.org/10.1007/s11434-009-0703-4

Zhang, R., & Somasundaran, P. (2006). Advances in adsorption of surfactants and their mixtures at solid/solution interfaces. *Advances in Colloid and Interface Science*, *123–126*(SPEC. ISS.), 213–229. https://doi.org/10.1016/j.cis.2006.07.004

Zhao, F., He, J., Li, X., Bai, Y., Ying, Y., & Ping, J. (2020). Smart plant-wearable biosensor for in-situ pesticide analysis. *Biosensors and Bioelectronics, 170.* https://doi.org/10.1016/j.bios.2020.112636

Zheng, L., Hong, F., Lu, S., & Liu, C. (2005). Effect of nano-TiO2 on strength of naturally aged seeds and growth of spinach. *Biological Trace Element Research, 104*(1), 83–91. https://doi.org/10.1385/bter:104:1:083

Zulfiqar, F., Navarro, M., Ashraf, M., Akram, N. A., & Munné-Bosch, S. (2019). Nanofertilizer use for sustainable agriculture: Advantages and limitations. *Plant Science, 289.* https://doi.org/10.1016/j.plantsci.2019.110270

12 Nanotechnology-Enabled Food Packaging and Its Antimicrobial Potential

Subhankar Das, Manjula Ishwara Kalyani, Yugal Kishore Mohanta, and Majid Sharifi-Rad
Corresponding author: Subhankar Das

12.1 INTRODUCTION

Food packaging is a vital component in delivering food in a hygienic way and preserving its nutritional value for a longer period of time. Food spoilage is caused by the cumulative actions of various environmental factors as well as microbial contamination [1]. Moreover, the rise in food-spoiling microorganisms has motivated researchers to develop efficient packaging to protect the food and its nutritional components [2]. Nanotechnology has thereby been proven to be useful in food packaging because it prevents the growth of microorganisms that cause food spoilage and further increases the shelf life of the packed food. Richard Feynman first proposed the idea of nanotechnology in 1959, while the term nanotechnology was coined by Norio Taniguchi in 1974. Nanoparticles are defined as particles having a size between 1 and 100 nanometers. Nanoparticles, due to their small size and large surface area, provide unique characteristic properties that can be explored in various applications [3,4].

Compared to traditional food packaging, the newer version of food packaging has a number of improved functions, such as better antimicrobial properties, antibiofilm activity, and improved biodegradability and biocompatibility. Modern food packaging also involves better barrier properties that allow permeability to essential molecules such as oxygen, moisture, and carbon dioxide in a controlled manner. Furthermore from postharvest to market, food at various stages comes into contact with different contaminants, whereas efficient food packaging protects the food against contaminations. The use of nanoparticles in food packaging not only prevents spoilage but also prevents the development of any pathogenic microbes that might cause serious illness to consumers. The integration of nanoparticles into food packaging prevents food from spoiling prematurely and thereby increases the shelf life of packaged food products [5–8,13].

Proper packaging will also prevent the wastage of food, subsequently reducing the impact on the environment. Due to food spoilage, food ends up in landfills that release harmful and toxic greenhouse gases into the environment and contribute to global warming [9]. The antimicrobial compounds can be broadly classified into inorganic, organic, and natural materials. The organic materials consist of halogenated compounds, phenols, quaternary ammonium salts, and organic acids, as well as natural materials that include chitosan and chitin. On the other hand, inorganic materials include metal and metallic oxide nanoparticles such as silver, gold, nickel oxide, zinc oxide, titanium oxide, and calcium oxide. These materials are excellent in combating a broad range of biocidal activities against various food-borne pathogens [10–14]. The antimicrobial properties of the nanoparticles are greatly dependent on their small size, high reactivity, and large surface

 DOI: 10.1201/9781003389408-12

area, which enable the nanoparticles to penetrate the bacterial cell. However, the exact mechanisms by which the nanoparticles exhibit the antimicrobial properties are still under investigation [15].

Nanoparticles can be integrated into food packaging materials of both biodegradable and nonbiodegradable types. However, biodegradable food packaging has recently become popular due to being environmentally friendly, less hazardous, and easily degradable. In contrast, nonbiodegradable polymers are persistent in nature, do not degrade easily, and cause environmental pollution, including air pollution, upon burning [16,8].

This chapter delves into the different nanoparticles that have been added to food packaging, their antimicrobial effects against pathogens and microorganisms that cause food to spoilage, and the various mechanisms exhibited by the nanoparticles embedded in food packaging against food-borne pathogens. We will also briefly explore the toxicity associated with nanoparticles integrated into food packaging.

12.2 DEVELOPMENT OF DIFFERENT FOOD PACKAGING AND INCORPORATION OF NANOTECHNOLOGY

In recent years, there has been a growing desire among a diverse range of individuals for leading a healthy lifestyle and for an intake of food that is high in nutrients. Because of this, there is an urge to develop improved and novel forms of food packaging that can retain the nutritional contents and at the same time protect the packed food. The primary function of implementing nanotechnology as an antimicrobial agent in food packaging is to combat the growth of microorganisms while also enhancing the functions of traditional food packaging. These functions include the extension of the product's shelf life, the preservation of its quality, freshness, and nutritional value, and the guarantee of its safety. However, the polymeric matrix used to embed the desired nanoparticles also varies and is composed of bio-based, fossil fuel-based materials. However, recently, emphasis has been given to the biodegradable polymeric matrices for their easy degradability and nontoxic and ecofriendly properties, as shown in Table 12.1. Nanoparticles also have the ability to improve the physio-mechanical properties of the packaging, which include enhanced stress tolerance and water vapor permeability. The polymeric matrix that is used to insert the appropriate nanoparticles might vary (as shown in Figure 12.1), and its components can be based on biodegradable or fossil fuel-based materials. However, there has been a shift in focus towards biodegradable polymeric matrices due to the fact that these matrices degrade quickly, are nontoxic, and are favorable to the environment [17,18].

12.2.1 Nanoparticles Enabled with Starch-Based Food Packaging

Starch, which occurs naturally and is a widely available polymer, is mainly found in plants such as cereals, tubers of various plant roots, and fruits. The plants, through the photosynthesis process using carbon dioxide, produce glucose, which later, through polymerization, produces starch. Starch is mainly composed of linear and branched polysaccharides. The appealing nature of the starch polymer in the food packaging sector is due to its easy degradability, low cost, ecofriendliness, and ease of availability. However, the integration of nanoparticles into the starch-based polymeric matrix further influences the physio-mechanical properties of the polymer.

A study conducted by Ni et al. in 2018 reported the fabrication of starch-based flexible coatings for the purpose of food packaging. The induction of zinc oxide nanoparticles (ZnONPs) was capped using carboxymethyl cellulose (CMC), which facilitated the miscibility between the starch and the ZnONPs. This led to an excellent improvement in the film, enhancing its hydrophobicity as well as its thermal properties. It showed antimicrobial activity against *E. coli*, and the migration of the nanoparticles was also reported to be low. With pure starch, the disadvantage is that the water vapor transmittance rate is high, whereas with the inclusion of ZnONPs, the transmittance rate decreased.

TABLE 12.1
List of nanoparticle-integrated antimicrobial polymers for food packaging applications

Type of nanomaterial	Polymeric matrix used	Other additives	Antimicrobial activity	Reference
Zinc oxide nanoparticles	Ethylcellulose/gelatin nanofibers	—	*Escherichia coli*, *Staphylococcus aureus*	[7]
TiO_2 nanoparticles	Soluble soy bean polysaccharide	Curcumin	*P. aeruginosa*, *Staphylococcus aureus*	[13]
Silver nanoparticles	Corn starch	—	Bacterial strain: *Escherichia coli*, *Staphylococcus aureus*, *Salmonella spp.* Fungal strains: *Penicillium spp.*	[24]
AgNPs, ZnONPs oxide, and CuONPs	Starch	—	*S. aureus*, *E. coli*	[26]
ZnO nanoparticles	Chitosan	—	*Escherichia coli*, *Bacillus subtilis*	[28]
Copper nanoparticles	Chitosan	—	*Staphylococcus aureus*, *Salmonella enterica*	[32]
ZnONP-loaded gallic acid	Chitosan	—	*B. subtilis*, *E. coli*	[33]
TiO_2 nanoparticles	Chitosan	—	*Escherichia coli*, *Salmonella*, *Staphylococcus aureus*, *Listeria monocytogenes*	[34]
Zinc oxide nanoparticles	Chicken skin gelatin/tapioca starch	—	*Escherichia coli*, *Staphylococcus aureus*	[46]
Titanium (Ti)-doped CuO nanoparticles	Gelatin	—	*Staphylococcus aureus*, *Escherichia coli*	[81]
Nanocellulose; starch-capped AgNPs	Polyvinyl alcohol (PVA)	—	*Staphylococcus aureus* (MRSA), *Escherichia coli* (DH5-alpha)	[82]
Zinc oxide nanoparticles	Low-density polyethylene (LDPE)	—	*Escherichia coli*	[83]
Silver nanoparticles	Low-density polyethylene (LDPE)	—	*Escherichia coli*, *Salmonella typhi*, *Bacillus cereus*, *Pseudomonas aeruginosa*	[84]
Copper nanoparticles	Low-density polyethylene (LDPE)	—	*Escherichia coli*, *Staphylococcus aureus*	[85]
Silver nanoparticles	Agar	—	*Aeromonas hydrophilla*	[86]
Silver nanocluster	Zein	—	*E. coli*	[87]
Tyrosine-capped (AgNPTyr); tryptophan-capped silver nanoparticles (AgNPTrp)	Agar polymer	—	*Listeria monocytogenes*, *Escherichia coli*	[88]
Silver nanoparticles	Polyvinyl chloride (PVC),	—	Bacterial strain: *Bacillus subtilis*, Fungal strains: *Aspergillus niger*, *Fusarium solani*,	[89]
Aluminum-doped zinc oxide nanoparticles	Polylactide (PLA)	—	*Escherichia coli*	[90]
Iron nanoparticles	Poly(lactic acid) PLA	—	Bacteria strains: *Bacillus subtilis*, *Escherichia coli*, *Staphylococcus epidermidis* Fungal strains: *Geotrichum candidum*, *Rhodotorula rubra*	[91]

Titanium dioxide and silver nanoparticles	Polylactide (PLA)	—	*E. coli*, *Listeria monocytogenes*	[92]
SiO_2 nanoparticles	Poly(butylene adipate-co-terephthalate) (PBAT)	—	*E. coli*, *S. aureus*	[93]
Zinc oxide nanoparticles	Poly (butylene adipate-co-terephthalate) PBAT	—	*Staphylococcus aureus*, *Pseudomonas aeruginosa*, *Bacillus subtilis*	[94]
Silver nanoparticles	Chitosan	Anthocyanin-rich purple corn extract	*E. coli*, *Salmonella*, *S. aureus*, *L. monocytogenes*	[95]
Zinc oxide nanoparticles	Fish protein isolate; fish skin gelatin	Basil leaf essential oil	*L. monocytogenes*, *P. aeruginosa*	[96]
Silver nanoparticles	—	Tragacanth; hydroxypropyl methylcellulose; beeswax	*Bacillus cereus*, *Staphylococcus aureus*, *Streptococcus pneumoniae*, *Listeria monocytogenes*, *Escherichia coli*, *Salmonella typhimurium*, *Pseudomonas aeruginosa*, *Klebsiella pneumoniae*	[97]
Silver nanoparticles	Starch-PVA	—	Bacterial strain: *Listeria innocua*, *Escherichia coli* Fungal strains: *Aspergillus niger*, *Penicillium expansum*	[98]
Silver nanoparticles	poly(3-hydroxybutyrate-co-3-hydroxyvalerate) PHBV	—	*Salmonella enterica*, *Listeria monocytogenes*	[99]
Zinc oxide nanoparticles	Poly(3-hydroxybutyrate-co-3-hydroxyvalerate) (PHBV)	—	*Salmonella enterica*, *Listeria monocytogenes*	[100]
Ti-doped ZnO nanoparticles	Starch-coated polyethylene film	—	*Escherichia coli*	[101]

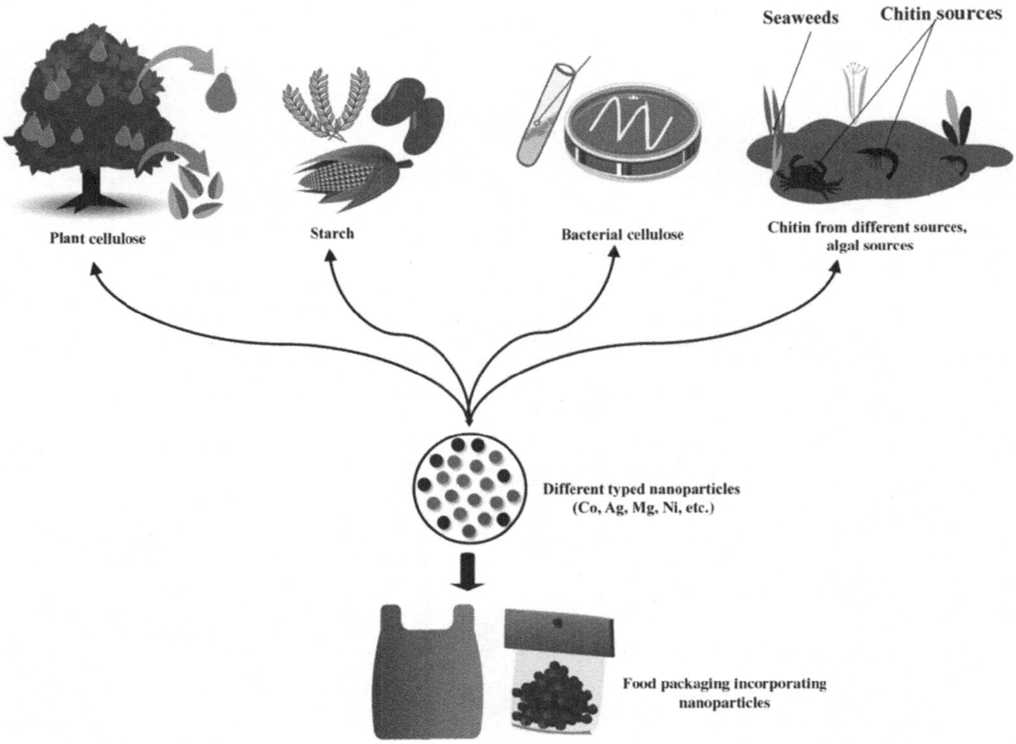

FIGURE 12.1 Various biological sources used to develop biodegradable packaging material embedded with nanoparticles.

Also, the film was impenetrable by water molecules due to the ZnONPs coating that increased the tortuosity, making the film hydrophobic [19–21].

Furthermore, nanoparticles participate in producing synergistic effects with the polymer matrix and its other components to provide enhanced antibacterial activity in addition to other improved film qualities. A study conducted by Babapour et al. in 2021 revealed the synergistic effect of a film that was prepared using potato starch with the addition of ZnONPs and fennel essential oil. The result of the study revealed a decrease in the solubility of the film, its barrier properties against water vapor, as well as its oxygen and tensile strength. However, the study further emphasized the antimicrobial activity against bacterial strains such as *Staphylococcus aureus*, *Escherichia coli*, and the fungal strain *Aspergillus flavus*, thus ensuring the protection of packed food from any microbial spoilage or contamination [22]. In another study, silver nanoparticles (AgNPs) were embedded into starch matrix obtained from the sugar palm plant. AgNPs are a well-known antimicrobial agent, and coating food packaging with AgNPs will enable it to last longer, increasing its shelf life, and will prevent the spoilage of the packed food by any bacterial strains. Incorporating AgNPs into the polymer also reduces the moisture content of the packed food [23].

Since the size and shape of the nanoparticles are among the crucial requirements for their antibacterial action, one approach that uses chemical reducing agents for the synthesis of AgNPs with precisely regulated shape and size. Nevertheless, the chemical approaches are inherently hazardous and harmful to the environment. Instead of using harmful solvents and reducing agents, green nanoparticle synthesis relies on more benign substances to produce uniform nanoparticles of the required size and shape. In a study conducted by Ortega et al. in 2017, the synthesis of AgNPs was carried out by introducing silver nitrate into the gelatinized filmogenic suspension. The prepared nanocomposite films showed high antibacterial activity against *Salmonella* spp., *E. coli*, and *Penicillium* spp. [24,25].

Additionally, it has been proposed that a blend of metallic nanoparticles in low concentration can further boost antibacterial activity of a film below the toxicity limit compared to the antibacterial activity displayed by single metallic nanoparticles. A study by Peighambardoust et al. in 2019 incorporated three different types of nanoparticles, namely silver, copper oxide, and zinc oxide, into starch films. The results revealed that the blend of metallic nanoparticles in the polymeric matrix had developed better surface contact. Including ZnONPs alone resulted in an enhanced polymer because of the powerful interaction of the nanoparticles with the starch chains. Including the three nanoparticles in the starch matrix improved the material's tensile strength and Young's modulus. Synergistic enhancement of antibacterial activity was also observed when all three nanoparticles were present in the starch at low concentrations [26]. Beyond their antibacterial effect against food-borne microbes, titanium nanoparticles can provide a number of other benefits, including the prevention of odors, allergens, and degradation in stored foods. According to a study by Malathi et al. in 2019, titanium oxide can be successfully integrated into a rice starch polymer. The study further emphasized the antibacterial potential of the titanium dioxide nanoparticles on *E. coli* reactive oxygen species (ROS) upon illuminated titanium dioxide nanoparticles by inactivating the microorganisms [27]. Therefore starch, with all its benefits, is a conveniently accessible material that is not hazardous in nature and is able to hold nanoparticles for the purpose of food packaging.

12.2.2 Nanoparticles Incorporated into Chitosan-Based Food Packaging

Chitosan is a type of cationic polysaccharide produced from chitin by the process of alkaline deacetylation. Both chitosan and chitin are linear polysaccharides that contain $\beta(1\rightarrow4)$ linked residues of N-acetyl-2-amino-2-deoxy-D-glucose and 2-amino-2-deoxy-D-glucose. Chitin can be extracted from the exoskeleton of crustaceans, the cuticles of insects, the cell walls of algae, and the mycelium of fungi. On the other hand, chitosan in nature can be found in some fungal species. Moreover, primary amine protonation makes chitosan soluble in acidic aqueous environments. In addition, chitosan is itself an antimicrobial biopolymer. When nanoparticles are placed in a chitosan matrix, they impart antibacterial action in a synergistic manner [28,29,30]. A study conducted by Garavand et al. in 2022 explored the antimicrobial potential of chitosan. Corn starch along with polyvinyl alcohol-based films were produced and incorporated with chitosan nanoparticles. The chitosan nanoparticles revealed their inhibitory action against the bacterial strains *E. coli*, *Salmonella typhimurium*, and *S. aureus*. In addition, gram-positive bacteria strains had more inhibitory effects from chitosan nanoparticles than gram-negative bacterial strains [31].

Food packaging has benefited from the conjugation of several metallic and metal oxide nanoparticles with the chitosan polymeric matrix. As described in a study by Cárdenas et al. (2009), where colloidal copper nanoparticles were incorporated with the film produced by chitosan. The composite film prepared from copper oxide nanoparticles (CuONPs) and chitosan demonstrated high barrier properties against UV radiation and low oxygen as well as water vapor permeability. Furthermore, the composite film showed enhanced antimicrobial activity against the *S. aureus* and *Salmonella enterica serovar Typhimurium* bacterial strains and have revealed the disintegration of the cell wall of these bacterial strains. The stress on the bacterial cell increased due to the perforation in the cell wall and as a result of the internal turgor, leading to the release of internal components of the cells. In addition, there have been observations of blebs in the cell wall of the bacterial strain, suggesting stress on the cell [32]. In a similar study reported by Yadav et al. in 2020, chitosan-based zinc nanoparticle films were prepared and loaded with gallic acid. ZnONPs, a well-known metal oxide nanoparticle with a broad range of antibacterial activity, are widely used as a safe ingredient in food, drug, and cosmetic products and have been classified as Generally Recognized as Safe

(GRAS). Due to the presence of ZnO in the chitosan film, it showed enhanced antibacterial activity against *E. coli* and *B. subtilis*. The antibacterial action of the film was carried by the ZnONPs and the gallic acid. The gallic acid and the Zn^{2+} ions combined their actions to target the negatively charged outer cell wall, leading to the release of intracellular components and ultimately the death of the bacterial cell [33].

Titanium dioxide is another excellent metallic oxide nanoparticle that has an important role in the food packaging sector. TiO has the ability to block UV radiation, biocompatible, hydrophilic properties, ethylene scavenging, and antimicrobial activities. In a study reported by Zhang et al. in 2019, titanium dioxide nanoparticles were incorporated into the chitosan filled with plum peel extracts, which are a rich source of anthocyanin. The prepared food packaging films showed antibacterial activity against food-borne pathogens that included *E. coli*, *S. aureus*, *Salmonella*, and *Listeria monocytogenes*. The combination of synergistic actions of chitosan, TiO, and plum peel extracts resulted in the highest antibacterial activity when compared with only chitosan film, chitosan TiO conjugate film, and chitosan/plum peel extract conjugate film [34].

12.2.3 Nanoparticles Incorporated into Cellulose-Based Food Packaging

Cellulose, which is the most inexhaustible biopolymer present on the planet, is a major component of the cell wall in many plant, algal, bacterial, fungal, and sea organisms like tunicates and oomycete species. Cellulose, due to its intriguing structure and properties, is used in a number of products that include textiles, foods, and pharmaceuticals. There has been a recent shift towards the development and use of cellulose-based composite materials because of their potential for easy availability, easy degradation, environmental friendliness, and low cost. Nanoparticles integrated into cellulose-based food packaging have given the food industry a way to overcome problems with conventional food packaging. CMC is an anionic water-soluble cellulose derivative also known as cellulose gum or mucoadhesive. In a study reported by Ebrahimi et al. in 2019, CMC-based nanobiocomposite films were made using a blend of nanoparticles that included AgNPs, CuONPs, and ZnONPs. The study also showed that the nanobiocomposite lacked antibacterial properties until the nanoparticles were introduced, which clearly showed the antimicrobial roles of the AgNPs, ZnONPs, and CuONPs. Both *S. aureus* and *E. coli* were inhibited by nanobiocomposite films comprised of Ag and ZnO nanoparticles, whereas CuONPs inhibited the growth of *E. coli* only. However, the AgNPs showed the highest activity against bacterial strains [35,36,14].

Ethyl cellulose, also an ethyl ether of cellulose, has the advantage of being nonreactive due to its nonionic nature, and is therefore extensively used in the pharmaceutical industry as a coating agent, binder, drug transporter, and stabilizer. In a study, ethyl cellulose was blended with gelatin to produce nanofibers with ZnONPs embedded into it to achieve better functionality in the composite polymer. The addition of ZnONPs revealed that the antibacterial effects on bacterial strains such as *E. coli* and *S. aureus* were dose-dependent [7,57]. In another study, AgNPs were incorporated into CMC as well as cellulose nanocrystals (CNC). The biocomposite film coated with AgNPs showed antibacterial activity against *E. coli* and *S. aureus*. However, the effect of the antibacterial activity on *S. aureus* was higher than that on *E. coli* due to their differences in cell wall morphology. Since the gram-negative bacterial strains have a thick phospholipid bilayer as well as a lipopolysaccharide layer, these layers create obstacles for the AgNPs to enter the cytoplasm of the bacterial cell and, to some extent, impact the antibacterial activity of the nanoparticles. Also, the mobilization of the AgNPs was reduced due to the presence of CNC, which lowers their release [37,54].

Bacterial strains can also make what is considered to be the purest form of cellulose. The food sector is one of the prime users of this high-purity cellulose [38]. In a study reported by Atta et al. in 2021, bacterial cellulose-based food packaging was developed with embedded AgNPs. In addition, the nanocomposite film exhibited antibacterial activity against bacterial strains including *S. aureus*,

Pseudomonas aeruginosa, and *E. coli*, as well as antifungal activity against strains like *Candida albicans* and *Trichosporon spp*. The results showed nanocomposite films provided significant protection to the texture and prolonged the storage period of tomato and orange. In terms of toxicity, the AgNPs were tested against NIH-3T3 fibroblasts, which revealed that the nanoparticles were nontoxic and biocompatible [39].

Thereby, cellulose provides a suitable base for the induction of nanoparticles for the purpose of food packaging.

12.2.4 Nanoparticles Incorporated into Gelatin-Based Food Packaging

Gelatin is a natural biopolymer that is obtained from collagen by the hydrolysis process. The protein gelatin has a wide range of applications due to its biodegradable, widely available, cost-efficient, and biocompatible nature. Gelatin is a polyampholyte due to the fact that it possesses both cationic and anionic groups. In a study reported by Sooch et al. in 2021, gelatin-based food packaging polymeric film was embedded with titanium-doped CuONPs. Because of the inclusion of these nanoparticles in gelatin-based active food packaging, the shelf life of *Solanum lycopersicum* (tomatoes) was increased to up to 18 days at a temperature of 40°C±3°C. The nanoreinforced film also exhibited antifungal action against *Botrytis cineria* and *Penicillium expansum*, and antibacterial activity against *S. aureus* and *E. coli*, which are typical food pathogens [80,81]. In a similar study reported by Lee et al. in 2020, chicken skin gelatin along with tapioca starch biocomposite and ZnONPs were blended to create a bionanocomposite. The incorporation of 5% ZnO into the composite film revealed a high zone of inhibition against *S. aureus* and *E. coli*. The antibacterial activity of the nanoparticles increased proportionally with their concentration. Direct interaction between nanoparticles and bacteria can destroy the bacterial outer membrane, and entry of the Zn ions further leads to ROS. The production of intracellular ROS inhibits all metabolic processes inside the bacterium [46].

A study was published where a bionanocomposite film was prepared using gelatin and cellulose nanofibers with different amounts of zinc nanoparticles and/or selenium nanoparticles. The nanocomposite film was tested against food-borne bacterial strains, *P. fluorescens*, *E. coli* O157:H7, *S. aureus*, and *L. monocytogenes*. However, an interesting observation was reported in the study whereby the gram-negative *E. coli* was more susceptible to ZnONPs compared to the gram-positive *S. aureus*, which is contrary to the reports in many research publications that the gram-positive bacterial strains are more susceptible than the gram-negative due to their morphological differences. However, one reason that has been given in the study for *E. coli* being more susceptible to ZnO than *S. aureus* is the presence of a large quantity of carotenoid pigments, which are antioxidants that protect the bacterial strain from oxidative stress caused by the ZnONPs [47,37].

Thus, gelatin provides a great matrix to enable the incorporation of metal and metal oxide nanoparticles for the specific application like food packaging.

12.2.5 Nanoparticles Incorporated into Pectin-Based Food Packaging

Pectin is a hydrocolloid derived from plants, and was initially isolated from tamarind by Louis Nicolas Vauquelin in the year 1790. In 1825, Henri Braconnot was the first to use the name pectin, derived from the Greek *pektikos*, which means solidified or congealed. Pectin is a linear macromolecule made up of (1-4)-linked D-galacturonic acid. Pectin is used for a wide variety of purposes, including the production of food items, medicine, and edible films. Pectin, therefore, has been used to make ecofriendly composite films that are both biodegradable and biocompatible, in addition to having a high gelling capacity. Due to its hydrophobic nature, however, pectin exhibits negative effects in terms of food packaging. Therefore, various nanomaterials, including nanoclay, metal oxide nanoparticles, and nanocellulose, that improve the interaction between the polymeric matrix and its components have been used. This addition of nanomaterials also enhances the antimicrobial and antioxidant properties of the film [104,105,106].

A study by Sani et al. reported in 2021 used a composite film prepared using starch from potatoes and pectin from apple peels, enabled with microencapsulated essential oil from *Zataria multiflora* and further embedded with zirconium oxide (ZrO_2) nanoparticles. In addition to its antibacterial potential, the biocomposite film extended the shelf life of packed meat. The antibacterial capabilities of the film were significantly enhanced by the incorporation of encapsulated essential oil and ZrO_2 nanoparticles [107]. In another study, functional polymeric films were developed based on pectin embedded with curcumin and sulfur nanoparticles. The polymeric composite showed enhanced thermal as well as water contact angles, and the composite film revealed antibacterial activity against food-borne pathogenic bacterial strains like *E. coli* and *L. monocytogenes*. Furthermore, the polymeric film had color-changing properties upon any change in pH [108].

12.2.6 Nanoparticles Incorporated into Protein-Based Food Packaging

Proteins are a kind of heteropolymer that is thermoplastic. They are made up of different polar and nonpolar alpha amino acids. Because amino acids are capable of forming a diverse range of intermolecular connections, these molecules can participate in a wide variety of interactions. Interestingly, protein-based films or coatings, owing to their chain-to-chain interaction, have been shown to be especially potent in blocking oxygen, even at low relative humidity (RH), in spite of the fact that protein-based films have a hydrophilic characteristic. There are a number of proteins, for instance, gelatin, collagen, soy protein, gluten, maize zein, keratin, and casein. Moreover, proteins have extremely interesting qualities of polymers like flexibility, shear and tensile strength, and elasticity [109,110,111].

The findings of a study by Huang et al. (2021) revealed that they had created biodegradable food packaging film containing soy protein isolate with the addition of mangosteen peel extract and ZnO nanoparticles. The biodegradable film provided further evidence of the strong hydrogen bonding between soy protein isolate, mangosteen peel extract, and ZnONPs. The composite film also had exceptional UV-blocking, antioxidant, and antibacterial activities against bacterial strains such as *E. coli* and *S. aureus* [112].

Another study reported the preparation of flexible polymeric films comprising caprolactam casein and ZnONPs coatings. When ZnONPs were combined with caprolactam casein, they exhibited significantly greater antibacterial activity against *S. aureus* and *E. coli*, with an incubation time of two days. Additionally, ZnONPs had a lesser effect on gram-negative bacteria like *E. coli* than they did on gram-positive bacteria like *S. aureus*. One of the likely causes of this sensitivity was the composition of the cell wall. The peptidoglycan multilayers that make up *S. aureus* are porous, and the Zn^+ ions or reactive species can easily be breached, making the bacteria more vulnerable to cell lysis. *E. coli*, in contrast, has been reported to be less susceptible to the assault of reactive species because its cell wall is relatively thin and predominantly composed of peptidoglycan, the outer layer consisting of lipopolysaccharide, lipoprotein, and phospholipids [113].

12.2.7 Other Polymeric Composites Embedded with Nanotechnology for Food Packaging

Food packaging can also include other polymeric matrices to incorporate nanomaterials for extended functionality. The food packaging polymers range from fossil-based polymers that include polyvinyl chloride (PVC), polyethylene (PE), polyethylene terephthalate (PET), and polypropylene (PP), to biodegradable and bio-based polymers such as starch, chitosan, and cellulose. Yet, increased environmental concerns have led to the use of biodegradable and bio-based polymers due to their nontoxic and quickly biodegradable nature [40,41,42]. However, the introduction of nanotechnology into such polymers has improved their physio-mechanical characteristics as well as their capacity to function as an antibacterial film against a variety of food-borne microorganisms [43,44].

A study conducted by Francis et al. in 2022 used a blend of polymeric matrices comprised of PVA, starch, and glycerol. The polymeric blend was further integrated with ZnONPs and CuONPs. The water barriers and hydrophobicity properties of CuO and ZnO blended nanocomposites were improved. It was suggested that the antibacterial action of the composite polymer was caused by the release of Cu^{2+} and Zn^{2+} from the polymeric nanocomposite film. The nanocomposites and polymeric films exhibited antifungal activity in strains such as *Aspergillus niger*, *Aspergillus calidoustus*, and *Penicillium chrysogenum*, and antibacterial activity involving strains like *Klebsiella pneumoniae* and *S. aureus*, thus improving the shelf life of the packed food [45]. In a similar study, PVA was blended with nanocellulose and AgNPs, as reported by Sarwar et al. in 2017. The study found that the nanocomposite film had strong antimicrobial activity against *S. aureus* (MRSA) and *E. coli* (DH5-alpha) whereby the antibacterial activity increased with the increase in the amount of nanoparticles in the PVA. The cytotoxic evaluation in the study revealed the nontoxic nature of the film, which had no cytotoxic effects against HepG2, with viable cells more than 90% [82].

In a study reported by Mathew et al. in 2019, PVA was blended with rice starch that had been fabricated with AgNPs. However, the combination of the components led to enhanced mechanical and optical characteristics with lower sensitivity to water. The polymer film exhibited excellent barrier properties against bacterial strains such as gram-positive *S. aureus* and gram-negative *S. typhimurium*. In gram-positive bacteria, AgNPs had more difficulty in penetrating the cell due to the complexity of the cell wall structure and the thick peptidoglycan layer, while in gram-negative bacteria, AgNPs had easier penetration in the cell due to the thin peptidoglycan layer. However, it was emphasized in the study that antimicrobial compounds are also known to interfere with the degradation process, which slows the breakdown of the polymer composite containing AgNPs [102].

The drawbacks associated with PVA films that cause them to be fragile have been mitigated with fillers in several attempts, as in a study by Rahmadiawan et al. in 2022 where nanocellulose from white ginger was added to reinforce the PVA and zinc nanoparticle biocomposite film. The film exhibited improved thermal resistance, and its antimicrobial potential increased against both bacterial and fungi strains including *S. aureus*, *B. subtilis*, *E. coli*, *P. aeruginosa*, and *C. albicans* [103].

Polyethylene is a popularly used material in the food packaging sector due to its chemically resistive nature, high strength, cost-effectiveness, and transparency. Low-density polyethylene (LDPE) has been reported to have been integrated with nanoparticles in an attempt to improve their antibacterial characteristics. A study by Chougule et al. in 2020 reported the integration of AgNPs synthesized biologically using leaf extract of *Moringa oleifera* into the LDPE polymer. The composite polymer exhibited strong antibacterial activity against *E. coli*. The AgNPs' tiny size enabled their penetration into bacterial cells, where they attacked sulfur- and phosphorus-containing molecules, leading to cell death, and subsequently inactivated the key enzymes, causing disruption in the respiratory chain. Furthermore, the initiation of oxidative stress within the bacterial cell triggered a cascade of events that ultimately resulted in cell death [83,84].

A literature survey revealed that polymers such as agar, zein, and soybean have also been used in conjugation with different types of nanoparticles in developing antimicrobial food packaging. In a study by Salarbashi et al. in 2020, soybean polysaccharide conjugated with curcumin was used along with TiO_2 nanoparticles, since curcumin is a natural polyphenolic compound that has the properties of antimicrobial activity. The TiO_2 with curcumin synergistically showed strong antimicrobial activity against *E. coli* and *P. aeruginosa* [13,86,87,88].

Biodegradable thermoplastic such as poly(lactic acid) PLA is a linear aliphatic polyester and can be produced by bacterial fermentation supplemented with nutrients from corn starch or sugar beets. Since they are renewable and easily degradable, they were used in studies to incorporate nanoparticles such as aluminum-doped zinc oxide, titanium dioxide, and iron nanoparticles. The studies revealed enhanced antimicrobial activity against both gram-positive and gram-negative bacterial strains [90,91,92].

In an investigation by Lu et al. in 2020, the PLA matrix was incorporated with clove essential oil that had been immobilized in mesoporous silica nanoparticles. The prepared film successfully inhibited the growth of *E. coli* and *S. aureus*. With the increase in the amount of mesoporous silica nanoparticles, there was an increase in the antibacterial activity of the film. This study's findings also revealed that the mesoporous silica carrier has the potential to significantly reduce clove essential oil release [114]. Since PLA is biodegradable and has favorable physiochemical qualities, it has found widespread application in food packaging; yet, it has a very limited capacity to shield against UV radiation and serve as an antioxidant. In an investigation by Boarino et al. in 2022, PLA was blended with lignin nanoparticles to overcome all the limitations and further improve the PLA film properties [115]. Similarly, in a study reported by Ramji et al. in 2020, a biocomposite was prepared which consisted of nickel oxide supported on silica nanoparticles embedded in the PLA matrix. According to the findings presented in the study, the antibacterial capability was effective against gram-positive (*L. monocytogenes*) and gram-negative (*Salmonella*). It is widely recognized that materials containing metal oxides endow polymers with outstanding antibacterial properties, and it is also common knowledge that reinforcing the polymer matrix with nanoscale of these elements significantly increases film antibacterial properties [118].

Poly(butylene adipate-co-terephthalate) (PBAT) is a biodegradable large-scale polymer mainly used to manufacture films, fibers, and packaging materials due to its flexible, elastic, and water-resistant characteristics. PBAT is also used as a hardener for biodegradable and bio-based polymeric materials like PLA and PHBV. PBAT is reported to have been incorporated with zinc oxide and nanosilica that enabled the composite film with enhanced antibacterial activity [93,94]. Arumugam et al. in 2022 described the development of a PBAT biocomposite film employing the solution-casting technique, which includes the incorporation of natural grape seed essential oil (GEO) as a plasticizer and silica nanoparticles. Incorporating silica nanoparticles into the PBAT/GEO blend had several benefits, such as boosting the flim's tensile strength, thermal stability, and antibacterial properties. In addition, the inclusion of silica nanoparticles, which induced antibacterial activity against *E. coli* and *S. aureus*, further contributed to lengthening the period of time, so that perishable foods remained fresh after being packaged. According to literature studies, the surface area of silica nanoparticles was shown to be directly related to the antibacterial efficacy. Greater antibacterial activity was seen with smaller particles with a greater surface-to-volume ratio [116]. According to the findings of research published in 2016 by Venkatesan et al., a nanocomposite film that was created utilizing PBAT and ZnONPs showed improved antibacterial activity. In addition, nanocomposites composed of PBAT and ZnONPs demonstrated antibacterial efficacy against *E. coli* and *S. aureus*. Research has also shown that an increase in the percentage of zinc oxide in the load causes a widening of the zone of inhibition, which goes from 1 to 10 weight percent [117].

Hence, there is a requirement for expanded research and development in the food packaging domain, specifically in the area of comprehending the interaction between polymers and the nanoparticles embedded inside them for the purpose of food packaging.

12.3 MECHANISMS BEHIND THE ANTIMICROBIAL ACTIVITY OF FOOD PACKAGING

Nanoparticles are well known to possess excellent antimicrobial properties that enable them to perform a critical role in improving food packaging as well as act as a biocidal agent for various microorganisms. However, nanomaterials used in food packaging consist of metal and metal oxide nanoparticles. Metal and metal oxide nanoparticles, due to their unique physiochemical properties, which include shape, size, surface area, and particle charge, exhibit enhanced antimicrobial characteristics. Therefore, nanoparticles embedded in food packaging are a great addition to deter the growth of any undesirable microorganisms [7].

However, studies suggest that to be a potent antimicrobial agent, nanoparticles have to have a specific size and shape along with other physiochemical properties. Furthermore, with smaller size, the surface-to-volume ratio of NPs increases, thereby ensuring proper contact with the microbial outer membrane. This is the likely reason for their superior antibacterial activity in comparison to their counterparts. Also, smaller NPs dissolve more rapidly because their surface area to volume ratio is higher [25,48,49,50]. As stated in a study conducted by Darvishi et al. in 2019 where ZnONPs were synthesized using the leaf extract of *Juglans regia*, the spherical ZnONPs had a higher surface-to-volume ratio and a smaller size, making them more effective in penetrating bacteria cell walls than the flower-shaped ZnONPs [51]. The results of this study emphasize the significance of nanoparticle shape in demonstrating its antibacterial activities.

In a separate study by Ramji et al. in 2020, nanocomposites made of nickel oxide were supported by silica nanoparticles that were embedded in a PLA matrix. The silica nanoparticles showed antibacterial activity against both gram-negative (*Salmonella*) and gram-positive bacterial strains (*L. monocytogenes*). It is well known that antibacterial compounds are often more effective against gram-negative bacteria than against gram-positive bacteria. In contrast to gram-positive bacteria, whose cell walls are composed of many layers of thick peptidoglycan, gram-negative bacteria have only a single layer, making them more susceptible to these antibacterial compounds. Although it was noted in the study that the specific mechanisms of the silica nanoparticles were unknown to the researchers, several plausible explanations have been offered in relation to the antibacterial activity demonstrated by nickel oxide supported on silica nanoparticles. These possibilities include the integrity of the bacterial cell wall being destroyed when it comes into direct contact with it, and the silica nanoparticles' ion release interfering with DNA replication and protein synthesis. Since heavy metals are more attracted to protein molecules, they attach to the functional groups of proteins and cause denaturation of the proteins, resulting in bacteria death. In another study, the preparation of a flexible polymeric film composed of caprolactam casein and ZnONP coatings was reported. In this study, it was determined that ZnONPs had a lower impact on *E. coli* (gram-negative) bacterial strains than on *S. aureus* (gram-positive) bacteria strains. However, in this study, one possible reason for that outcome was provided: the peptidoglycan multilayers that compose *S. aureus* are porous, making the bacteria more susceptible to cell lysis when exposed to Zn^{+} ions or reactive species [113,118].

A study conducted by Pal et al. in 2007 revealed that the ability to bind with molecules increases with a decrease in size. It was also shown that truncated triangle-shaped AgNPs displayed much higher antibacterial activity than either spherical or rod-shaped AgNPs. Nanoparticles' charge is also a key factor in determining the degree to which they will exhibit antibacterial properties. Furthermore, the efficiency of the nanoparticles as a bactericidal agent varies depending on the structural and morphological construction of a bacterial strain. The peptidoglycan layer of gram-positive bacterial strains is thicker than that of gram-negative bacteria. Hence, the destructive effect upon physical contact of nanoparticles with the outer covering of gram-negative bacteria is more detrimental than for gram-positive bacterial strains. In addition, the affinity of nanoparticles for gram-negative bacterial strains is high due to the presence of lipopolysaccharide in their outer membrane, which is negatively charged and thereby attracts and facilitates the uptake of positive ions from the nanoparticles, leading to damage inside the bacterial cell. Although both gram-positive and gram-negative bacterial cells are negatively charged, they demonstrate an affinity for positively charged nanoparticles. The surface charge of the synthesized nanoparticles can be modified using a variety of capping agents. When the positive charge of nanoparticles increases, the electrostatic barriers are lowered, and as a result, the antibacterial activity is enhanced. Furthermore, the positively charged nanoparticles disrupt the electron transport chain in the bacterial cell [52,53,50,82].

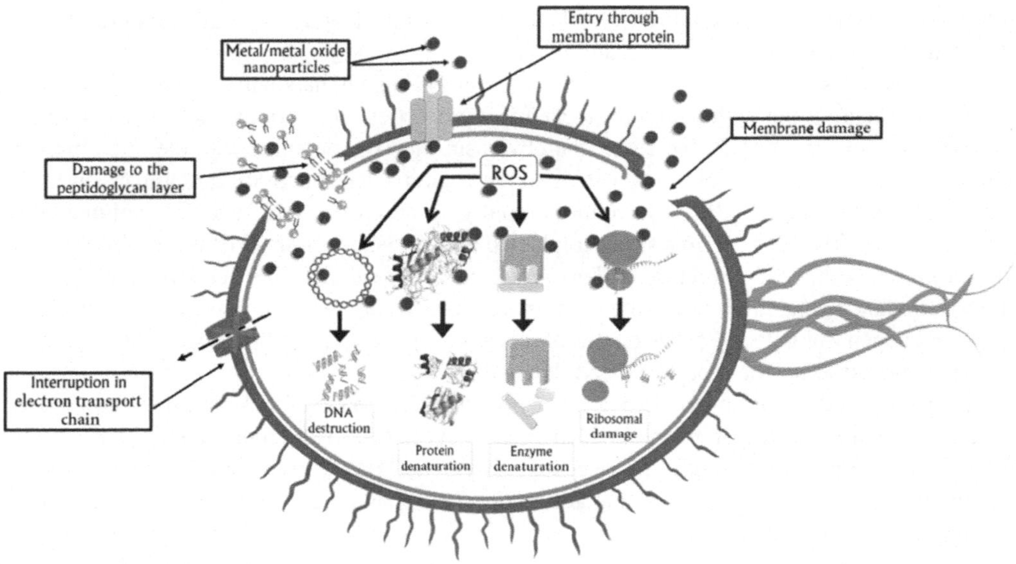

FIGURE 12.2 Antimicrobial mechanisms by metal/metal oxide nanoparticles on a bacterial cell.

Prominent mechanisms (as shown in Figure 12.2) are exhibited by the nanoparticles to act against microorganisms. The entry of the nanoparticles into the microbial cell can be attributed to three basic mechanisms: the destruction of the outer membrane, transport through a membrane such as the β-barrel porin channel (nonspecific diffusion), and specific uptake [55]. The bacterial outer membrane is equipped with ion channels known as porins. The basic functions of these ion channels are communication between cells and intake of ions and nutrients by cells. Nanoparticles have the ability to cross the porin channel to enter into the intracellular environment. The passive diffusion of the nanoparticles through the porin changes the structure of the porin channel and facilitates the entry of small nanoparticles into the intracellular region of the bacterial cell [50,56]. A study by Jin et al. in 2016 reported the entry of the ultrasmall silver nanoclusters into bacterial strain *E. coli* DH 5α cells by distorting the outer membrane. The study further revealed that in the bacterial strain *E. coli* DSM 4230 the entry of the ultrasmall nanoclusters took place across the porin channel [58]. A transmission electron microscope study by Sinha et al. in 2011 reported the destruction of the cell membrane of the bacterial cell and the presence of ZnO nanoparticles accumulated inside cytoplasm. The four most probable mechanisms related to the antimicrobial potential of ZnONPs were also proposed: Zn cations being released from ZnONPs; interaction between ZnONPs and microbial cells; cell membrane penetration by ZnONPs; and generation of reactive oxygen species (ROS) [59,83]. Once within the bacterial cell, the nanoparticles trigger a series of events, including the production of ROS, which damage internal components such as DNA and mitochondria and increase the cell's oxidative stress. In addition, Zn^{2+} ions disrupt cell enzyme systems by dislodging Mg^{2+} ions, which have an essential role in bacterial enzyme activity [60,61,94].

Li et al. in 2009 studied the mechanisms demonstrated by AgNPs on *E. coli*. The results showed that the AgNPs induced damage to the permeability of the bacterial outer membrane, which caused the leakage of reducing sugars and proteins and promoted the inactivation of the respiratory chain dehydrogenases [62]. In terms of protein interaction, as Meghana et al. showed in 2012 with CuONPs, the toxicity inside the bacterial cell was elevated through the introduction of free radicals by the CuONPs. The CuONPs caused damage to the intracellular proteins of the bacterial cell [63]. Similarly, in terms of fungus strains, the outer membrane is first damaged, followed by disruption in membrane potential and release of intracellular components [64]. A study conducted by Li et al. in 2010 reported that AgNPs caused condensation of the DNA, halting its replication activity and further cell reproduction. In addition, the AgNPs were suggested to have the ability to impact the proteins. Recombinase A protein, an important protein, is required to repair any damage to the plasmid and chromosomal DNA as part of the SOS response, and its damage affects translation, transcription, and post-translational activity. AgNPs have the potential to affect the activity of enzymes such as respiratory chain dehydrogenase [65].

Therefore, food packaging embedded with nanoparticles is active in protecting the food from any unwanted microbial strains that could spoil it. As a result, they are beginning to be used in modern food packaging materials.

12.4 NEGATIVE IMPACTS OF NANOPARTICLES IN FOOD PACKAGING

The implementation of nanotechnology in food packaging has greatly influenced the food sector, given the importance of getting rid of various food-spoiling microorganisms and extending the shelf life, aroma, and nutritional content of the packed food. However, these nanoparticles could also induce toxicity in humans, plants, animals, and other living organisms upon exposure. Among the metallic nanoparticles, AgNPs, which are known for their wide range of applications, including antimicrobial activity, are highly toxic in their ionic (Ag^+) state. Silver nanoparticles are able to release Ag^+ ions when they come into contact with water. As the study suggests, AgNPs cause toxicity due to their negative impact on mitochondrial activity, which in turn affects cellular energy metabolism. In addition, the AgNPs' toxicity leads to the generation of ROS and induces oxidative stress. A study conducted by Kim et al. in 2009 evaluated the toxicity of AgNPs to human hepatoma cells, and found that Ag^+ ions were released less when the cells were treated with AgNPs but had the tendency to release Ag^+ ions. Another interesting finding is that the cytotoxicity of AgNPs cannot be correlated with the release of Ag^+ ions, but AgNPs are themselves able to induce intrinsic toxicity, comparable to copper nanoparticles, which are more cytotoxic than copper ions. Furthermore, such nontoxicity is exhibited across all varieties of metallic and metal oxide nanoparticles [66–69]. In terms of food packaging, migration of nanoparticles into the packed food occurs, and it has therefore become imperative to analyze the patterns of nanoparticle migration into various types of food and their impact on nature and humans. In a study conducted by Echegoyen et al. in 2013, the migration of AgNPs was reported from three different commercially available sources under acidic conditions at a temperature of 40 °C for a duration of 40 days. In addition, the study revealed that the values of nanoparticle migration increased dramatically during heating in the microwave as compared to a conventional oven [70]. Similar findings have been reported in terms of copper nanoparticle migration into acidic foods. In a study, when stimulated by 3% acid at 60 °C for 10 days, the migration of copper from food packaging material was higher than that of AgNPs [66].

Nanoparticles produce considerable toxicity in humans through a number of different routes, including respiratory inhalation, dermal absorption, and gastrointestinal uptake, upon exposure to such nanoparticles. Human exposure can be estimated by taking into account three variables: the rate of NP migration, the frequency at which the food containing the migratory NPs is consumed, and the individual's weight [67, 68, 71].

For instance, a study conducted by Yun et al. in 2014 reported the toxicity exhibited by silicon dioxide (SiO_2) nanoparticles, AgNPs, and iron oxide (Fe_2O_3) nanoparticles, which, on being administered orally to rats, led to several organ toxicity in the rats. However, the effects of SiO_2 and Fe_2O_3 at the highest doses did not show any relevant treatment-related changes. In contrast, the AgNPs caused considerable damage, including increases in white blood cell count and alkaline phosphatase. Moreover, lymphocyte infiltration into the liver as well as portal inflammation were detected in the histopathological analysis, and the kidneys of the rat were reported to be affected by the toxicity of the AgNPs. The study also revealed gastrointestinal absorption of the AgNPs and their later excretion through feces [72]. In similar studies reported by Johari et al. in 2013, the toxicity of AgNPs was investigated on *Oncorhynchus mykiss* fish, commonly known as rainbow trout. The trout were highly sensitive to the colloidal AgNPs in the early stages of life compared to the later stages [73]. An investigation by Yu et al. in 2019 revealed the toxicity effect of titanium dioxide nanoparticle nanocomposites incorporated in the PVA matrix. The study further reported that titanium dioxide nanoparticles were able to penetrate human Caco-2 cells, mostly through epidermal growth factor receptor (EGFR)-associated endocytosis, and trigger the cellular inflammatory response. The epithelial integrity and survival of human colon cells were not compromised by titanium dioxide nanoparticles at low doses (10 and 100 g/mL), although the nanoparticles did change the microvillar organization and raise intracellular-free calcium levels. In contrast, titanium dioxide nanoparticles at concentrations higher than 200 ng/mL significantly reduced the survival of human colon cells by disrupting cell membranes and degrading intracellular components. The investigation concluded that the photoactivation of titanium dioxide nanoparticles enhanced their toxicity to colon cells because they induced greater amounts of ROS, which injure cells more severely [108]. Despite the fact that the effect of nanoparticles on humans remains to be explored, the topic has received relatively little investigation. A study reported by Dabrunz et al. in 2011 explained the toxic impact of nanosized titanium dioxide compared with that of non-nanosized titanium dioxide on *Daphnia magna*. The study further revealed that nanosized titanium dioxide is more toxic to *Daphnia magna* than its counterpart, which is non-nanosized titanium dioxide. The impact further decreased the rate of molting up to 10% compared to the control, which was 100%. In long-term exposure, up to 96 hours, this led to an excessive level of immobility and mortality [74].

Ecotoxicology is another perspective on nanotechnology-enabled food packaging and is also of concern when packaging will be disposed in soil or landfill or will come into contact with water bodies. The biodegradation process of such food packaging material embedded with nanoparticles thus poses a risk for the environment and habitats [17,75,76].

Inadequacies exist in terms of the analysis and characterization of nanoparticles that are present in the materials used for food packaging, and it has become absolutely crucial to appropriately evaluate the risk associated with the exposure and migration of nanoparticles towards the environment, living organisms, and humans.

12.5 CONCERNS RELATED TO THE RISE IN THE NANOPARTICLE RESISTANCE AMONG MICROORGANISMS

Because of the increasing use of nanoparticles in a wide variety of products, there is a risk that microbes will evolve resistance towards nanoparticles. In addition, the use of nanoparticles in food packaging has increased the risk of developing resistance to a variety of nanoparticles. A wide variety of mechanisms, including efflux pump, biofilm formation, enzymatic detoxifications, genetic material manipulation, and electrostatic repulsion, are used by different microbial strains to contribute significantly to nanoparticle resistance. However, the development of resistance against nanoparticles is likely to be slower than the development of resistance by microorganisms against antibiotics. This is because of nanoparticles' ability to target bacterial strains in multiple ways compared to antibiotics. However, there are reports of various bacteria strains having developed resistance to silver, gold, and copper nanoparticles. A study by Li et al. reported in 2019 stated that long-term exposure to sulfided AgNPs can result in phenotypic as well as genotypic alterations in *E. coli* that promote silver tolerance; however, acute exposure of the *E. coli* showed no change [77]. As microbial strains come into contact with the environment, further adaptation may enable them to develop resistance. Food packaging embedded with nanoparticles may also give rise to such resistance upon disposal in open landfills [78].

A study reported by Kotchaplai et al. in 2017 found that the nanoscale zerovalent iron tolerance on bacterial strains is different for long- and short-duration exposure. In extended periods of exposure to the nanoscale zerovalent iron, the *P. putida* F1 bacterial strain was reported to have developed resistance, causing the cells to adapt and permanently alter their phenotype by appearing morphologically as a small colony. It is generally understood that one of the adaptive responses of bacterial cells to environmental stressors is the phenotypic response [79].

Consequently, more research into the environmental effects of nanoparticle-enabled food packaging and its potential consequences upon disposal is warranted. To avoid upsetting the natural order, we also need a system for properly disposing of food packaging that contains nanoparticles.

12.6 CONCLUSIONS AND FUTURE PERSPECTIVES

The implementation of nanotechnology in various fields is enormous, and nanoparticles play a multifarious role in the food packaging sector. With the emergence of various food-borne diseases and food-spoiling microbes, nanoparticles have an important role in the prevention, control, and eradication of such microorganisms. Furthermore, the incorporation of metallic and metal oxide nanoparticles lowers the chances of food wastage due to microbial spoilage. Using nanotechnology in food packaging has many benefits, including extending the product's freshness, increasing its shelf life, and preserving its nutritional value. However, caution should be exercised when using metal and metal oxide nanoparticles in food packaging, as there is a possibility of some of these particles migrating into the packaged food and then entering the body of the consumer, causing toxic effects. The use of metal and metallic oxide nanoparticles necessitates further investigation into the assessment of associated risks.

REFERENCES

(1) Alfei, S., Marengo, B. and Zuccari, G., 2020. Nanotechnology application in food packaging: A plethora of opportunities versus pending risks assessment and public concerns. *Food Research International*, 137, pp.109664.

(2) Jafarizadeh-Malmiri, H., Sayyar, Z., Anarjan, N., Berenjian, A., Jafarizadeh-Malmiri, H., Sayyar, Z., Anarjan, N. and Berenjian, A., 2019. Nanobiotechnology in food packaging. In: *Nanobiotechnology in food: Concepts, applications and perspectives*; Springer: Cham, 69–79.

(3) Durán, N. and Marcato, P.D., 2013. Nanobiotechnology perspectives. Role of nanotechnology in the food industry: A review. *International Journal of Food Science & Technology*, 48(6), pp.1127–1134.

(4) Momin, J.K., Jayakumar, C. and Prajapati, J.B., 2013. Potential of nanotechnology in functional foods. *Emirates Journal of Food and Agriculture*, 25(1), pp.10–19.

(5) Mei, L. and Wang, Q., 2020. Advances in using nanotechnology structuring approaches for improving food packaging. *Annual Review of Food Science and Technology*, 11, pp.339–364.

(6) Sharma, C., Dhiman, R., Rokana, N. and Panwar, H., 2017. Nanotechnology: An untapped resource for food packaging. *Frontiers in Microbiology*, 8, pp.1735.

(7) Liu, Y., Li, Y., Deng, L., Zou, L., Feng, F. and Zhang, H., 2018. Hydrophobic ethylcellulose/gelatin nanofibers containing zinc oxide nanoparticles for antimicrobial packaging. *Journal of Agricultural and Food Chemistry*, 66(36), pp.9498–9506.

(8) Ahmed, S., Sameen, D.E., Lu, R., Li, R., Dai, J., Qin, W. and Liu, Y., 2022. Research progress on antimicrobial materials for food packaging. *Critical Reviews in Food Science and Nutrition*, *62*(11), pp.3088–3102.

(9) Gondal, A.H. and Tayyiba, L., 2022. Prospects of using nanotechnology in agricultural growth, environment and industrial food products. *Reviews in Agricultural Science*, 10, pp.68–81.

(10) Correa, M.G., Martínez, F.B., Vidal, C.P., Streitt, C., Escrig, J. and de Dicastillo, C.L., 2020. Antimicrobial metal-based nanoparticles: A review on their synthesis, types and antimicrobial action. *Beilstein Journal of Nanotechnology*, 11(1), pp.1450–1469.

(11) Ramesh, P., Saravanan, K., Manogar, P., Johnson, J., Vinoth, E. and Mayakannan, M., 2021. Green synthesis and characterization of biocompatible zinc oxide nanoparticles and evaluation of its antibacterial potential. *Sensing and Bio-Sensing Research*, 31, pp.100399.

(12) Hoseinnejad, M., Jafari, S.M. and Katouzian, I., 2018. Inorganic and metal nanoparticles and their antimicrobial activity in food packaging applications. *Critical Reviews in Microbiology*, 44(2), pp.161–181.

(13) Salarbashi, D., Tafaghodi, M. and Heydari-Majd, M., 2020. Fabrication of curcumin-loaded soluble soy bean polysaccharide/TiO_2 nanoparticles bio-nanocomposite for improved antimicrobial activity. *Nanomedicine Journal*, 7(4), pp.85–92.

(14) Ebrahimi, Y., Peighambardoust, S.J., Peighambardoust, S.H. and Karkaj, S.Z., 2019. Development of antibacterial carboxymethyl cellulose-based nanobiocomposite films containing various metallic nanoparticles for food packaging applications. *Journal of Food Science*, 84(9), pp.2537–2548.

(15) Rajeshkumar, S. and Malarkodi, C., 2014. In vitro antibacterial activity and mechanism of silver nanoparticles against foodborne pathogens. *Bioinorganic Chemistry and Applications*, 2014, 581890.

(16) Sharma, D. and Dhanjal, D.S., 2016. Bio-nanotechnology for active food packaging. *Journal of Applied Pharmaceutical Science*, 6(9), pp.220–226.

(17) Souza, V.G.L. and Fernando, A.L., 2016. Nanoparticles in food packaging: Biodegradability and potential migration to food—A review. *Food Packaging and Shelf Life*, 8, pp.63–70.

(18) Espitia, P.J.P., Soares, N.D.F.F., Coimbra, J.S.D.R., de Andrade, N.J., Cruz, R.S. and Medeiros, E.A.A., 2012. Zinc oxide nanoparticles: Synthesis, antimicrobial activity and food packaging applications. *Food and Bioprocess Technology*, 5, pp.1447–1464.

(19) Ogunsona, E., Ojogbo, E. and Mekonnen, T., 2018. Advanced material applications of starch and its derivatives. *European Polymer Journal*, 108, pp.570–581.

(20) Ojogbo, E., Ogunsona, E.O. and Mekonnen, T.H., 2020. Chemical and physical modifications of starch for renewable polymeric materials. *Materials Today Sustainability*, 7, pp.100028.

(21) Ni, S., Zhang, H., Dai, H. and Xiao, H., 2018. Starch-based flexible coating for food packaging paper with exceptional hydrophobicity and antimicrobial activity. *Polymers*, 10(11), pp.1260.

(22) Babapour, H., Jalali, H. and Mohammadi Nafchi, A., 2021. The synergistic effects of zinc oxide nanoparticles and fennel essential oil on physicochemical, mechanical, and antibacterial properties of potato starch films. *Food Science & Nutrition*, 9(7), pp.3893–3905.

(23) Rozilah, A., Jaafar, C.A., Sapuan, S.M., Zainol, I. and Ilyas, R.A., 2020. The effects of silver nanoparticles compositions on the mechanical, physiochemical, antibacterial, and morphology properties of sugar palm starch biocomposites for antibacterial coating. *Polymers*, 12(11), pp. 2605.

(24) Ortega, F., Giannuzzi, L., Arce, V.B. and García, M.A., 2017. Active composite starch films containing green synthetized silver nanoparticles. *Food Hydrocolloids*, 70, pp.152–162.

(25) Shanmuganathan, R., MubarakAli, D., Prabakar, D., Muthukumar, H., Thajuddin, N., Kumar, S.S. and Pugazhendhi, A., 2018. An enhancement of antimicrobial efficacy of biogenic and ceftriaxone-conjugated silver nanoparticles: Green approach. *Environmental Science and Pollution Research*, 25, pp.10362–10370.

(26) Peighambardoust, S.J., Peighambardoust, S.H., Pournasir, N. and Pakdel, P.M., 2019. Properties of active starch-based films incorporating a combination of Ag, ZnO and CuO nanoparticles for potential use in food packaging applications. *Food Packaging and Shelf Life*, 22, pp.100420.

(27) Malathi, A.N. and Singh, A.K., 2019. Antimicrobial activity of rice starch based film reinforced with titanium dioxide (TiO2) nanoparticles. *Agricultural Research Journal*, 56, pp.111.

(28) Priyadarshi, R. and Negi, Y.S., 2017. Effect of varying filler concentration on zinc oxide nanoparticle embedded chitosan films as potential food packaging material. *Journal of Polymers and the Environment*, 25, pp.1087–1098.

(29) Mittal, H., Ray, S.S., Kaith, B.S., Bhatia, J.K., Sharma, J. and Alhassan, S.M., 2018. Recent progress in the structural modification of chitosan for applications in diversified biomedical fields. *European Polymer Journal*, 109, pp.402–434.

(30) Aranaz, I., Alcántara, A.R., Civera, M.C., Arias, C., Elorza, B., Heras Caballero, A. and Acosta, N., 2021. Chitosan: An overview of its properties and applications. *Polymers*, 13(19), pp. 3256.

(31) Garavand, Y., Taheri-Garavand, A., Garavand, F., Shahbazi, F., Khodaei, D. and Cacciotti, I., 2022. Starch-polyvinyl alcohol-based films reinforced with chitosan nanoparticles: Physical, mechanical, structural, thermal and antimicrobial properties. *Applied Sciences*, 12(3), pp. 1111.

(32) Cárdenas, G., Díaz V, J., Meléndrez, M.F., Cruzat C, C. and García Cancino, A., 2009. Colloidal Cu nanoparticles/chitosan composite film obtained by microwave heating for food package applications. *Polymer Bulletin*, 62, pp. 511–524.

(33) Yadav, S., Mehrotra, G.K. and Dutta, P.K., 2021. Chitosan based ZnO nanoparticles loaded gallic-acid films for active food packaging. *Food Chemistry*, 334, pp. 127605.
(34) Zhang, X., Liu, Y., Yong, H., Qin, Y., Liu, J. and Liu, J., 2019. Development of multifunctional food packaging films based on chitosan, TiO2 nanoparticles and anthocyanin-rich black plum peel extract. *Food Hydrocolloids*, 94, pp.80–92.
(35) Shaghaleh, H., Xu, X. and Wang, S., 2018. Current progress in production of biopolymeric materials based on cellulose, cellulose nanofibers, and cellulose derivatives. *RSC Advances*, 8(2), pp.825–842.
(36) Ergun, R., Guo, J. and Huebner-Keese, B. 2016. Cellulose. In: *Encyclopedia of Food and Health*; Oxford University Press: Oxford, pp.694–702.
(37) He, Y., Li, H., Fei, X. and Peng, L., 2021. Carboxymethyl cellulose/cellulose nanocrystals immobilized silver nanoparticles as an effective coating to improve barrier and antibacterial properties of paper for food packaging applications. *Carbohydrate Polymers*, 252, pp.117156.
(38) Lahiri, D., Nag, M., Dutta, B., Dey, A., Sarkar, T., Pati, S., Edinur, H.A., Abdul Kari, Z., Mohd Noor, N.H. and Ray, R.R., 2021. Bacterial cellulose: Production, characterization, and application as antimicrobial agent. *International Journal of Molecular Sciences*, 22(23), pp.12984.
(39) Atta, O.M., Manan, S., Ul-Islam, M., Ahmed, A.A.Q., Ullah, M.W. and Yang, G., 2021. Silver decorated bacterial cellulose nanocomposites as antimicrobial food packaging materials. *ES Food & Agroforestry*, 6, pp.12–26.
(40) Das, S. and Kalyani, M.I., 2022. From trash to treasure: Review on upcycling of fruit and vegetable wastes into starch based bioplastics. *Preparative Biochemistry & Biotechnology*, 53(7), pp.713–727.
(41) Motelica, L., Ficai, D., Ficai, A., Oprea, O.C., Kaya, D.A. and Andronescu, E., 2020. Biodegradable antimicrobial food packaging: Trends and perspectives. *Foods*, 9(10), pp.1438.
(42) Yadav, A., Mangaraj, S., Singh, R., Kumar, N. and Arora, S., 2018. Biopolymers as packaging material in food and allied industry. *International Journal of Chemical Studies*, 6(2),pp.2411–2418.
(43) Eleftheriadou, M., Pyrgiotakis, G. and Demokritou, P., 2017. Nanotechnology to the rescue: Using nano-enabled approaches in microbiological food safety and quality. *Current Opinion in Biotechnology*, 44, pp.87–93.
(44) Ashfaq, A., Khursheed, N., Fatima, S., Anjum, Z. and Younis, K., 2022. Application of nanotechnology in food packaging: Pros and cons. *Journal of Agriculture and Food Research*, 7, pp. 100270.
(45) Francis, D.V., Thaliyakattil, S., Cherian, L., Sood, N. and Gokhale, T., 2022. Metallic nanoparticle integrated ternary polymer blend of PVA/Starch/Glycerol: A promising antimicrobial food packaging material. *Polymers*, 14(7), pp.1379.
(46) Lee, S.W., Said, N.S. and Sarbon, N.M., 2021. The effects of zinc oxide nanoparticles on the physical, mechanical and antimicrobial properties of chicken skin gelatin/tapioca starch composite films in food packaging. *Journal of Food Science and Technology*, 58, pp.4294–4302.
(47) Ahmadi, A., Ahmadi, P., Sani, M.A., Ehsani, A. and Ghanbarzadeh, B., 2021. Functional biocompatible nanocomposite films consisting of selenium and zinc oxide nanoparticles embedded in gelatin/cellulose nanofiber matrices. *International Journal of Biological Macromolecules*, 175, pp.87–97.
(48) Lee, N.Y., Ko, W.C. and Hsueh, P.R., 2019. Nanoparticles in the treatment of infections caused by multidrug-resistant organisms. *Frontiers in Pharmacology*, 10, pp.1153.
(49) Bajpai, V.K., Kamle, M., Shukla, S., Mahato, D.K., Chandra, P., Hwang, S.K., Kumar, P., Huh, Y.S. and Han, Y.K., 2018. Prospects of using nanotechnology for food preservation, safety, and security. *Journal of Food and Drug Analysis*, 26(4), pp.1201–1214.
(50) Slavin, Y.N., Asnis, J., Hńfeli, U.O. and Bach, H., 2017. Metal nanoparticles: Understanding the mechanisms behind antibacterial activity. *Journal of Nanobiotechnology*, 15, pp.1–20.
(51) Darvishi, E., Kahrizi, D. and Arkan, E., 2019. Comparison of different properties of zinc oxide nanoparticles synthesized by the green (using Juglans regia L. leaf extract) and chemical methods. *Journal of Molecular Liquids*, 286, pp.110831.
(52) Pal, S., Tak, Y.K. and Song, J.M., 2007. Does the antibacterial activity of silver nanoparticles depend on the shape of the nanoparticle? A study of the gram-negative bacterium Escherichia coli. *Applied and Environmental Microbiology*, 73(6), pp.1712–1720.
(53) Fan, X., Yahia, L.H. and Sacher, E., 2021. Antimicrobial properties of the Ag, Cu nanoparticle system. *Biology*, 10(2), pp.137.

(54) Dairi, N., Ferfera-Harrar, H., Ramos, M. and Garrigós, M.C., 2019. Cellulose acetate/AgNPs-organoclay and/or thymol nano-biocomposite films with combined antimicrobial/antioxidant properties for active food packaging use. *International Journal of Biological Macromolecules*, 121, pp.508–523.
(55) Białas, N., Sokolova, V., van der Meer, S.B., Knuschke, T., Ruks, T., Klein, K., Westendorf, A.M. and Epple, M., 2022. Bacteria (E. coli) take up ultrasmall gold nanoparticles (2 nm) as shown by different optical microscopic techniques (CLSM, SIM, STORM). *Nano Select*, 3(10), pp.1407–1420.
(56) Agarwal, H., Menon, S., Kumar, S.V. and Rajeshkumar, S., 2018. Mechanistic study on antibacterial action of zinc oxide nanoparticles synthesized using green route. *Chemico-Biological Interactions*, 286, pp.60–70.
(57) Wasilewska, K. and Winnicka, K., 2019. Ethylcellulose–A pharmaceutical excipient with multidirectional application in drug dosage forms development. *Materials*, 12(20), pp.3386.
(58) Jin, J.C., Wu, X.J., Xu, J., Wang, B.B., Jiang, F.L. and Liu, Y., 2017. Ultrasmall silver nanoclusters: Highly efficient antibacterial activity and their mechanisms. *Biomaterials Science*, 5(2), pp.247–257.
(59) Sinha, R., Karan, R., Sinha, A. and Khare, S.K., 2011. Interaction and nanotoxic effect of ZnO and Ag nanoparticles on mesophilic and halophilic bacterial cells. *Bioresource Technology*, 102(2), pp.1516–1520.
(60) Kim, S.H., Lee, H.S., Ryu, D.S., Choi, S.J. and Lee, D.S., 2011. Antibacterial activity of silver-nanoparticles against Staphylococcus aureus and Escherichia coli. *Microbiology and Biotechnology Letters*, 39(1), pp.77–85.
(61) Singh, R., Cheng, S. and Singh, S., 2020. Oxidative stress-mediated genotoxic effect of zinc oxide nanoparticles on Deinococcus radiodurans. *3 Biotech*, 10, pp.1–13.
(62) Li, W.R., Xie, X.B., Shi, Q.S., Zeng, H.Y., Ou-Yang, Y.S. and Chen, Y.B., 2010. Antibacterial activity and mechanism of silver nanoparticles on Escherichia coli. *Applied Microbiology and Biotechnology*, 85, pp.1115–1122.
(63) Meghana, S., Kabra, P., Chakraborty, S. and Padmavathy, N., 2015. Understanding the pathway of antibacterial activity of copper oxide nanoparticles. *RSC Advances*, 5(16), pp.12293–12299.
(64) Kim, K.J., Sung, W.S., Suh, B.K., Moon, S.K., Choi, J.S., Kim, J.G. and Lee, D.G., 2009. Antifungal activity and mode of action of silver nano-particles on Candida albicans. *Biometals,* 22, pp.235–242.
(65) Li, W.R., Xie, X.B., Shi, Q.S., Duan, S.S., Ouyang, Y.S. and Chen, Y.B., 2011. Antibacterial effect of silver nanoparticles on Staphylococcus aureus. *Biometals*, 24, pp.135–141.
(66) Hannon, J.C., Kerry, J.P., Cruz-Romero, M., Azlin-Hasim, S., Morris, M. and Cummins, E., 2016. Human exposure assessment of silver and copper migrating from an antimicrobial nanocoated packaging material into an acidic food simulant. *Food and Chemical Toxicology*, 95, pp.128–136.
(67) Chang, Y.N., Zhang, M., Xia, L., Zhang, J. and Xing, G., 2012. The toxic effects and mechanisms of CuO and ZnO nanoparticles. *Materials*, 5(12), pp.2850–2871.
(68) Kim, S., Choi, J.E., Choi, J., Chung, K.H., Park, K., Yi, J. and Ryu, D.Y., 2009. Oxidative stress-dependent toxicity of silver nanoparticles in human hepatoma cells. *Toxicology in Vitro*, 23(6), pp.1076–1084.
(69) Sardoiwala, M.N., Kaundal, B. and Choudhury, S.R., 2018. Toxic impact of nanomaterials on microbes, plants and animals. *Environmental Chemistry Letters*, 16, pp.147–160.
(70) Echegoyen, Y. and Nerín, C., 2013. Nanoparticle release from nano-silver antimicrobial food containers. *Food and Chemical Toxicology*, 62, pp.16–22.
(71) Chau, C.F., Wu, S.H. and Yen, G.C., 2007. The development of regulations for food nanotechnology. *Trends in Food Science & Technology*, 18(5), pp.269–280.
(72) Yun, J.W., Kim, S.H., You, J.R., Kim, W.H., Jang, J.J., Min, S.K., Kim, H.C., Chung, D.H., Jeong, J., Kang, B.C. and Che, J.H., 2015. Comparative toxicity of silicon dioxide, silver and iron oxide nanoparticles after repeated oral administration to rats. *Journal of Applied Toxicology*, 35(6), pp.681–693.
(73) Johari, S.A., Kalbassi, M.R., Soltani, M. and Yu, I.J., 2013. Toxicity comparison of colloidal silver nanoparticles in various life stages of rainbow trout (Oncorhynchus mykiss). *Iranian Journal of Fisheries Science*, 12(1), pp.76–95.
(74) Dabrunz, A., Duester, L., Prasse, C., Seitz, F., Rosenfeldt, R., Schilde, C., Schaumann, G.E. and Schulz, R., 2011. Biological surface coating and molting inhibition as mechanisms of TiO2 nanoparticle toxicity in Daphnia magna. *PLoS One*, 6(5), pp.e20112.

(75) Cushen, M., Kerry, J., Morris, M., Cruz-Romero, M. and Cummins, E.J.F.C., 2013. Migration and exposure assessment of silver from a PVC nanocomposite. *Food Chemistry*, 139(1-4), pp.389–397.
(76) Primožič, M., Knez, Ž. and Leitgeb, M., 2021. (Bio) Nanotechnology in food science—Food packaging. *Nanomaterials*, 11(2), p.292.
(77) Li, M., Li, J., Sun, J., He, Y., Chen, P. and Zhang, C., 2019. Is sulfidation a true detoxification process for silver nanoparticles?: From the perspective of chronic exposure. *Environmental Science: Nano*, 6(12), pp.3611–3624.
(78) Kotsilkov, S., Ivanov, E. and Vitanov, N.K., 2018. Release of graphene and carbon nanotubes from biodegradable poly (lactic acid) films during degradation and combustion: Risk associated with the end-of-life of nanocomposite food packaging materials. *Materials*, 11(12), pp.2346.
(79) Kotchaplai, P., Khan, E. and Vangnai, A.S., 2017. Membrane alterations in Pseudomonas putida F1 exposed to nanoscale zerovalent iron: Effects of short-term and repetitive nZVI exposure. *Environmental Science & Technology*, 51(14), pp.7804–7813.
(80) Yasmin, R., Shah, M., Khan, S.A. and Ali, R., 2017. Gelatin nanoparticles: A potential candidate for medical applications. *Nanotechnology Reviews*, 6(2), pp.191–207.
(81) Sooch, B.S. and Mann, M.K., 2021. Nanoreinforced biodegradable gelatin based active food packaging film for the enhancement of shelf life of tomatoes (Solanum lycopersicum L.). *Food Control*, 130, pp.108322.
(82) Sarwar, M.S., Niazi, M.B.K., Jahan, Z., Ahmad, T. and Hussain, A., 2018. Preparation and characterization of PVA/nanocellulose/Ag nanocomposite films for antimicrobial food packaging. *Carbohydrate Polymers*, 184, pp.453–464.
(83) Rojas, K., Canales, D., Amigo, N., Montoille, L., Cament, A., Rivas, L.M., Gil-Castell, O., Reyes, P., Ulloa, M.T., Ribes-Greus, A. and Zapata, P.A., 2019. Effective antimicrobial materials based on low-density polyethylene (LDPE) with zinc oxide (ZnO) nanoparticles. *Composites Part B: Engineering*, 172, pp.173–178.
(84) Chougule, S.S., Gurme, S.T., Jadhav, J.P., Dongale, T.D. and Tiwari, A.P., 2021. Low density polyethylene films incorporated with biosynthesised silver nanoparticles using Moringa oleifera plant extract for antimicrobial, food packaging, and photocatalytic degradation applications. *Journal of Plant Biochemistry and Biotechnology*, 30, pp.208–214.
(85) Lomate, G.B., Dandi, B. and Mishra, S., 2018. Development of antimicrobial LDPE/Cu nanocomposite food packaging film for extended shelf life of peda. *Food Packaging and Shelf Life*, 16, pp.211–219.
(86) Basumatary, K., Daimary, P., Das, S.K., Thapa, M., Singh, M., Mukherjee, A. and Kumar, S., 2018. Lagerstroemia speciosa fruit-mediated synthesis of silver nanoparticles and its application as filler in agar based nanocomposite films for antimicrobial food packaging. *Food Packaging and Shelf Life*, 17, pp.99–106.
(87) Mei, L., Teng, Z., Zhu, G., Liu, Y., Zhang, F., Zhang, J., Li, Y., Guan, Y., Luo, Y., Chen, X. and Wang, Q., 2017. Silver nanocluster-embedded zein films as antimicrobial coating materials for food packaging. *ACS Applied Materials & Interfaces*, 9(40), pp.35297–35304.
(88) Shankar, S. and Rhim, J.W., 2015. Amino acid mediated synthesis of silver nanoparticles and preparation of antimicrobial agar/silver nanoparticles composite films. *Carbohydrate Polymers*, 130, pp.353–363.
(89) Braga, L.R., Rangel, E.T., Suarez, P.A.Z. and Machado, F., 2018. Simple synthesis of active films based on PVC incorporated with silver nanoparticles: Evaluation of the thermal, structural and antimicrobial properties. *Food Packaging and Shelf Life*, 15, pp.122–129.
(90) Valerini, D., Tammaro, L., Di Benedetto, F., Vigliotta, G., Capodieci, L., Terzi, R. and Rizzo, A., 2018. Aluminum-doped zinc oxide coatings on polylactic acid films for antimicrobial food packaging. *Thin Solid Films*, 645, pp.187–192.
(91) Ligaj, M., Tichoniuk, M., Cierpiszewski, R. and Foltynowicz, Z., 2020. Efficiency of novel antimicrobial coating based on iron nanoparticles for dairy products' packaging. *Coatings*, 10(2), pp.156.
(92) Li, W., Zhang, C., Chi, H., Li, L., Lan, T., Han, P., Chen, H. and Qin, Y., 2017. Development of antimicrobial packaging film made from poly (lactic acid) incorporating titanium dioxide and silver nanoparticles. *Molecules*, 22(7), pp.1170.
(93) Venkatesan, R. and Rajeswari, N., 2019. Preparation, mechanical and antimicrobial properties of SiO 2/poly (butylene adipate-co-terephthalate) films for active food packaging. *Silicon*, 11, pp.2233–2239.

(94) Seray, M., Skender, A. and Hadj-Hamou, A.S., 2021. Kinetics and mechanisms of Zn 2+ release from antimicrobial food packaging based on poly (butylene adipate-co-terephthalate) and zinc oxide nanoparticles. *Polymer Bulletin*, 78, pp.1021–1040.
(95) Qin, Y., Liu, Y., Yuan, L., Yong, H. and Liu, J., 2019. Preparation and characterization of antioxidant, antimicrobial and pH-sensitive films based on chitosan, silver nanoparticles and purple corn extract. *Food Hydrocolloids*, 96, pp.102–111.
(96) Arfat, Y.A., Benjakul, S., Prodpran, T., Sumpavapol, P. and Songtipya, P., 2014. Properties and antimicrobial activity of fish protein isolate/fish skin gelatin film containing basil leaf essential oil and zinc oxide nanoparticles. *Food Hydrocolloids*, 41, pp.265–273.
(97) Bahrami, A., Mokarram, R.R., Khiabani, M.S., Ghanbarzadeh, B. and Salehi, R., 2019. Physico-mechanical and antimicrobial properties of tragacanth/hydroxypropyl methylcellulose/beeswax edible films reinforced with silver nanoparticles. *International Journal of Biological Macromolecules*, 129, pp.1103–1112.
(98) Cano, A., Cháfer, M., Chiralt, A. and González-Martínez, C., 2016. Development and characterization of active films based on starch-PVA, containing silver nanoparticles. *Food Packaging and Shelf Life*, 10, pp.16–24.
(99) Castro-Mayorga, J.L., Fabra, M.J. and Lagaron, J.M., 2016. Stabilized nanosilver based antimicrobial poly (3-hydroxybutyrate-co-3-hydroxyvalerate) nanocomposites of interest in active food packaging. *Innovative Food Science & Emerging Technologies*, 33, pp.524–533.
(100) Castro-Mayorga, J.L., Fabra, M.J., Pourrahimi, A.M., Olsson, R.T. and Lagaron, J.M., 2017. The impact of zinc oxide particle morphology as an antimicrobial and when incorporated in poly (3-hydroxybutyrate-co-3-hydroxyvalerate) films for food packaging and food contact surfaces applications. *Food and Bioproducts Processing*, 101, pp.32–44.
(101) Tankhiwale, R. and Bajpai, S.K., 2012. Preparation, characterization and antibacterial applications of ZnO-nanoparticles coated polyethylene films for food packaging. *Colloids and Surfaces B: Biointerfaces*, 90, pp.16–20.
(102) Mathew, S., Jayakumar, A., Kumar, V.P., Mathew, J. and Radhakrishnan, E.K., 2019. One-step synthesis of eco-friendly boiled rice starch blended polyvinyl alcohol bionanocomposite films decorated with in situ generated silver nanoparticles for food packaging purpose. *International Journal of Biological Macromolecules*, 139, pp.475–485.
(103) Rahmadiawan, D., Abral, H., Yesa, W.H., Handayani, D., Sandrawati, N., Sugiarti, E., Muslimin, A.N., Sapuan, S.M. and Ilyas, R.A., 2022. White ginger nanocellulose as effective reinforcement and antimicrobial polyvinyl alcohol/ZnO hybrid biocomposite films additive for food packaging applications. *Journal of Composites Science*, 6(10), pp.316.
(104) Chandel, V., Biswas, D., Roy, S., Vaidya, D., Verma, A. and Gupta, A., 2022. Current advancements in pectin: Extraction, properties and multifunctional applications. *Foods*, 11(17), pp.2683.
(105) Malathi, A.N., Santhosh, K.S. and Nidoni, U., 2014. Recent trends of biodegradable polymer: Biodegradable films for food packaging and application of nanotechnology in biodegradable food packaging. *Current Trends in Technology and Science*, 3(2), pp.73–79.
(106) Ezati, P. and Rhim, J.W., 2020. pH-responsive pectin-based multifunctional films incorporated with curcumin and sulfur nanoparticles. *Carbohydrate Polymers*, 230, pp.115638.
(107) Sani, I.K., Geshlaghi, S.P., Pirsa, S. and Asdagh, A., 2021. Composite film based on potato starch/apple peel pectin/ZrO2 nanoparticles/microencapsulated Zataria multiflora essential oil; investigation of physicochemical properties and use in quail meat packaging. *Food Hydrocolloids*, 117, pp.106719.
(108) Yu, Z., Wang, W., Sun, L., Kong, F., Lin, M. and Mustapha, A., 2020. Preparation of cellulose nanofibril/titanium dioxide nanoparticle nanocomposites as fillers for PVA-based packaging and investigation into their intestinal toxicity. *International Journal of Biological Macromolecules*, 156, pp.1174–1182.
(109) Gupta, V., Biswas, D. and Roy, S., 2022. A comprehensive review of biodegradable polymer-based films and coatings and their food packaging applications. *Materials*, 15(17), pp.5899.
(110) Vroman, I. and Tighzert, L., 2009. Biodegradable polymers. *Materials*, 2(2), pp.307–344.
(111) Mangaraj, S., Yadav, A., Bal, L.M., Dash, S.K. and Mahanti, N.K., 2019. Application of biodegradable polymers in food packaging industry: A comprehensive review. *Journal of Packaging Technology and Research*, 3, pp.77–96.

(112) Huang, X., Zhou, X., Dai, Q. and Qin, Z., 2021. Antibacterial, antioxidation, uv-blocking, and biodegradable soy protein isolate food packaging film with mangosteen peel extract and ZnO nanoparticles. *Nanomaterials*, 11(12), pp.3337.
(113) Wang, Y., Ma, J., Xu, Q. and Zhang, J., 2017. Fabrication of antibacterial casein-based ZnO nanocomposite for flexible coatings. *Materials & Design*, 113, pp.240–245.
(114) Lu, W., Cui, R., Zhu, B., Qin, Y., Cheng, G., Li, L. and Yuan, M., 2021. Influence of clove essential oil immobilized in mesoporous silica nanoparticles on the functional properties of poly (lactic acid) biocomposite food packaging film. *Journal of Materials Research and Technology*, 11, pp.1152–1161.
(115) Boarino, A., Schreier, A., Leterrier, Y. and Klok, H.A., 2022. Uniformly dispersed poly (lactic acid)-grafted lignin nanoparticles enhance antioxidant activity and UV-barrier properties of poly (lactic acid) packaging films. *ACS Applied Polymer Materials*, 4(7), pp.4808–4817.
(116) Arumugam, S., Kandasamy, J., Thiyaku, T. and Saxena, P., 2022. Effect of low concentration of SiO2 nanoparticles on grape seed essential oil/PBAT composite films for sustainable food packaging application. *Sustainability*, 14(13), pp.8073.
(117) Venkatesan, R. and Rajeswari, N., 2017. ZnO/PBAT nanocomposite films: Investigation on the mechanical and biological activity for food packaging. *Polymers for Advanced Technologies*, 28(1), pp.20–27.
(118) Ramji, V. and Vishnuvarthanan, M., 2022. Influence of NiO supported silica nanoparticles on mechanical, barrier, optical and antibacterial properties of polylactic acid (PLA) bio nanocomposite films for food packaging applications. *Silicon*, 14(2), pp.531–538.

13 Nano Seed Priming

A Sustainable and Innovative Technology for Seed Quality Enhancement

Bhaswati Sarmah, Meghali Barua, Mira Landep Widiastuti, and Vishwa Jyoti Baruah
Corresponding author: Vishwa Jyoti Baruah

13.1 INTRODUCTION

Nanotechnology refers to the understanding, manipulation, and control of matter at the nanoscale, which is typically defined as the scale ranging from 1 to 100 nanometers. At this scale, materials exhibit unique properties and behaviors that can be utilized in various fields such as electronics, medicine, energy, and materials science (Tiwari et al., 2012). Working at the nanoscale provides opportunities to create, manipulate, and characterize materials with precise control over their structure and properties with prospects to develop new materials with enhanced functionalities, improved performance, and novel applications. This field has been of interest to scientists since the mid-twentieth century, when Nobel laureate Richard Feynman gave a lecture on the subject in 1959 (Feynman, 1960). Since then, researchers have made significant strides in producing materials at the nanoscale, including nanoparticles (NPs) (Laurent et al., 2010). The physiochemical properties of NPs are heavily influenced by their size and shape, which has made them valuable for varied applications. One interesting property of NPs is their ability to exhibit unique optical properties based on their size and shape (Dreaden et al., 2012) that exhibit characteristic colors. These color changes are due to differences in the physiological characteristics of the NPs, and an ability to tune these optical properties based on their size and shape has made them useful for a wide range of applications, like bioimaging (Khan et al., 2019). By using NPs with specific absorption properties, scientists can selectively label and visualize different structures and molecules in biological samples. Overall, the study of NPs and their properties has opened up new avenues for research and innovation in fields such as medicine, electronics, and energy.

While nanoparticles (NPs) can have complex structures, it is important to note that not all nanoparticles necessarily consist of three distinct layers as described. The structure and composition of nanoparticles can vary depending on their synthesis method and intended application. Nanoparticles can be categorized into different types based on their composition, such as metal nanoparticles (e.g., gold, silver, iron), metal oxide nanoparticles (e.g., titanium dioxide, zinc oxide), semiconductor nanoparticles (e.g., quantum dots), carbon-based nanoparticles (e.g., carbon nanotubes, graphene), and many others. In some cases, nanoparticles may have a core–shell structure, where a core material is surrounded by a shell layer of a different material (Shin et al., 2016). This core–shell architecture can provide enhanced properties or functionalities to the nanoparticles. Regarding surface functionalization, it is indeed common to modify the surface of nanoparticles with various molecules. This process is known as surface functionalization or surface modification. The surface layer of nanoparticles can be functionalized with molecules

DOI: 10.1201/9781003389408-13

such as small molecules, polymers, surfactants, ligands, or biomolecules to alter their surface properties, enhance stability, enable targeted delivery, or facilitate interactions with specific environments or biological systems (Nam et al., 2013). It is important to note that while some nanoparticles may exhibit the three-layered structure described, nanoparticle structures can be highly diverse and can vary based on their composition, shape, and intended application (Mašek et al., 2017).

The uniqueness of NPs has attracted the interest of researchers in various fields. NPs can be used for drug delivery systems, where drugs are encapsulated or attached to nanoparticles to improve their solubility, stability, and targeted delivery to specific cells or tissues. The surface functionalization of nanoparticles can enable controlled release of drugs, enhance their circulation time in the body, and facilitate targeted drug delivery to tumor cells or specific organs (Lee et al., 2011). Functionalized NPs can be used as sensing elements to interact with target molecules, resulting in measurable changes in properties like electrical conductivity, fluorescence, or surface plasmon resonance, which can be detected and quantified. By functionalizing the surface of NPs with specific materials or molecules, they can selectively interact with target gases, leading to measurable changes in electrical, optical, or chemical properties (Barrak et al., 2016). This enables the detection and monitoring of gases in diverse applications, including environmental monitoring, industrial safety, and healthcare (Khan et al., 2017). Certain nanomaterials, such as metal-organic frameworks (MOFs) or porous nanoparticles, can have high surface areas and specific adsorption properties, making them effective for capturing and storing CO_2. These NPs can be designed with tailored structures and functionalized surfaces to enhance their CO_2 adsorption capacity and selectivity, contributing to efforts in carbon capture and storage (CCS) technologies (Ramacharyulu et al., 2015). Overall, NPs are a promising class of materials with diverse and multidimensional prospects in various fields (Shaalan et al., 2016).

Priming is a seed enhancement technique which helps to initiate the initial stages of germination, such as imbibition (the absorption of water by the seed) and the activation of metabolic processes, without allowing the seed to fully emerge from the seed coat. This partial germination prepares the seed for faster and more uniform germination when sown in the field. It is a technique commonly used to improve the performance of seeds by providing favorable conditions for germination while avoiding complete sprouting which, in turn, enhances the seed vigor and performance in the field (Finch-Savage and Bassel, 2016). During seed priming, the seeds are soaked in water or a priming solution for a specific period of time. The duration of priming can vary depending on the seed species, but it is typically a relatively short period, usually ranging from a few hours to a couple of days. The controlled hydration during priming triggers biochemical and physiological changes within the seed, such as the breakdown of dormancy mechanisms, the activation of enzymes, and the synthesis of growth-promoting substances. These changes help to enhance germination performance by improving seed vigor, increasing the speed and percentage of germination, and promoting more uniform seedling emergence. It is important to note that the duration and conditions of seed priming should be carefully determined, as excessive or inadequate hydration can have negative effects on seed viability and subsequent germination. Different seed species may require specific priming protocols tailored to their characteristics and requirements.

Nano seed priming is a sustainable and innovative technology that has emerged as a promising approach to improve seed quality (Shelar et al., 2021). Nano seed priming takes this concept to the next level by using nanotechnology to enhance the potency and efficiency of seed priming. The applications of nanotechnology in seed preparation involve controlled application of nanomaterials to seeds to improve seed performance (Rai-Kalal and Jajoo, 2021). Being minute in size, these particles have unique physical and chemical properties that are highly effective in improving seed quality. They readily penetrate the seed coat to reach the interior of the seed, where they activate or inhibit specific biochemical pathways, improving seed germination, seedling growth, and stress tolerance (Nile et al., 2022; Shelar et al., 2021).

13.2 APPLICATIONS OF NANO SEED PRIMING

Nano seed priming has many advantages over traditional seed priming methods (Shelar et al., 2021). The first and the most important advantage is that it requires very little water, making it very sustainable and environmentally friendly (Acharya et al., 2020). Nanomaterials have been found to be so efficient and effective that they can improve the seed quality with minimal effort. Nano seed primers also have the potential to improve crop yield and quality, potentially providing significant economic benefits to farmers and food producers (do Espirito Santo Pereira et al., 2021). Another promising application of nano seed priming is promoting plant germination and early growth under adverse environmental conditions like salinity and drought (Shelar et al., 2021; Rhaman et al., 2022; Sabagh et al., 2021). It has also been shown to improve plant resistance to pests and diseases, thus reducing the need for chemical pesticides and herbicides (Shukla et al., 2019). Other potential uses include improving the nutritional quality of crops like increased vitamin and mineral content.

Nano seed priming is both a sustainable and an innovative technology with a great prospect to transform seed quality enhancement approaches (Shelar et al., 2021). Using nanomaterials to improve the effectiveness and efficiency of seed preparation can improve crop performance and reduce reliance on chemical additives (Abbasi Khalaki et al., 2021). The technology continues to evolve, and we can expect even more exciting uses and benefits in the future. It has been shown to have a significant impact on many aspects of seed quality and performance. The main applications of nanopriming are discussed in the following subsections.

13.2.1 Enhanced Seed Germination

Nanopriming greatly enhances seed germination rate, resulting in faster and more uniform seedling emergence (Maroufi et al., 2011). This is achieved by stimulating various metabolic processes within the seed, such as increased respiratory activity. Nanopriming can enhance the water uptake and retention capacity of seeds and seedlings, potentially leading to improved survival and growth in water-restricted conditions (Wu and Li, 2022). This is accomplished through several mechanisms viz-à-viz aquaporin activation, modification of cell wall properties, and production of Osmo protectants. Nanopriming can activate aquaporins, proteins that facilitate the movement of water across cell membranes (Nile et al., 2022). It can improve the water-holding capacity of seeds and seedlings under water-deficient conditions by enhancing the expression and activity of aquaporins (Saha et al., 2022). Nanopriming can also modify cell wall properties, which can affect water uptake and retention (Santo Pereira et al., 2021) by increasing the production of pectin and other polysaccharides and improve the water-holding capacity of cell walls. Nanopriming has also been demonstrated to stimulate the production of osmoprotectants such as proline and glycine betaine, helping plants maintain water balance under stressful conditions (Chakraborti et al., 2022). These molecules act as osmolytes, preventing water loss from plant tissues and maintaining cellular integrity. Nanopriming can help plants survive and thrive in water-constrained conditions by enhancing the water uptake and retention capacity of seeds and seedlings (Shelar et al., 2021). This is especially important in drought-prone areas or areas with limited access to water resources.

13.2.2 Improved Seedling Growth

Nanopriming improves seedling growth by promoting cell division and elongation, resulting in larger, healthier seedlings (Nile et al., 2022). When nanoparticles are applied to seeds during nanopriming, they can penetrate the seed coat and reach the interior of the seed. Once inside, nanoparticles can interact with the components of the cell and trigger various physiological and biochemical responses in the seedlings. This stimulates cell division, leading to increased cell numbers in the growing tissues of the seedling which, in turn, contributes to the overall growth and development of the seedlings. Additionally, nanoparticles have shown to improve seedling growth by promoting cell division and elongation, resulting in longer and more vigorous seedlings (Chakraborti et al., 2022). This is achieved by activating specific genes and enzymes that regulate plant growth and development. Furthermore, nanoparticles used in nanopriming can also influence various physiological processes in seeds and seedlings. They can regulate gene expression, activate signaling pathways, and modulate the levels of hormones and enzymes involved in growth and development. These molecular-level interactions contribute to the observed improvements in seedling growth. Nanopriming stimulates the production and activity of growth-promoting hormones such as auxins and cytokinin, which are key regulators of cell division and elongation (Nile et al., 2022). By increasing the levels of these hormones in seedlings, nanopriming can promote the growth and development of plant tissues. Nanopriming can also increase the activity of enzymes involved in cell wall modification, such as cellulases and pectinases (Tung et al., 2022). These enzymes break down cell wall components, allowing cells to expand and stretch. By enhancing the activity of these enzymes, nanopriming promotes plant tissue growth. Moreover, nanopriming can stimulate the production of stress-response proteins such as heat shock proteins and LEA proteins (enriched in late embryogenesis). This protects plant cells from environmental stressors and promotes seedling growth. These proteins help maintain cellular integrity and prevent damage from abiotic stressors, allowing seedlings to grow and develop under challenging conditions. Promoting cell division and elongation, nanopriming improves seedling growth and development, resulting in larger, healthier plants (Arun et al., 2022; Aswathi et al., 2021). This can have important impacts on crop productivity and yield, especially in areas with limited resources or unfavorable environmental conditions.

13.2.3 Improved Stress Tolerance

Nanopriming can increase the ability of plants to withstand a variety of abiotic and biotic stressors such as drought, salinity, and disease. It does this by increasing the production of stress-sensitive proteins and antioxidants that protect plants from damage caused by stressors. There was tremendous improvement in plant stress tolerance and the ability of the plants to withstand a variety of abiotic and biotic stressors, including drought, salinity, and disease increases after nanopriming. These may be attributed to activation of stress-response genes that are involved in stress-response pathways such as encoding for antioxidant enzymes, chaperones, and osmoprotectants. These genes can help plants cope with environmental stressors by scavenging reactive oxygen species (ROS), preventing protein misfolding and maintaining cellular integrity. Nanopriming can also induce the production of secondary metabolites, which are chemical compounds that are not directly involved in the growth and development of plants but play important roles in defense against various stresses, attraction of pollinators, and other ecological functions.

When plants are exposed to NPs during nanopriming, it can trigger signaling pathways and metabolic responses that lead to the synthesis and accumulation of secondary metabolites. Flavonoids and phenolic compounds are among the secondary metabolites that have been observed to increase in plants subjected to nanopriming. Flavonoids are a diverse group of compounds that have antioxidant and antimicrobial properties. They are involved in plant defense mechanisms against pathogens, UV radiation, and oxidative stress. The induction of flavonoid production through nanopriming can enhance the plants' ability to defend themselves against biotic stresses and environmental challenges. Similarly, phenolic compounds are known for their antioxidant and antimicrobial activities. They have been associated with various beneficial effects, such as enhanced plant growth, improved nutrient uptake, and increased tolerance to abiotic stresses like drought and salinity. Nanopriming can stimulate the production of phenolic compounds, thereby contributing to the plants' stress tolerance and overall health.

The induction of secondary metabolite production through nanopriming holds potential for various applications, including enhancing plant resistance to diseases and pests, improving crop productivity, and increasing the nutritional value of plants. However, it is important to consider the dosage, timing, and specific nanomaterials used in nanopriming, as excessive or improper application of nanoparticles could have adverse effects on plant growth and development. These compounds also have antibacterial properties and help protect plants from pathogen infections. Changes in membrane properties were observed due to nanopriming which alters the properties of plant membranes, making them more resistant to environmental stressors such as drought and salinity. Nanopriming enhances the production of lipids and other membrane components and helps maintain membrane fluidity and integrity under stress conditions. It can help improve plant productivity and yield by improving plant stress tolerance, especially in areas with harsh environmental conditions or where plants are prone to disease. It also helps reduce the need for chemicals such as pesticides and fertilizers that can harm the environment.

13.2.4 Improved Nutrient Uptake

Nanopriming improves the uptake and utilization of essential nutrients from the soil and subsequently increasing the plant growth and yield. This is achieved by stimulating the activity of enzymes involved in nutrient absorption and transport. It has been shown to improve the uptake and utilization of nutrients such as nitrogen, phosphorus, and potassium, increasing plant growth and yield. Some of the ways nanopriming can improve nutrient absorption are enhanced root growth and development, nutrient transporter activation, and increased efficiency of nutrient utilization. The increased expression and activity of transporters facilitate the movement of nutrients across cell membranes, allowing for more effective nutrient uptake by the roots. This, in turn, can enhance nutrient availability for various metabolic processes and support plant growth and development. Nanopriming stimulates root growth and development, resulting in a larger root system that can more effectively absorb nutrients from the soil. This may increase the uptake and utilization of nutrients by plants. Nanopriming activates nutrient transporters in plant cells, allowing nutrients to be more easily absorbed and transported to other parts of the plant. This can improve the efficiency of nutrient absorption and utilization. It can also enhance the efficiency with which plants utilize nutrients by enhancing the activity of enzymes involved in nitrogen metabolism, allowing plants to utilize nitrogen more effectively. This can increase plant growth and yield. By improving nutrient uptake and utilization, nanopriming can help plants grow, especially in areas where soil nutrient availability is limited or where plants are grown under suboptimal conditions. It helps improve the productivity and yield of your crop. It also helps reduce the need for chemical fertilizers, which can have a negative impact on the environment. Stimulating enzyme activity may be one way nanopriming improves nutrient uptake, but it is actually not the only mechanism involved. It can improve nutrient absorption and utilization.

13.3 INFLUENCE OF NANOPRIMING IN PRIMARY PLANT METABOLISM

To understand the effect of nanopriming on the seed quality, it is necessary to understand various processes related to seed food reserve accumulation, health, and quality, like primary metabolism, photosynthesis, respiration, assimilation, and mobilization of photosynthates (starch metabolism/seed filling).

13.3.1 Primary Metabolism

Metabolism includes all chemical reactions that occur in cells that help in growth and development of the organism, be it plant or animal. These chemical reactions may be catabolic (breakdown) or anabolic (synthesis) in nature. Primary metabolism includes biosynthesis and breakdown of primary food reserves such as sugars or polysaccharides (starch, hemicellulose, etc.), proteins, lipids or fatty acids, and nucleotides which are essential for the survival of the organism. In plants, it includes photosynthesis and respiration, whereas secondary metabolism includes biosynthesis and breakdown of smaller compounds which are not essential for survival but certainly give some edge to the organism for survival, including phenolics, alkaloids, and saponins.

When nanomaterials are applied during seed priming, they can influence various metabolic pathways and enzymatic activities, leading to modifications in primary plant metabolism. Nano seed priming can affect processes such as photosynthesis, respiration, protein synthesis, carbohydrate metabolism, and hormone regulation. One of the primary effects of nano seed priming on metabolism is the modulation of enzymatic activity. Nanomaterials can interact with enzymes and alter their activity levels, leading to changes in metabolic pathways. This can result in enhanced energy production, nutrient utilization, and biosynthesis of essential molecules in plants. Nano seed priming has been reported to promote photosynthesis by improving chlorophyll content, increasing the efficiency of light absorption, and enhancing the activity of photosynthetic enzymes. This leads to increased carbon assimilation and improved plant growth. Furthermore, it can positively influence respiratory metabolism by enhancing mitochondrial activity and ATP production. This increased energy availability can support various cellular processes, including nutrient uptake, biosynthesis, and stress tolerance. It can also affect protein synthesis and carbohydrate metabolism. It has been shown to enhance the synthesis of enzymes and structural proteins, leading to improved cellular functions and growth. Additionally, nano seed priming can modulate carbohydrate metabolism, resulting in optimized energy allocation, storage, and utilization within plants. Hormone regulation is another aspect of primary plant metabolism that can be influenced by nano seed priming. Nanomaterials applied during seed priming can affect the production, transport, and signaling of plant hormones such as auxins, cytokinins, gibberellins, and abscisic acid. This hormonal modulation can have profound effects on various physiological processes, including seed germination, root development, flowering, and stress responses.

13.3.2 Photosynthesis

By photosynthesis, green plants, algae, and cyanobacteria synthesize food from water, carbon dioxide (CO_2), and sunlight in the chloroplast whose end products are glucose and oxygen (O_2). In other words, through photosynthesis, the plant converts sunlight energy into chemical energy. The higher the photosynthetic rate, the higher the accumulation of biomass in the plant, thereby not only enhancing the health status of the plant but also increasing the productivity. Photosynthesis converts CO_2 to O_2, and thus photosynthesis directly or indirectly sustains all forms of life in this world. CO_2 from the atmosphere enters plant cells through stomata and through diffusion reaches mesophyll, which contains chloroplast. The photosynthetic process utilizes water absorbed by the plant, which acts as a source of oxygen. Light with wavelength range of 400 nm (blue region) to 700 nm (red region) is considered as photosynthetically active radiation (Liu and Van Iersel, 2021). Chloroplast consists of granum, which is a stack of thylakoids, and stroma. Chlorophyll, a green pigment and the main component that captures solar energy during photosynthesis, is located on the thylakoid membrane. Most plants have two types of chlorophyll: chlorophyll a and chlorophyll b. Chlorophyll b captures the solar energy and transfers it to chlorophyll a.

When nanomaterials are applied during seed priming, they can positively influence various aspects of photosynthesis, resulting in improved photosynthetic efficiency. Here are some ways in which nano seed priming can enhance photosynthesis:

(1) Increased chlorophyll content: Nano seed priming can lead to higher chlorophyll content in plant leaves. Chlorophyll is the pigment responsible for capturing light energy during photosynthesis. With increased chlorophyll levels, plants can absorb more light, thereby enhancing the overall efficiency of photosynthesis.

(2) Enhanced light absorption: Nanoparticles used in seed priming can enhance light absorption by plant leaves. They can alter the optical properties of leaves, allowing for improved light penetration and absorption. This increased light absorption provides plants with more energy for photosynthesis.
(3) Activation of photosynthetic enzymes: Nano seed priming can stimulate the activity of photosynthetic enzymes, such as RuBisCo (ribulose-1,5-bisphosphate carboxylase/oxygenase) and ATP synthase. These enzymes play crucial roles in the carbon fixation and energy production processes of photosynthesis. Activation of these enzymes leads to improved photosynthetic efficiency.
(4) Regulation of stomatal conductance: Nanomaterials applied during seed priming can affect stomatal conductance, which is the opening and closing of stomata (pores) on leaves. Stomata control the exchange of gases, including carbon dioxide for photosynthesis and water vapor for transpiration. Nano seed priming can optimize stomatal conductance, ensuring an adequate supply of carbon dioxide for photosynthesis while minimizing water loss.
(5) Increased electron transport rate: Nano seed priming can enhance the rate of electron transport in the photosynthetic machinery. Electron transport is a series of reactions that occur during photosynthesis, leading to the generation of ATP (adenosine triphosphate) and NADPH (nicotinamide adenine dinucleotide phosphate), which are energy-rich molecules crucial for carbon fixation and sugar production.

By enhancing photosynthesis, nano seed priming can improve plant growth, biomass production, and crop yield. It can also contribute to the plant's ability to tolerate environmental stresses such as drought, heat, and high light intensity.

13.3.3 Assimilation and Mobilization of Photosynthates (Starch Metabolism/Seed Filling)

The glucose produced during photosynthesis in the leaf can be transported to other parts of the plants through phloem in the form of sucrose only, hence glucose must be converted into sucrose. After reaching the designated destination, sucrose is either converted into a higher form of food reserve such as starch, protein, or lipids for storage, or it may be utilized to release energy for conducting various plant functions. Seed stores food reserve mainly as proteins, lipids, and carbohydrates. Sugar produced through photosynthesis is utilized for synthesis of fats, proteins, complex carbohydrates, and other compounds, which is known as assimilation. The processes that convert complex, nonmobile, and localized food reserve to simple, translocatable forms that are made available to germinating seed are known as mobilization of reserve. These processes are highly sensitive to the external environment (Yin et al., 2020).

The germination process starts with water imbibition, which starts the metabolic processes like respiration that provide energy for the development of the embryo into a seedling. Rehydration also activates numerous enzymes involved in protein or lipid synthesis. During germination, seed reserves which are stored in endosperm, cotyledons, or perisperm are mobilized, such that they are brought into the vicinity of the embryo.

Starch is the most common reserve carbohydrate in endospermic seeds, such as rice, wheat, and maize; others include hemicellulose and mucilage. Imbibed seed stimulates synthesis of gibberellic acid in the scutellum, which then diffuses to the aleurone layer of the seed and induces synthesis of α-amylase and other hydrolytic enzymes. The α-amylase then enters the endosperm, where starch is broken down into simpler sugar through a series of enzyme-mediated reactions, and then transported to the developing embryo to provide energy until the seedling starts photosynthesis.

13.3.4 Influence of Nanopriming in Primary Plant Metabolism

The growth and development processes of plants are highly influenced by both biotic and abiotic factors. In today's world, due to a number of developmental programs, various nanoparticles are released into the environment, which may affect plants adversely or favorably (Mohamed and Kumar, 2016), and nanoparticles also show greater penetration into the cells (Poddar et. al., 2020). Nanoparticles can enter plant cells by diffusion through the plasma membrane (Auría-Soro et al., 2019), by endocytosis (Foroozandeh and Aziz, 2018), or through transmembrane proteins (Li et al., 2020).

Crop production can be elevated by enhancing the efficiency of photosynthesis, but under abiotic stress conditions photosynthetic processes may be hampered due by excessive production of the ROS scavenging mechanism, which may lead to reduced crop production. Photosynthetic activity can be enhanced and protected using engineered nanoparticles which can disperse the oxidative stress, enhance the light-harvesting molecules, or block the electron transport system (Sonkar et. al., 2023). However, the effect of nanoparticles on the effectiveness of the photosynthetic process is not constant, as it varies from plant to plant within the same species, and with the nanoparticle itself. For example, single-walled C nanotubes (SWCNTs) can enhance photosynthesis by three times, but iron oxide NPs and silver NPs hamper photosynthesis due to their toxic effects, affecting crop production adversely (Sonkar et. al., 2023; Poddar et. al., 2020). The threefold increase in photosynthesis by SWCNTs is due to speedy electron transport and enhanced plant signaling activities like nitric acid (Poddar et. al., 2020). Nanoparticles of silica, titanium oxide (TiO_2), and so on enhanced photosynthetic rate by elevating O_2 transport or by absorbing more light (TiO_2 being a good light absorber), transferred energy to electrons more efficiently, and improved RuBisCo activities, an important enzyme for C-assimilation. An adverse effect of AgNPs was reported, which may be due to its damaging effect on primary root cells (root apical meristem) that leads to improper growth and development of the plants, thus affecting the photosynthetic process also (Pokhrel and Dubey, 2013).

Nanopriming not only influences photosynthesis but has also been found to break seed dormancy and enhance seed germination, affecting primary and secondary metabolite production there by enhancing seed vigor (Nile et al., 2022), thus influencing the overall growth and development of the plant. Seed dormancy is an adaptive mechanism in many crops which helps them to prevail over unfavorable environmental conditions or preharvest sprouting, but it becomes a nuisance when we want our seed to germinate. In nanoprimed seed, especially with micronutrients, better water uptake was observed (Anand et al., 2019), thus enhancing seed germination and vigor, which was achieved due to enhanced seed cell membrane integrity, phosphorylation activity (Acharya et al., 2020), and phytohormone response to nanoparticles (Yang et al., 2017). The enhanced seed germination and vigor in nanoprimed seed may be due to nanopore development on the seed coat, induction of ROS up to the level that favors germination, and/or higher activation of starch-hydrolyzing enzymes (Singh et al., 2020).

Nanoparticles may be absorbed directly by the leaves or may be received from root through transpirational flow when soil is also exposed to nanoparticles, which affects the total content of primary metabolites without affecting secondary metabolite content (Poddar et al., 2020). It also alters the saturated and unsaturated fatty acid composition in both roots and leaves. These effects were more profound when leaves, soil, and water supply were all exposed to CdONPs. With varied doses, the effect of nanoparticles on plant cell activities may vary. Seeds treated with $TiSiO_4$NPs showed a beneficial effect on photosynthetic processes without compromising the germinative property of lettuce (*Lactuca sativa L.*) seed (Mariz-Ponte et al., 2020). $TiSiO_4$NPs of size <50 nm, at higher dose (100mg/L), boosted the photosynthetic rate by increasing chlorophyll, enhancing Fv/Fm and electron transport rate more compared to when seeds were treated with a lower dose of $TiSiO_4$NPs (50mg/L), although net photosynthetic rate and Calvin cycle components like RuBisCo were not affected. The enhanced photosynthetic rate at lower doses was due to stimulating pigment accumulation, whereas at higher doses it was due to enhanced efficiency of photosystem II, ETA, and other gas exchange components. However, soluble sugar and starch content was not upset, indicating nontoxicity of $TiSiO_4$NPs. Thus, $TiSiO_4$ can be utilized to boost plant growth and development.

Starch metabolism is not only essential for plant growth and development but also an important factor for seed germination. Seed priming with 5 ppm and 10 ppm photosynthesized silver nanoparticles (AgNPs) synthesized through a green route using kaffir lime leaf extract enhanced not only germination but also vigor more in the aged seeds of rice. This was due to amplified α-amylase activities, which is an important enzyme for starch metabolism, thus promoting higher concentration of readily usable sugar to be utilized by germinating seed and growing seedlings (Mahakham et al., 2017). It was also suggested that photosynthesized AgNP-primed seed led to better promotion of aquaporin gene expression (needed for water uptake), ROS production, activation of antioxidant activities (SOD, CAT), creation of nanopores for improved water uptake, increased dehydrogenase activities (for improved root activities), and loosening of the cell wall by producing hydroxyl radicals that lead to improved germination rate and vigor (Mahakham et al., 2017) compared to hydropriming, $AgNO_3$, and unprimed seed. Aquaporin was also found to favor ROS diffusion across the cell membranes (Bienert and Chaumont, 2014) and regulation of plant development by managing ROS (Maurel et al., 2015; Cordeiro, 2015). Seed with low ROS or aged seeds with high ROS led to poor germination; however, seed priming with nanoparticles optimized the ROS level, thus improving germination (Pereira et al., 2021). The activity of α-amylase, needed for starch metabolism, was found to be imitated by FeS_2NPs in spinach seeds (Srivastava et al., 2013). Many other nanoparticles were found to enhance α-amylase activity (Ma et al., 2010; Laware and Raskar, 2014) that led to a high degree of starch degradation, ultimately improving germination and vigor, which again led to higher production and productivity.

13.3.5 Aging

Nano seed priming has the potential to delay or reduce the aging process in plants to some extent. Aging in plants, also known as senescence, refers to the physiological and biochemical changes that occur as plants mature and approach the end of their life cycle. These changes can lead to decreased vigor, reduced productivity, and increased susceptibility to stresses (Pagano et al., 2023). Nano seed priming can help mitigate the effects of aging by enhancing seed germination, promoting early seedling growth, and improving overall plant health. By providing beneficial nanomaterials during seed priming, the seeds are equipped with additional resources and protective mechanisms that can delay the onset of senescence (Salehi et al., 2023). One way in which nano seed priming can affect aging is by enhancing antioxidant defense systems. Nanomaterials used in seed priming can increase the activity of antioxidant enzymes and boost the production of antioxidant molecules in plants. This helps to counteract the harmful effects of ROS, which are generated during various metabolic processes and can accelerate aging. By reducing oxidative damage, nano seed priming can slow down the aging process in plants (Ragab and Saad-Allah, 2021).

Additionally, nano seed priming can improve nutrient uptake and nutrient use efficiency in plants. Nutrients play a vital role in plant growth and development, and their availability becomes limited as plants age. By enhancing nutrient absorption and utilization, nano seed priming can help sustain the plant's physiological functions and delay the onset of senescence (Tondey et al., 2021). Furthermore, nano seed priming can enhance stress tolerance in plants. As plants age, they become more susceptible to various stresses, such as drought, salinity, and diseases. Nano seed priming can enhance the plant's ability to cope with these stresses by activating stress-responsive genes, modulating hormone levels, and improving the overall stress defense mechanisms. By reducing the impact of stress, nano seed priming can slow down the aging process and maintain plant vitality (Johnson and Puthur, 2021; Mostofa et al., 2021). While nano seed priming shows promise in delaying aging, it is important to note that aging is a natural and complex process influenced by multiple factors. The specific effects of nano seed priming on aging may vary depending on various factors, including the type of nanomaterials used, their concentration, the plant species, and environmental conditions (do Espirito Santo Pereira et al., 2021).

13.4 INFLUENCE OF NANOPRIMING IN ABIOTIC STRESS TOLERANCE

Nanopriming has been shown to improve plant abiotic stress tolerance. This is one of the most important applications of this technology in agriculture. Abiotic stresses such as drought, salinity, and temperature extremes can severely limit crop growth and yield, resulting in significant economic losses for farmers. However, nanopriming can mitigate these effects and improve plant productivity even under unfavorable environmental conditions. Below are some of the ways nanopriming improves abiotic stress tolerance in plants:

- Increased water absorption and retention: Nanopriming enhances the water absorption and retention of seeds and seedlings, improving survival and growth in water-limited conditions. It does this by increasing the production of aquaporins and other water transporters that regulate the movement of water across cell membranes.
- Enhanced antioxidant activity: Nanopriming can enhance the activity of antioxidant enzymes such as superoxide dismutase (SOD) and catalase (CAT), protecting plant cells from oxidative damage caused by abiotic stressors. This helps maintain cell homeostasis and prevent cell death.
- Enhanced photosynthesis: Nanopriming can increase the efficiency of plant photosynthesis, resulting in increased biomass production and yield even under low light and high temperature conditions. It does this by increasing the production of chlorophyll and other photosynthetic pigments, as well as the activity of key enzymes involved in carbon fixation.
- Regulation of stress-response genes: Nanopriming activates specific genes involved in abiotic stress responses, leading to the production of stress-response proteins and other protective molecules. This mitigates the effects of stressors such as salinity and drought and improves plant growth and survival in harsh environmental conditions.

Nanopriming has the potential to significantly improve crop abiotic stress tolerance, leading to increased agricultural productivity and sustainability. Using nanotechnology to improve seed quality and performance can help ensure food security in a changing climate and help develop more resilient and adaptive crop systems.

13.5 INFLUENCE OF NANOPRIMING IN BIOTIC STRESS TOLERANCE

Nanopriming has also been shown to improve plant tolerance to biological stressors such as pests and diseases. Some ways that nanopriming can improve biological stress tolerance are discussed in the following subsections.

13.5.1 Activation of Defense Mechanisms

Nanopriming activates plant defense mechanisms, such as the production of plant hormones and pathogenic proteins, and helps protect plants from biological stressors. It activates plant defense mechanisms against biological stressors. When plants are exposed to biological stressors such as pests and pathogens, various defense mechanisms are activated, including the production of plant hormones and pathogenesis-related (PR) proteins. These defense mechanisms help plants recognize and respond to biological stressors and initiate effective defensive responses.

Nanopriming may facilitate the activation of these defense mechanisms and increase the production of plant hormones and PR proteins. Phytohormones such as salicylic acid and jasmonic acid play important roles in regulating plant defense responses. They have antibacterial properties and can induce the expression of PR proteins that help protect plants from pathogens. Nanopriming helps improve the ability of plants to defend themselves against biological stressors by increasing the production of plant hormones and PR proteins.

In addition to activating defense mechanisms, nanopriming helps strengthen the plant's immune system, making it more responsive to future biological stressors, allowing for the protection of the plant in the long term.

13.5.2 Induction of Secondary Metabolites

Similar to abiotic stress tolerance, nanopriming can induce the production of secondary metabolites such as phenolic compounds that have antimicrobial properties and help protect plants from pathogen infection. Nanopriming can also induce plants to produce secondary metabolites that have antimicrobial properties and help protect plants from pathogenic infections. Secondary metabolites are organic compounds produced by plants that are not directly involved in growth and development but perform specialized functions such as defense against herbivores and pathogens.

Phenolic compounds are a class of secondary metabolites that have been shown to have antimicrobial properties. They help protect plants from pathogenic microorganisms by disrupting cell membranes and inhibiting growth. Nanopriming induces the production of phenolic compounds in plants and increases their tolerance to biological stressors. In addition to phenolic compounds, other secondary metabolites such as flavonoids, alkaloids and terpenoids may also be induced by nanopriming, resulting in enhanced plant defense against biological stressors. By inducing the production of these secondary metabolites, nanopriming can provide plants with a natural defense mechanism against pests and diseases, reducing the need for chemical pesticides and fungicides

13.5.3 Altering Plant–Microbe Interactions

Nanopriming alters the interaction between plants and beneficial microorganisms such as mycorrhizal fungi and rhizobia, improving plant growth and health and providing protection from pathogenic microorganisms.

Mycorrhizal fungi are a type of beneficial fungi that form symbiotic bonds with plant roots, helping to improve nutrient uptake and water uptake by plants. Nanopriming of plant roots by mycorrhizal fungi has been shown to promote colonization of plants and improve plant growth and stress tolerance. Rhizosphere bacteria are another type of beneficial microorganisms that live in the rhizosphere (soil surrounding plant roots) and help protect plants from pathogenic microorganisms. Nanopriming has been shown to promote the growth of certain species of rhizobia, to improve plant growth and health, and to provide protection from pathogenic microorganisms. By altering the interactions between plants and beneficial microbes, nanopriming helps improve plant growth and health, provides protection from biological stressors, and reduces reliance on fungicides, making it a more sustainable and environmentally friendly approach to crop cultivation.

13.5.4 Enhancement of plant immunity

Nanopriming can enhance the innate immune response of plants by enhancing detection of pathogenic microorganisms and activation of defense responses. It can enhance a plant's innate immune responses by enhancing detection of pathogenic microorganisms and activation of defense responses. Plants have an innate immune system that allows them to recognize and respond to pathogenic microorganisms. This system involves recognition of pathogen-associated molecular patterns (PAMPs) by plant cell-surface receptors that trigger the activation of defense responses. Nanopriming can enhance this process by enhancing PAMP recognition and immune response activation. Nanopriming has been shown to enhance the expression of genes involved in plant immune responses, including genes encoding pathogenesis-related (PR) proteins involved in plant defense against pathogens. This helps improve plant tolerance to biological stressors such as pathogens and pests.

Furthermore, nanopriming can enhance systemic acquired resistance (SAR) responses in plants. SARs are a class of induced resistance that occur in response to pathogen infection and involve the generation and transport of mobile signaling molecules that activate defense responses in uninfected parts of plants. Nanopriming amplifies SAR responses by increasing the production and transport of these signaling molecules, enhancing plant immunity against pathogens. Nanopriming enhances the plant's innate immune response, resulting in enhanced plant immunity and protection against biological stressors. By improving crop tolerance to biological stressors, nanopriming reduces crop losses to pests and diseases and reduces the need for chemical pesticides and fungicides that can have negative environmental impacts.

13.5.5 Regulation of Gene Expression

Nanopriming can modulate the expression of genes involved in stress-response pathways, resulting in improved plant tolerance to biological stressors. Under stressful conditions, plants undergo a complex network of molecular and biochemical changes to adapt and survive. These changes are regulated by the expression of specific genes involved in stress-response pathways, including those associated with defense mechanisms, signaling, and metabolic processes. It can regulate the expression of these genes, resulting in improved plant tolerance to biological stressors. For example, nanopriming can induce expression of genes encoding enzymes involved in the synthesis of osmoprotectants such as proline and glycine betaine, which are useful in protecting plants from biological stressors such as pathogens and pests.

Nanopriming can also induce the expression of genes involved in the production of plant hormones such as salicylic acid, jasmonic acid, and ethylene, which play important roles in plant defense responses to biological stressors. By regulating the expression of these genes, nanopriming enhances the ability of plants to respond and adapt to biological stressors, improving stress tolerance and plant health and leading to more sustainable and environmentally friendly approaches to agriculture.

13.6 CONCLUSIONS

Nanotechnology encompasses a broad range of techniques and approaches, including the synthesis and fabrication of nanomaterials, nanodevices, and nanosystems, as well as the investigation and understanding of their properties and interactions. It involves working with nanoparticles, nanotubes, nanowires, nanocomposites, and other nanostructured materials. The field of nanotechnology has the potential to revolutionize various industries by enabling advances in areas such as electronics and computing, medicine and healthcare, energy generation and storage, environmental remediation, and many others. However, as with any emerging technology, it is crucial to consider the ethical, environmental, and safety implications associated with the production and use of nanomaterials. Seed priming is a valuable technique in agriculture and horticulture to improve the germination and early growth of seeds, leading to more successful and uniform crop establishment. Nanopriming involves the application of nanomaterials, such as metal and metal oxide nanoparticles, to seeds prior to germination. Nanopriming has been studied and applied in different plant species, including crops and horticultural plants, to enhance seed germination, early growth, and stress tolerance. By promoting cell division and elongation, nanopriming can improve the establishment of seedlings, increase biomass production, and enhance plants' ability to withstand biotic and abiotic stresses, such as drought, salinity, and disease. It is worth noting that the specific effects of nanopriming can vary depending on the type of nanomaterials used, their concentration, the timing and duration of application, as well as the specific plant species and environmental conditions. Nevertheless, the impact of nanopriming on seed quality and performance is very promising, and this technology has the potential to revolutionize the way we approach seed treatment and crop production. We can expect even more exciting applications and benefits of nanoprimers in the years to come.

REFERENCES

Abbasi Khalaki, M., Moameri, M., Asgari Lajayer, B., & Astatkie, T. (2021). Influence of nano-priming on seed germination and plant growth of forage and medicinal plants. *Plant Growth Regulation,* 93, 13–28.

Acharya, P., Jayaprakasha, G. K., Crosby, K. M., Jifon, J. L., & Patil, B. S. (2020). Nanoparticle mediated seed priming improves germination, growth, yield, and quality of watermelons (*Citrullus lanatus*) at multi-locations in Texas. *Scientific Reports*, 10, 5037. https://doi.org/10.1038/s41598-020-61696-7

Anand, A., Kumari, M., & Koul, T. A. (2019). Hydrogen peroxide signaling integrates with phytohormones during the germination of magneto-primed tomato seeds. *Scientific Reports*, 9, 8814. https://doi.org/10.1038/s41598-019-45102-5

Arun, M. N., Hebbar, S. S., Senthivel, T., Nair, A. K., Padmavathi, G., Pandey, P., & Singh, A. (2022). *Seed priming: The way forward to mitigate abiotic stress in crops*. London, UK: IntechOpen.

Aswathi, K. R., Kalaji, H. M., & Puthur, J. T. (2021). Seed priming of plants aiding in drought stress tolerance and faster recovery: A review. *Plant Growth Regulation*, 97(2), 235–253.

Auría-Soro, C., Nesma, T., Juanes-Velasco, P., Landeira-Viñuela, A., FidalgoGomez, H., Acebes-Fernandez, V., Gongora, R., Almendral Parra, M. J., Manzano-Roman, R., & Fuentes, M. (2019). Interactions of nanoparticles and biosystems: Microenvironment of nanoparticles and biomolecules in nanomedicine. *Nanomaterials*, 9(10), 1365. https://doi.org/10.3390/nano9101365

Barrak, H., Saied, T., Chevallier, P., Laroche, G., M'nif, A., & Hamzaoui, A. H. (2019). Synthesis, characterization, and functionalization of ZnO nanoparticles by N-(trimethoxysilylpropyl) ethylenediamine triacetic acid (TMSEDTA): Investigation of the interactions between Phloroglucinol and ZnO@ TMSEDTA. *Arabian Journal of Chemistry*, 12(8), 4340–4347.

Bienert, G. P., & Chaumont, F. (2014). Aquaporin-facilitated transmembrane diffusion of hydrogen peroxide. *BBA-Gen. Subjects*, 1840, 1596–1604.

Chakraborti, S., Bera, K., Sadhukhan, S., & Dutta, P. (2022). Bio-priming of seeds: Plant stress management and its underlying cellular, biochemical and molecular mechanisms. *Plant Stress*, 3, 100052.

Cordeiro, R. M. (2015). Molecular dynamics simulations of the transport of reactive oxygen species by mammalian and plant aquaporins. *BBA-Gen. Subjects,* 1850, 1786–1794.

do Espirito Santo Pereira, A., Caixeta Oliveira, H., Fernandes Fraceto, L., & Santaella, C. (2021). Nanotechnology potential in seed priming for sustainable agriculture. *Nanomaterials,* 11(2), 267.

Dreaden, E. C., Alkilany, A. M., Huang, X., Murphy, C. J., & El-Sayed, M. A. (2012). The golden age: Gold nanoparticles for biomedicine. *Chemical Society Reviews*, 41(7), 2740–2779.

Feynman, R. (2018). There's plenty of room at the bottom. In *Feynman and computation* (pp. 63–76). CRC Press.

Finch-Savage, W. E., & Bassel, G. W. (2016). Seed vigour and crop establishment: Extending performance beyond adaptation. *Journal of Experimental Botany*, 67(3), 567–591.

Foroozandeh, P., & Aziz, A. A. (2018). Insight into cellular uptake and intracellular trafficking of nanoparticles. *Nanoscale Research Letters*, 13(1), 1–12. https://doi.org/10.1186/s11671-018-2728-6

Johnson, R., & Puthur, J. T. (2021). Seed priming as a cost effective technique for developing plants with cross tolerance to salinity stress. *Plant Physiology and Biochemistry*, 162, 247–257.

Khan, I., Saeed, K., & Khan, I. (2019). Nanoparticles: Properties, applications and toxicities. *Arabian Journal of Chemistry*, 12(7), 908–931.

Khan, I., Yamani, Z. H., & Qurashi, A. (2017). Sonochemical-driven ultrafast facile synthesis of SnO2 nanoparticles: Growth mechanism structural electrical and hydrogen gas sensing properties. *Ultrasonics Sonochemistry*, 34, 484–490.

Laurent, S., Forge, D., Port, M., Roch, A., Robic, C., Vander Elst, L., & Muller, R. N. (2008). Magnetic iron oxide nanoparticles: synthesis, stabilization, vectorization, physicochemical characterizations, and biological applications. *Chemical Reviews*, 108(6), 2064–2110.

Laware, S. L., & Raskar, S. (2014). Effect of titanium dioxide nanoparticles on hydrolytic and antioxidant enzymes during seed germination in onion. *International Journal of Current Microbiol Applied Science*, 3(7), 749–60. www.ijcmas.com/vol-3-7/

Lee, J. E., Lee, N., Kim, T., Kim, J., & Hyeon, T. (2011). Multifunctional mesoporous silica nanocomposite nanoparticles for theranostic applications. *Accounts of Chemical Research*, 44(10), 893–902.

Li, J. H, Santos-Otte, P., Au, B., Rentsch, J., Block, S. & Ewers, H. (2020). Directed manipulation of membrane proteins by fluorescent magnetic nanoparticles. *Nature Communications*, 11(1), 4259. https://doi.org/10.1038/s41467-020-18087-3

Liu, J. & Van Iersel, M. W. (2021). Photosynthetic physiology of blue, green, and red light: Light intensity effects and underlying mechanisms. *Frontiers in Plant Science*, 12. https://doi.org/10.3389/fpls.2021.619987

Ma, X., Geiser-Lee, J., Deng, Y., & Kolmakov, A. (2010). Interactions between engineered nanoparticles (ENPs) and plants: Phytotoxicity, uptake and accumulation. *Science of the Total Environment*, 408(16), 3053–61. https://doi.org/10.1016/j.scitotenv.2010.03.031

Mahakham, W., Sarmah, A. K., Maensiri, S., & Theerakulpisut, P. (2017). Nanopriming technology for enhancing germination and starch metabolism of aged rice seeds using phytosynthesized silver nanoparticles. *Scientific Reports*, 7, 8263. https://doi.org/10.1038/s41598-017-08669-5

Mariz-Ponte, N., Sario, S., Mendes, R. J., Correia, C. V., Moutinho-Pereira, José, Correia, C. M., & Santos, C. (2020). Silicon titanium oxide nanoparticles can stimulate plant growth and the photosynthetic pigments on lettuce crop. *Agriculture (Poľnohospodárstvo)*, 66(4), 148–160. https://doi.org/10.2478/agri-2020-0014

Maroufi, K., Farahani, H. A., & Moradi, O. (2011). Evaluation of nano priming on germination percentage in green gram (*Vigna radiata L.*). *Advances in Environmental Biology*, 5(11), 3659–3664.

Mašek, J., Lubasova, D., Lukáč, R., Turanek-Knotigova, P., Kulich, P., Plockova, J., ... & Turanek, J. (2017). Multi-layered nanofibrous mucoadhesive films for buccal and sublingual administration of drug-delivery and vaccination nanoparticles-important step towards effective mucosal vaccines. *Journal of Controlled Release,* 249, 183–195.

Maurel, C. et al. (2015). Aquaporins in plants. *Physiological Reviews*, 95, 1321–1358.

Mostofa, M. G., Rahman, M. M., Ansary, M. M. U., Keya, S. S., Abdelrahman, M., Miah, M. G., & Phan Tran, L. S. (2021). Silicon in mitigation of abiotic stress-induced oxidative damage in plants. *Critical Reviews in Biotechnology*, 41(6), 918–934.

Nam, J., Won, N., Bang, J., Jin, H., Park, J., Jung, S., ... & Kim, S. (2013). Surface engineering of inorganic nanoparticles for imaging and therapy. *Advanced Drug Delivery Reviews*, 65(5), 622–648.

Nile, S. H., Thiruvengadam, M., Wang Yao, Samynathan, R., Shariati, M. A., Rebezov, M., Nile, A., Sun, M., Venkidasamy, B., Xia, J., & Kai, G. (2022). Nano-priming as emerging seed priming technology for sustainable agriculture: Recent developments and future perspectives. *Journal of Nanobiotechnology*, 20(1), 1–31. https://doi.org/10.1186/s12951-022-01423-8

Pagano, A., Macovei, A., & Balestrazzi, A. (2023). Molecular dynamics of seed priming at the crossroads between basic and applied research. *Plant Cell Reports*, 42(4), 657–688.

Pereira, A. E. S., Oliveira, H. C., Fraceto, L. F. & Santaella, C. (2021). Nanotechnology potential in seed priming for sustainable agriculture. *Nanomaterials*, 11(2), 267. https://doi.org/10.3390/nano11020267

Poddar, K., Sarkar, D, & Sarkar, A. (2020). Nanoparticles on photosynthesis of plants: Effects and role. In: *Green Nanoparticles: Synthesis and Biomedical Applications*, 273–287. Springer: Cham.

Pokhrel, L. R., & Dubey, B. (2013). Evaluation of developmental responses of two crop plants exposed to silver and zinc oxide nanoparticles. *Science of the Total Environment*, 452–453, 321–332.

Ragab, G., & Saad-Allah, K. (2021). Seed priming with greenly synthesized sulfur nanoparticles enhances antioxidative defense machinery and restricts oxidative injury under manganese stress in Helianthus annuus (L.) seedlings. *Journal of Plant Growth Regulation,* 40, 1894–1902.

Rai-Kalal, P., & Jajoo, A. (2021). Priming with zinc oxide nanoparticles improve germination and photosynthetic performance in wheat. *Plant Physiology and Biochemistry,* 160, 341–351.

Ramacharyulu, P. V. R. K., Muhammad, R., Kumar, J. P., Prasad, G. K., & Mohanty, P. (2015). Iron phthalocyanine modified mesoporous titania nanoparticles for photocatalytic activity and CO_2 capture applications. *Physical Chemistry Chemical Physics*, 17(39), 26456–26462.

Rhaman, M. S., Tania, S. S., Imran, S., Rauf, F., Kibria, M. G., Ye, W., … & Murata, Y. (2022). Seed priming with nanoparticles: An emerging technique for improving plant growth, development, and abiotic stress tolerance. *Journal of Soil Science and Plant Nutrition*, 22(4), 4047–4062.

Sabagh, A. E., Çiğ, F., Seydoşoğlu, S., Battaglia, M. L., Javed, T., Iqbal, M. A., & Awad, M. (2021). Salinity stress in maize: Effects of stress and recent developments of tolerance for improvement. *Cereal Grains,* 1, 213.

Saha, D., Choyal, P., Mishra, U. N., Dey, P., Bose, B., Prathibha, M. D., … & Singhal, R. K. (2022). Drought stress responses and inducing tolerance by seed priming approach in plants. *Plant Stress*, 22, 100066.

Salehi, H., Cheheregani Rad, A., Raza, A., Djalovic, I., & Prasad, P. V. (2023). The comparative effects of manganese nanoparticles and their counterparts (bulk and ionic) in Artemisia annua plants via seed priming and foliar application. *Frontiers in Plant Science*, 13, 1098772.

Shaalan, M., Saleh, M., El-Mahdy, M., & El-Matbouli, M. (2016). Recent progress in applications of nanoparticles in fish medicine: A review. *Nanomedicine: Nanotechnology, Biology and Medicine*, 12(3), 701–710.

Sheikh, M. M., & Sakthi, K. D. (2016). Effect of nanoparticles on plants with regard to physiological attributes. In: *Plant Nanotechnology: Principles and Practices*, 119–153. Cham: Springer.

Shelar, A., Singh, A. V., Maharjan, R. S., Laux, P., Luch, A., Gemmati, D., … & Patil, R. (2021). Sustainable agriculture through multidisciplinary seed nanopriming: Prospects of opportunities and challenges. *Cells*, 10(9), 2428.

Shin, W. K., Cho, J., Kannan, A. G., Lee, Y. S., & Kim, D. W. (2016). Cross-linked composite gel polymer electrolyte using mesoporous methacrylate-functionalized SiO_2 nanoparticles for lithium-ion polymer batteries. *Scientific Reports*, 6(1), 26332.

Shukla, P., Chaurasia, P., Younis, K., Qadri, O. S., Faridi, S. A., & Srivastava, G. (2019). Nanotechnology in sustainable agriculture: Studies from seed priming to post-harvest management. *Nanotechnology for Environmental Engineering,* 4, 1–15.

Singh, Y., Kaushal, S., & Sodhi, R. S. (2020). Biogenic synthesis of silver nanoparticles using cyanobacterium Leptolyngbya sp. WUC 59 cell-free extract and their effects on bacterial growth and seed germination. *Nanoscale Advances*, 2, 3972–82. https://doi.org/10.1039/D0NA00357C

Sonkar, S., Sarangi, P. K., Pandey, B., Prakash, A., & Singh, A. K. (2023). Response of plant photosynthesis to nanomaterials. In: *Nanomaterial Interactions with Plant Cellular Mechanisms and Macromolecules and Agricultural Implications*, 49–67. Cham: Springer.

Srivastava, P., Bragança, J., Ramanan, S. R., Kowshik, M. (2013). Synthesis of silver nanoparticles using haloarchaeal isolate Halococcus salifodinae BK3. *Extremophiles*, 17(5), 821–31. https://doi.org/10.1007/s00792-013-0563-3

Tiwari, J. N., Tiwari, R. N., & Kim, K. S. (2012). Zero-dimensional, one-dimensional, two-dimensional and three-dimensional nanostructured materials for advanced electrochemical energy devices. *Progress in Materials Science*, 57(4), 724–803.

Tondey, M., Kalia, A., Singh, A., Dheri, G. S., Taggar, M. S., Nepovimova, E., … & Kuca, K. (2021). Seed priming and coating by nano-scale zinc oxide particles improved vegetative growth, yield and quality of fodder maize (Zea mays). *Agronomy,* 11(4), 729.

Tung, H. T., Nguyen, P. L. H., Van Lich, T., Ngan, H. T. M., Luan, V. Q., Khai, H. D., … & Nhut, D. T. (2022). Enhanced shoot and plantlet quality of Gerbera (Gerbera jamesonii Revolution Yellow) cultivar on medium containing silver and cobalt nanoparticles. *Scientia Horticulturae,* 306, 111445.

Wu, H., & Li, Z. (2022). Recent advances in nano-enabled agriculture for improving plant performance. *The Crop Journal*, 10(1), 1–12.

Yang, J., Cao, W., & Rui, Y. (2017). Interactions between nanoparticles and plants: Phytotoxicity and defense mechanisms. *Journal of Plant Interactions*, 12, 158–69. https://doi.org/10.1080/17429145.2017.1310944

Yin, S., Li, P., Xu, Y., Liu, J., Yang, T., Wei, J., Xu, S., Yu, J., Fang, H., Xue, L., Hao, D., Yang, Z., & Xu, C. (2020). Genetic and genomic analysis of the seed-filling process in maize based on a logistic model. *Heredity*, 124, 122–134. https://doi.org/10.1038/s41437-019-0251-x

14 Decoding of Nanoneutraceuticals and Their Delivery

A Modern Therapeutics in Healthcare

Amol D. Gholap, Navnath T. Hatvate, Pankaj R. Khuspe, Trushali A. Mandhare, Pooja Kashid, and Lalitkumar K Vora
Corresponding author: Amol D. Gholap

14.1 INTRODUCTION

Since its inception more than 3000 years ago, the idea of nutraceuticals has undergone substantial development. The founder of medicine, Hippocrates, commanded, "Let food be thy medicine and medicine be thy food," emphasizing the importance of nutrition for good health and disease prevention. The words "nutrition" and "pharmaceutics" are the roots of the term "nutraceutic." The word is used to describe goods that are isolated from herbal products, dietary supplements (nutrients), certain diets, and processed meals such as cereals, soups, and drinks that are also utilized as medicine in addition to providing nutrition [1,2]. A food ingredient or component thereof that has therapeutic or preventative health advantages is referred to as a nutraceutical, and as a natural means of achieving therapeutic objectives [3,4].

Due to growing consumer interest in naturally derived compounds over the past few years, there has been an increase in the consumption of nutraceuticals among consumers. In recent years, changes in lifestyle have induced disease like type 2 diabetes along with cardiovascular diseases, which are becoming more common among the younger population. At the same time, people are becoming more aware of the link between eating habits and these deadly diseases. As a result, people are paying closer attention to the quality of the food they consume and showing an increased interest in foods that can help improve their health and delay the onset of harmful conditions. All of these factors have encouraged research into the potential health benefits of nutraceuticals and their mechanisms of action while also igniting the market to create cutting-edge products that can attract customers' attention [5–9].

14.1.1 CONCEPT OF NUTRACEUTICALS

Nutraceuticals include characteristics of general biological therapies that are used to stop cancerous processes, promote wellbeing, and manage a number of disease symptoms. Nutraceuticals' significant success and strong demand can be attributed to their ability to reveal the right therapeutic potential while minimizing potential adverse effects associated with conventional medications used to treat a variety of illnesses.

DOI: 10.1201/9781003389408-14

Although the definition of the phrase varies from country to country, it is typically understood to mean a product that has been separated from foods and is typically sold in medical forms unrelated to food. A substance that has physiological benefits or offers protection against chronic diseases may be referred to as a nutraceutical product. Nutraceuticals can be used to promote energy, delay aging, avoid chronic diseases, and improve health [10].

Nutraceuticals have a huge range of health advantages. To name a few, there are a number of nutraceuticals that can be used to reduce the risk of heart disease and various cancers but also to treat or prevent fatigue, hypertension, diabetes, obesity-related conditions, arthritis, osteoporosis, menopausal symptoms, macular degeneration, cataracts, deteriorated memory and concentration, neurodegenerative diseases, constipation and digestive distress, and higher cholesterol levels.

The idea of nutraceuticals is a strategy for preventing and treating such conditions, and it can be conceptualized as follows: Researchers have been intrigued by the potential physiological and health benefits of plant-based nutrients such as beta-carotene and polyphenols (curcumin, resveratrol, and rutin), which are found in vegetables and fruits [11,12].

14.1.2 Classification of Nutraceuticals

Various categories of nutraceuticals exist depending on how they are understood and used. Nutraceuticals can be divided into two broad categories more broadly: potential nutraceuticals and established nutraceuticals. A potential nutraceutical is one that promises to provide certain medical or health benefits. Once there is enough clinical evidence to support the claimed health benefit, it might become an established one. Folic acid, for instance, was first classified as a prospective nutraceutical before sufficient clinical research data were produced to elevate it to a proven nutraceutical for the prevention of neural tube abnormalities. Because there is a dearth of thorough studies and clinical data, many nutraceutical items are still classified as potential products.

Another method of categorizing nutraceuticals is into three main categories: herbal/natural products (herbs or plant products as concentrates and extracts), dietary supplements (reagents derived from other sources aiding specific functions, such as sports nutrition), and nutrients (compounds with established nutritional functions, such as vitamins, minerals, amino acids, and fatty acids). Nutraceuticals are also divided into three groups according to the source of their ingredients: plant-based (lutein, lycopene, gallic acid, etc.), animal-based (conjugated linoleic acids, choline, lecithine, etc.), and microbes (*Lactobacillus acidophilus*, *Bifidobacterium bifidum*, *Saccharomyces boulardii*, etc.). Nutraceuticals can be categorized into five categories based on their mode or mechanism of action: anticancer (curcumin, capsaicin, genestein, etc.), antioxidant (ascorbic acid, beta-carotene, tocopherols, etc.), anti-inflammatory (linolenic acid, curcumin, quercetin, etc.), osteogenic (soy protein, calcium, daidzein, etc.), and positive influence on blood profile (beta glucan, resveratrol, tannins, etc.).

Nutraceuticals can also be divided into molecular or elemental groupings depending on their chemical nature. They consist of the following compounds: isoprenoid/terpenoid derivatives (carotenoids, tocopherols, saponins, etc.), phenolic compounds (coumanins, tannins, flavonols, etc.), protein/amino-based (amino acids, folate, capsaicinoids, etc.), carbohydrate derivatives (oligosaccharides, nonstarch polysaccharides, ascorbic acid, etc.), minerals (Ca, Se, K, etc.), fatty acids and structural lipids (lecithin, sphingolipids, n-3 PUFA, etc.), and microbiological (prebiotics, probiotics, and synbiotics).

Based on their sources, nutraceuticals or functional foods can be divided into two categories: natural (or traditional) and unnatural (or nontraditional). Products derived from plants, animals, minerals, or microorganisms fall within the category of natural sources. Traditional nutraceuticals are the name for this category. They can be subdivided into chemical constituents (nutrients, herbals, phytochemicals), nutraceutical enzymes (chemical constituents), and probiotic microorganisms. Biotechnologically produced nutritional supplements, also known as nontraditional nutraceuticals,

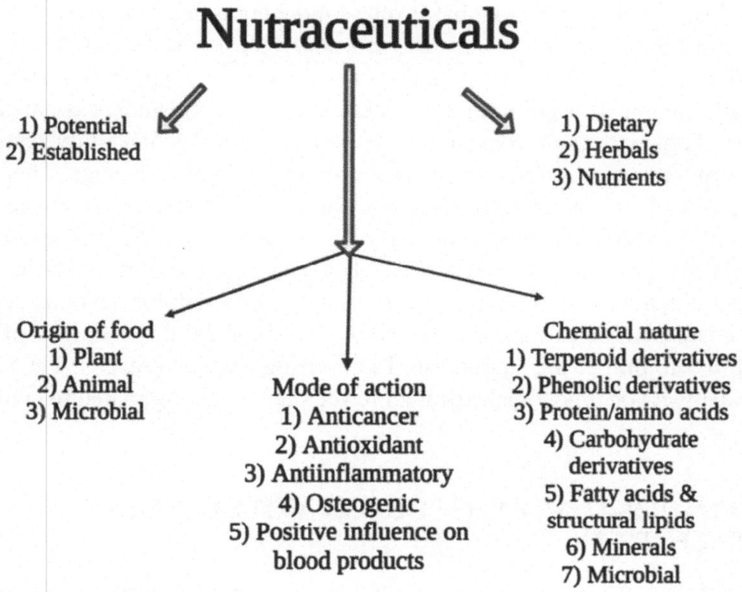

FIGURE 14.1 Classification of nutraceuticals depending on several sources.

are presented in Figure 14.1. They can be divided into recombinant and fortified nutraceuticals [13–15].

14.1.3 Nutraceuticals in Chronic Disease

Lifelong ailments with a significant influence on quality of life are chronic diseases such as type 2 diabetes mellitus (T2DM), cardiovascular disease (CVD), and immune-based illnesses such as rheumatoid arthritis, osteoarthritis, and Crohn's disease. The incidence of many of these illnesses is on the rise, especially those that are regarded as lifestyle diseases (T2DM, CVD), along with the risk factors that accompany them (obesity, hyperglycemia, hypertension, and hypercholesterolemia) [16,17]. Increasing physical activity and making dietary changes to reduce risk factors are now the primary therapeutic techniques for these disorders; as the condition worsens, pharmaceutical therapies become necessary. In particular, when knowledge of dietary sources of nutrients/nonnutrients is limited and nutraceuticals present a promising alternative, nutraceuticals with good safety profiles have become increasingly appealing alternatives to reduce some of the risk factors and symptoms linked to these chronic conditions [18,19].

14.2 NUTRACEUTICALS FOR HUMAN HEALTH

Over the years, the nutraceuticals sector has aimed for betterment of human health with excellent positive effects on the same. There are many applications in which nutraceuticals can be used for the treatment of inflammation, arthritis, high cholesterol, and diabetes. One of the natural gifts used widely for human use in Indian households is honey. Natural honey is used in feeding newborns owing to its good nutritional value and other health benefits. Natural honey is an excellent source of enzymes and antioxidants that help with digestion. It can be applied to the treatment of ulcers, the lowering of risk factors linked to metabolic and cardiovascular disorders, and the management of ocular illnesses. Additionally, honey has potent antibacterial properties and can be used to clean infected wounds [20,21].

The food industry, working at a global level, is inducing a great volume of food waste, and much research aims to exploit this food waste as a source of nutraceuticals. The implementation of the same will have a significant effect on the environment. One type of food waste, citrus peel, contains phenolic chemicals used for effective antioxidant action. Numerous positive impacts on human health are known to result from natural substances' antioxidant activity. Antioxidants safeguard the body by scavenging free radicals and avoiding oxidative damage. Their effectiveness in reducing the risk of cancer and cardiovascular illness, their capacity to reduce inflammation, and their capacity to manage and prevent Alzheimer's disease are all well documented. Citrus peel efficacy in the treatment of type-2 diabetes has been supported by scientific evidence. The phenolic content obtained from shaddock peels shows good inhibitory activity against amylase along with glucosidase. The extract can also be implemented for the treatment of hypertension, owing to its ability to inhibit the angiotensin I-converting enzyme (ACE1). Thus shaddock peel has shown good effects on type 2 diabetes with fewer side effects compared to conventional therapies [22,23].

14.3 CONVENTIONAL MODE OF DESIGN AND DELIVERY OF NUTRACEUTICALS

Nutraceuticals are marketed either in single formulations or in combinations in their concentrated forms. Many dosage form options are available, including capsules, powders for oral delivery, and pills. From the topical administration point of view, tinctures are given. Suppositories containing food bioactive components as nutraceuticals are used for effective treatment of patients. One of the major challenges associated with this formulation is lesser bioavailability because of the low solubility along with reduced permeability of the bioactive components. The stability of the bioactive compounds also causes less absorption of the drug compared to the gastrointestinal tract, thus inducing only partial absorption and reduced biological activity. In the conventional mode of delivery of nutraceuticals, the majority of the supplied nutraceutical dose is either excreted or causes nonspecific toxicity and unpleasant side effects due to undesirable biodistribution. This is a serious issue for researchers, who have used the concepts of nanotechnology for the efficient delivery of nutraceuticals, strengthening their biological activity by stabilizing and improving the solubility and permeability of poorly soluble nutraceuticals to overcome these problems. Effective encapsulation and administration using a nanoparticle-based delivery system could increase efficacy and decrease nonspecific toxicity.

14.3.1 Overview of Problems Associated with Nutraceuticals

Everything we eat affects how long we live. Only if the ingredients in food are able to enter the systemic circulation do they have the power to change biological processes. Low aqueous solubility and bioavailability are two typical issues with the majority of the natural ingredients utilized in nutraceuticals.

14.3.2 Issues of Solubility and Bioavailability in Nutraceuticals

Whether a bioactive substance is absorbed orally depends on how it interacts with the digestive system. Since water makes up the majority of gastrointestinal fluid, providing a medication, vitamin, bioactive, or other nutraceutical with poor aqueous solubility serves no purpose. In actuality, the majority of nutraceuticals lie within BCS Classes II and IV, which represent low aqueous solubility and low membrane permeability, respectively. Bioactives' aqueous solubility is closely tied to their release behavior, which is connected to their bioavailability. Table 14.1 lists some of the issues with frequently used nutraceuticals.

TABLE 14.1
Examples of nutraceuticals and their health benefits and issues

Nutraceutical	Food source	Health impacts	Problems
Omega-3	Mackerel, salmon, cod liver oil, trout, tuna, eggs, milk, yogurt, margarine, cereal, flaxseed, oatmeal, canola oil, mustard oil	Used to decrease heart disease risk	Less water solubility, less chemical stability, high lipid oxidation
Silymarin	Seeds of *Silybum marianum*	Acts as an antioxidant through free radical scavenger mechanism; used as a hepatoprotective. agent	Issues with water solubility, chemical stability, and gastric degradation, leading to lesser bioavailability
β-carotene, α-carotene	Orange/red vegetables, carrots, tomatoes, broccoli, citrus fruits, berries, cantaloupe, pumpkin, paprika, green vegetables, dandelion	Used as antioxidants and antiaging agents	Issues with water solubility and oxidation, leading to stability issues
Naringenin	Grapefruit, oranges, tomatoes	Acts as an antioxidant, anti-inflammatory, and hepatoprotective agent	Issues with water solubility, leading to reduced bioavailability
Lycopene	Tomatoes and tomato products	Acts as an antioxidant action and a prostate-protecting agent	Issues with water solubility, leading to reduced bioavailability
Lutein	Bananas, egg yolk, green vegetables (kale, spinach, lettuce, broccoli, peas)	Used as an antioxidant agent	Issues with water solubility, leading to reduced bioavailability
Puerarin	Kudzu roots	Used as an antioxidant agent	Poor water solubility
Capsaicin	Chili peppers	Used as an anti-inflammatory agent and known for its neurophysiological effects	Poor water solubility
Curcumin	Turmeric	Used in households and the pharmaceutical industry for many purposes: anti-inflammatory, antioxidant, anticancer, protective agent against several chronic conditions (HIV infection, neurological, cardiovascular, and skin diseases)	Issues with water solubility, leading to reduced bioavailability
Vitamin A (retinol)	Fish liver oils, eggs, chicken	Acts as an antioxidant and promotes immune health	Issues with water solubility, leading to poor bioavailability

By creating nanoformulations, these agents could be successfully delivered in vivo, minimizing the drawbacks that come with them. The bioactive substance may not degrade due to thermal stability, and photodegradation can be avoided. Vitamins, minerals, and lipids, which make up a large portion of food, are all sensitive to oxidation, which can result in an unpleasant taste and disintegration of food, but all of these substances can be encapsulated inside innovative carrier systems to prevent these issues. By using liposomes, solid lipid nanoparticles, and other colloidal systems, the release behavior in vivo may be altered [24–26].

14.3.3 Interaction with Food and Gastrointestinal Constituents

The fluid in the gastrointestinal tract is extremely rich in constituents. The components that make up gastrointestinal fluid come from two different sources: first, the body's own synthesized enzymes, mucus, minerals, bile salt, phospholipids, and other substances; and second, proteins, dietary fibers, surfactants, and other substances. In these conditions, the components in nutraceuticals are likely to combine to create complexes, which could reduce the amount of components available for systemic circulation. Due to the many chemical interactions that take place in the fluid of the gastrointestinal tract, many bioactive medicines have decreased bioavailability. Chelating substances found in the fluid of the gastrointestinal tract can change the further absorption and digestion of various mineral ions. The gastrointestinal tract contains a variety of charged species, including enzymes, bile salts, phospholipids, and others, to which electrically charged proteins (biopolymers) may bind through electrostatic interactions. Nutraceuticals may degrade as a result of other factors, such as pH changes, ionic strength, enzyme breakdowns, and mechanistic motilities. As a result, delivery techniques with protective mechanisms, such as solid lipid nanoparticles, nanostructured lipid carriers, and encapsulated formulations, may help to stop the degradation or loss of biological components such as vitamins and minerals as well as increase their effectiveness [27–29].

14.4 DIFFERENT METHODS OF SYNTHESIS AND FORMULATION OF NANOPARTICLES WITH NUTRACEUTICALS

There are several methods for the synthesis and formulation of nanoparticles with nutraceuticals, including the following:

- Physical methods: These methods include techniques such as high-pressure homogenization, ultrasonication, and spray drying, which are used to reduce the particle size of nutraceuticals to the nanoscale [30,31].
- Chemical methods: These methods involve the use of chemical agents to create nanoparticles, such as surfactants, polymers, and other agents, to stabilize and control the size of the particles [32].
- Biomedical methods: These methods involve the use of biological systems, such as bacteria or cells, to create nanoparticles.
- Combination methods: This includes a combination of any of the above methods to improve the characteristics of the nanoparticles, such as the rate of release, bioavailability, and stability. The choice of method will depend on the nutraceutical being used, the desired properties of the nanoparticles, and the intended application [33].

14.4.1 Physical Methods

Mechanical energy is applied to reduce the particle size of the nutraceuticals to that of the nanoscale with the help of physical methods. The synthesis, along with formulation of the nanoparticles with nutraceuticals, is carried out using several physical methods [12,13]. These methods include the following:

- High-pressure homogenization: This method uses high pressure to homogenize a mixture of the nutraceutical and a liquid, such as water or an organic solvent. The high pressure causes the mixture to pass through a small nozzle, leading to the formation of small droplets that solidify upon release of pressure, resulting in the formation of nanoparticles.
- Ultrasonication: This method uses high-frequency sound waves to generate cavitation bubbles in a mixture of nutraceutical and liquid. The rapid expansion and contraction of these bubbles

leads to the formation of small droplets, which solidify upon release of pressure, resulting in the formation of nanoparticles.
- Spray drying: This method involves atomizing a solution or suspension of the nutraceutical into a stream of hot air, which causes the droplets to dry and solidify, forming nanoparticles [34,35].

These methods can be efficient in creating nanoparticles and can provide good control of particle size and shape; however, they may lead to potential damage to nutraceuticals, and it is important to consider the specific nutraceutical used, the desired properties of the nanoparticles, and the intended application before choosing a method.

14.4.2 Chemical Methods

Chemical methods for the synthesis and formulation of nanoparticles with nutraceuticals involve the use of chemical agents to create nanoparticles. These methods include the following:

14.4.2.1 Sol-gel Methods

This method uses a chemical solution called a "sol" that contains the nutraceutical and a precursor for the nanoparticle's matrix. The sol is then gelated to form the nanoparticles.

- Emulsion-based methods: These involve mixing the nutraceutical with a liquid, such as water or an organic solvent, and then emulsifying the mixture using an emulsifying agent, such as a surfactant. The emulsion droplets solidify upon release of pressure, resulting in the formation of nanoparticles.
- Polymer-based methods: These involve using a polymer, such as polyethylene glycol (PEG), to stabilize and control the size of the nutraceutical particles. The nutraceutical is mixed with the polymer and then processed to form nanoparticles.
- Coprecipitation: This method involves mixing a nutraceutical with a salt solution and then adding a precipitating agent to cause the formation of nanoparticles.
- Interfacial deposition: This method involves depositing a nutraceutical onto the surface of an existing nanoparticle, such as a liposome.

Chemical methods can be efficient in creating nanoparticles and can provide good control of particle size and shape; however, they may lead to potential damage to nutraceuticals, and it is important to consider the specific nutraceutical used, the desired properties of the nanoparticles, and the intended application before choosing a method.

14.4.2.2 Biochemical Methods

Biochemical methods for the synthesis and formulation of nanoparticles with nutraceuticals involve the use of biological systems, such as bacteria or cells, to create nanoparticles. These methods include the following:

- Microorganism-based methods: These involve using microorganisms, such as bacteria or fungi, to produce nanoparticles by secreting extracellular polymeric substances (EPS) around nutraceuticals.
- Cell-based methods: These involve using cells, such as plant or animal cells, to produce nanoparticles by secreting polymers or other substances around nutraceuticals.
- Enzymatic-based method: This method involves using enzymes to modify nutraceuticals and form nanoparticles.

Biochemical methods can be efficient in creating nanoparticles and can provide good control of particle size and shape; however, they may lead to potential damage to nutraceuticals, and it is important to consider the specific nutraceutical used, the desired properties of the nanoparticles, and the intended application before choosing a method. Additionally, it is important to consider the safety and potential toxicity of the biological system used [36,37].

14.4.3 Combination Methods

Combination methods for the synthesis and formulation of nanoparticles with nutraceuticals involve the use of a combination of different methods to improve the characteristics of the nanoparticles. These methods include the following:

- Physical-chemical methods: These involve using a combination of physical and chemical methods to create nanoparticles. The surfactants are used for the stabilization of the nanoparticles, and high-pressure homogenization ensures that the particles are the correct size.
- Physical-biomedical methods: These involve using a combination of physical and biomedical methods to create nanoparticles. For example, ultrasonication can be used to reduce the particle size of nutraceuticals, and then bacteria can be used to produce extracellular polymeric substances (EPS) around nutraceuticals.
- Chemical-biomedical methods: These involve using a combination of chemical and biomedical methods to create nanoparticles, for example, using a polymer to stabilize and control the size of the nutraceutical particles and then using enzymes to modify the nutraceuticals.

Combination methods can provide more control over the characteristics of the nanoparticles, such as the rate of release, bioavailability, and stability. It can also improve the efficiency of creating nanoparticles. However, it is important to consider the specific nutraceutical used, the desired properties of the nanoparticles, and the intended application before choosing a method [38–40].

14.5 NANOTECHNOLOGICAL BOOSTER TOOLS FOR NUTRACEUTICALS

Nanotechnology can provide various tools and techniques that can be used to enhance nutraceutical performance. These tools include the following:

- Nanoencapsulation: This is a technique that involves enclosing a nutraceutical within a nanoparticle, such as a liposome or a polymeric nanoparticle. This can improve the solubility, stability, and bioavailability of nutraceuticals.
- Nanoparticle formulation: This involves the use of various methods to create nanoparticles from nutraceuticals, such as high-pressure homogenization, ultrasonication, and spray drying. This can improve the solubility, stability, and bioavailability of nutraceuticals.
- Surface modification: This is a technique that involves modifying the surface of nutraceutical nanoparticles to improve their properties, for example by adding a coating of a material such as polyethylene glycol (PEG). This approach helps to improve the stability and bioavailability of nutraceuticals.
- Nanoparticle targeting: This is a technique that involves attaching specific molecules to the surface of nutraceutical nanoparticles. This approach helps the particles to target specific cells or tissues in the body.
- Nanoparticle tracking analysis (NTA): This is a technique that allows visualization and quantification of nanoparticles. It can be used to determine the size, shape, concentration, and stability of nutraceutical nanoparticles.

- DLS: This is the technique that allows us to determine the size of nanoparticles in a solution, and it can be used to determine the size distribution of nutraceutical nanoparticles[41,42].

14.5.1 From Nanopharmaceuticals to Nanonutraceuticals

Nanopharmaceuticals are defined as drugs or drug delivery systems that use nanotechnology to improve their performance. They can include nanoparticles, liposomes, dendrimers, and other types of nanostructures that are used to improve the bioavailability, stability, and targeted delivery of drugs. Nanonutraceuticals, on the other hand, are defined as dietary supplements or functional foods that use nanotechnology to improve their performance. They can include nanoparticles, liposomes, dendrimers, and other types of nanostructures that are used to improve the bioavailability, stability, and targeted delivery of nutraceuticals. Both nanopharmaceuticals and nanonutraceuticals have the potential to improve the performance of their respective products. They can improve bioavailability, stability, and targeted delivery, which can lead to better treatment outcomes for patients and more effective dietary supplements for consumers. Notably, the development of nanopharmaceuticals and nanonutraceuticals is a complex process that can take several years to complete. The process also requires experts in various fields, including chemistry, pharmacology, toxicology, and regulatory affairs [43,44].

14.5.2 Nanoparticles in Nutraceuticals

Nanoparticles in nutraceuticals refer to the use of nanotechnology to create small particles from nutraceuticals that can improve their performance. The use of nanoparticles in nutraceuticals can improve their solubility, stability, and bioavailability, which can lead to better treatment outcomes for patients and more effective dietary supplements for consumers. By reducing the particle size of nutraceuticals to the nanoscale, their surface area increases, leading to an increase in the rate of dissolution and absorption in the body. Encapsulating nutraceuticals in a nanoparticle, such as a liposome or a polymeric nanoparticle, can help to protect them from degradation and oxidation. By attaching specific molecules to the surface of nutraceutical nanoparticles, they can be targeted to specific cells or tissues in the body, which can lead to more effective treatment outcomes. Some of the methods used to create nanoparticles from nutraceuticals include physical methods such as high-pressure homogenization, ultrasonication and spray drying, chemical methods such as sol-gel methods, emulsion-based methods, and polymer-based methods, and biochemical methods such as microorganism-based methods, cell-based methods, and enzymatic-based methods. Nanoparticles in nutraceuticals can provide many benefits, and it is important to thoroughly test their safety and potential toxicity before they are used in any application [45].

14.5.3 Solid Lipid Nanoparticles and Nanostructured Lipid Carriers

Solid lipid nanoparticles (SLNs) along with nanostructured lipid carriers (NLCs) are implemented for the improvement of the nutraceutical performance. They are composed of lipids, which are naturally occurring compounds that can help to improve the solubility, stability, and bioavailability of nutraceuticals.

SLNs are composed of a solid lipid core, which can encapsulate a nutraceutical. The solid lipid core helps to protect the nutraceutical from degradation and oxidation, and the small particle size increases the surface area available for dissolution, leading to improved bioavailability. NLCs act similar to SLNs, but they contain lipid components in liquid as well as solid form. The liquid lipid component is added for better solubility and bioavailability of the nutraceuticals. The drug loading capacity of NLCs is greater than that of SLNs, and they are more suitable for hydrophobic drugs, which are not soluble into the solid lipids. Various methods are implemented to produce SLNs and NLCs, including high-pressure homogenization, microfluidization, and ultrasonification. These methods are useful for the generation of the narrow size distribution type of the nanoparticles to improve their performance [46].

14.5.4 Liposomes for Nutraceuticals

Liposomes are spherical structures composed of a phospholipid bilayer that can encapsulate a nutraceutical. They are similar to cells in that they have an outer phospholipid membrane that surrounds an internal aqueous compartment. Liposomes can help improve the solubility of hydrophobic nutraceuticals, which are not soluble in water or other polar solvents, by encapsulating them within the hydrophobic core of the liposome. Liposomes can help to protect nutraceuticals from degradation and oxidation by encapsulating them within the phospholipid bilayer. For the better attachment of molecules like antibodies, carbohydrates along with peptides and the engineered liposomes can be used. These are helpful for the effective targeting of specific cells or tissues inside the body. The liposomes are designed for effective release of the nutraceutical payload at a specific rate into the target site with reference to body response conditions like pH and temperature changes. Various methods are used for the production of the liposomes, including the thin film hydration method, the reverse phase evaporation method, and the extrusion method. Liposomes can be made in different sizes, and it is important to choose the right size and properties of the liposomes to suit the intended application. Liposomes have been widely studied in the field of drug delivery and have shown promising results in improving the solubility, stability, and targeted delivery of nutraceuticals. However, it is important to conduct proper testing to determine the safety and potential toxicity of liposomes before they are used in any nutraceutical [47].

14.5.5 Nanoemulsions for Nutraceuticals

Nanoemulsions are a type of emulsion in which the droplet size is in the range of 20 to 200 nanometers. Nanoemulsions can help improve the solubility of hydrophobic nutraceuticals, which are not soluble in water or other polar solvents, by encapsulating them within the hydrophobic core of the emulsion droplets. Nanoemulsions can help to protect nutraceuticals from degradation and oxidation by encapsulating them within emulsion droplets. Nanoemulsions can be designed to release their nutraceutical payload at a specific rate or in response to specific conditions, such as pH or temperature changes. The increment in the rate of dissolution and the absorption of the nutraceuticals into the system can be achieved with the help of the high surface area to volume ratio and by monitoring the droplet size of the nanoemulsions. Nanoemulsions can be produced using various methods, such as high-pressure homogenization, ultrasonication, and microfluidization. These methods can create emulsion droplets with a narrow size distribution, which can help to improve their performance [48].

14.5.6 Nanogels for Nutraceuticals

Nanogels are a type of hydrogel that are composed of a network of polymers and have a particle size in the nanometer range. Nanogels can help improve the solubility of hydrophobic nutraceuticals, which are not soluble in water or other polar solvents, by encapsulating them within the hydrophobic core of the nanogel. Nanogels can help to protect nutraceuticals from degradation and oxidation by encapsulating them within the nanogel matrix. Nanogels can be designed to release their nutraceutical payload at a specific rate or in response to specific conditions, such as pH or temperature changes, by controlling the properties of the polymers and the loading of the nutraceuticals. Nanogels have a high surface area to volume ratio, which enhances the rate of dissolution and the absorption of the nutraceuticals into the body. Different methods are used for the preparation of the

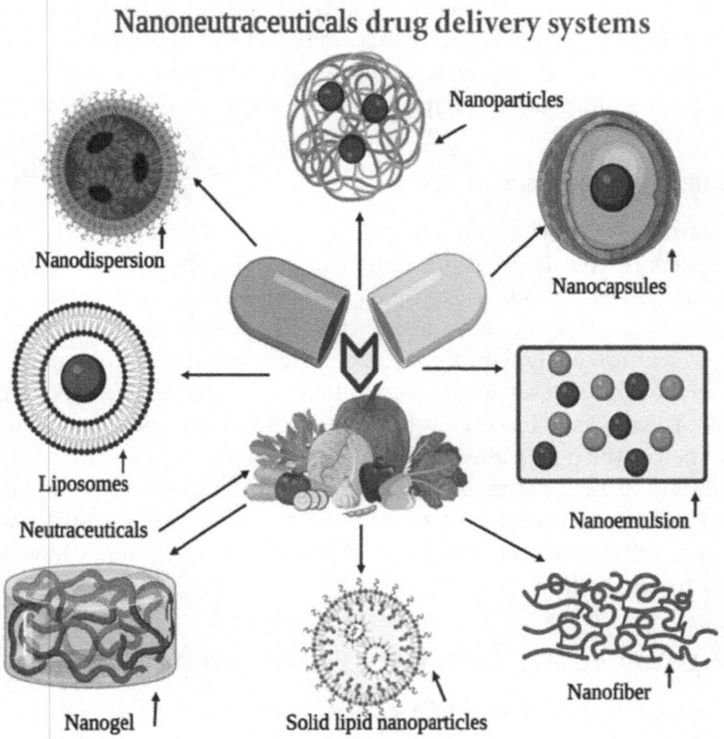

FIGURE 14.2 Nanonutraceuticals drug delivery systems.

Includes nanocarriers involving nanodispersion, nanoparticles, liposomes, nanogels, solid lipid nanoparticles, nanoemulsion, nanocapsules, and nanofibers.

nanogel, including inverse emulsion polymerization, self-assembly, and nanoprecipitation. These methods can create nanogels with a narrow size distribution, which can help to improve their performance [49]. Like other nanocarriers, nanogels are used for nutraceutical delivery, as presented in Figure 14.2.

14.5.7 Nanodispersions and Nanoencapsulates

Nanodispersions and nanoencapsulates are forms of delivery systems used in nutraceuticals to improve the bioavailability and stability of active ingredients. Nanodispersions are suspensions of nanoparticles in a liquid medium, while nanoencapsulates are particles that have a coating or shell around them, which can protect the active ingredient from degradation and improve its solubility. Enhancement of efficacy can be achieved by using these systems for better targeting of specific cells or tissues into the body. However, it should be noted that more research is needed to fully understand the potential health and safety implications of using these systems for nutraceuticals [50].

14.5.8 Nanolaminates for Nutraceuticals

Nanolaminates are a type of delivery system that consists of multiple layers of material, each with different properties, that are used to encapsulate and protect active ingredients in nutraceuticals. These systems can be composed of a variety of materials, such as polymers, lipids, or proteins, and can be used to improve the stability, bioavailability, and targeted delivery of the active ingredient.

The nanolaminates are implemented for the protection of the sensitive ingredients against the environment, including light and oxygen. These are also implemented to control the release rate of the active moieties inside the body. However, it should be noted that more research is needed to fully understand the potential health and safety implications of using these systems for nutraceuticals [51].

14.5.9 Nanofibers for Nutraceuticals

The delivery of nutraceuticals can be carried out by using nanofibers, which are very thin fibers. The size of these fibers is typically in the nanometer range. The fibers can be made from a variety of materials, such as polymers, proteins, or lipids, and can be used to improve the stability, bioavailability, and targeted delivery of active ingredients. Nanofibers can be produced using various techniques, such as electrospinning, which uses an electric field to pull a polymer solution into thin fibers, or self-assembly, which uses natural interactions between molecules to form fibers. One of the advantages of using nanofibers as a delivery system is that they can mimic the extracellular matrix of the body, which increases the chances of active ingredients being taken up by cells. Nanofibers provide an excellent surface area to volume ratio, which enhances the interaction of the active moieties with cells and controls the release action. However, it should be noted that more research is needed to fully understand the potential health and safety implications of using these systems for nutraceuticals [52].

14.6 NANONUTRACEUTICALS IN CANCER TREATMENT

Much research is exploring the potential of nanonutraceuticals in the treatment of disease. A size reduction tool is used for preparation of the delivery system including nanoemulsions, solid lipid nanoparticles and nanocomposites. Nanospheres and carbon nanotubes are also used, but monitoring of crucial factors like solubility, bioavailability, targeted delivery, and stability should be considered. Pharmaceutical nanotechnology has helped the delivery of nanosized nutraceuticals in the nonpharmaceutical domain for effective delivery of the added bioactives required to benefit the patient's health. More investigations are needed to determine the applications of different nanocarriers for delivery of nutraceuticals to increase their bioavailability, distribution, and effects. The creation of nanonutraceuticals, a new class of novel products, results from this. Particularly for individuals who are ineligible for standard pharmaceutical therapy, nanonutraceutical formulations offer a valuable method for managing medical disorders. Nutraceuticals are significantly impacted by nanopharmaceuticals and the significant transformation of the pharmaceutical business.

Agarwal et al. have explored the patent and approval scenario for nanopharmaceuticals. The effective application of nanopharmaceuticals for biomedical relies on excellent knowledge of manufacturing processes and safety considerations. The enhancement of their function, as well as excellent nutrient bioavailability, is crucial for beneficial impacts on human health, including the treatment of cancer, cardiovascular disease, and neurological disorders. Nutraceuticals include characteristics of general biological therapies that are used to stop cancerous processes, promote wellbeing, and manage a number of disease symptoms. Nutraceuticals' significant success and strong demand can be attributed to their ability to reveal the right therapeutic potential while minimizing the adverse effects associated with conventional medications used to treat a variety of illnesses. According to previous studies, the combination of curcumin and magnetic nanoparticles significantly reduced the number of cancer cells in vitro when used in conjunction with magnetic hyperthermia. Although there are many anticancer drugs on the market, acquired drug resistance and the severe side effects of these therapies are the main obstacles to effective chemotherapy.

It is proposed that newer, more creative pharmaceuticals need to be rationally created with fewer side effects. Phytochemicals and nutritional dietary components have a long and storied history, as well as significant uses in the field of contemporary medicine. Nutritional supplements can control

DNA transcription and the elements that cause DNA damage in tumor cells. The natural components have showed their potential to resensitize drug-resistant cancer by means of their ability to influence many signaling pathways, including the AMPK, EGF-mediated, and NF-kB signaling pathways. Nutraceuticals act on their molecular targets to target malignant cells on numerous levels, causing cell cycle arrest or apoptosis in the process. They also decrease angiogenesis, invasion, and metastasis. By obstructing the signaling pathways required for cancer progression, they halt the spread of cancer. Since the advent of nanomedicine, the scientific community has learned how to use this discipline to overcome some of the most fundamental problems with conventional medicine, such as drug side effects, nonspecific cell targeting, and drug instability. The rapid development of innovative formulations known as nanopharmaceuticals and nanonutraceuticals is a result of these breakthroughs. Due to its potential to reduce the aforementioned limitations of conventional therapy, nanomedicine is at the forefront of the fight against cancer. Tumor cells with an efflux pump contribute to the release of chemotherapeutic agents. The development of resistance to conventional therapy and the inadequate targeting of cancer cells are listed as major hurdles for conventional approaches to treatment. Drug-carrying nanoparticles can overcome resistance by targeting tumor cells precisely while still carrying out their intended function, as presented in Figure 14.3. The history and existing knowledge of the nanopharmaceuticals and nanonutraceuticals are required for further exploration of the anticancer properties of nutraceuticals.

The treatment of numerous chronic and harmful conditions, such as Alzheimer's disease, Parkinsonism, cancer, diabetes, inflammation, and obesity, is provided by well-established therapies in the healthcare industries through the incorporation of nanotechnology in the form of potent nutraceuticals and cutting-edge nanocarriers. Nanotechnology's diverse uses in the fields of healthcare and cosmetics expand the functional benefits that can be delivered by nutraceutical delivery. The emergence of nanotechnology, which enables bioactive substances to cross the blood–brain barrier, has helped nutraceuticals overcome their obstacles. The fundamental way that nutraceuticals differ from nonpharmaceuticals is that they supply nutritional supplements for therapeutic reasons rather than a pharmacologically active substance. The use of phytochemicals in nutraceuticals has increased dramatically. Many plant-based phytoconstituents contain biological functional traits like plant growth or defenses, and they have been investigated for their diverse effects and excellent potential as medicines for the betterment of human life. Oxidative stress usually induces DNA damage, which is addressed by the antioxidant characteristics of the phytochemicals, and the same is required for the chemoprevention of the cancer. Indoles commonly found in the cabbage can be used to lessen the effects of estrogen, reducing the risk of breast cancer [43–45,53]. Some of the curcumin nutraceutical patents for human benefit are listed in Table 14.2.

14.7 REGULATORY CONCERNS ABOUT NANONUTRACEUTICALS

Like other pharmaceuticals, nanonutraceuticals require clinical data to show their safety and effectiveness in the treatment of health issues. One of the major issues is the lack of regulatory framework. The nanopharmaceutical aspect of nutraceuticals is related to the nanoversion of the selected nutraceuticals, like nanoparticles, for better human health. Their development is primarily motivated by the desire to enhance their physicochemical properties (e.g. solubility) and pharmacokinetic properties (tmax, Cmax, and area under curve properties), with the ultimate goal of reducing the dose necessary to observe the therapeutic or nutritional outcome and, consequently, the potential risk of toxicity. Therefore, factors such as efficiency, quality, and safety should be taken into account.

Further research is therefore required to address the regulatory difficulties surrounding nanopharmaceuticals. According to already existing legislation and without any additional testing, the FDA has cleared all of the nanopharmaceuticals that are now on the market. As a result, the

TABLE 14.2
Patent details for few examples of curcumin-loaded nanoformulation for human health

No.	Patent ID	Patent title	Assignee	Inventor
1	CA-2732635-A1	Curcumin nanoparticles and methods of producing the same	Santosh Kumar Kar, Feroz Akhtar, Gopesh Ray, Atul Kumar Pandey	Santosh Kumar Kar, Feroz Akhtar, Gopesh Ray, Atul Kumar Pandey
2	WO-2013175507-A1	Novel highly bioavailable, water soluble and sustained release nanoformulations of hydrophobic plant derived compounds and extracts	Laila Pharmaceuticals Pvt. Ltd.	Ravichandran Sripathy, Venkata Narasimha Siva Rama Raju Mandapati, Gopaal Ajay, Nirvanashetty Somashekara, Ramchand Nanappan Chaniyilparampu, Rama Raju Gokaraju, Ganga Raju Gokaraju, Kiran Bhupathiraju, Dwarakanath Anjana
3	US-2015283239-A1	Graphene derivative-based composition for drug delivery and preparation method thereof	Research & Business Foundation Sungkyunkwan University	Hyoyoung Lee, Surajit Some
4	WO-2011101859-A1	A novel water soluble curcumin loaded nanoparticulate system for cancer therapy	Institute Of Life Sciences	Kumar Sahoo Sanjeeb, Mohanty Chandana
5	US-10420731-B1	Method of synthesizing lignin-based nanocompositions	King Saud University	Mohammed Saeed A. Alqahtani, Ali Saeed Althabit Alqahtani, Rabbani Syed S. Baji
6	EP-3582757-A1	Ayurvedic encapsulated gold nanoparticles, fabrication methods and cancer therapeutic methods	The Curators of the University of Missouri	Kattesh Katti, Menka Khhobchandani, Kavita Katti, Alsam Khan, Chintamani Joshi, Vinay Mutalik
7	US-8333994-B2	Stabilized, biocompatible gold nanoparticles and enviro-friendly method for making same	The Curators of the University of Missouri	Kattesh V. Katti, Raghuraman Kannan, Kavita K. Katti, Satish Kumar Nune
8	US-2021113554-A1	Nutraceuticals having sustained release for improved bioavailability and method of production	Bioactives Llc	Doddabele L. Madhavi, Daniel I. Kagan
9	EP-3057604-B1	Nanoparticle compositions and methods as carriers of nutraceutical factors across cell membranes and biological barriers	Nanosphere Health Sciences Inc.	Richard Clark Kaufman
10	US-2014056981-A1	Method for producing drug-loaded polymeric nanoparticles by polymerization in presence of drugs	Centro de Investigacion en Quimica Aplicada	Raúl Guillermo López-Campos, Hened Saade Caballero

TABLE 14.2 (Continued)
Patent details for few examples of curcumin-loaded nanoformulation for human health

No.	Patent ID	Patent title	Assignee	Inventor
11	WO-2016013026-A1	Curcumin-sophorolipid complex	Council of Scientific & Industrial Research	Pradeep Kumar Singh, Asmita Ashutosh Prabhune, Satishchandra Balkrishna Ogale
12	WO-2016113762-A1	Modified cyclodextrin coated magnetite nanoparticles for targeted delivery of hydrophobic drugs	Council of Scientific & Industrial Research	Pattayil Joy, Jayaprabha Kunnoth Naduvili
13	WO-2006038552-A1	Composition containing fine particles and process for producing the same	Eisai R&D Management Co., Ltd.	Yosuke Ueki
14	CA-2665343-C	Water-dispersible oral, parenteral, and topical formulations for poorly water soluble drugs using smart polymeric nanoparticles	The Johns Hopkins University, Anirban Maitra, Georg Feldmann, Savita Bisht	Anirban Maitra, Georg Feldmann, Savita Bisht
15	EP-2346492-A2	A process for nanoemulsification of curcumin and derivatives of curcumin	Laila Pharmaceuticals Pvt. Ltd.	Anitha Krishnan Nair, Ramchand Nanappan Chaniyilparampu, Bhupathiraju Kiran, Rama Raju Gokaraju, Ganga Raju Gokaraju, Trimurtulu Golakoti
16	ES-2718225-B2	Inclusion complex to improve the biodisponibility of biologically active hydrosolubble compounds	University of Valladolid	Gil Jesus Martin, Petruta Mihaela Matei, Lebena Eduardo Perez
17	CA-3128653-A1	Nanoemulsion compositions comprising saponins for increasing bioavailability	Readymix Foods Corp., Iraj Mehrnia, Thomas Hielscher	Iraj Mehrnia, Thomas Hielscher
18	EP-4119126-A1	Nanoformulation with diverse functional molecules from turmeric and process for preparation of the same	Zeroharm Sciences Private Limited	Swetha Ravikanti, Lakshmanan Chitrabhanu
19	EP-2814332-B1	Nanoparticles comprising a vegetable hydrophobic protein and a water miscible non-volatile organic solvent and uses thereof	Bionanoplus, S.L.	Hesham H.A. Salman, Izaskun Goñi Azcárate, Irene Esparza Catalán
20	WO-2014001050-A1	Colloidal particles comprising hydrophobic polymer, curcuminoid, and blue colourant	Unilever Plc, Unilever N.V., Conopco, Inc., D/B/A Unilever	Ashok Ranchodbhai Patel, Krassimir Petkov Velikov
21	US-2013149381-A1	Absorption method for entrapment of drugs in polymeric nanoparticles	Centro de Investigacion en Quimica Aplicada	Raúl Guillermo López Campos, Hened Saade Caballero
22	EP-3405304-B1	Gold nanoparticles and ecological method of production	Torskal, Morel, Anne-laure	Anne-Laure Morel
23	AU-2022202259-A1	Novel cannabinoid formulations	Yissum Research Development Company of the Hebrew University of Jerusalem Ltd.	Abraham J. Domb, Amnon Hoffman, Dvora Izgelov

(continued)

TABLE 14.2 (Continued)
Patent details for few examples of curcumin-loaded nanoformulation for human health

No.	Patent ID	Patent title	Assignee	Inventor
24	EP-3609612-B1	Coated powders having high photostability	Nanophase Technologies Corporation	Harry W. Sarkas, Abigail R. Hooper, Nathan H. Hoffmann
25	CA-2992879-C	Delivery of bioactive, nanoencapsulated antioxidants	Board of Supervisors of Louisiana State University and Agricultural and Mechanical College, Iuliu Hatieganu University of Medicine and Pharmacy, Christina M. Sabliov, Carlos E. Astete, Ede Bodoki, Oliviu Vostinaru, Ovidiu Ciprian Samoila, Elena Dinte	Christina M. Sabliov, Carlos E. Astete, Ede Bodoki, Oliviu Vostinaru, Ovidiu Ciprian Samoila, Elena Dinte
26	US-2013337075-A1	Nanomedicines for early nerve repair	Purdue Research Foundation	Ji-Xin Cheng
27	EP-2587940-B1	Particles comprising hydrophobic polymer and hydrophobic phenolic compound	Unilever N.V., Unilever PLC	Ashok Ranchodbhai Patel, Jyoti Kumar Tiwari, Krassimir Petkov Velikov
28	US-2016310620-A1	Curcumin coated magnetite nanoparticles for biomedical applications	Council of Scientific & Industrial Research	Alias Joy Pattayil, Kunnoth Naduvilidam Jayaprabha
29	US-2016193159-A1	Liposomal curcumin for treatment of cancer	Board of Regents, The University of Texas System	Razelle Kurzrock, Lan Li, Kapil Mehta, Bharat Aggarwal
30	US-2018318217-A1	Liposomal Curcumin for Treatment of Diseases	Board of Regents, The University of Texas System, Signpath Pharma Inc.	Razelle Kurzrock, Lan Li, Kapil Mehta, Bharat Bhushan Aggarwal, Lawrence Helson
31	US-10416167-B2	Nanoparticles for mitochondrial trafficking of agents	University of Georgia Research Foundation, Inc.	Shanta Dhar, Sean M. Marrache
32	US-9555011-B2	Formulation of active agent loaded activated PLGA nanoparticles for targeted cancer nano-therapeutics	University of North Texas Health Science Center at Fort Worth	Arthur R. C. Braden, Jamboor K. Vishwanatha
33	US-2020390710-A1	Method for the Production of Commercial Nanoparticle and Microparticle Powders	Iceutica Pty Ltd.	Aaron Dodd, H William Bosch, Felix Meiser, Marck Norret, Adrian Russell
34	US-2020188311-A1	Edible plant-derived microvesicle compositions for diagnosis and treatment of disease	University of Louisville Research Foundation, Inc.	Huang-Ge Zhang
35	US-2020095553-A1	Formulation comprising extracellular vesicles, method for producing the same, and uses thereof	Mackay Memorial Hospital	Kuen-Der Yang, Yeou-Ping Tsao, Chie-Pein Chen

TABLE 14.2 (Continued)
Patent details for few examples of curcumin-loaded nanoformulation for human health

No.	Patent ID	Patent title	Assignee	Inventor
36	US-10799457-B2	Exosomal compositions and methods for the treatment of disease	University of Louisville Research Foundation, Inc.	Huang-Ge Zhang
37	ES-2885052-T3	Multifunctional formulation composed of natural ingredients and its preparation / manufacturing method	Datt Life Science Pvt Ltd	Rajan Datt, Ramadhar Kumar, Siddharth Pandey, Pallavi Shrivastava
38	US-2019374467-A1	Milk-derived microvesicle compositions and related methods	University of Louisville Research Foundation, Inc.	Ramesh C. Gupta, Radha Munagala, Farrukh Aqil, Jeyaprakash Jeyabalan
39	US-8192718-B1	Pharmaceutical composition of nanoparticles	Gp Medical, Inc., National Tsing Hua University	Hsing-Wen Sung, Kiran Sonaje, Ho-Ngoc Nguyen, Er-Yuan Chuang, Hosheng Tu
40	US-9895322-B2	Pharmaceutical compositions for the delivery of substantially water-insoluble drugs	Mewa Singh, Maninder Sandhu	Mewa Singh, Maninder Sandhu
41	US-2018000962-A1	Cysteine engineered antibodies and conjugates	Genentech, Inc.	Charles W. Eigenbrot, Jr., Jagath Reddy Junutula, Henry Lowman, Helga E. Raab, Richard Vandlen
42	US-2020316546-A1	Production of encapsulated nanoparticles at high volume fractions	Iceutica Pty Ltd.	Aaron Dodd, Felix Meiser, Marck Norret, Adrian Russell, H. William Bosch
43	US-2018169662-A1	Production of encapsulated nanoparticles at commercial scale	Iceutica Pty Ltd.	Aaron Dodd, Felix Meiser, Marck Norret, Adrian Russell, H. William Bosch
44	US-9138411-B2	Curcumin-ER, a liposomal-PLGA sustained release nanocurcumin for minimizing QT prolongation for cancer therapy	University of North Texas Health Science Center at Fort Worth, Signpath Pharma, Inc.	Amalendu Prakash Ranjan, Anindita Mukerjee, Jamboor K. Vishwanatha, Lawrence Helson
45	EP-3003281-B1	Novel core-shell nanoparticles for oral drug delivery	South Dakota State University	Omathanu Perumal, Mohammed Saeed A. Alqahtani
46	DK-2729123-T3	Cannabinoidreceptorbindende funds and formations and methods	Patchtek Inc, Jonathan Bevier	Jonathan Bevier
47	US-2020069598-A1	Cholestosome vesicles for incorporation of molecules into chylomicrons	Therasyn Sensors, Inc.	Jerome J. Schentag, Mary P. McCourt, Lawrence Mielnicki, Julie Hughes
48	US-10682055-B1	Fluorescent imaging on a head-mountable device	Verily Life Sciences Llc	Adrian Wong, Philipp Helmut Schmaelzle, Andrew Jason Conrad, Samuel D'Amico
49	US-2023113351-A1	Compositions and methods for managing disorders	Diamond Therapeutics Inc.	Judith Blumstock, William J. TYLER
50	US-9557271-B2	Optical biomodule for detection of diseases	Mohammad A. Mazed, Sayeeda Mazed	Mohammad A. Mazed, Sayeeda Mazed
51	US-10414826-B2	Monomethylvaline compounds capable of conjugation to ligands	Seattle Genetics, Inc.	Svetlana O. Doronina, Peter D. Senter, Brian E. Toki, Toni Beth Kline

(continued)

TABLE 14.2 (Continued)
Patent details for few examples of curcumin-loaded nanoformulation for human health

No.	Patent ID	Patent title	Assignee	Inventor
52	US-2020397807-A1	Nicotinyl riboside compounds and their uses	MitoPower, LLC	Gautham Tumkur PRANESH, Gangadhara Ganapati, Atignal Shankara Rao Arvind, G. Mani Subramanian
53	ES-2885868-T3	Composition, particulate materials and methods to manufacture particulate materials	University of Queensland	Chengzhang Yu, Meihua Yu, Hongwei Zhang, Yusilawati Ahmad Nor, Hao Song, Neena Mitter
54	US-10905770-B2	Topical delivery of therapeutic agents using cell-penetrating peptides for the treatment of age-related macular degeneration and other eye diseases	Macregen, Inc.	Keith Roizman, John Jacob Requard, III, Felicity Jane De Cogan
55	US-11147800-B2	Combinatorial methods to improve the therapeutic benefit of bisantrene and analogs and derivatives thereof	Race Oncology Ltd.	William J. Garner, Arnie Franklin, John Rothman
56	CA-2914583-C	Coated particles and compositions comprising same	Vyome Biosciences Pvt. Ltd., Suresh Chawral, Sudhanand Prasad, Kirti Bajaj, Vyome Therapeutics Limited	Suresh Chawral, Sudhanand Prasad, Kirti Bajaj
57	US-10278922-B2	Hydrogel encapsulated cells and anti-inflammatory drugs	Massachusetts Institute of Technology, The Children's Medical Center Corporation	Daniel G. Anderson, Robert S. Langer, Tram T. Dang
58	US-9663462-B2	Anti-parasitic methods and compositions utilizing diindolylmethane-related indoles	Bioresponse, L.L.C.	Michael A. Zeligs
59	US-2018042549-A1	Methods for making controlled delivery devices having zero order kinetics	Board of Regents, The University of Texas System	Paul S. Ho, Salomon Stavchansky, Phillip Bowman, Zhiquan Luo, Zhuojie Wu, Ashish Rastogi
60	US-2019321293-A1	Liposomal mitigation of drug-induced long QT syndrome and potassium delayed-rectifier current	Signpath Pharma, Inc.	Lawrence Helson
61	US-2022396796-A1	Rna for cancer therapy	Curevac Ag	Mariola Fotin-Mleczek, Aleksandra Kowalczyk, Regina Heidenreich, Ulrike Gnad-Vogt, Ute Klinkhardt, Katja Fiedler
62	WO-2015006735-A1	Nanoparticles and methods of producing the same	The Board of Trustees of the University of Illinois	Ying Liu, Z. Jim Wang

TABLE 14.2 (Continued)
Patent details for few examples of curcumin-loaded nanoformulation for human health

No.	Patent ID	Patent title	Assignee	Inventor
63	EP-1993365-B1	Bioavailable curcuminoid formulations for treating Alzheimer's disease and other age-related disorders	The Regents of the University of California, the US Government represented by the Department of Veterans Affairs	Sally A. Frautschy, Greg M. Cole
64	US-2019367529-A1	Use of the fl-one hundred eighteen core chemical structure platform to generate fl-one hundred eighteen derivatives for treatment of human disease	Canget Biotekpharma, Llc	Fengzhi Li, Xiang Ling
65	US-2019224193-A1	Compositions of natural extracts and use thereof in methods for preventing or treating diseases	Christopher Brian Reid	Christopher Brian Reid, Que T. Collins
66	ES-2739194-T3	Formulation and method to increase oral bioavailability of drugs	Yissum Res Dev Co of Hebrew Univ Jerusalem Ltd	Amnon Hoffman, Abraham J Domb, Irina Cherniakov

Source: Patent data collected from https://patents.google.com/?q=(carotenoids+nanoparticle)&oq=carotenoids+nanoparticle)

FDA's regulatory principles have been questioned by the approval of novel pharmaceuticals. The pharmacodynamics and pharmacokinetic characteristics of nutraceuticals are intended to be changed by nanocarriers. Furthermore, the nanoformulation of nutraceuticals has a number of benefits, including enhanced bioavailability, solubility, entrapment effectiveness, stability, target specificity, and controlled or sustained release of bioactive chemicals. Due to a lack of information and the potential health risks associated with nanoparticles, more toxicological studies are advised to determine the safety and regulatory status of nanoengineered particles. The development of innovative nanoformulations, for which clearance from federal agencies such as the FDA is required to be adopted for safety regulations and efficacy, was the nutraceutical industry's most significant benefit [41,54,55].

14.8 SAFETY PERSPECTIVES FOR NANONUTRACEUTICALS

Concerns regarding potential health and environmental implications have been raised by the use of nanoparticles in foods and food packaging. Due to their small size, nanoparticles may go undetected in food and are more prone to accumulate in tissues and organs. According to studies, human cell-sized nanoparticles can be inhaled and may cause damage where they are deposited in the lungs. Additionally, the particles could enter the bloodstream or go to other organs, posing a risk to health. It has been revealed that nanoparticles as small as 50 nm can infiltrate cells and that particles 30 nm in size can pass through blood. The liver, spleen, heart, brain, and other organs may be impacted by their deposition upon translocation from the blood circulatory system. The introduction of foreign particles into the bloodstream is also likely to occur with nanoparticles, which are designed to transport dietary supplements. Because of their distinctive size, nanoparticles will also be more readily absorbed and bioavailable, boosting their concentration in cells. It is unclear what these

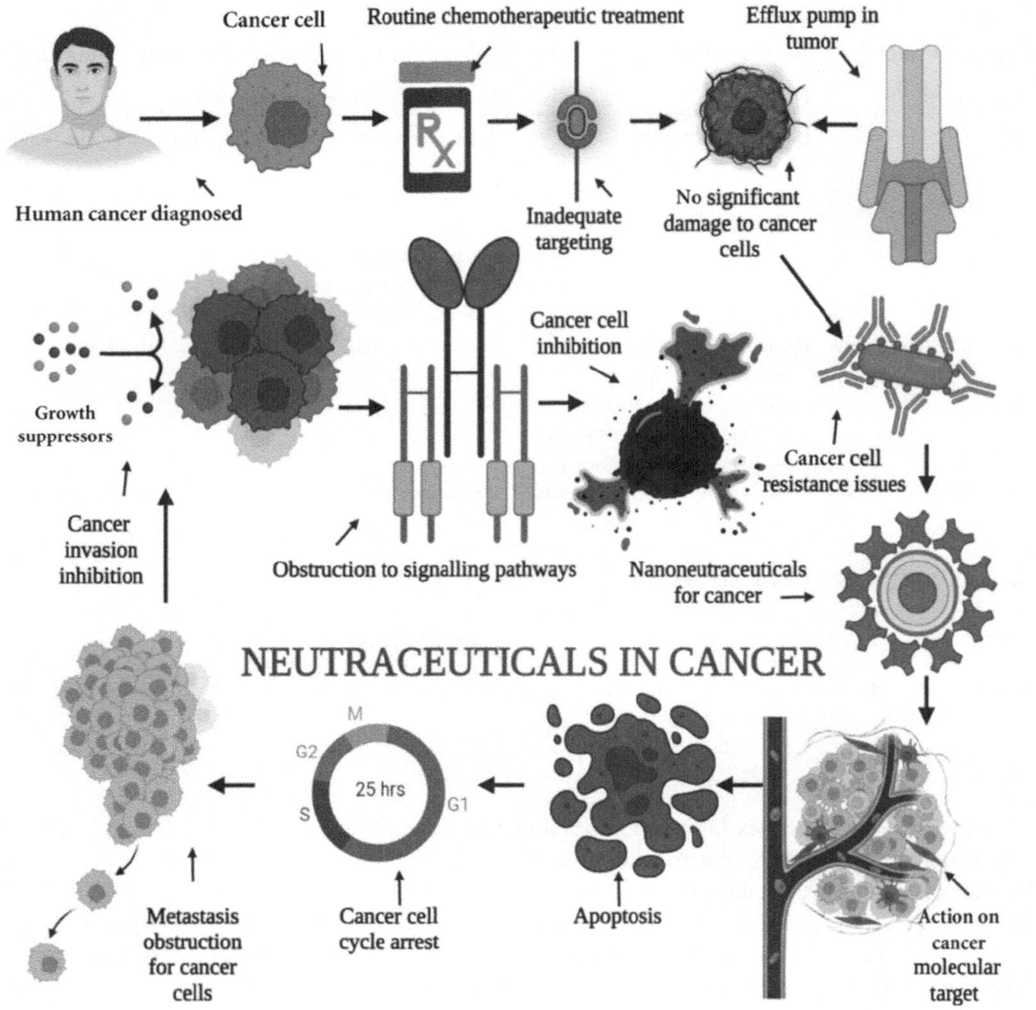

FIGURE 14.3 Role of nanonutraceuticals in cancer treatment.

Conventional chemotherapeutic agents suffer inadequate targeting and also involve the presence of an efflux pump inside the tumor, which further develops resistance. Nanonutraceuticals are used to address these issues, providing targeted action on cancer cells and inducing cell cycle arrest and apoptosis. The results of chili pepper studies on the potential anticancer activities of nutraceuticals made from phytochemicals are encouraging [43–45,53].

higher quantities will do to the environment. The particles may also interact with proteins and other nutrients, which may change how well they are absorbed and cause them to end up in the gastro-intestinal system.

This could have a number of unfavorable outcomes. The characteristics of nanoparticles, such as their size, size distribution, rate of agglomeration, structure, chemical composition, surface area, surface charge, and behavior when they aggregate, are anticipated to play a significant role in determining the toxicity of the nanomaterial when used in foods. Colon cancer, obesity, food allergies, diabetes, colitis, and immunological dysfunction have all been linked to the presence of nanoparticles in food. Environmental issues could arise from the disposal of packaging materials

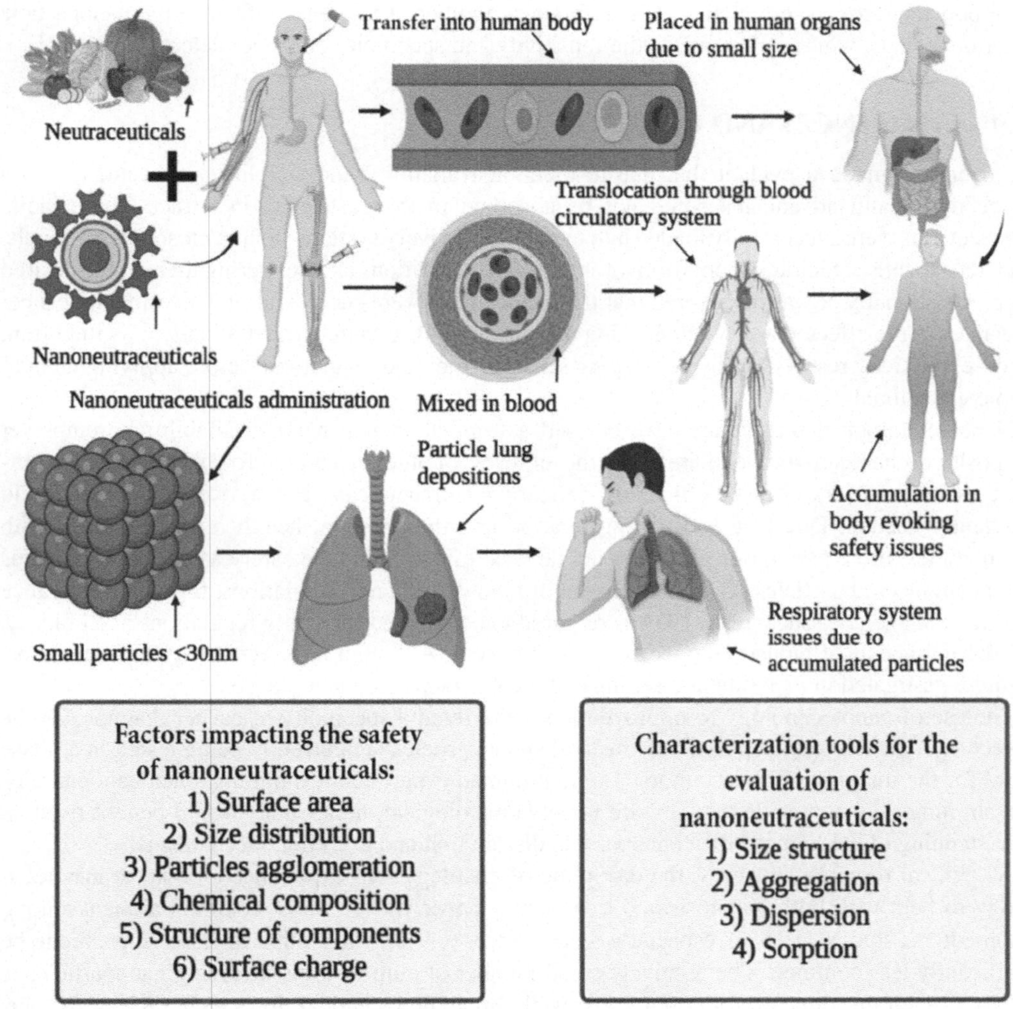

FIGURE 14.4 Safety issues of delivery for nanonutraceuticals related to factors such as surface charge and size.

The inhalation of smaller particles (<30 nm) produces lung accumulation of the particles, which creates respiratory system damage. The particles get mixed in the blood, and their translocation through the blood circulatory system produces safety issues due to longer periods of accumulation inside the body.

with incorporated nanoparticles. Consuming fish or other plant or animal commodities produced using such contaminated water or suspected pollution of water sources carry a significant danger, as presented in Figure 14.4.

Although analytical methods such as spectroscopy and image processing are now available to study the properties of nanoparticles such as size, structure, aggregation, dispersion, and sorption, the variety of these particles makes it necessary to combine various methods to fully understand the safety aspects of nanomaterial. To promote technological advances to ensure food safety, a detailed understanding of the implications of nanoscale applications is necessary.

Consideration must be given to how cost-effectively nanopharmaceuticals and nanonutraceuticals are developed. A properly thought-out and well-designed manufacturing process will be necessary to support their cost of production. For instance, a number of variables influence the nanoparticle detection signals, which in turn affect the sensitivity and specificity of cancer detection [56–58].

14.9 CHALLENGES AND OUTLOOK

The overview made it evident that due to their bioavailability and stability, nutraceuticals with prospective health advantages could not be absorbed in the gastrointestinal tract. These labile nutraceuticals were successfully loaded into nanobased delivery systems such as niosomes, nanogels, and nanocapsules, shielding them from a variety of pH conditions and delivering them to the desired place. Additionally, it was discovered that the nanocarriers were successful in delivering anticancer bioactives to the affected area while leaving the surrounding area unharmed. Finally, it is important to address toxicity restrictions by developing sensible rules and regulations before applying them to people and animals.

Nanotechnological techniques have crossed a frontier, increasing bioavailability to empower the positive characteristics and improve the efficacy of nutraceuticals. To examine the significance of nanocarriers in more efficiently delivering nutraceuticals, a new overview of this field is urgently needed. Due to a lack of information and the potential health risks associated with nanoparticles, more toxicological studies are advised to determine the safety and regulatory status of nanoengineered particles. The development of innovative nanoformulations, for which clearance from federal agencies such as the FDA is required to be adopted for safety regulations and efficacy, was the nutraceutical industry's most significant benefit. Although nanotechnology has been thoroughly investigated in experimental settings, clinical data are frequently absent.

The use of nanotechnology to improve the sensitivity and specificity of cancer diagnosis is the subject of numerous ongoing trials. Ultrasmall silica particles are currently being tested in a phase 1 trial for the imaging of brain tumors. The first human experiment examining silica nanoparticles in brain tumors is now underway. Future targeted oncological therapeutics would benefit from an understanding of the human silica nanoparticle distribution and excretion mechanisms.

As clinical research advances, the discipline of oncology will experience a boom in nanotechnology to help with both diagnosis and treatment. Cancer was the most common ailment among nanomedicine studies (19.3%), especially breast cancer (6.9%), while other illnesses appeared to be significantly less common. The relatively small number of clinical trials involving nanoparticles is consistent with the situation for curcumin, a well-known nutraceutical, for which 3.8% of the pertinent literature consisted of clinical trials, as well as for the ethnopharmacology literature, in which 1.3% of the studies were clinical trials. As a reward for these efforts, we anticipate that further clinical studies for nanopharmaceuticals and nanoneutraceuticals will be conducted in the near future and receive high numbers of citations [2,59].

14.10 CONCLUSIONS

The treatment of numerous chronic and harmful conditions, such as Alzheimer's disease, Parkinsonism, cancer, diabetes, inflammation, and obesity, is provided by well-established therapies in the healthcare industries through the incorporation of nanotechnology into potent nutraceuticals and cutting-edge nanocarriers. Nanotechnology's diverse uses in the fields of healthcare and cosmetics expand the functional benefits that can be delivered by nutraceutical delivery. The emergence of nanotechnology, which enables bioactive substances to cross the blood–brain barrier, has helped nutraceuticals overcome their obstacles. The pharmacodynamics and pharmacokinetic characteristics of nutraceuticals are intended to be changed by nanocarriers. Furthermore, the nanoformulation of nutraceuticals has a number of benefits, including enhanced bioavailability, solubility, entrapment effectiveness, stability, target specificity, and controlled or sustained release of bioactive chemical for the same. One of the great development during the last decade is intervention by means of nanotechnology-based tools for effective delivery in the food and nutrient sector. The exact mechanism of the potential benefits is under scrutiny, and many explorations are still required to ensure their safety and to meet regulatory standards. The same types of studies and assessments are also required for nanoprobiotics and nanoprebiotics. Many regulatory authorities in different countries are evaluating the possible risks in relation to nanotechnology-based food safety and supply chain scenarios. To address all these types of issues, additional evidence-based research is required, along with evaluations of the safety, improved bioavailability, and efficacy of nutraceuticals for the betterment of the human health in the near future.

REFERENCES

1 Girija, A. R.; Sakthi Kumar, D. Novel Paradigm of Design and Delivery of Nutraceuticals with Nanoscience and Technology. In *Nutraceuticals*; Elsevier, 2016; pp. 343–385.

2 Paolino, D.; Mancuso, A.; Cristiano, M. C.; Froiio, F.; Lammari, N.; Celia, C.; Fresta, M. Nanonutraceuticals: The New Frontier of Supplementary Food. *Nanomaterials*, 2021, *11* (3), 792. https://doi.org/10.3390/nano11030792

3 Gholap, A.D.; Hatvate, N.T.; Sawant, S.D.; Khemka, P.N.; Vora, L.K. Unlocking the Potential of Nanobiohybrids to Combat Environmental Pollution. In *Nanobiohybrids for Advanced Wastewater Treatment and Energy Recovery*; IWA Publishing, 2023; pp. 169–190.

4. Gholap, A.D.; Sawant, S.D.; Sayyad, S.F.; Hatvate, N.T.; Chavan, M.; Rojekar, S.; Faiyazuddin, M. Food Biopackaging for Human Benefits: Status and Perspectives. In *Biobased Packaging Materials*; Springer Nature Singapore: Singapore, 2023; pp. 1–35.

5. Perera, K.Y.; Jaiswal, A.K.; Jaiswal, S. Biopolymer-Based Sustainable Food Packaging Materials: Challenges, Solutions, and Applications. *Foods* 2023, *12*, 2422, doi:10.3390/foods12122422.

6. Versino, F.; Ortega, F.; Monroy, Y.; Rivero, S.; López, O.V.; García, M.A. Sustainable and Bio-Based Food Packaging: A Review on Past and Current Design Innovations. *Foods* 2023, *12*, 1057, doi:10.3390/foods12051057.

7. Gholap, A.D.; Bhowmik, D.D.; Deshmukh, A.Y.; Hatvate, N.T. Quintessential Impact of Dendrimer Bioconjugates in Targeted Drug Delivery. In 2023; pp. 257–302.

8. Gholap, A.D.; Gupta, J.S.; Kamandar, P.A.; Banchhod, G. V.; Hatvate, N.T. Antibody-Drug Conjugates for Cancer Therapy: An up-to-Date Review on the Chemistry and Pharmacology. In; 2023; pp. 105–190.

9. Gholap, A.D.; Said, R.P.; Pawar, R.D.; Ambore, G.S.; Hatvate, N.T. Importance of Carbohydrate-Drug Conjugates in Vaccine Development: A Detailed Review. In; 2023; pp. 191–256.

10. Maurya, A.P.; Chauhan, J.; Yadav, D.K.; Gangwar, R.; Maurya, V.K. Nutraceuticals and Their Impact on Human Health. In *Preparation of Phytopharmaceuticals for the Management of Disorders*; Elsevier, 2021; pp. 229–254.

11. Mali, S.; Rathod, S.; Kale, N.; Shinde, N. Overview of Nutraceuticals. *Asian J. Pharm. Res.* 2022, 61–70, doi:10.52711/2231-5691.2022.00010.

12. Reque, P.M.; Brandelli, A. Encapsulation of Probiotics and Nutraceuticals: Applications in Functional Food Industry. *Trends Food Sci. Technol.* 2021, *114*, 1–10, doi:10.1016/j.tifs.2021.05.022.

13. AlAli, M.; Alqubaisy, M.; Aljaafari, M.N.; AlAli, A.O.; Baqais, L.; Molouki, A.; Abushelaibi, A.; Lai, K.-S.; Lim, S.-H.E. Nutraceuticals: Transformation of Conventional Foods into Health Promoters/Disease Preventers and Safety Considerations. *Molecules* 2021, *26*, 2540, doi:10.3390/molecules26092540.

14. De, S.; Gopikrishna, A.; Keerthana, V.; Girigoswami, A.; Girigoswami, K. An Overview of Nanoformulated Nutraceuticals and Their Therapeutic Approaches. *Curr. Nutr. Food Sci.* 2021, *17*, 392–407, doi:10.2174/1573401316999200901120458.

15 Nwosu, O.K.; Ubaoji, K.I. Nutraceuticals: History, Classification and Market Demand. In *Functional Foods and Nutraceuticals*; Springer International Publishing: Cham, 2020; pp. 13–22.

16. Rai, R.H.; Goyal, R.K.; Singh, R.B.; Handjiev, S.; Singh, J.; Darlenska, T.H.; Smail, M.M.A. Vitamins and Minerals as Nutraceuticals in Cardiovascular Diseases and Other Chronic Diseases. In *Functional Foods and Nutraceuticals in Metabolic and Non-Communicable Diseases*; Elsevier, 2022; pp. 651–670.

17. Allam, V.S.R.R.; Chellappan, D.K.; Jha, N.K.; Shastri, M.D.; Gupta, G.; Shukla, S.D.; Singh, S.K.; Sunkara, K.; Chitranshi, N.; Gupta, V.; et al. Treatment of Chronic Airway Diseases Using

Nutraceuticals: Mechanistic Insight. *Crit. Rev. Food Sci. Nutr.* 2022, *62*, 7576–7590, doi:10.1080/10408398.2021.1915744.
18. Daimary, U.D.; Parama, D.; Rana, V.; Banik, K.; Kumar, A.; Harsha, C.; Kunnumakkara, A.B. Emerging Roles of Cardamonin, a Multitargeted Nutraceutical in the Prevention and Treatment of Chronic Diseases. *Curr. Res. Pharmacol. Drug Discov.* 2021, *2*, 100008, doi:10.1016/j.crphar.2020.100008.
19. Sachdeva, V.; Roy, A.; Bharadvaja, N. Current Prospects of Nutraceuticals: A Review. *Curr. Pharm. Biotechnol.* 2020, *21*, 884–896, doi:10.2174/1389201021666200130113441.
20. Durazzo, A.; Lucarini, M.; Santini, A. Nutraceuticals in Human Health. *Foods* 2020, *9*, 370, doi:10.3390/foods9030370.
21. Choudhary, B.; Chauhan, O.P.; Mishra, A. Edible Seaweeds: A Potential Novel Source of Bioactive Metabolites and Nutraceuticals With Human Health Benefits. *Front. Mar. Sci.* 2021, *8*, doi:10.3389/fmars.2021.740054.
22. Carrizzo, A.; Izzo, C.; Forte, M.; Sommella, E.; Di Pietro, P.; Venturini, E.; Ciccarelli, M.; Galasso, G.; Rubattu, S.; Campiglia, P.; et al. A Novel Promising Frontier for Human Health: The Beneficial Effects of Nutraceuticals in Cardiovascular Diseases. *Int. J. Mol. Sci.* 2020, *21*, 8706, doi:10.3390/ijms21228706.
23. Damián, M.R.; Cortes-Perez, N.G.; Quintana, E.T.; Ortiz-Moreno, A.; Garfias Noguez, C.; Cruceño-Casarrubias, C.E.; Sánchez Pardo, M.E.; Bermúdez-Humarán, L.G. Functional Foods, Nutraceuticals and Probiotics: A Focus on Human Health. *Microorganisms* 2022, *10*, 1065, doi:10.3390/microorganisms10051065.
24. Fathi, M.; Mozafari, M.R.; Mohebbi, M. Nanoencapsulation of Food Ingredients Using Lipid Based Delivery Systems. *Trends Food Sci. Technol.* 2012, *23*, 13–27, doi:10.1016/j.tifs.2011.08.003.
25. Wahlang, B.; Pawar, Y.B.; Bansal, A.K. Identification of Permeability-Related Hurdles in Oral Delivery of Curcumin Using the Caco-2 Cell Model. *Eur. J. Pharm. Biopharm.* 2011, *77*, 275–282, doi:10.1016/j.ejpb.2010.12.006.
26. Ali, A.; Ahmad, U.; Akhtar, J.; Badruddeen; Khan, M.M. Engineered Nano Scale Formulation Strategies to Augment Efficiency of Nutraceuticals. *J. Funct. Foods* 2019, *62*, 103554, doi:10.1016/j.jff.2019.103554.
27. McClements, D.J. Food Hydrocolloids: Application as Functional Ingredients to Control Lipid Digestion and Bioavailability. *Food Hydrocoll.* 2021, *111*, 106404, doi:10.1016/j.foodhyd.2020.106404.
28. Swallah, M.S.; Fu, H.; Sun, H.; Affoh, R.; Yu, H. The Impact of Polyphenol on General Nutrient Metabolism in the Monogastric Gastrointestinal Tract. *J. Food Qual.* 2020, *2020*, 1–12, doi:10.1155/2020/5952834.
29. Shahidi, F.; Pan, Y. Influence of Food Matrix and Food Processing on the Chemical Interaction and Bioaccessibility of Dietary Phytochemicals: A Review. *Crit. Rev. Food Sci. Nutr.* 2022, *62*, 6421–6445, doi:10.1080/10408398.2021.1901650.
30. Gholap, A.D.; Kapare, H.S.; Pagar, S.; Kamandar, P.; Bhowmik, D.; Vishwakarma, N.; Raikwar, S.; Garkal, A.; Mehta, T.A.; Rojekar, S.; et al. Exploring Modified Chitosan-Based Gene Delivery Technologies for Therapeutic Advancements. *Int. J. Biol. Macromol.* 2024, *260*, 129581, doi:10.1016/j.ijbiomac.2024.129581.
31. Gholap, A.D.; Rojekar, S.; Kapare, H.S.; Vishwakarma, N.; Raikwar, S.; Garkal, A.; Mehta, T.A.; Jadhav, H.; Prajapati, M.K.; Annapure, U. Chitosan Scaffolds: Expanding Horizons in Biomedical Applications. *Carbohydr. Polym.* 2024, *323*, 121394, doi:10.1016/j.carbpol.2023.121394.
32. Helal, N.A.; Eassa, H.A.; Amer, A.M.; Eltokhy, M.A.; Edafiogho, I.; Nounou, M.I. Nutraceuticals' Novel Formulations: The Good, the Bad, the Unknown and Patents Involved. *Recent Pat. Drug Deliv. Formul.* 2019, *13*, 105–156, doi:10.2174/1872211313666190503112040.
33. Puri, V.; Nagpal, M.; Singh, I.; Singh, M.; Dhingra, G.A.; Huanbutta, K.; Dheer, D.; Sharma, A.; Sangnim, T. A Comprehensive Review on Nutraceuticals: Therapy Support and Formulation Challenges. *Nutrients* 2022, *14*, 4637, doi:10.3390/nu14214637.
34. Mohammady, M.; Mohammadi, Y.; Yousefi, G. Freeze-Drying of Pharmaceutical and Nutraceutical Nanoparticles: The Effects of Formulation and Technique Parameters on Nanoparticles Characteristics. *J. Pharm. Sci.* 2020, *109*, 3235–3247, doi:10.1016/j.xphs.2020.07.015.
35. Jamali, S.N.; Assadpour, E.; Jafari, S.M. Formulation and Application of Nanoemulsions for Nutraceuticals and Phytochemicals. *Curr. Med. Chem.* 2020, *27*, 3079–3095, doi:10.2174/0929867326666190620102820.

36. Samanta, P.; Dey, S.; Ghosh, A.R.; Kim, J.K. Nanoparticle Nutraceuticals in Aquaculture: A Recent Advances. *Aquaculture* 2022, *560*, 738494, doi:10.1016/j.aquaculture.2022.738494.
37. Patil, S.G.; Gattani, S.G.; Patil, K.; Bagade, S.; Mahapatra, D.K. Fabrication Methods of Therapeutic Nanoparticles: Potential Opportunities and Challenges in Drug Delivery and Nutraceutical Delivery. *Nutraceutical Deliv. Syst.* 2022, 91–131.
38. Srivastava, N.; Choudhury, A.R. Microbial Polysaccharide-Based Nanoformulations for Nutraceutical Delivery. *ACS Omega* 2022, *7*, 40724–40739, doi:10.1021/acsomega.2c06003.
39. Koul,B.;Poonia,A.K.;Yadav,D.;Jin,J.-O.Microbe-MediatedBiosynthesisofNanoparticles:Applications and Future Prospects. *Biomolecules* 2021, *11*, 886, doi:10.3390/biom11060886.
40. Khan, Y.; Sadia, H.; Ali Shah, S.Z.; Khan, M.N.; Shah, A.A.; Ullah, N.; Ullah, M.F.; Bibi, H.; Bafakeeh, O.T.; Khedher, N. Ben; et al. Classification, Synthetic, and Characterization Approaches to Nanoparticles, and Their Applications in Various Fields of Nanotechnology: A Review. *Catalysts* 2022, *12*, 1386, doi:10.3390/catal12111386.
41. Aguilar-Pérez, K.M.; Ruiz-Pulido, G.; Medina, D.I.; Parra-Saldivar, R.; Iqbal, H.M.N. Insight of Nanotechnological Processing for Nano-Fortified Functional Foods and Nutraceutical—Opportunities, Challenges, and Future Scope in Food for Better Health. *Crit. Rev. Food Sci. Nutr.* 2023, *63*, 4618–4635, doi:10.1080/10408398.2021.2004994.
42. Hussien, E.T.; Yuvaraj, M.; Sivaji, M.; Thilagavathi, T. Development of Nutraceuticals Using Nanotechnological Tools: Current Scenario And Future Prospects. In *Nanotechnology Horizons in Food Process Engineering*; Apple Academic Press, 2023; pp. 335–352.
43. Salama, L.; Pastor, E.; Stone, T.; Mousa, S. Emerging Nanopharmaceuticals and Nanonutraceuticals in Cancer Management. *Biomedicines* 2020, *8*, 347, doi:10.3390/biomedicines8090347.
44. Patra, J.K.; Das, G.; Fraceto, L.F.; Campos, E.V.R.; Rodriguez-Torres, M. del P.; Acosta-Torres, L.S.; Diaz-Torres, L.A.; Grillo, R.; Swamy, M.K.; Sharma, S.; et al. Nano Based Drug Delivery Systems: Recent Developments and Future Prospects. *J. Nanobiotechnology* 2018, *16*, 71, doi:10.1186/s12951-018-0392-8.
45. Maha Z Rizk; Hanan Farouk Aly. Nutraceuticals and Their Nanotechnology-Based Therapeutic Applications in Different Diseases. *Int. J. Biol. Med.* 2020, 15–32, doi:10.36811/ijbm.2020.110020.
46. Ghasemiyeh, P.; Mohammadi-Samani, S. Solid Lipid Nanoparticles and Nanostructured Lipid Carriers as Novel Drug Delivery Systems: Applications, Advantages and Disadvantages. *Res. Pharm. Sci.* 2018, *13*, 288, doi:10.4103/1735-5362.235156.
47. Subramani, T.; Ganapathyswamy, H. An Overview of Liposomal Nano-Encapsulation Techniques and Its Applications in Food and Nutraceutical. *J. Food Sci. Technol.* 2020, *57*, 3545–3555, doi:10.1007/s13197-020-04360-2.
48. Zhang, R.; Zhang, Z.; McClements, D.J. Nanoemulsions: An Emerging Platform for Increasing the Efficacy of Nutraceuticals in Foods. *Colloids Surfaces B Biointerfaces* 2020, *194*, 111202, doi:10.1016/j.colsurfb.2020.111202.
49. Silva, M.P.; Fabi, J.P. Food Biopolymers-Derived Nanogels for Encapsulation and Delivery of Biologically Active Compounds: A Perspective Review. *Food Hydrocoll. Heal.* 2022, *2*, 100079, doi:10.1016/j.fhfh.2022.100079.
50. Prieto, C.; Lagaron, J.M. Nanoencapsulation and Nanocoating of Bioactives of Application Interest in Food, Nutraceuticals and Pharma. *Nanomaterials* 2024, *14*, 313, doi:10.3390/nano14030313.
51. Manohar, M.; Paladhi, A.; Vallinayagam, S.; Rudrapal, M. Nanotechnological Approach in Nutraceuticals. In *Polyphenols*; Wiley, 2023; pp. 67–81.
52. Jurić, M.; Donsì, F.; Maslov Bandić, L.; Jurić, S. Natural-Based Electrospun Nanofibers: Challenges and Potential Applications in Agri-Food Sector. *Food Biosci.* 2023, *56*, 103372, doi:10.1016/j.fbio.2023.103372.
53. El-Far, A.; Godugu, K.; Salaheldin, T.; Darwish, N.; Saddiq, A.; Mousa, S. Nanonutraceuticals: Anti-Cancer Activity and Improved Safety of Chemotherapy by Costunolide and Its Nanoformulation against Colon and Breast Cancer. *Biomedicines* 2021, *9*, 990, doi:10.3390/biomedicines9080990.
54. Magne, T.M.; Alencar, L.M.R.; Carneiro, S.V.; Fechine, L.M.U.D.; Fechine, P.B.A.; Souza, P.F.N.; Portilho, F.L.; de Barros, A.O. da S.; Johari, S.A.; Ricci-Junior, E.; et al. Nano-Nutraceuticals for Health: Principles and Applications. *Rev. Bras. Farmacogn.* 2022, *33*, 73–88, doi:10.1007/s43450-022-00338-7.

55. Gupta, C.; Prakash, D. Nanonutraceuticals for Drug Delivery. In *Advances in Novel Formulations for Drug Delivery*; Wiley, 2023; pp. 521–540.
56. Gupta, R.C.; Lall, R.; Srivastava, A. *Nutraceuticals: Efficacy, Safety and Toxicity*; Academic Press, 2021.
57. Dasgupta, N.; Ranjan, S.; Ramalingam, C.; Kumar, A. Nanonutraceuticals: Are They Safe? In *Nanotechnology in Nutraceuticals*; CRC Press, 2016; pp. 339–366.
58. Bhavin, S.E.; Gajjar, A.K. Nanonutraceuticals. In *Nanotechnology in Medicine*; Wiley, 2021; pp. 281–298.
59. Dubey, A.K.; Chaudhry, S.K.; Singh, H.B.; Gupta, V.K.; Kaushik, A. Perspectives on Nano-Nutraceuticals to Manage Pre and Post COVID-19 Infections. *Biotechnol. Reports* 2022, *33*, e00712, doi:10.1016/j.btre.2022.e00712.

15 Nanoemulsions for Fat Reduction

Vanga Sri Varsha, Tanmayi Boreda, Sanjana Reddy Pailla, Kiran Yellappa Vajanthri, Bhushan Vishal, Yashasvi Kambhampati, T Gourav, and Bishwambhar Mishra
Corresponding author: Bishwambhar Mishra

15.1 INTRODUCTION

Nanotechnology has transformed several sectors, including food, by manipulating materials at the nanoscale (1 to 100 nm) (Gupta et al., 2016). Nanotechnology has the potential to modify food qualities such as texture, color, taste, and stability, resulting in the development of innovative food items. Nanoemulsions, which include droplets as small as 100 nm, have emerged as a critical use of nanotechnology in food. These emulsions, which are generally made up of oil, water, and emulsifiers, improve the water-solubility, bioavailability, and sensory properties of active substances such as flavoring agents and antioxidants (Fryd and Mason, 2012). Researchers have investigated both low-energy and high-energy techniques of producing nanoemulsions, with implications ranging from medicine delivery to enhancing food digestibility to synthesizing complex materials. Reduced fat consumption in food production also has environmental and health benefits. It has the potential to reduce greenhouse gas emissions and land use while also decreasing the risk of cardiovascular disease and obesity (Fryd and Mason, 2012). To minimize fat content in food items, several tactics such as employing plant-based oils, low-fat components, and functional additives such as inulin and polyphenols are used (Aswathanarayan and Vittal, 2019). This chapter delves further into nanoemulsions, their formulation, stability, and critical role in tackling fat-related issues in food. It also discusses future developments and problems in this subject, as well as the wide applications of nanoemulsions in many food categories, providing insights via real-world case studies (Silva et al., 2012; Koroleva and Yurtov, 2012).

15.2 FORMULATION ASPECTS AND METHODS FOR PREPARATION

15.2.1 High-energy Methods

High-energy techniques break down water and oil phases into minute drops and encourage homogeneous mixing by applying strong mechanical forces or external energy inputs. For several reasons, like enhanced stability and increased bioavailability of active components, high-energy techniques are essential in the creation of nanoemulsions (McClements, 2011).

15.2.1.1 Microfluidization

Micro fluidization is a high-energy method for nanoemulsion preparation. High shear forces are produced by forcing the emulsion through microchannels. This method produces a stable and finely dispersed nanoemulsion by dissolving the water and oil phases into nanosized drops. Improved homogeneity, increased stability, and effective encapsulation of bioactive substances are benefits of

DOI: 10.1201/9781003389408-16

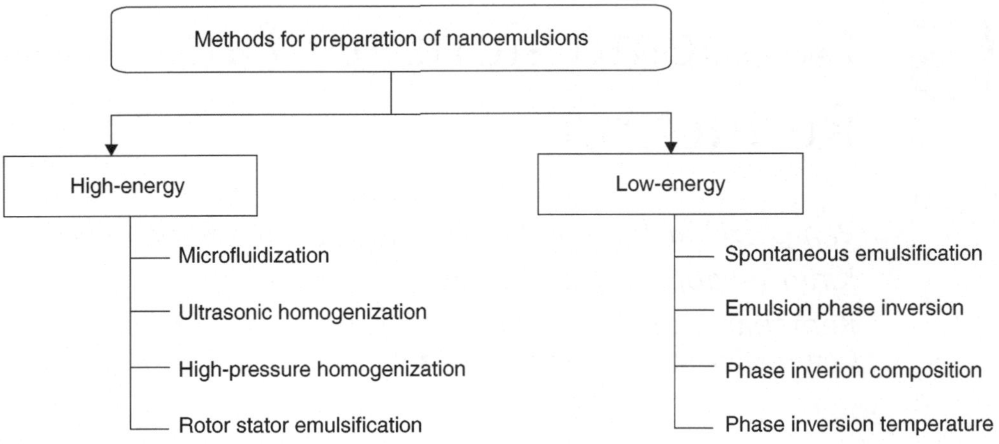

FIGURE 15.1 Overview of methods for preparation of nanoemulsions.

microfluidization. It has uses in the pharmaceutical, cosmetic, and food sectors, since fine control over droplet size and dispersion is essential to the effectiveness and quality of a product in these fields (McClements, 2012).

15.2.1.2 Ultrasonic Homogenization

Ultrasonic homogenization is a high-energy method for creating nanoemulsions, using ultrasonic waves to break water and oil phases into nanosized drops. This method creates microbubbles in the emulsion by using an ultrasonic probe to produce strong cavitation. High shear pressures produced by these implosions result in smaller droplet sizes and consistent mixing (Che Marzuki et al., 2019). For creating stable nanoemulsions with improved bioavailability, which are perfect for medicines and nutraceuticals, ultrasonic homogenization has been recognized for its effectiveness. This technique is frequently used in the food and cosmetics industries because it produces formulations that are transparent and nonsticky, which enhances sensory qualities and consumer acceptance (Wang et al., 2015).

15.2.1.3 High-pressure Homogenization

The process of high-pressure homogenization (HPH) is very effective for producing nanoemulsions. It involves applying high pressure to force the water and oil phases through a narrow opening, creating strong shear forces that reduce droplets to the nanoscale. HPH is extremely useful in the food, cosmetics, and pharmaceutical industries because it produces consistent, stable, and bioavailable nanoemulsions. HPH enhances product quality, sensory qualities, and shelf life by reducing droplet size. Due to its effectiveness in creating nanoemulsions, this technology is now widely used, and ongoing research is being done to make it better for other purposes (Modarres-Gheisari et al., 2019).

15.2.1.4 Rotor-stator Emulsification

In the rotor-stator emulsification technique, a rotor and stator system is used, and while the rotor quickly rotates inside the stator, strong mechanical shear forces are generated at the intersection of the water and oil phases. With the help of these forces, the phases break into nanoscale droplets, creating a stable nanoemulsion with fine dispersion (Gazolu-Rusanova et al., 2020). This method is highly efficient, offering close control over droplet size and distribution. In the food, cosmetics, and pharmaceutical industries, rotor-stator emulsification is frequently used to create nanoemulsions with increased stability and bioactive component delivery (Shi et al., 2022).

15.2.2 Low-Energy Methods

Low-energy methods for making nanoemulsions use techniques that reduce the amount of mechanical energy or external forces needed to produce fine oil-in-water or water-in-oil droplets at nanoscale. In comparison to high-energy technologies, these techniques are successful, affordable, and used in the pharmaceutical, food, and cosmetic industries (Safaya and Rotliwala, 2020).

15.2.2.1 Spontaneous Emulsification

A simple and fast technique for creating nanoemulsions without the use of external energy inputs is spontaneous emulsification. This process involves vigorously mixing an oil phase, an emulsifying agent, and a water phase. As a result of the decreased interfacial tension, nanosized droplets are spontaneously formed. With droplet sizes in the nanoscale range, this procedure produces stable, fine emulsions. The pharmaceutical, cosmetics, and food sectors can all benefit from the use of spontaneous emulsification because it is a quick, affordable, and effective way to create nanoemulsions for a variety of formulations and delivery methods (Akram et al., 2021).

15.2.2.2 Emulsion Phase Inversion

A powerful technique for producing nanoemulsions is the emulsion phase inversion method, in which the formulation spontaneously switches from water-in-oil (W/O) and oil-in-water (O/W) emulsions. The concentrations of surfactant and co-surfactant as well as the oil-to-water ratio must be carefully adjusted during this procedure. A W/O emulsion first forms, and if the water content or surfactant ratios are steadily increased, a phase inversion takes place, leading to the creation of stable nanoemulsions with tiny droplet sizes (Gao et al., 2019). For improving solubility, bioavailability, and product performance, this method's tremendous adaptability and ability to precisely adjust nanoemulsion properties make it important in the pharmaceutical, cosmetic, and food sectors (Akram et al., 2021; Feng et al., 2020).

15.2.2.3 Phase Inversion Composition

For preparation of nanoemulsions, the phase inversion composition (PIC) approach provides a flexible process. To create finely dispersed nanometer-sized droplets, a phase inversion can also be induced by altering the composition of the emulsion's constituent parts. The system goes through a phase transition, often from water-in-oil (W/O) to oil-in-water (O/W) or vice versa, which results in the development of a stable nanoemulsion by carefully regulating the ratios of oil, water, and emulsifiers. Since the PIC approach allows for precise droplet size control, it is advantageous in sectors like medicines and cosmetics (Feng et al., 2020).

15.2.2.4 Phase Inversion Temperature

The phase inversion temperature (PIT) method depends on the temperature-induced transition from the water-in-oil (W/O) to the water-in-oil (W/O) emulsion phases or vice versa. This technique induces the development of nanosized droplets during the phase inversion process by carefully controlling temperature and emulsifying agents. PIT technique is useful for various applications in the cosmetic, pharmaceutical, and food sectors due to its ease and scalability. It provides control over droplet size and stability, enhancing the bioavailability of medicinal formulations and enhancing product quality, as well as improving the sensory qualities of cosmetics and food products (Ren et al., 2019; Gao et al., 2019).

Figure 15.1 depicts nanoemulsion preparation in two categories: high-energy methods with microfluidization, ultrasonic homogenization, high-pressure homogenization, and rotor-stator emulsification; and low-energy methods with spontaneous emulsification, emulsion phase inversion, phase inversion composition, and phase inversion temperature.

15.2.3 Factors Affecting Nanoemulsion Stability

Stability in nanoemulsions is affected by many variables. The selection of surfactants and cosurfactants plays an important role in stabilizing the emulsion by reducing interfacial tension. Droplet size and dispersion are also impacted by emulsification techniques such as high-pressure homogenization or sonication. Stability is impacted by the composition of the emulsion, including the oil and water phases. Phase separation can be caused by temperature and pH changes in the interfacial characteristics. Stability may be impacted by the presence of ions, additives, or pollutants. For creating and maintaining stable nanoemulsions proper consideration of these factors is necessary (Oca-Ávalos et al., 2017).

15.3 ROLE OF NANOEMULSIONS IN FAT REDUCTION

Nanoemulsions play a crucial role in the food sector, particularly in reducing fat content in food items. They are a sort of emulsion in which nanometer-sized droplets of one liquid are scattered within another liquid. In numerous ways, nanoemulsions can be utilized to lower the fat content of food products such as fat mimetics, reduced calorie density, controlled release, and reduced saturation (Meral et al., 2022).

15.3.1 Fat Reduction Challenges in Food Products

Reducing fat content in food products while preserving their flavor, texture, and overall quality presents several formidable challenges. Firstly, fat plays a pivotal role in many recipes by carrying and enhancing flavors. A reduction in fat content can potentially lead to a loss of flavor, resulting in bland or unappealing food. Secondly, fat contributes significantly to the creamy, silky, and rich texture of various dishes. Altering or decreasing fat levels can alter the mouthfeel of a product, making it less appealing to consumers. Additionally, fat acts as a barrier to moisture and oxygen, thus extending the freshness and shelf life of many food products (Sneha and Kumar, 2022). A reduction in fat content may compromise product stability and result in a reduced shelf life. Moreover, in recipes that rely on emulsification or fat as a binding agent, such as salad dressings, mayonnaise, or baked goods, reducing fat can disrupt these processes, making it challenging to maintain the desired product consistency. Finally, while consumers often opt for lower-fat options for health reasons, completely eliminating fat can raise concerns about the nutritional integrity of the food, as fat contains essential fatty acids and fat-soluble vitamins (Silva et al., 2012).

15.3.2 Nanoemulsions Addressing the Challenges of Fat Reduction

Nanoemulsions are liquid-in-liquid dispersions with droplet diameters of around 100 nanometers. They have several potential uses in the food business, including providing bioactive lipophilic chemicals; increasing food quality, nutritional value, functional characteristics, and shelf life; and improving food product stability and safety. However, there are several limitations to using nanoemulsions for fat reduction, such as the following:

- High manufacturing costs: Nanoemulsions necessitate the use of costly solvents, emulsifiers, and functional additives, which might have an impact on the end product's quality, stability, and functioning. Vegetable oils or plant oils, for example, are often employed as solvents for nanoemulsions, but they may be less soluble or have a greater viscosity than typical petroleum-based solvents. Furthermore, natural, or biodegradable emulsifiers like lecithin, chitosan, alginate, or carrageenan may be less stable or higher toxicity than synthetic or animal-derived emulsifiers (Aswathanarayan and Vittal, 2019).

- Low solubility: Because of the hydrophobic character of the nanoemulsion droplets, some lipophilic substances in water may not dissolve effectively in nanoemulsions. As a result, the lipophilic chemicals in the food product may have low bioavailability or perform poorly. Some antioxidants and vitamins, for example, may not be properly distributed to target cells or tissues in the body if they are encapsulated in nanoemulsions (Thiruvengadam et al., 2018).
- Low stability: Because of the high surface tension and interfacial surface tension between the water and oil phases, nanoemulsions are prone to phase separation and coalescence. This can shorten the shelf life of the nanoemulsion system by affecting its homogeneity and uniformity. Some nanoemulsions, for example, may create enormous droplets that can split or coalesce larger droplets that can lose their stability (Dasgupta and Ranjan, 2018).
- Low bioavailability: Because of their low solubility and permeability in nanoemulsions, several lipophilic substances in food may not be efficiently absorbed across the intestinal mucosa. As a result, the lipophilic chemicals in the food product may have inadequate bioavailability or effectiveness. Some medications or nutrients, for example, may not reach their target cells or tissues in the body if encapsulated in nanoemulsions (McClements, 2021).

By overcoming these obstacles, nanoemulsions have the potential to be a viable component for fat reduction in food business. By encapsulating lipophilic chemicals inside a stable and biocompatible framework, nanoemulsions can help lower the fat content and calories in meals. By boosting flavor, color, texture, shelf life, and bioavailability, nanoemulsions can also assist in improving the sensory characteristics and nutritional value of meals. Nanoemulsions can also help to preserve the health advantages of lipophilic substances by avoiding degradation and increasing absorption in the gastrointestinal system (Naseema et al., 2021; Sneha and Kumar, 2022).

15.3.3 Potential Benefits of using Nanoemulsions in Fat Reduction

Nanoemulsions provide a holistic solution to the challenges connected with reducing fat content in food items. First and foremost, they have the extraordinary capacity to improve the texture and mouthfeel of low-fat or reduced-fat products (Thiruvengadam et al., 2018). Nanoemulsions contribute to a creamier and more pleasant sensory experience by providing a finely dispersed, uniform distribution of fat droplets, making these goods more desirable to customers. Furthermore, nanoemulsions help to increase shelf life. They resist phase separation and preserve thermodynamic stability, effectively increasing the shelf life of various food products. This stability, in turn, serves to avoid fat crystallization and oil separation, ensuring that the products retain their quality over time (Aswathanarayan and Vittal, 2019; Silva et al., 2012).

One of the fundamental benefits of nanoemulsions is their ability to reduce fat content in food items while maintaining the perception of a fatty mouthfeel and flavor. This distinguishing feature enables the development of healthier alternatives that do not sacrifice taste or texture, thus meeting the growing demand for healthy solutions. Nanoemulsions can be used to increase nutrition delivery in low-fat goods in addition to fat reduction (Dasgupta and Ranjan, 2018; Naseema et al., 2021). They act as efficient transporters for fat-soluble vitamins and bioactive compounds, enhancing bioavailability and nutritional value. The adaptability of nanoemulsions makes them particularly appealing in the field of fat reduction. They may be smoothly integrated into an extensive range of food applications, including sauces, beverages, dressings, bread items, and dairy products, providing a versatile solution to the food industry's diversified needs. It is vital to emphasize that the successful use of nanoemulsions for fat reduction is based on factors such as emulsifier choice, processing conditions, and product-specific requirements. Proper formulation and careful optimization are essential to achieve the desired sensory and nutritional properties in reduced-fat food items (McClements, 2021; Ozogul et al., 2022).

15.4 FORMULATION OF NANOEMULSIONS FOR FAT REDUCTION

Nanoemulsions are colloidal suspensions of two immiscible liquids, usually water and oil, that are stable thermodynamically and are often stabilized using an interfacial layer of surfactants or emulsifying agents. The food industry has been interested in them because they can lower the fat level while keeping the desired sensory qualities.

15.4.1 Selection of Emulsifiers and Surfactants

It is essential for the proper formulation of a nanoemulsion to choose the appropriate emulsifiers and surfactants. These substances aid in maintaining the stability of the emulsion by inhibiting droplet coalescence at the interface between the oil and water phases. Table 15.1 indicates the emulsifiers and surfactants, including lecithin, Tween series, Span series, glycerol monostearate, xanthan gum, guar gum, carrageenan, sodium caseinate, and cellulose derivatives, each sourced and designed for specific applications, which play essential roles in the food and pharmaceutical industries, contributing to emulsion stability, texture, and functionality.

TABLE 15.1
Overview of properties of surfactants and emulsifiers with their applications

Emulsifier/ surfactant	Source	Properties	Applications	Reference
Lecithin (e.g., soy lecithin)	Usually made from soybeans	Natural, GRAS, good emulsification characteristics	Oil-in-water emulsions	Smykov, 2020
Tween series (e.g., Tween 20, Tween 80)	Sorbitol and oleic acid-based polysorbates	Nonionic surfactants, effective in stabilizing oil-in-water emulsions	Food and pharmaceutical sectors	Meral et al., 2022
Span series (e.g., Span 80, Span 85)	Esters of sorbitan fatty acids	Lipophilic surfactants, suitable for water-in-oil emulsions	Used with hydrophilic emulsifiers	Silva et al., 2012
Glycerol monostearate (GMS)	Ester of stearic acid in glycerol	Thickening agent, emulsifier, culinary and medicinal applications	Various culinary and medicinal uses	Akram et al., 2021
Xanthan gum	Produced by *Xanthomonas campestris* fermentation of carbohydrates	Hydrocolloid stabilizer, improves emulsion stability	Stabilizing hydrocolloids, food industry	Dhull et al., 2020
Guar gum	Derived from guar beans	Hydrocolloid stabilizer, increases viscosity and stability	Food industry (like xanthan gum)	Fryd and Mason, 2012
Carrageenan	Extracted from certain red seaweed species	Hydrocolloid with superior emulsifying qualities	Desserts, dairy products	Gazolu-Rusanova et al., 2020
Sodium caseinate	Milk-derived protein	Protein-based emulsifier, good stability in acidic environments	Dairy products, acidic environments	Gupta et al., 2016
Cellulose derivatives (methylcellulose, hydroxypropyl methylcellulose)	Chemically altered cellulose	Stabilizing, emulsifying, and thickening qualities	Various food and industrial applications	Aswathanarayan and Vittal, 2019

15.4.2 Choice of Oil Phase and Water Phase

When producing a nanoemulsion for a particular purpose, the oil phase and water phase selection are critical. The decision is made based on many elements, including the flavor profile, desired product attributes, and the planned use of the nanoemulsion (Duncan, 2011). Here are some factors to consider when selecting the oil phase:

- Reduced-fat oils: Start with reduced-fat oils if you want to make a nanoemulsion that will help weight loss. These may include vegetable oils with low fat content, such as canola, sunflower, or olive oil. Compared to conventional oils, these oils have a lower fat content while still offering some appealing sensory qualities.
- Flavor profile: Consider the oil's flavor profile, because it can have a big effect on how the finished product tastes. Some oils have characteristic flavors that may or may not be appealing in the final product.
- Nutritional profile: Examine the oil's dietary characteristics. You may use oils with good nutritional profiles for items aimed at health-conscious consumers, including those high in omega-3 fatty acids or unsaturated fats.
- Stability: Make sure the oil you choose will work with the emulsifiers and surfactants you are using. Some oils might need particular emulsifiers in order to create stable nanoemulsions (Che Marzuki et al., 2019).

Here are some factors to consider when selecting the water phase:

- Water purity: For the aqueous phase, use only high-quality, filtered water. The shelf life and stability of a nanoemulsion can be influenced by water quality. Adding substances to the water phase may be necessary, depending on the application. Obtain the desired flavor, texture, and functioning of the finished product; these additives may contain salt, sugar, acids, or other substances (Sneha and Kumar, 2022).
- Functional components: Think about whether you need to add any functional components to the water phase, including vitamins, minerals, or flavor enhancers. These can boost the product's sensory qualities or provide nutritional value.
- Acidity or pH: When working with certain components or applications, the pH of the water phase is crucial. If required, change the pH to achieve the best flavor and stability.

In the end, the aims of your product and the intended sensory and nutritional qualities should guide your decisions about the water and oil phases. The ideal ratio of oil and water phases for your unique nanoemulsion formulation can be found by conducting stability tests and sensory assessments (Walker et al., 2015).

15.4.3 Role of Stabilizers and Additives

In the creation of nanoemulsions, stabilizers and additives are essential. They support the finished product's general stability, texture, and sensory qualities. Here is a list of the roles played by stabilizers:

- Phase separation: This can be avoided by using stabilizers to keep immiscible phases (such as water and oil) evenly dispersed throughout the nanoemulsion. They provide a shield around the droplets to keep them from splitting and consolidating (Oca-Ávalos et al., 2017).
- Enhancing emulsion stability: Stabilizers increase the nanoemulsion's long-term stability. They ensure that the emulsion is consistent and stable throughout time by reducing creaming (the rise of oil droplets) and coalescence. Stabilizers have the ability to modify the size of the

drops in a nanoemulsion. They also support the maintenance of a narrow droplet size distribution by offering electrostatic repulsion or steric between the droplets.
- Enhancing viscosity: Some stabilizers, including hydrocolloids such as xanthan gum and guar gum, can make an emulsion more viscous. In applications where a heavier texture is preferred, this can be useful.
- Sedimentation prevention: Stabilizers aid in preventing the sedimentation of solid particles or denser phases at the bottom of the emulsion. This is crucial for formulations that contain suspended particles (Dickinson, 2020).

Here is a list of the roles played by additives:

- Flavor enhancers: To improve the flavor profile of the nanoemulsion, additives like herbs, spices, and flavoring agents can be used. They can transmit particular tastes or cover up undesirable flavors from certain components.
- Nutritional supplements: To strengthen the nanoemulsion with particular nutrients, additives like vitamins, minerals, or functional substances may be used. The finished product's nutritional value may improve as a result (Jamali et al., 2020; Dima et al., 2020).
- Preservatives: By preventing oxidation and microbiological development, additives like antioxidants or antimicrobial agents can also be used to improve the shelf life of the nanoemulsion.
- pH adjusters: For flavor or stability purposes, some formulae may need the pH level adjusted. The appropriate pH range can be achieved using bases or acids.
- Colorants and pigments: To improve or alter the visual look of the nanoemulsion, additives like natural or synthetic colorants may be used.
- Texture modifiers: To change the finished product's mouthfeel and texture, thickeners or gelling agents might be added.

It is critical to remember that the choice and concentration of stabilizers and additives should consider the precise objectives of the nanoemulsion formulation as well as regulatory standards for food or medicinal goods. The best mixture and concentrations of stabilizers and additives can be found by conducting stability tests and sensory assessments (Dickinson, 1998, 2020; Duncan, 2011).

15.5 APPLICATIONS OF NANOEMULSIONS IN FOOD PRODUCTS

Nanoemulsions have been proved to have various positive effects on the food products in the market (McClements, 2011). These emulsions help in enhancing the quality and, in some cases, the flavor of the food. Also, they contribute to increasing the shelf life of these products (Chung et al., 2016). Some very common applications are discussed in the following subsections.

15.5.1 Low-fat and Reduced-fat Dairy Products

Nanoemulsions help in replacing fats in dairy products such as low-fat yogurt, cheese, and ice cream. They help in retaining the creamy texture of the product by incorporating the oil droplets in nanoscale. They can also mimic the texture of full-fat products. Flavors and odors can be encapsulated by nanoemulsions and disseminated evenly throughout the product (Dhull et al., 2020; Chung et al., 2016). This helps to retain the product's taste and sensory qualities, compensating for any flavor loss due to the lower fat content. Nanoemulsions can be utilized for encasing fat-soluble vitamins along with other bioactive substances, improving nutrient absorption and nutritional profile in low-fat dairy products (Mahalakshmi et al., 2022). They can also improve the shelf life of dairy products by controlling lipid oxidation. By substitution of water and oil nanoemulsions for some of the fat content, low-calorie cheese products can be created that retain the texture and melting characteristics of cheese (Dhull et al., 2020).

15.5.2 Salad Dressings and Sauces

By decreasing phase separation and sedimentation, nanoemulsions can increase the stability of salad dressings and sauces. The small droplet size ensures homogeneous distribution of oil, water, and other substances, resulting in a longer shelf life (Smykov, 2020). The caloric content can also be elevated while retaining the texture. Similarly, nanoemulsions can be used to modify texture, deliver nutrients effectively, and reduce lipid oxidation (Jafari et al., 2020; Huang et al., 2010). By minimizing the need for stabilizers and emulsifiers, nanoemulsions can be utilized to generate "clean-label" salad dressings and sauces. This corresponds to consumer preferences for natural and less processed goods (Dima et al., 2020).

15.5.3 Baked Goods and Confectionery

The fat in baked goods can be replaced by nanoemulsions while keeping the texture intact, as in the products mentioned above. They also help in enhancing the textural component of baked goods with dietary restrictions. Nanoemulsions can help baked foods retain moisture, keeping them from drying out and getting stale; this increases freshness and shelf life. Nanoemulsions can also help to ensure consistent quality and uniform distribution of components such as antioxidants, vitamins, and natural colorants throughout the products (Meral et al., 2022).

15.5.4 Meat and Meat Analogues

The qualities most improved by the use of nanoemulsions are texture, fat reduction, flavor, distribution of nutrients, and shelf life. Some applications unique to meat products are improved binding and reduced allergenicity (Snoussi et al., 2022; Liang et al., 2020). The incorporation of nanoemulsions can make the binding of meat analogues similar to that of real meat. Nanoemulsions also help in encapsulating allergens present in analogs to serve people with allergies (McClements, 2011).

15.5.5 Beverages

Nanoemulsions help in increasing the bioavailability of certain water-insoluble compounds through beverages, by improving their absorptivity and solubility in the gastrointestinal tract. They also improve the stability of water-based beverages. They can be used to minimize the turbidity or cloudiness of some beverages, such as fruit juices and veggie drinks, by preventing insoluble particles or oils from segregating (McClements et al., 2021). They can be used to improve the stability of elements such as vitamins, antioxidants, and extracts of plants in functional beverages such as energy drinks and sports drinks (McClements, 2015; Jaiswal et al., 2015). By emulsifying fat and preventing phase separation, nanoemulsions can improve the texture of plant-based milk replacements such as almond milk and soy milk. Nanoemulsions are also employed in the manufacturing of drinks infused with cannabidiol (CBD) and tetrahydrocannabinol (THC) to improve cannabinoid solubility and stability, resulting in enhanced product uniformity (Dickinson, 2020).

15.5.6 Nutraceutical and Functional Food Products

Lipophilic substances such as vitamins E and D, carotenoids, and coenzyme Q10 can be encapsulated in nanoemulsions, increasing their solubility and bioavailability. Nanoemulsions can act as carriers

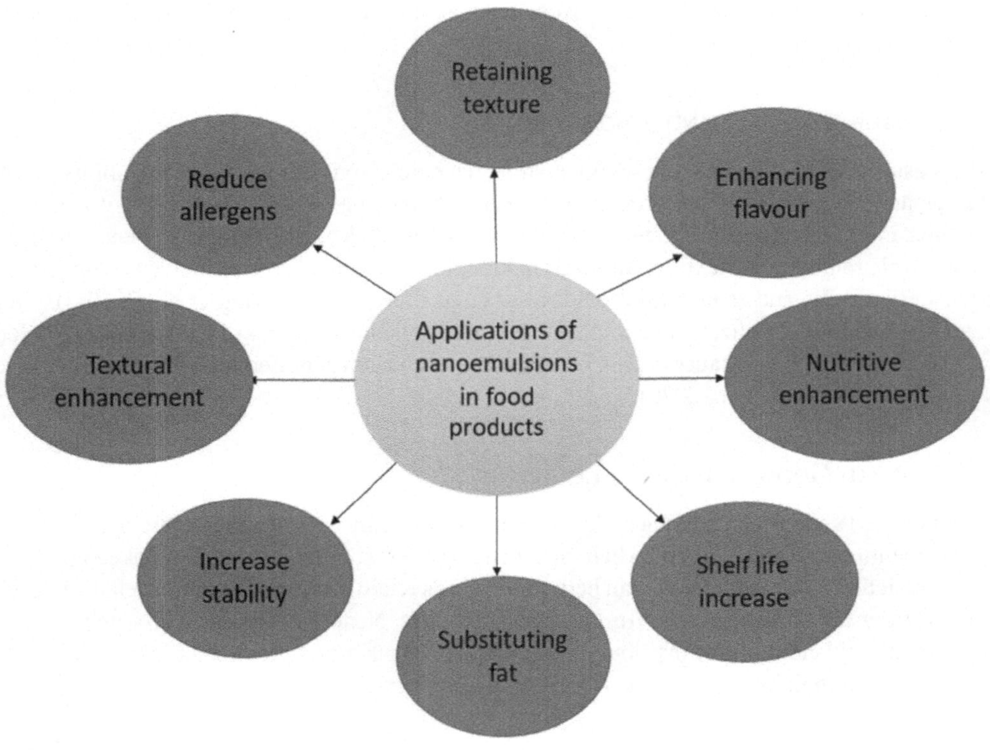

FIGURE 15.2 Applications of nanoemulsions in various food products.

for nutraceuticals, preventing degradation and allowing controlled release, which is especially advantageous for sensitive substances such as probiotics and peptides (Jamali et al., 2020; Huang et al., 2010). By delivering bioactive substances in food products, nanoemulsions may reduce the gap between pharmaceuticals and food, resulting in functional foods with health advantages. The texture and aesthetic of functional meals can be improved with nanoemulsions, making them more attractive to consumers (Jamali et al., 2020). Figure 15.2 illustrates the various applications of nanoemulsions in the field of food industry such as retaining texture, enhancing flavor, nutritive enhancement, and increasing shelf life and stability.

15.6 SENSORY AND TEXTUAL ASPECTS

15.6.1 Mimicking Sensory Properties of Fat

The appearance of a food product significantly influences consumers' initial sensory impressions and the overall acceptability of the product. The color of a food product may be measured both mathematically and experimentally using tristimulus coordinates and a colorimeter. Interactions between volatile substances emitted from food or beverages and nasal cavity receptors determine aroma. Volatile flavors may be quantified via headspace analysis (Dickinson, 2020). Specific types of nonvolatile molecules present or dissolved in saliva during mastication affect the flavor of a food. Taste molecules in emulsions partition between the oil and water phases, which is characterized by their temporal dependency in the water phase. Nonvolatile flavor concentrations can be measured using static or dynamic approaches within a food product or mouth during mastication (Chung et al., 2016).

15.6.2 Influence of Fat on Mouthfeel and Texture

Mouthfeel is an important element of food items since it influences their entire flavor. It refers to feelings caused by interactions between consumed food and oral receptors. Creaminess, firmness, hardness, thickness, thinness, smoothness, richness, sliminess, wateriness, and astringency are all mouthfeel qualities of food emulsions. The kind, concentration, and interactions of fat droplets considerably impact the perceived mouthfeel of food emulsions. Perception of creaminess has been linked to other sensory qualities, such as thickness, smoothness, and slipperiness, using mathematical models. Sensory perception is also influenced by the physical condition and phase transitions of the fat phase within the mouth (Chung et al., 2016; Walker et al., 2015).

15.6.3 Consumer Perception of Nanoemulsion-based Fat Reduction

There is no universally applicable technique for developing low-fat food items, since each category has distinct obstacles depending on the product's composition, physicochemical qualities, and sensory features. Food producers must determine the best fat-reduction approach for their product, considering criteria such as cost of use, ease of use, toughness, and label-friendliness. Cost of use refers to the additional costs associated with changes in ingredients or processing operations, while ease of use refers to the ease of implementing changes without significant changes in equipment or processes (Jaiswal et al., 2015). Concerns about label-friendliness, such as reporting unwanted elements like allergies and genetically modified products, may restrict the kind of ingredients that may be employed to build successful low-fat methods.

To prevent wasting time and money throughout the research and development process, these criteria should be addressed in the early phases of establishing a viable method to minimize the fat content of certain emulsion-based products (Thiruvengadam et al., 2018). Figure 15.3 shows the various properties of food that influence the oral processing, sensory perception, and biological responses of individuals, as well as perceptual response due to alteration of properties of bolus.

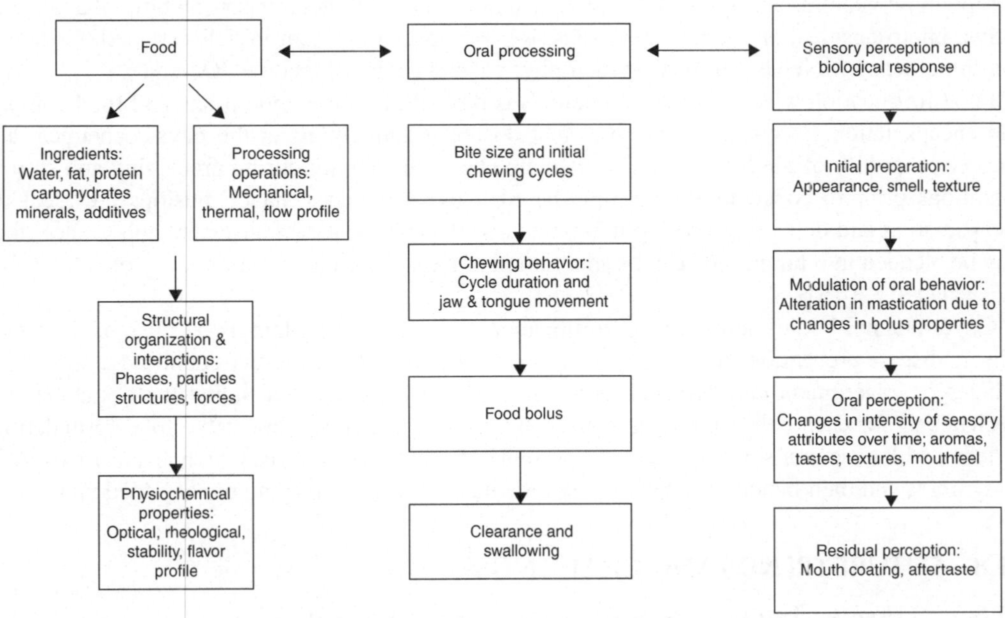

FIGURE 15.3 Overview of the relationships among food properties, oral processing, sensory perception, and biological response of food emulsions.

15.7 CASE STUDIES AND EXAMPLES

15.7.1 Real-world Exemplification of Successful Fat Reduction using Nanoemulsions

Various ways for producing reduced-fat goods with the same flavor, taste, texture, and mouthfeel as full-fat products have been developed. Ingredients such as hydrocolloid thickening agents (proteins and gums), fat mimetics (Olestra and SALATRIM), particulates (starch granules, and hydrocolloid thickening agents, proteins, and gums), and protein microspheres (hydrogel beads, and air bubbles) are examples of these. Innovative processing methods and food structuring techniques have also been employed (Chung et al., 2016).

Olestra is a sucrose polyester that was created to substitute digestible fats in items such as fried dishes, snacks, breads, and fillings. Because of its steric hindrance properties, it is a zero-calorie fat alternative. However, it may impair fat-soluble vitamin absorption and induce unpleasant gastrointestinal symptoms. The FDA restricts the amount of olestra and the types of foods that can include it (Walker et al., 2015).

SALATRIM, a set of structured triacylglycerols, is another fat mimetic that is used to lower calorie content in some meals. It has identical physicochemical, organoleptic, and functional qualities to conventional fats but roughly half the calories. For many years, SALATRIM has been utilized as a fat mimetic in low-calorie food items and has been advocated as an effective fat replacer due to its ability to suppress appetite and improve fullness (Chung et al., 2016).

15.7.2 Nanoemulsion Food Fortification using Lipophilic Vitamins

The purposeful addition of a specific nutrient to a food vehicle that is extensively eaten by people at risk of deficiency is referred to as food fortification. The regulations of the Codex Alimentarius provide the general conditions for commercializing fortified foods, which require fortified foods to be inexpensive, stable throughout processing, storage, and transportation, and not contain excessive or negligible quantities of vital nutrients. Malnutrition in many populations, particularly in poor nations, has highlighted the need for dietary fortification with lipophilic vitamins A, D, and E. Lipophilic vitamins are lipid-soluble bioactive molecules with poor water-solubility that are oxidative, photosensitive, and temperature-sensitive. Encapsulating them in delivery methods boosts their bioefficacy while also preserving their physical and chemical features (Öztürk, 2017).

Food fortification with bioactive compounds is difficult, because appropriate and food compatible encapsulation systems must be used that do not negatively affect the physicochemical and sensory properties of the food, and that maintain chemical stability during processing and storage conditions prior to consumption. As lipid-based delivery vehicles, nanoemulsions are ideal for encapsulating and delivering lipid-soluble vitamins. Because of their translucent appearance, they may be blended into liquids and can be spray dried or freeze dried to create a variety of solid dishes (Jaiswal et al., 2015).

Governmental bodies support food fortification with lipophilic vitamins because of their vital roles in disease prevention. However, regulations may be different for every country due to varying policies of government and deficiencies in different nutrients. Understanding the destabilization mechanisms for each delivery system could pave the way for producing stable food formulations enriched with nanoemulsions (Öztürk, 2017). Table 15.2 indicates various vitamins that are fortified using lipids and their benefits along with the food matrix used and the amount of added vitamins.

15.8 FUTURE TRENDS AND CHALLENGES

15.8.1 Emerging Trends in Nanoemulsions for Fat Reduction

Emerging trends in the use of nanoemulsions for fat reduction include the use of clean and natural label ingredients, precision fat reduction, customized nanoemulsions, functional ingredients, plant-based

TABLE 15.2
Examples of lipophilic vitamin-fortified foods

Vitamin	Formulation	Food matrix	Amount of added vitamin	Benefits	Reference
A	Water-dispersible spray dried form (Roche SD250)	Pandesal bread	2.8 mg/kg food	Increased liver store in children	Öztürk, 2017
D	nd	Milk	15.63 nmol/L	Increased 25-hydroxyvitamin D [25(OH)D] serum levels in human	Chung et al., 2016
D	nd	Orange juice	1000 IU/240 mL 10.4 ug/100g food	Increased serum 25(OH)D concentration	Walker et al., 2015
D	Dry vitamin D3 type 100, CWS (cold water-soluble)	Wheat and rye bread	10 ug/100g food	Increased serum 25(OH)D concentration	Duncan, 2011
D	Conventional emulsion	Cheddar cheese	111 IU/g milk for cheese	Increased serum 25(OH)D concentration	Jamali et al., 2020
E	a-Tocopherol acetate in oil-in-water emulsion	Ready-to-eat-breakfast cereals	30 and 400 IU	Increased bioavailability	Oca-Ávalos et al., 2017
D	Vitamin D3	Yogurt drink	5 ug/100 food	Increased serum 25(OH)D level, improved glycemic control in patients	Jaiswal et al., 2015
A, E, and D3	Oil-in-water emulsion	Milk for cheesemaking	3 ug/mL a-tocopherol acetate, 0.6 ug/mL retinyl palmitate, 10 ug/mL vitamin D3	Increased retention in curd, maintaining bioactivity and chemical stability of cheese during storage	Huang et al., 2010

and alternative proteins, sustainability, regulatory frameworks, consumer education, health and wellness, and global expansion. Natural and clean-label ingredients, such as plant-based emulsifiers like lecithin, are being explored to enhance the acceptability of reduced-fat products. Researchers are also working on fine-tuning nanoemulsions to achieve precise fat reduction while preserving the desired sensory attributes of foods (Che Marzuki et al., 2019). Customized nanoemulsions are being developed to meet individual dietary preferences and health goals. Functional ingredients, such as antioxidants and vitamins, are being incorporated to promote health and wellness. As the demand for plant-based and alternative protein products grows, nanoemulsions are being explored to reduce fat content while maintaining taste and texture. Sustainability is a key driver in product development, with regulatory bodies developing clear guidelines and standards for the use of nanoemulsions in food products. Consumer education is also crucial, with transparent labeling and communication of technology's advantages building consumer trust (Fryd and Mason, 2012).

15.8.2 Ongoing Research and Development

Nanoemulsions are being researched for their potential in various fields, including the food industry, pharmaceuticals, cosmetics, personal care, agriculture, biotechnology, medical imaging, energy storage, environmental remediation, and biomedical research. These nanoemulsions are being studied for their potential in reducing fat content in food products, improving drug delivery systems, and enhancing the bioavailability and stability of bioactive compounds. They are also being explored for their role in cell culture and bioprocessing, facilitating the delivery of nutrients and growth factors to cells, and improving the efficiency of bioproduction processes (Koroleva and Yurtov, 2012).

In energy storage, researchers are exploring the use of nanoemulsions for energy-dense and environmentally friendly fuel formulations, enhancing combustion efficiency, and reducing emissions. They are also being studied for their potential in environmental remediation, such as the remediation of oil spills. In biomedical research, nanoemulsions are used as model systems to study phenomena related to emulsification and droplet dynamics, contributing to a deeper understanding of colloidal systems. Safety and regulatory aspects are also being considered in the development of nanoemulsions, with a focus on understanding potential toxicity and establishing safety guidelines (Gupta et al., 2016).

15.8.3 Challenges and Areas for Improvement

Nanoemulsions are dispersions of two immiscible phases with nanosized particle sizes (d 200 nm). They can be oil-in-water (o/w) or water-in-oil (w/o) nanoemulsions and are thermodynamically unstable but kinetically stable. By encapsulating lipophilic bioactive molecules in the oil core, oil-in-water nanoemulsions have the ability to preserve, stabilize, and transport lipophilic bioactive substances such as lipophilic vitamins or omega-3. Emulsifiers are surface-active chemicals that absorb droplet oil-water interfaces, lowering interfacial tension and creating smaller droplets. Depending on the processing parameters and emulsion composition, many types of food-grade surfactants and emulsifiers can be employed to create nanoemulsions. Because of their hydrophilic groups, high molecular weight biopolymers, such as amphiphilic proteins and polysaccharides, can function as effective emulsifiers and hydrophobic moieties (Shi et al., 2022). Amphiphilic compounds include small molecule surfactants, phospholipids, quillaja saponins, and sucrose esters. Because of the growing desire for clean-label goods in the global food industry, natural emulsifiers have a lot of promise for usage in food applications. Nanoemulsions may be made using either high-energy or low-energy approaches (Gao et al., 2019).

Nanoemulsions are a promising solution for fat reduction, but their long-term stability and sensory attributes need to be improved. They can alter food products' taste, texture, and appearance, but balancing fat reduction with sensory qualities is a complex task. Increasing the nutritional content of nanoemulsion-based products is also a major challenge. Escalating production from lab scale to the industrial level requires advanced technology and process control. Cost-effectiveness is crucial, as specialized equipment and costly raw materials are required. Regulatory bodies are increasingly focusing on the safety and potential health impacts of nanoemulsions. Raising consumer awareness about the benefits and safety of nanoemulsions is essential for their successful adoption (Meral et al., 2022).

15.9 CONCLUSIONS

In conclusion, the role of nanoemulsions in fat reduction within the realm of food products is an evolving and promising field, underscored by the challenges presented by the high fat content in many consumer goods. Factors affecting nanoemulsion stability play a pivotal role in achieving successful fat-reduction strategies, while formulation intricacies, such as the selection of emulsifiers, oil and water phases, stabilizers, and additives, require careful consideration. These nanoemulsions are finding applications in a wide array of food products, ranging from dairy to baked goods and beverages, offering low-fat and reduced-fat options that strive to mimic the sensory properties of fat (Aswathanarayan and Vittal, 2019). Real-world case studies provide evidence of successful fat reduction using nanoemulsions, showcasing their potential in practical food formulations. Looking ahead, emerging trends in this field, such as natural ingredients and personalized nutrition, highlight the commitment to healthier and more sustainable products. However, ongoing developmental and research efforts are essential to deal with stability, scalability, and regulatory challenges. As we navigate these complexities, the use of nanoemulsions for fat reduction continues to hold promise in reshaping the landscape of healthier and more indulgent food choices, catering to evolving consumer preferences and nutritional needs (Meral et al., 2022).

REFERENCES

Akram, Salman, Nicolas Anton, Ziad Omran, and Thierry Vandamme. 2021. "Water-in-Oil Nano-Emulsions Prepared by Spontaneous Emulsification: New Insights on the Formulation Process." *Pharmaceutics* 13 (7): 1030. https://doi.org/10.3390/pharmaceutics13071030

Aswathanarayan, Jamuna Bai, and Ravishankar Rai Vittal. 2019. "Nanoemulsions and Their Potential Applications in Food Industry." *Frontiers in Sustainable Food Systems* 3 (November). https://doi.org/10.3389/fsufs.2019.00095

Che Marzuki, Nur Haziqah, Roswanira Abdul Wahab, and Mariani Abdul Hamid. 2019. "An Overview of Nanoemulsion: Concepts of Development and Cosmeceutical Applications." *Biotechnology & Biotechnological Equipment* 33 (1): 779–97. https://doi.org/10.1080/13102818.2019.1620124

Chung, Cheryl, Gordon Smith, Brian Degner, and David Julian McClements. 2016. "Reduced Fat Food Emulsions: Physicochemical, Sensory, and Biological Aspects." *Critical Reviews in Food Science and Nutrition* 56 (4): 650–85. https://doi.org/10.1080/10408398.2013.792236

Dasgupta, Nandita, and Shivendu Ranjan. 2018. *An Introduction to Food Grade Nanoemulsions*. Singapore: Springer Singapore. https://doi.org/10.1007/978-981-10-6986-4

Dhull, Smith, Degner, and McClements. 2020. *Nanotechnological Approaches in Food Microbiology*. Edited by Sanju Bala Dhull, Prince Chawla, and Ravinder Kaushik. CRC Press. https://doi.org/10.1201/9780429342776

Dickinson, Eric. 1998. "Structure, Stability and Rheology of Flocculated Emulsions." *Current Opinion in Colloid & Interface Science* 3 (6): 633–38. https://doi.org/10.1016/S1359-0294(98)80092-7

Dickinson, Eric. 2020. "Advances in Food Emulsions and Foams: Reflections on Research in the Neo-Pickering Era." *Current Opinion in Food Science* 33 (June): 52–60. https://doi.org/10.1016/j.cofs.2019.12.009

Dima, Cristian, Elham Assadpour, Stefan Dima, and Seid Mahdi Jafari. 2020. "Bioavailability of Nutraceuticals: Role of the Food Matrix, Processing Conditions, the Gastrointestinal Tract, and Nanodelivery Systems." *Comprehensive Reviews in Food Science and Food Safety* 19 (3): 954–94. https://doi.org/10.1111/1541-4337.12547

Duncan, Timothy V. 2011. "Applications of Nanotechnology in Food Packaging and Food Safety: Barrier Materials, Antimicrobials and Sensors." *Journal of Colloid and Interface Science* 363 (1): 1–24. https://doi.org/10.1016/j.jcis.2011.07.017

Feng, Jin, Carlos Rodríguez-Abreu, Jordi Esquena, and Conxita Solans. 2020. "A Concise Review on Nanoemulsion Formation by the Phase Inversion Composition (PIC) Method." *Journal of Surfactants and Detergents* 23 (4): 677–85. https://doi.org/10.1002/jsde.12414

Fryd, Michael M., and Thomas G. Mason. 2012. "Advanced Nanoemulsions." *Annual Review of Physical Chemistry* 63 (1): 493–518. https://doi.org/10.1146/annurev-physchem-032210-103436

Gao, Qingsheng, Wenbiao Zhang, Zhangping Shi, Lichun Yang, and Yi Tang. 2019. "Structural Design and Electronic Modulation of Transition-Metal-Carbide Electrocatalysts toward Efficient Hydrogen Evolution." *Advanced Materials* 31 (2). https://doi.org/10.1002/adma.201802880

Gazolu-Rusanova, Dilek, Ivan Lesov, Slavka Tcholakova, Nikolai Denkov, and Badreddine Ahtchi. 2020. "Food Grade Nanoemulsions Preparation by Rotor-Stator Homogenization." *Food Hydrocolloids* 102 (May): 105579. https://doi.org/10.1016/j.foodhyd.2019.105579

Gupta, Ankur, Burak H. Eral, Alan T. Hatton, and Patrick S. Doyle. 2016. "Nanoemulsions: Formation, Properties and Applications." *Soft Matter* 12 (11): 2826–41. https://doi.org/10.1039/C5SM02958A

Huang, Qingrong, Hailong Yu, and Qiaomei Ru. 2010. "Bioavailability and Delivery of Nutraceuticals Using Nanotechnology." *Journal of Food Science* 75 (1): R50–57. https://doi.org/10.1111/j.1750-3841.2009.01457.x

Jafari, Seid Mahdi, Ali Sedaghat Doost, Maryam Nikbakht Nasrabadi, Sareh Boostani, and Paul Van der Meeren. 2020. "Phytoparticles for the Stabilization of Pickering Emulsions in the Formulation of Novel Food Colloidal Dispersions." *Trends in Food Science & Technology* 98 (April): 117–28. https://doi.org/10.1016/j.tifs.2020.02.008

Jaiswal, Manjit, Rupesh Dudhe, and P. K. Sharma. 2015. "Nanoemulsion: An Advanced Mode of Drug Delivery System." *3 Biotech* 5 (2): 123–27. https://doi.org/10.1007/s13205-014-0214-0

Jamali, Seyedeh Narges, Elham Assadpour, and Seid Mahdi Jafari. 2020. "Formulation and Application of Nanoemulsions for Nutraceuticals and Phytochemicals." *Current Medicinal Chemistry* 27 (18): 3079–95. https://doi.org/10.2174/0929867326666190620102820

Koroleva, Yu M, and Evgenii V Yurtov. 2012. "Nanoemulsions: The Properties, Methods of Preparation and Promising Applications." *Russian Chemical Reviews* 81 (1): 21–43. https://doi.org/10.1070/RC2012v081n01ABEH004219

Liang, Xiuping, Cuicui Ma, Xiaojia Yan, Huanhuang Zeng, David Julian McClements, Xuebo Liu, and Fuguo Liu. 2020. "Structure, Rheology and Functionality of Whey Protein Emulsion Gels: Effects of Double Cross-Linking with Transglutaminase and Calcium Ions." *Food Hydrocolloids* 102 (May): 105569. https://doi.org/10.1016/j.foodhyd.2019.105569

Mahalakshmi, L., K.S. Yoha, Jeyan Arthur Moses, and C. Anandharamakrishnan. 2022. "Nano Delivery Systems for Food Bioactives." In *Food, Medical, and Environmental Applications of Nanomaterials*, 205–30. Elsevier. https://doi.org/10.1016/B978-0-12-822858-6.00008-X

McClements, David Julian. 2011. "Edible Nanoemulsions: Fabrication, Properties, and Functional Performance." *Soft Matter* 7 (6): 2297–2316. https://doi.org/10.1039/C0SM00549E

McClements, David Julian. 2012. "Nanoemulsions versus Microemulsions: Terminology, Differences, and Similarities." *Soft Matter* 8 (6): 1719–29. https://doi.org/10.1039/C2SM06903B

McClements, David Julian. 2015. Food Emulsions. CRC Press. https://doi.org/10.1201/b18868

McClements, David Julian. 2021. "Advances in Edible Nanoemulsions: Digestion, Bioavailability, and Potential Toxicity." *Progress in Lipid Research* 81 (January): 101081. https://doi.org/10.1016/j.plipres.2020.101081

McClements, David Julian, Arun K. Das, Pubali Dhar, Pramod Kumar Nanda, and Niloy Chatterjee. 2021. "Nanoemulsion-Based Technologies for Delivering Natural Plant-Based Antimicrobials in Foods." *Frontiers in Sustainable Food Systems* 5 (February). https://doi.org/10.3389/fsufs.2021.643208

Meral, Raciye, Mehmet Mustafa Ekin, Nazan Kutlu, and Yagmur Erim Kose. 2022. "The Nanoemulsions: A New Design and Fat-reducing Strategy for the Bakery Industry. Their Effects on Some Quality Attributes of Fat-reduced Cakes." *Journal of Food Processing and Preservation* 46 (12). https://doi.org/10.1111/jfpp.17160

Modarres-Gheisari, Seyed Mohammad Mohsen, Roghayeh Gavagsaz-Ghoachani, Massoud Malaki, Pedram Safarpour, and Majid Zandi. 2019. "Ultrasonic Nano-Emulsification — A Review." *Ultrasonics Sonochemistry* 52 (April): 88–105. https://doi.org/10.1016/j.ultsonch.2018.11.005

Naseema, A., Lohitesh Kovooru, Alok Kumar Behera, K.P. Pramodh Kumar, and Priyanka Srivastava. 2021. "A Critical Review of Synthesis Procedures, Applications and Future Potential of Nanoemulsions." *Advances in Colloid and Interface Science* 287 (January): 102318. https://doi.org/10.1016/j.cis.2020.102318

Oca-Ávalos, Juan Manuel Montes de, Roberto Jorge Candal, and María Lidia Herrera. 2017. "Nanoemulsions: Stability and Physical Properties." *Current Opinion in Food Science* 16 (August): 1–6. https://doi.org/10.1016/j.cofs.2017.06.003

Ozogul, Yesim, Gokcem Tonyali Karsli, Mustafa Durmuş, Hatice Yazgan, Halil Mecit Oztop, David Julian McClements, and Fatih Ozogul. 2022. "Recent Developments in Industrial Applications of Nanoemulsions." *Advances in Colloid and Interface Science* 304 (June): 102685. https://doi.org/10.1016/j.cis.2022.102685

Öztürk, Bengü. 2017. "Nanoemulsions for Food Fortification with Lipophilic Vitamins: Production Challenges, Stability, and Bioavailability." *European Journal of Lipid Science and Technology* 119 (7): 1500539. https://doi.org/10.1002/ejlt.201500539

Ren, Gaihuan, Zhicheng Sun, Zengzi Wang, Xiaoyang Zheng, Zhenghe Xu, and Dejun Sun. 2019. "Nanoemulsion Formation by the Phase Inversion Temperature Method Using Polyoxypropylene Surfactants." *Journal of Colloid and Interface Science* 540 (March): 177–84. https://doi.org/10.1016/j.jcis.2019.01.018

Safaya, Malvika, and Yogesh Chandrakant Rotliwala. 2020. "Nanoemulsions: A Review on Low Energy Formulation Methods, Characterization, Applications and Optimization Technique." *Materials Today: Proceedings* 27: 454–59. https://doi.org/10.1016/j.matpr.2019.11.267

Shi, Yameng, Min Zhang, Kai Chen, and Mingqi Wang. 2022. "Nano-Emulsion Prepared by High Pressure Homogenization Method as a Good Carrier for Sichuan Pepper Essential Oil: Preparation, Stability, and Bioactivity." *LWT* 154 (January): 112779. https://doi.org/10.1016/j.lwt.2021.112779

Silva, Hélder Daniel, Miguel Ângelo Cerqueira, and António A. Vicente. 2012. "Nanoemulsions for Food Applications: Development and Characterization." *Food and Bioprocess Technology* 5 (3): 854–67. https://doi.org/10.1007/s11947-011-0683-7

Smykov, Igor T. 2020. "Nanotechnology in the Dairy Industry." In *The ELSI Handbook of Nanotechnology*, 223–75. Wiley. https://doi.org/10.1002/9781119592990.ch11

Sneha, K., and Ashwani Kumar. 2022. "Nanoemulsions: Techniques for the Preparation and the Recent Advances in Their Food Applications." *Innovative Food Science & Emerging Technologies* 76 (March): 102914. https://doi.org/10.1016/j.ifset.2021.102914

Snoussi, Ahmed, Moncef Chouaibi, Hayet Ben Haj Koubaier, and Nabiha Bouzouita. 2022. "Encapsulation of Tunisian Thyme Essential Oil in O/W Nanoemulsions: Application for Meat Preservation." *Meat Science* 188 (June): 108785. https://doi.org/10.1016/j.meatsci.2022.108785

Thiruvengadam, Muthu, Govindasamy Rajakumar, and Ill-Min Chung. 2018. "Nanotechnology: Current Uses and Future Applications in the Food Industry." *3 Biotech* 8 (1): 74. https://doi.org/10.1007/s13205-018-1104-7

Walker, Rebecca, Eric A. Decker, and David Julian McClements. 2015. "Development of Food-Grade Nanoemulsions and Emulsions for Delivery of Omega-3 Fatty Acids: Opportunities and Obstacles in the Food Industry." *Food & Function* 6 (1): 41–54. https://doi.org/10.1039/C4FO00723A

Wang, Zheng, Marcos A. Neves, Hiroko Isoda, and Mitsutoshi Nakajima. 2015. "Preparation and Characterization of Micro/Nano-Emulsions Containing Functional Food Components." *Japan Journal of Food Engineering* 16 (4): 263–76. https://doi.org/10.11301/jsfe.16.263

Index

Note: Page numbers in *italic* refer to figures.